Rasch Models

Foundations, Recent Developments, and Applications

Gerhard H. Fischer
Ivo W. Molenaar
Editors

Rasch Models

Foundations, Recent Developments, and Applications

With 19 Illustrations

Springer-Verlag
New York Berlin Heidelberg London Paris
Tokyo Hong Kong Barcelona Budapest

Gerhard H. Fischer
Institut für Psychologie
Universität Wien
Liebiggasse 5
A-1010 Vienna
Austria

Ivo W. Molenaar
Vakgroep Statistiek en Meettheorie
Rijksuniversiteit Groningen
Grote Kruisstraat 2/I
9712 TS Groningen
The Netherlands

Library of Congress Cataloging-in-Publication Data

Rasch models : foundations, recent developments, and applications /
Gerhard H. Fischer, Ivo W. Molenaar, editors.
p. cm.
Papers originally presented at a workshop held at the University of Vienna, Feb. 25-27, 1993.
Includes bibliographical references and indexes.
ISBN 0-387-94499-0
1. Psychology—Mathematical models—Congresses. 2. Psychometrics—Congresses. 3. Rasch, G. (Georg), 1901-1980. I Fischer, Gerhard H. II. Molenaar, Ivo W.
BF39.R28 1995
150'.1'5195—dc20 95-3740

Printed on acid-free paper.

Production managed by Laura Carlson; manufacturing supervised by Jacqui Ashri.
Camera-ready copy prepared by the editors.
Printed and bound by Edwards Brothers, Inc., Ann Arbor, MI.
Printed in the United States of America.

9 8 7 6 5 4 3 2 1

ISBN 0-387-94499-0 Springer-Verlag New York Berlin Heidelberg

Preface

Thirty-two years after the publication of the legendary 'Rasch book' (Rasch, 1960), the rich literature on the Rasch model and its extensions was scattered in journals and many less accessible sources, including 'grey' literature. When asked by students or junior researchers for references to the Rasch model, it was a typical reaction on the part of the editors to state that it was difficult to name one, or just a few; actually, only a whole list of references differing in notation and level of formal abstraction seemed to meet the request in most cases. Therefore, in 1992 the editors decided to invite a number of outstanding authors in the field of Rasch modeling to contribute to a book presenting the current state of knowledge about Rasch models.

The aim was not just to collect a number of papers on the subject, rather to produce a well-organized monograph. To this end, a workshop was held in Vienna from 25 to 27 February 1993 in which, after a process of mutual reviewing, drafts of all chapters were read and discussed by all authors, leading to a more systematic organization of the topics treated in unified notation and terminology. (The workshop was sponsored by the University of Vienna; here, the editors would like to express their thanks, in the name of all contributors, both for the financial support and for the hospitality granted.) In spite of the attempts to present a homogeneous monograph, however, it is natural that some individual variations of views and styles of presentation remain, giving expression to the fact that researchers are individuals, and research is a lively subject.

It was clear from the beginning that the volume would be published in English, the language of international scientific communication. Since none of the editors and authors is a native English speaker, in spite of all the efforts to produce a correct English text there is no hope that the result will be perfect. We apologize for the lack of idiomatic English, or of elegance of style, wherever it should be missed by the native English reader. This is the price to be paid for the more important goal of reaching the interested audience world-wide.

The scholars of item response theory will decide whether the authors and editors of this book should be praised or criticized. Here, however, we would like to mention those who have contributed immensely to the present book project and who might otherwise be forgotten: Ms. Karin Waldherr has typed five and corrected many of the 21 chapters expertly in TeX, and has helped with the list of references and indexes, to say nothing of her many other organizational duties and of her extensive correspondence with the authors. She even pointed out to the editors many errors of all kinds, thus helping to improve the final manuscripts. We would like to thank her heartily for her invaluably loyal service. Thanks also go to Ms. Heidi Glatzmeier, who has spent much time on revisions and authorial corrections of many chapters and on inserting index labels in the

LaTeX files; and to Mag. Elisabeth Seliger for proofreading several chapters. Without the assistance of these ladies it would have been impossible to produce the complicated typescript from which the book is reproduced. Finally, thanks go to the Department of Psychology of the University of Vienna for making the 'ladypower' available to the project.

Gerhard H. Fischer
Vienna

Ivo W. Molenaar
Groningen

December 12, 1994

Contents

Part I
The Dichotomous Rasch Model

1

Some Background for Item Response Theory and the Rasch Model

Ivo W. Molenaar[1]

ABSTRACT The present chapter offers a general introduction to item response theory as a measurement model, with a discussion of the sources of random variation in this model. Next, the Rasch model is briefly introduced. Finally, an outline of the subsequent chapters is presented. The chapter emphasizes issues of interpretation; formal definitions are often postponed until later chapters.

1.1 Introduction

In a psychological test or attitude scale, one tries to measure the extent to which a person possesses a certain property. The social and behavioral sciences often deal with properties like intelligence, arithmetic ability, neuroticism, political conservatism, or manual dexterity. It is easy to find examples of observable human behavior indicating that a person has more or less of such a general property, but the concept has a surplus value, in the sense that no specific manifest behavior fully covers it. This is the reason why such properties are called *latent traits*.

The use of a test or scale presupposes that one can indirectly infer a person's position on a latent trait from his/her responses to a set of well-chosen items, and that this also allows us to predict his/her behavior when confronted with other items from the same domain. A statistical model of the measurement process should allow us to make such predictions. Moreover, it should allow generalization to other persons taking the same test, and address the question of generalizability over test taking occasions.

Throughout this book the terminology of persons, items, and answers will be used. This is because most applications of the theory deal with human individuals who are asked to provide the correct answer to a set of problems (an achievement scale), or are asked to express agreement or disagreement with a set of statements (an attitude scale). The models to be discussed, however, can be used in a wider class of measurement problems. In Stokman (1977), for example, the 'persons' are resolutions regarding decolonization, the 'items' are delegations at the General Assembly of the United Nations, and the 'answers' are votes in favor or against. The 'persons' may also be laboratory animals, the 'items' may

[1]Rijksuniversiteit Groningen, Vakgroep Statistiek en Meettheorie, Grote Kruisstraat 2/I, 9712 TS Groningen; e-mail W.MOLENAAR@PPSW.RUG.NL

be tasks, and an 'answer' may be the successful or unsuccessful completion of the task as scored by a human expert.

The history of scales for human achievements and attitudes begins with *measurement by fiat*, i.e., a scale was used because a domain expert had made it and claimed that it was good. *Formal measurement models* are a major step forward, because they incorporate assumptions about the measurement process. Their fulfillment imposes restrictions on the observed data. This permits some form of empirical assessment of the quality of measurement. A further discussion of this concept 'quality of measurement' is postponed to Section 1.3. The use of formal measurement models begins with factor analysis and the Thurstone models for attitude scaling. From roughly 1940 to 1980, the leading model was classical test theory with its decomposition of the total test score into a latent true score and an error component.

It is now widely recognized that classical test theory has some deficiencies which render it more desirable to model measurement processes with *item response theory (IRT)*. Briefly stated, IRT can do the same things better and can do more things, when it comes to modeling existing tests, constructing new ones, applying tests in non-standard settings, and above all interpreting the results of measurement. For a review of the reasons for the transition and a brief history of IRT, see, e.g., Fischer (1974), Hambleton & Swaminathan (1985), or Baker (1992). One of the main lines of development of IRT models originates in Rasch's (1960) 'structural model for the items of a test', now known as the *Rasch model* (RM). Chapter 21 gives some historical background for this model.

IRT is built around the central idea that the probability of a certain answer when a person is confronted with an item, ideally can be described as a simple function of the person's position on the latent trait plus one or more parameters characterizing the particular item. For each item, the probability of a certain answer as a function of the latent trait value is called the *item characteristic curve (ICC)* or *item response function (IRF)*.

Given the answers of n persons to k items intended to measure the same latent trait, such person and item parameters can be estimated and the assumptions underlying the IRT model can be tested. This will help a researcher to assess the quality of the test as a measurement instrument, and to predict its performance in future applications (other items, other persons, other occasions). Where required, IRT can also assist in improving the quality of the test, e.g., by indicating which items are inappropriate and should be changed, deleted, or replaced. After this stage, the test may be used as a standard instrument to measure similar persons.

The RM, which is the topic of the present book, is a rather simple but at the same time very powerful IRT model for measurement. Section 1.3 and Chapters 2 through 14 deal with its *dichotomous* form, in which each answer can be scored in only two ways: positive (correct, agree, indicative of a high position on the latent trait), or negative (wrong, disagree, indicative of a low amount of the property being measured). Later chapters consider the *polytomous* RM, in which more than two answer categories are modeled.

Many other IRT models have been proposed, see Hambleton & Swaminathan (1985) or Baker (1992) for an overview. They mostly make less stringent assumptions, and are therefore easier to use as a model for an existing test. On the other hand, they typically pose more problems during parameter estimation, fit assessment, and interpretation of results. Whenever possible, it is thus recommended to find a set of items that satisfies the RM, rather than find an IRT model that fits an existing item set. The present book deals only with the RM and its extensions that share its set of desirable properties with respect to statistical procedures and foundation of measurement.

1.2 Sources of Randomness

This section discusses two different views on the types of random variation in an IRT model. Suppose, for example, that a group of pupils, of mixed gender, has taken a spelling test that satisfies the RM. Say that the outcomes suggest that the girls are better spellers than the boys. In assessing the extent to which this conclusion is generalizable, one may ask whether it remains valid

- for measurement of the same children with the same test on another day;
- for other boys and girls from the same population;
- for other items measuring spelling ability.

When statistical methods are applied to the actual data obtained, it is recommended to reflect on the sources of randomness to be considered. Three aspects are generally distinguished.

1.2.1 Within Subjects Within Items

Many IRT researchers adhere to the so-called *stochastic subject view* (Holland, 1990) in which it is meaningful to say, e.g., that a fixed person S_v has a probability of .6 to answer positively to a fixed item I_i. This probability would then refer to small variations in the circumstances of the person or the test taking situation.

Suppose that we could brainwash the person after each test session. Then our statement would mean that in 60% of repeated person-item confrontations the result would be positive and in 40% it would be negative. Lord and Novick (1968) use the term *propensity distribution* for similar variations at the level of the total test score in classical test theory. They also give a quote (p. 29–30) from Lazarsfeld (1959) about the variation at the item level (repeatedly asking Mr. Brown whether he is in favor of the United Nations).

The proponents of the *random sampling view*, however, maintain that each subject either has always a positive or always a negative reply, and interpret the probability as the proportion of subjects with a fixed value θ on the latent trait who give the positive answer. In this view the latent trait, with its probability distribution across subjects, serves the purpose of explaining away the global dependence between items at the population level into local independence given θ. In this case there is no within-person within-item probability other than zero or one.

In the latter view, a person belongs to the latent class with value θ; within this class, each item of the test has a fixed success probability, and answers for different items are statistically independent. When the number of such classes is low, this leads to the latent class models discussed in Chapter 13. One could argue that in this view it is not meaningful to speak about an individual parameter for each person; the person is fully described by his/her answer pattern of zeroes and ones which indicate the items to which the answer is always negative and those to which the answer is always positive. Before we pass to a more thorough discussion of the two views, we should consider the other sources of randomness.

1.2.2 Sampling of Persons

The proponents of both views seem to agree that it is useful to consider the generalization to other persons from the same population of intended respondents. Especially at the stage when a test or questionnaire is developed, the researcher will often use a much smaller group of respondents than will be measured at a later stage. The smaller group is used to estimate the item parameters, and to assess or improve the test quality. Ideally it is a random sample from the population that one intends to measure at a later stage, but in practice this is not always possible. IRT offers somewhat better possibilities for the intended generalization to a wider group of persons than classical test theory, whose concepts like proportion correct, item-test correlation, and reliability are rather strongly dependent on the composition of the group of respondents actually observed.

Nevertheless it should be emphasized that also in IRT the item parameters, or the fit of the model, may change when a drastically different group of respondents is considered. Differences in culture, or in gender, should be examined before one may widely generalize IRT results. Ideally the test is not affected by such differences, but such a claim should be empirically established for any difference in respondent group for which it is reasonable to assume that it might endanger the generalizability. Research in this area uses concepts like *local homogeneity* (Ellis & Van den Wollenberg, 1993), *item or test bias* or *differential item or test functioning* (DIF or DTF, e.g., Shealy & Stout, 1993; Fischer, 1993b).

For the stochastic subject view, any application of IRT with this aim of measuring other subjects should consider sampling of persons as a second source of randomness. For the random sampling view, it is so far the first source. Although the latent trait values are unknown and therefore an actual sample from the subpopulation of persons with the same latent trait value cannot be drawn, the proportion of persons in this subpopulation who answer positively to an item is the core concept in this latter view.

1.2.3 Sampling of Items

It has already been argued that measurement of a latent trait often implies the goal that similar conclusions about the model and about the persons should hold for a test consisting of different items from the same domain. Actual random sampling of items, however, is only found in a small minority of testing situations, e.g., in individualized instruction when each examinee is confronted with a random sample of items drawn from a previously calibrated bank of items. Even when such a bank is available, the use of a stratified sample (with respect to content area or difficulty level), or the selection of an optimal test (see Chapter 7), is usually preferred to the random sampling of items. Although Lord (1974, p. 249) refers to "the probability that the examinee will give the right answer to a randomly chosen item having the same ICC as item i", and random sampling of items has been considered by a few others as well, most authors (in both views) seem to favor an IRT analysis treating items as fixed and persons as random. The possibilities of generalization to other items are then part of a more qualitative or indirect analysis (based on a careful choice of the fixed item set) rather than on statistical generalization to a universe of similar items.

1.2.4 The Two Views Discussed

Summarizing, it appears that there is a certain freedom to incorporate and to interpret sources of randomness in IRT models, dependent on the goal of the research and the data collection method. In particular, the probability of a positive answer on a fixed item given a latent trait value has two different interpretations, the stochastic subject view and the random sampling view.

It is remarkable that this difference in interpretation is often ignored, not only in applications of IRT but even in papers and books wholly devoted to IRT. Did the authors tacitly adhere to one of the two views? Or did they tacitly defend the opinion that the view difference did hardly matter, neither for choosing analysis methods (for parameter estimation or tests of fit) nor for interpretation of the statistical results?

There is no firm consensus on such issues among IRT researchers, not even among the authors of this book. The present author feels inclined to favor the stochastic subject view for a vast majority of IRT applications. For attitude items it is quite plausible that the answer of the subject depends on small variations in mood and context that the researcher does not want to measure, even if (s)he could. For items measuring an ability or achievement there is more ground to argue that a fixed item-person pair either always leads to a positive answer or always to a negative one. Here too, however, on closer scrutiny the 'always or never' view is often too simple for most person-item confrontations. Respondents may sometimes miss the correct answer during test taking although they would know it in a more relaxed situation. On the other hand they may improvise or guess the correct answer (not only for multiple choice questions!) when they might well have missed it when tested the next day.

The random sampling view has been relaxed by not strictly requiring that each single individual has a deterministic answer pattern. It then simply refrains from asking by what mechanism the members of a group with the same latent trait value may produce different answer patterns to the set of items under consideration. Indeed, one does not need an answer to this question for formulating an IRT model in which the so-called manifest probabilities, i.e., the probabilities of specific answer patterns for persons randomly sampled from the population, are modeled.

Suppose that n persons have taken a test consisting of k dichotomous items. Let the k-vector $\boldsymbol{x} = (x_1, x_2, \ldots, x_k)$ denote an answer pattern, where one puts $x_i = 1$ for a positive answer to item I_i, and $x_i = 0$ for a negative answer. Add the usual assumption of *local independence*, i.e., all items measure the same latent trait to be denoted by θ, and given θ the answers to the k items are statistically independent (this property is discussed more fully in Section 1.3). Then the answer vector $\boldsymbol{X}$ for a randomly sampled person has manifest probabilities of the form

$$P(\boldsymbol{X} = \boldsymbol{x}) = \int_{-\infty}^{\infty} \prod_{i=1}^{k} P(X_i = x_i|\theta) dG(\theta), \tag{1.1}$$

where G denotes the cumulative distribution function of the latent trait in the population considered. Holland (1990) observes that in this relaxed form of the random sampling view the latent trait is little more than a convenient tool to express the restriction on the manifest probabilities caused by local independence.

Such a modest interpretation of the concept of a latent trait goes at least back to Birnbaum (Lord & Novick, 1968, p. 358–359):

> "Much of psychological theory is based on a trait orientation, but nowhere is there any necessary implication that traits exist in any physical or physiological sense. It is sufficient that a person behaves as if he were in the possession of a certain amount of each of a number of relevant traits and that he behave as if these amounts substantially determined his behavior."

Note that the present book deals with the unidimensional case in which one latent trait θ suffices. It is easy to rewrite (1.1) for a vector $\boldsymbol{\theta}$ containing more than one latent trait, and such multidimensional versions of IRT are sometimes found in the literature. In most applications, however, it is desirable to find items that measure just one latent trait, and nothing else. The importance of local independence is exactly this: one observes global dependence between item scores, i.e., they correlate positively in the population of all respondents, but this dependence can be fully explained by the presence of one latent trait.

The last question is whether and where the difference in view leads to a different method and/or a different interpretation of the results of an IRT analysis. A final answer can only be given in the chapters to follow this one. In order to simplify such discussions, the reader may use the combination of the stochastic subject view, random sampling of persons, and fixed item choice as the basic paradigm, unless the contrary is explicitly stated. There will be applications in which the random sampling view is adequate. More often, however, inferences about specific persons are desired, e.g., for selection, placement or classification. Then the stochastic subject view appears to lead to more meaningful conclusions.

1.3 The Rasch Model

This section will discuss how the RM explains the occurrence of a data matrix containing the dichotomously scored answers of a sample of n persons (the subjects, denoted as $S_1, S_2, \dots, S_n$) to a fixed set of k items (denoted as $I_1, I_2, \dots, I_k$) that measure the same latent trait θ. It is assumed that each subject S_v has a real-valued person parameter θ_v denoting his/her position on the latent trait. Each item I_i has a real-valued item parameter β_i denoting the difficulty of that item.

Let $\boldsymbol{\beta}$ denote the k-vector of item parameters, and $\boldsymbol{\theta}$ the n-vector of person parameters. For each cell of the $n \times k$ data matrix $\boldsymbol{X}$, with elements x_{vi} equal to 0 or 1, one has

$$P(X_{vi} = x_{vi} \mid \theta_v, \beta_i) = \frac{\exp[x_{vi}(\theta_v - \beta_i)]}{1 + \exp(\theta_v - \beta_i)}. \tag{1.2}$$

One easily verifies that the probability of a positive answer $x_{vi} = 1$ equals .5 for $\theta_v = \beta_i$. For varying θ, one obtains the item reponse function as a logistic curve increasing from 0 (for $\theta = -\infty$) via .5 (for $\theta = \beta_i$) to 1 (for $\theta = \infty$). The importance of (1.2) is that it enables us not only to *explain* post hoc how this person responded to this item, but also to *predict* from a person and item parameter the probability distribution for the answer of this person to other items, and of other persons to this item, without actually observing the response to such person-item pairs.

In order to obtain the joint probability distribution of the whole matrix $\boldsymbol{X}$, we add to (1.2) the usual independence assumptions of IRT. First, we assume independence of answer vectors between persons. Note that this may be violated if a respondent knows the answers of other respondents and is influenced by that knowledge; such unfavorable measurement circumstances should be avoided, as we assume random sampling not only of persons, but also of answer vectors given by different persons.

Next, consider the answers of the same person to different items. In a group with mixed latent trait values, they will be positively correlated because subjects with higher latent trait values have a higher probability of giving a positive answer to any item than subjects with lower latent trait values. It will be assumed, however, that all systematic covariation between items is explained by the latent trait variation: given θ, the answers of a person to different items are stochastically independent. An interpretation of this so-called property of *local independence* is that not only must all items measure the same latent trait, but also they must each provide independent information on each person's position on that trait. If there would be subgroups of items that share other systematic variation than explained by θ, we would learn less about θ from such a subgroup than from the same number of items chosen differently. This has a relation with the item sampling discussed in Section 1.2, although local independence is found to hold in many contexts where items are not selected by sampling but by judgement. Summarizing, the items should be homogeneous (they all measure the same latent trait), but also in some sense heterogeneous (they have no more in common than the latent trait).

Combining (1.2) with the independence assumptions, one has, for any choice of 0 or 1 in the cells of the observed data matrix $\boldsymbol{x}$, a probability given by

$$P(\boldsymbol{X} = \boldsymbol{x} \mid \boldsymbol{\theta}, \boldsymbol{\beta}) = \prod_{v=1}^{n} \prod_{i=1}^{k} \frac{\exp[x_{vi}(\theta_v - \beta_i)]}{1 + \exp(\theta_v - \beta_i)}. \tag{1.3}$$

If all item parameters β_i and all person parameters θ_v were known, this would define a probability distribution on all $n \times k$ matrices with entries from the set $\{0,1\}$ in each cell. In practice such parameters are unknown, and methods to estimate them from the data will be discussed in Chapters 3 and 4. One restriction must be imposed on the parameters in order to obtain a unique solution: One must fix the origin of the scale, see Chapters 2 and 3.

It follows from (1.2) that all item response curves in the RM are logistic with the same slope; the only difference between items is the location of the curves. Other parametric IRT models either use more than one item parameter, or a different functional form than the logistic.

As will be explained more fully in subsequent chapters, the RM has several advantages not shared by its competitors. A major advantage is that the total number correct score of a person is a *sufficient statistic* for the unknown person parameter, and the item sum score across persons is a sufficient statistic for the unknown item parameter. This implies in both cases that one simply obtained number contains all statistical information about the parameter. Both for the actual estimation of the unknown parameters and for the interpretation of the data, this is important. Moreover, it becomes very easy to quantify the amount of available information on an unknown parameter in the form of the statistical *information function*, and this information is additive across the items or persons across which the sum is taken to obtain the sufficient statistics. The concept of statistical information will be discussed in more detail in later chapters.

1.3.1 Quality of Measurement

In Section 1 it was briefly mentioned that formal measurement models have the advantage of permitting empirical assessment of the quality of measurement. Although the author has the view that this is a correct statement, it is prudent to elaborate on it at the present stage. Now that IRT models have been introduced earlier in this chapter, some arguments can be presented why IRT models, and in particular the RM, are even more helpful than earlier measurement models in assessing, and sometimes improving, measurement quality.

Measurement can be described as the assigning of numbers to objects or individuals in a systematic way, as a means of representing a property of the objects or individuals (see Allen & Yen, 1979, p. 2). Although it will sometimes be evident that one such assignment method is better than another one, it will be equally evident that quality of measurement is not a unidimensional and simply operationalized concept. Indeed it is similar to the quality of many other things, e.g., the quality of a scientific book or the quality of a car, in the following sense. There are quite a few aspects on which higher and lower quality ratings can be given, some of them easily measured (such as price or weight), and some of them for which only a subjective evaluation can be given. Moreover, the weight of such aspects in an overall quality judgement may well differ between quality assessors, both because of differences of taste and because of differences in the kind of use that is intended.

A first aspect of quality of measurement that should be mentioned is the extent to which arithmetic properties of the numbers assigned can be related to the *empirical domain*. At the nominal level of measurement, only the distinction same versus different has this relation. At the ordinal level, the distinction between larger and smaller is added to the list of meaningful properties. It is desirable, however, that also certain metric properties of the numbers, such as differences or ratios, have an empirical counterpart. As is discussed, e.g., by Krantz, Luce, Suppes, and Tversky (1972, p. 123), most measurements in the social sciences do not have an empirical counterpart for addition in the form of an interpretation of the operation of concatenation.

They also argue, however, that this does not preclude fundamental measurement of the kind of latent properties to which our book is devoted. We have already seen that the difference between the person parameter and the item location parameter determines the probability of a positive reply in the dichotomous RM. The same holds in the two-parameter logistic model, see (2.15), after multiplication of the difference by the item discrimination parameter.

Moreover, consider as an example three subjects S_u, S_v, and S_w with person parameters $\theta_u = 0.5$, $\theta_v = 1.0$, and $\theta_w = 2.0$, respectively. We may wish to conclude that the difference on the latent trait between S_w and S_v is twice as large as that between S_v and S_u. In the same sense, if we obtain person parameters of the same persons at different time points, e.g., before and after a certain treatment, we may want to conclude that one treatment is twice as effective as the other.

Such conclusions are meaningful, in the sense that one may infer, for each item and also for the test as a whole, by how much the probabilities of certain answers do change, between two persons or between two time points. Such conclusions require that the latent scale have at least an *interval metric*, and are generalizable (ideally) across subjects and test items. In order to achieve this, we have to show that the interval scale property follows from empirically testable assumptions.

A strong case for conclusions of this type can only be made if (a) the scale level of the measurement has been rigorously established, (b) there exist adequate parameter estimation methods, where the degree of uniqueness of the estimates corresponds precisely to the scale level under (a), and (c) powerful tests of fit are available that help to verify the theoretical scale properties from the empirical structure of the data. It is an advantage of item response models, and in particular of the RM, that these three requirements can be satisfied remarkably well (see Chapters 2 through 6).

The quality aspect of *generalizability of conclusions* to other items and/or other persons has already been discussed in the present chapter. Another core concept for quality of measurement is *reliability*. In classical test theory this is defined as true score variance divided by observed score variance. Several properties of this parameter and several methods to estimate it are discussed, e.g., in Lord and Novick (1968). The index of subject separability is defined by Gustafsson (1977) as that part of the variance in estimated person parameters that is not due to estimation error. It is thus the exact counterpart of the classical reliability, with total score replaced by person parameter.

Item response theory, however, provides the user with means to express the reliability of person measurement in a much more refined way. This is a core theme in Chapters 5, 6, and 7. The unrealistic assumption of homoscedasticity (equal measurement error) in classical test theory is replaced by the consequence of the IRT assumptions that this error depends on a person's parameter value, and its size per person can be obtained via the information function. Moreover, it is possible to assess in advance how this would change if the person would take a different selection of items.

Validity of measurement is another important quality aspect. Although it is too dependent on scale content and research goal to be fully discussed here, it should always be taken into consideration. It suffices to mention here that this is at least as easy in IRT models as it is in older formal measurement models.

Ease of application is still another aspect of quality. John Tukey has once said that the practical power of a statistical method equals its theoretical power multiplied by its probability of being used. In this respect, IRT may present a barrier to novice users, unfamiliar with its concepts and its software. Now that more and more introductory texts and user-friendly software become available, and more and more researchers and institutions use IRT models, this barrier is expected to shrink.

A final aspect worth mentioning is the amount of *time and effort* required from the persons being measured. Here IRT offers good ways to achieve a maximum of information with a minimum number of items, to design a test such that information is highest for the interval of person parameter values for which it is most desired given the use of the test, and to offer different items to different persons (Chapter 7). Moreover, the use of other response formats than multiple choice scored correct/incorrect can increasingly be modeled (Hambleton, 1992).

Undoubtedly, there are still other aspects of quality of measurement than discussed here. Hopefully, however, this subsection makes clear why item response models, and in particular the RM, are in many respects superior to older efforts to formally model the process of measurement.

1.4 An Overview of the Book

Part I

In Chapter 2 the form of the RM is derived from several sets of basic assumptions, and its measurement properties are discussed. Chapters 3 and 4 discuss estimation methods for item and person parameters, respectively. Chapter 5 presents the ways in which one may check whether the assumptions of the RM hold for a given data matrix, and Chapter 6 discusses how one may assess 'person fit', i.e., the fit of the RM for one person at a time. Once a large set of items satisfying the RM has been obtained, Chapter 7 presents methods to select a good test of shorter length from that item bank.

Part II

Chapter 8 introduces the Linear Logistic Test Model (LLTM), an RM in which the item difficulties are governed by specific restrictions. In Chapter 9 variants of the RM occur that are useful for measuring change of the latent trait values of persons as a result of natural growth or intervention. Chapter 10 deals with dynamic RMs in which the person parameters change either during measurement or during time between test takings. Chapter 11 introduces methods for inference in which the RM as a measurement model is combined with a specific model for the latent trait values, in a way related to covariance structure analysis where measurement was modeled by factor analysis. The topic of Chapter 12 is an extension of the RM to items with different slopes of IRF's which still preserves the basic properties enabling separation of person and item parameters and conditional inference given certain sufficient statistics. Chapter 13 discusses the relation of the RM with a model in which persons belong to a fixed and moderate number of latent classes rather than each having their own real-valued person parameter. Chapter 14 introduces mixture models in which it is assumed that an RM with different item parameters may hold for different subgroups of the person population.

Part III

So far, all material dealt with dichotomously scored items. The third part of the book deals with the polytomous case in which more than two answer categories are offered and modeled. Parameter estimation is discussed in Chapter 15, derivation from sufficiency and from parameter separation in Chapter 16, and derivation as a generalized linear model with (quasi-)interchangeability in Chapter 17. Fit analysis for the polytomous RM is discussed in Chapter 18, and extensions that are useful for measurement of change in Chapter 19. Polytomous Rasch mixture models are the topic of Chapter 20. The final Chapter 21 gives some history of the RM and some thoughts on what Georg Rasch might have said about the present book if he were still alive.

2

Derivations of the Rasch Model

Gerhard H. Fischer[1]

ABSTRACT Chapters 1 and 3 through 6 describe a number of especially attractive features of the Rasch model (RM). These assets – in particular, sufficiency of the unweighted raw score, existence of conditional maximum likelihood estimators of the model parameters and of conditional likelihood ratio tests for hypothesis testing – suggest the question as to whether they are shared by a larger class of IRT models, or whether they are, within a certain framework, unique properties of the RM. In the latter case, we would have to conclude that the RM plays a rather singular role within IRT. As we shall see, this is actually so. The derivations in this chapter lay a foundation both for the RM and for the metric scale properties of Rasch measurement.

2.1 Derivation from Sufficiency

One outstanding property of the Rasch model (RM) is the sufficiency of the (unweighted) raw score $R_v = \sum_i X_{vi}$ for the ability parameter θ_v, given fixed item parameters β_i. The raw score statistic R_v as a means of 'measuring' θ_v is particularly appealing because, for many practical purposes, the item parameters β_i will no longer be required once the sufficiency of R_v has been established; then, as far as θ_v is concerned, a given response pattern $\boldsymbol{x}_v$ contains no information about θ_v beyond $\sum_i x_{vi} = r_v$, and hence sometimes need not even be recorded. In view of the practical advantages of the sufficiency of R_v, we shall investigate what family of IRT models is compatible with this property.

First, a notational convention has to be introduced: for reasons that will become apparent below, we have to use two different notations each for item characteristic curves (ICCs), person parameters, and item parameters. As long as no specific metric of the latent trait dimension has been established, person parameters will be denoted by ξ, item difficulty parameters by δ_i, and ICCs by $g_i(\xi)$. The ξ and δ_i parameters can be subjected to monotone scale transformations at any time, entailing changes of the form of the ICCs. Whenever person and item parameters have been transformed in such a way that the ICCs are of the form $g_i(\xi) = f(\phi(\xi) - \psi(\delta_i))$, however, which establishes (as we shall see) an interval metric for person and item measurement, we write $g_i(\xi) = f(\theta - \beta_i)$, with $\theta := \phi(\xi)$ and $\beta_i := \psi(\delta_i)$.[2]

---

[1]Department of Psychology, University of Vienna, Liebiggasse 5, A–1010 Vienna, Austria; e-mail: GH.FISCHER@UNIVIE.AC.AT

In this chapter, some algebraic properties of the conditional likelihood function of the Rasch model, described in Chapter 3, are used; readers not familiar with these properties may wish to read Chapter 3 prior to Chapter 2.

[2]The notation '$a := b$' or '$b =: a$' means that a is defined by b.

We shall start out by assuming a unidimensional IRT model for a given (finite or infinite) set of items I_i, satisfying the following assumptions:

(i) *continuous, strictly monotone increasing ICCs $g_i(\xi)$ of all I_i, for all latent trait parameters $\xi \in \mathbb{R}$;*

(ii) *lower limits $\lim_{\xi \to -\infty} g_i(\xi) = 0$ and upper limits $\lim_{\xi \to +\infty} g_i(\xi) = 1$* (in the following denoted as *no guessing* for short);

(iii) *the principle of 'local stochastic independence' of all responses,* i.e., *given a test of length k,*

$$P[(X_{v1} = x_{v1}) \wedge (X_{v2} = x_{v2}) \wedge \ldots \wedge (X_{vk} = x_{vk})]$$
$$= \prod_i g_i(\xi_v)^{x_{vi}} [1 - g_i(\xi_v)]^{1-x_{vi}}.$$

The following is the key assumption.

(iv) *Given a test of length k, the raw score statistic $R_v = \sum_{i=1}^k X_{vi}$ is sufficient for ξ_v.*

Upon these assumptions, the following result can be proved:

Theorem 2.1 *Assumptions (i), (ii), and (iii), together with the sufficiency of the raw score, (iv), imply that the item response model is equivalent to an RM with ICCs*

$$f_i(\theta) = \frac{\exp(\theta - \beta_i)}{1 + \exp(\theta - \beta_i)}. \tag{2.1}$$

(Remark: 'equivalent to an RM' means that there exists a strictly monotone function $\phi(\xi) =: \theta$ and constants β_i such that all ICCs $f_i(\theta)$ attain the form (2.1). Any model equivalent to an RM is empirically indistinguishable from an RM.)

Proof Dropping index v for convenience, sufficiency of R implies that, for any given response vector $\boldsymbol{x}$ and corresponding raw score r, the conditional likelihood function $L(\boldsymbol{x}|r)$ satisfies

$$L(\boldsymbol{x}|r) = \frac{L(\boldsymbol{x}|\xi)}{L(r|\xi)} = c(\boldsymbol{x}), \tag{2.2}$$

where $c(\boldsymbol{x})$ is a constant depending on $\boldsymbol{x}$ (and r) and on the set of items, but independent of ξ. Note that no assumptions are made on how $g_i(\xi)$ depends on item I_i; in particular, it is not assumed that items are characterized by just one scalar parameter each.

Let $\sum_{i=1}^k X_i = R$ be sufficient for ξ, i.e., (2.2) holds for any response vector with $\sum_i x_i = r$. Let $k \geq 2$ and I_i be any item with $2 \leq i \leq k$; moreover, define

$\boldsymbol{x}$ as a response vector with raw score r, satisfying $x_1 = 1$ and $x_i = 0$;

$\boldsymbol{x}^{(1,i)}$ as the partial response vector $(x_2, \ldots, x_{i-1}, x_{i+1}, \ldots, x_k)$, if $k > 2$;

$\boldsymbol{y}$ as a response vector with raw score r, satisfying $y_1 = 0$, $y_i = 1$, and $y_l = x_l$ otherwise; and

$\boldsymbol{y}^{(1,i)}$ as the partial response vector $(y_2, \ldots, y_{i-1}, y_{i+1}, \ldots, y_k)$, if $k > 2$.

According to assumptions, $L(\boldsymbol{x}|\xi) = L(\boldsymbol{x}^{(1,i)}|\xi)g_1(\xi)[1-g_i(\xi)]$, with $L(\boldsymbol{x}^{(1,i)}|\xi) := 1$ if $k = 2$, so that (2.2) becomes

$$L(\boldsymbol{x}|r) = \frac{L(\boldsymbol{x}^{(1,i)}|\xi)g_1(\xi)[1-g_i(\xi)]}{L(r|\xi)} = c(\boldsymbol{x}). \tag{2.3}$$

Similarly, for response vector $\boldsymbol{y}$,

$$L(\boldsymbol{y}|r) = \frac{L(\boldsymbol{y}^{(1,i)}|\xi)[1-g_1(\xi)]g_i(\xi)}{L(r|\xi)} = c(\boldsymbol{y}), \tag{2.4}$$

and dividing (2.4) into (2.3) yields that the quotient

$$\frac{L(\boldsymbol{x}^{(1,i)}|\xi)g_1(\xi)[1-g_i(\xi)]}{L(\boldsymbol{y}^{(1,i)}|\xi)[1-g_1(\xi)]g_i(\xi)} = \frac{c(\boldsymbol{x})}{c(\boldsymbol{y})} \tag{2.5}$$

is also independent of ξ. Observe that by definition,

$$L(\boldsymbol{x}^{(1,i)}|\xi) = L(\boldsymbol{y}^{(1,i)}|\xi),$$

so that (2.5) implies

$$\frac{g_1(\xi)[1-g_i(\xi)]}{[1-g_1(\xi)]g_i(\xi)} = \frac{c(\boldsymbol{x})}{c(\boldsymbol{y})} =: d_i(\boldsymbol{x}), \tag{2.6}$$

independent of ξ.

So far nothing has been assumed about the scale of the latent trait ξ. Therefore, a suitable monotone scale transformation $\xi \to \theta$ can be chosen, carrying the hitherto unspecified ICC of item I_1, $g_1(\xi)$, over into the logistic function

$$f_1(\theta) = \frac{\exp(\theta)}{1+\exp(\theta)}. \tag{2.7}$$

(The strict monotonicity of $g_1(\xi)$ implies that there exists a strictly monotone function $\phi(\xi) =: \theta$ such that $g_1(\xi)$ assumes the form (2.7).) Replacing $g_1(\xi)$ and $g_i(\xi)$ in (2.6) by $f_1(\theta)$ and $f_i(\theta)$, respectively, and inserting (2.7) in (2.6) yields, after a little algebra,

$$f_i = \frac{\exp(\theta-\beta_i)}{1+\exp(\theta-\beta_i)}, \qquad \text{with } \beta_i = \ln d_i(\boldsymbol{x}). \tag{2.8}$$

This shows that any model satisfying assumptions (i) through (iv) can be transformed into an equivalent RM with item parameters $\beta_1 = 0$ (normalization) and β_i for $i = 2, \ldots, k$. In other words, any model satisfying (i) through (iv) is empirically indistinguishable from an RM. □

This result was given by Fischer (1968, 1974), who employed complete induction for $k = 2, 3, \ldots$ That method of proof rested on the observation that, if $R = \sum_{i=1}^{k} X_i$ is a sufficient statistic of the test consisting of items $I_1, \ldots, I_k$, then $R' := \sum_{i=1}^{k-1} X_i$ is also sufficient within the test composed of $I_1, \ldots, I_{k-1}$. The intuitive reason for this plausible fact is that $R = R' + X_k$, so that the only way by which information from items $I_1, \ldots, I_{k-1}$ enters R is via R'; hence, R' has to be a sufficient summary of $X_1, \ldots, X_{k-1}$. Clearly, this result also follows from Theorem 2.1 above: if sufficiency of R implies the RM for test length k, then both R and R' must be sufficient for θ_v (cf. Chapter 3). However, a direct formal proof of the sufficiency of R' within the test $I_1, \ldots, I_{k-1}$, without use of Theorem 2.1, is not trivial. Since such a proof does not seem to exist in psychometric literature, we give one below.

Lemma 2.1 *Let assumptions (i), (ii), and (iii) hold. If, within test $I_1, \ldots, I_k$, raw score $R = \sum_{i=1}^{k} X_i$ is sufficient for ξ, then $R_S = \sum_{I_i \in S} X_i$ is also sufficient for ξ within test S, where S denotes some nonempty subset of the items.*

Proof Let $\boldsymbol{x}$ be an observed and $\boldsymbol{z}$ any response vector, both with raw score r, in the total test, $I_1, \ldots, I_k$. From the sufficiency of R it follows that

$$\frac{\prod_{i=1}^{k} g_i^{x_i}(1-g_i)^{1-x_i}}{\sum_{\boldsymbol{z}} \prod_{i=1}^{k} g_i^{z_i}(1-g_i)^{1-z_i}} = \frac{\pi_k \prod_{i=1}^{k} h_i^{x_i}}{\sum_{\boldsymbol{z}} \pi_k \prod_{i=1}^{k} h_i^{z_i}} = c(\boldsymbol{x}), \tag{2.9}$$

where $\pi_k = \prod_{i=1}^{k}(1-g_i)$ and $h_i = g_i/(1-g_i)$. The denominator of (2.9) can be written as $\pi_k \gamma_r(h_1, \ldots, h_k)$, where γ_r denotes the elementary symmetric function of order r of the h_i (for a definition of γ_r, see (3.10)). So (2.9) implies

$$\prod_{i=1}^{k} h_i^{x_i} = c(\boldsymbol{x})\gamma_r(h_1, \ldots, h_k). \tag{2.10}$$

Similarly, for any response vector $\boldsymbol{y}$ with the same raw score r,

$$\prod_{i=1}^{k} h_i^{y_i} = c(\boldsymbol{y})\gamma_r(h_1, \ldots, h_k). \tag{2.11}$$

Dividing (2.10) into (2.11) yields

$$\frac{\prod_{i=1}^{k} h_i^{y_i}}{\prod_{i=1}^{k} h_i^{x_i}} = \frac{c(\boldsymbol{y})}{c(\boldsymbol{x})}. \tag{2.12}$$

Equation (2.12) holds for all possible response vectors $\boldsymbol{x}$ and $\boldsymbol{y}$ with raw score r. Let, for instance, $x_1 = y_j = 1$ and $x_j = y_1 = 0$, and $x_l = y_l$ otherwise. Then it follows from (2.12) that

$$\frac{h_j(\xi)}{h_1(\xi)} = \frac{c(\boldsymbol{y})}{c(\boldsymbol{x})} =: d_j(\boldsymbol{x}),$$

or equivalently (with simplified notation),

$$h_j = d_j h_1, \tag{2.13}$$

for $j = 2, \dots, k$.

For subtest $I_1, \dots, I_{k-1}$ it follows that the conditional likelihood, given the raw score in that subtest, r', is

$$\frac{\prod_{i=1}^{k-1} g_i^{x_i}(1-g_i)^{1-x_i}}{\sum_{\boldsymbol{z}'} \prod_{i=1}^{k-1} g_i^{z_i}(1-g_i)^{1-z_i}} = \frac{\prod_{i=1}^{k-1} h_i^{x_i}}{\sum_{\boldsymbol{z}'} \prod_{i=1}^{k-1} h_i^{z_i}} = \frac{h_1^{r'} \prod_{i=1}^{k-1} d_i^{x_i}}{h_1^{r'} \sum_{\boldsymbol{z}'} \prod_{i=1}^{k-1} d_i^{z_i}} =$$

$$= \frac{\prod_{i=1}^{k-1} d_i^{x_i}}{\gamma_{r'}(d_1, \dots, d_{k-1})}, \tag{2.14}$$

where $\boldsymbol{z}'$ now denotes any response vector $(z_1, \dots, z_{k-1})$ with raw score r', and $d_1 = 1$ is defined for completeness. Since the right-hand term in (2.14) is independent of ξ, $R' = \sum_{i=1}^{k-1} X_i$ is sufficient for ξ. This holds – *ceteris paribus* – for any other subtest composed of $k-1$ items. – The same arguments can then be applied to all subtests of length $k-2$, etc. □

A result similar to that of Theorem 2.1 is due to Birnbaum (1968), who defined 'optimal weights' α_i such that the score $R = \sum_i \alpha_i X_i$ discriminates optimally at a given point ξ_0 of the latent continuum. Birnbaum asked for what family of IRT models these optimal weights are independent of ξ_0; he showed that such models must be equivalent to

$$f_i(\theta) = \frac{\exp[\alpha_i(\theta - \beta_i)]}{1 + \exp[\alpha_i(\theta - \beta_i)]}, \quad \alpha_i > 0. \tag{2.15}$$

This is usually referred to as the 'Birnbaum Model' or 'Two Parameter Logistic Model' (2PL). The α_i are called 'discrimination' parameters. Specializing Birnbaum's result by setting all $\alpha_i = 1$, the only class of models compatible with the optimality of equal and constant weights of all items at all points $\xi \in \mathbb{R}$ is models equivalent to the RM.

So far we have seen that any model satisfying (i) through (iv) *can* be transformed into an RM (2.1). This means that, after an appropriate transformation $\xi \to \theta$, the parameters θ and β_i enter f_i in subtractive form. Suppose we had decided to choose this additive concatenation of θ and β_i in the ICCs from the beginning, starting out with the following assumption (v).

(v) *There exists a universe of items with ICCs of the form $f(\theta - \beta)$ for all $\theta, \beta \in \mathbb{R}$, where β is the item difficulty, and $f(x)$ is a continuous bijective increasing (and hence strictly monotone) $\mathbb{R} \to (0,1)$ function.*

What is then the most general class of monotone scale transformations $k(\theta)$ and $l(\beta)$ compatible with the subtractive argument $\theta - \beta$ of f in (v)? This question is answered by a result of Pfanzagl (1971, p. 171; 1994, pp. 249–250).

Lemma 2.2 *Let the probabilities $P(+|\theta, \beta) = f(\theta - \beta)$ be fixed for all $\theta, \beta \in \mathbb{R}$. Then assumption (v) implies that the θ and β are unique up to linear transformations $a\theta + b_1$ and $a\beta + b_2$, $a > 0$, with arbitrary constants b_1 and b_2, i.e., the θ and β are measured on interval scales with a common measurement unit.*

(Remark: assumptions (i) and (ii) are now contained in (v) and hence need not be listed explicitly.)

Proof Assume a model with ICCs of the form $f(\theta - \beta)$. This implies that, for any individual with parameter $\theta \in \mathbb{R}$, and any item with difficulty $\beta \in \mathbb{R}$, the probability $P(+|\theta, \beta)$ is given. Suppose there exists another representation compatible with these probabilities $P(+|\theta, \beta)$, with new parameters $k(\theta)$ and $l(\beta)$, and with ICCs $m[k(\theta) - l(\beta)]$, such that

$$m[k(\theta) - l(\beta)] = f(\theta - \beta) \quad \text{for all } \theta, \beta \in \mathbb{R}, \tag{2.16}$$

where k and l are continuous strictly monotone increasing scale transformations on $\mathbb{R}$, and m is defined by (2.16). It then follows that

$$k(\theta) - l(\beta) = m^{-1} \circ f(\theta - \beta), \tag{2.17}$$

with '$\circ$' as the concatenation of functions. For $\beta = 0$, (2.17) implies

$$k(\theta) = m^{-1} \circ f(\theta) + l(0),$$

or denoting $m^{-1} \circ f$ by h and $l(0)$ by l_0,

$$k(\theta) = h(\theta) + l_0. \tag{2.18}$$

Similarly, for $\theta = 0$ (2.17) yields

$$l(\beta) = -h(-\beta) + k_0. \tag{2.19}$$

Inserting (2.19) and (2.18) in (2.17),

$$h(\theta) + l_0 + h(-\beta) - k_0 = h(\theta - \beta),$$

which for $\theta = \beta = 0$ yields $h(0) =: h_0 = k_0 - l_0$. Hence,

$$h(\theta) - h_0 + h(-\beta) - h_0 = h(\theta - \beta) - h_0. \tag{2.20}$$

Denoting $h(x) - h_0$ by $h^*(x)$ for short, (2.20) becomes

$$h^*(s) + h^*(t) = h^*(s + t), \tag{2.21}$$

with $\theta =: s$ and $-\beta =: t$. This is the 'functional equation of Cauchy' (Aczél, 1966, p. 34), which under the present monotonicity assumptions has the general solution $h^*(x) = ax$, $a > 0$. Remembering that $h^*(x) = h(x) - h_0$, we obtain $h(x) = ax + h_0$. Inserting this in (2.18), $k(\theta) = a\theta + h_0 + l_0$, and in (2.19), $l(\beta) = a\beta - h_0 + k_0$. Hence, the most general admissible transformations of θ and β are $a\theta + b_1$ and $a\beta + b_2$, with arbitrary constants $a > 0$, b_1, and b_2. □

Applying Lemma 2.2, we see that assumptions (i) through (iv), in combination with (v), yield a 'family of Rasch models',

$$P(+|S_v, I_i) = \frac{\exp[a(\theta_v - \beta_i) + b]}{1 + \exp[a(\theta_v - \beta_i) + b]}, \tag{2.22}$$

with arbitrary constants a and $b := b_1 - b_2$, where a corresponds to the common discrimination of all items, and b may be interpreted as a definition of the origin of the β-scale relative to that of the θ-scale. Obviously, any shift of the θ-scale can be compensated by a contrary shift of the β-scale; hence, we conclude that assumptions (i) through (v) determine the θ- and the β-scales up to positive linear transformations with the same multiplicative constant a, implying that these scales are *interval scales* with the same unit of measurement. We may formulate this consequence in the form of the following Theorem 2.2.

Theorem 2.2 *Assumptions (v), (iii), and (iv) imply that the model is an RM, and the parameters θ and β are unique up to positive linear transformations with a common multiplicative constant (i.e., they have interval scale properties with a common unit of measurement).*

(Remark: assumptions (i) and (ii) are contained in (v) and hence need not be stated explicitly.)

This result on the scale properties of the RM is in correspondence with the more general result of Lemma 2.2 (Pfanzagl, 1971, p. 171) on probabilistic measurement structures and in line with Hamerle (1979) and Fischer (1987a, 1988).

The assumption, however, that there exists a 'dense' universe of items with ICCs $f(\theta - \beta)$ for all $\theta, \beta \in \mathbb{R}$, is quite strong. It might be considered unrealistic. The following considerations should shed some light on the role of this assumption.

Suppose there exists only a finite universe of items, $I_1, \ldots, I_k$, with fixed parameters β_i. Then the functional equation (2.16) has to be replaced by a set of k equations, with constants b_i,

$$m(k(\theta) - b_i) = f(\theta - \beta_i) \quad \text{for } i = 1, \ldots, k \text{ and all } \theta \in \mathbb{R}. \tag{2.23}$$

Proceeding as above, with $h := m^{-1} \circ f$,

$$k(\theta) = h(\theta - \beta_i) + b_i, \tag{2.24}$$

which implies

$$h(\theta - \beta_i) + b_i = h(\theta - \beta_j) + b_j,$$

for any pair of items (I_i, I_j) with $\beta_i \neq \beta_j$, or

$$h(x) = h(x - (\beta_j - \beta_i)) + b_j - b_i. \tag{2.25}$$

Applying (2.25) to itself,

$$h(x) = h(x - 2(\beta_j - \beta_i)) + 2(b_j - b_i),$$

and repeating this step, we obtain more generally

$$h(x) = h(x - n(\beta_j - \beta_i)) + n(b_j - b_i), \tag{2.26}$$

for $n = 0, \pm 1, \pm 2, \ldots$ For another pair of items, say (I_p, I_q) with $\beta_p \neq \beta_q$, it follows from (2.25) and (2.26) that, for integers n and m,

$$h(x) = h(x - n(\beta_j - \beta_i) - m(\beta_q - \beta_p)) + n(b_j - b_i) + m(b_q - b_p),$$

which generalizes to

$$h(x) = h\Big(x - \sum_{i<j} n_{ji}(\beta_j - \beta_i)\Big) + \sum_{i<j} n_{ji}(b_j - b_i), \tag{2.27}$$

for $n_{ji} = 0, \pm 1, \pm 2, \ldots$

Suppose, first, that the quotients $(\beta_j - \beta_i)/(\beta_q - \beta_p)$ for all pairs (I_i, I_j) and (I_p, I_q) are rational. (From an empirical point of view, this assumption can be considered unproblematic, because it cannot be refuted empirically.) Let $\Delta > 0$ be the largest real number satisfying $\beta_j - \beta_i = w_{ji}\Delta$ for all (I_i, I_j), where the w_{ji} are integers. Then the greatest common divisor of the w_{ji} is obviously 1. Now (2.27) becomes

$$h(x) = h\Big(x - \sum_{i<j} n_{ji} w_{ji} \Delta\Big) + \sum_{i<j} n_{ji}(b_j - b_i). \tag{2.28}$$

From the definition of Δ and w_{ji} it follows that there exists a particular set of integers n_{ji}^* satisfying equation (2.29)[3],

[3]Equation (2.29) is equivalent to the diophantine equation $\sum_{i<j} n_{ji}^* w_{ji} = 1$, which has an integer solution (n_{ji}^*) iff the w_{ji} have no common divisor > 1, cf. Scholz and Schoeneberg (1955, p. 20, Theorem 2.12).

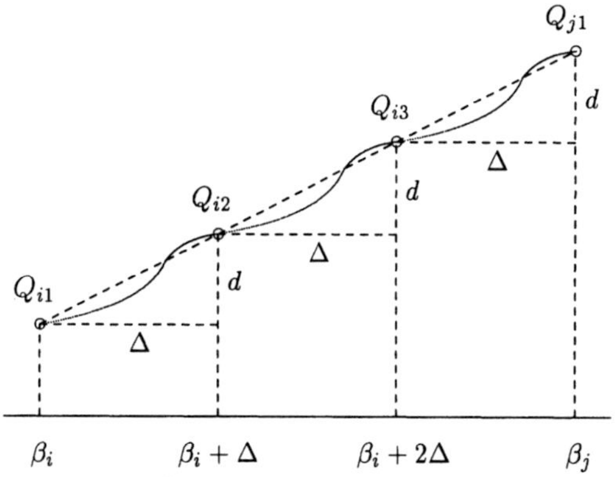

FIGURE 2.1. Equidistant points Q_{in} and a strictly monotone function $h(x)$ through these points

$$\sum_{i<j} n^*_{ji} w_{ji} \Delta = \Delta, \tag{2.29}$$

which implies that

$$h(x) = h(x - \Delta) + d, \tag{2.30}$$

with some constant d. From (2.30) it follows that

$$h(x) - \frac{d}{\Delta}x = h(x - \Delta) - \frac{d}{\Delta}(x - \Delta),$$

that is, $h^*(x) := h(x) - \frac{d}{\Delta}x$ is a periodic function with periodicity Δ; and $h(x)$ is the sum of $h^*(x)$ and a linear term, generating an infinite sequence of identical arcs as shown in Figure 2.1. The arcs connect points Q_{in} equally spaced on a straight line with slope

$$\frac{b_j - b_i}{\beta_j - \beta_i} = \frac{b_q - b_p}{\beta_q - \beta_p} = \frac{d}{\Delta} =: c, \tag{2.31}$$

for all pairs (I_i, I_j) and (I_p, I_q).

Hence, the function $h = m^{-1} \circ f$ in (2.24), and therefore also $k(\theta)$, is *not* fully specified any more. At the points Q_{in} we find the same linear relationship as we had in the proof of Lemma 2.2, but between the points Q_{in} the continuous transformation $k(\theta)$ is arbitrary, subject only to the monotonicity requirements.

If the spacing of the β_j is such that the period length Δ is small, there is little freedom left for $h = m^{-1} \circ f$ and $k(\theta)$, so that we may say θ is 'almost unique' except for linear transformations. Moreover, we shall show that, if there are at least 3 items, I_i, I_j, I_l with $\beta_i < \beta_j < \beta_l$, such that the quotient $(\beta_j - \beta_i)/(\beta_l - \beta_j)$

is irrational, then no $\Delta > 0$ exists, and h becomes a linear function $h(x) = cx + h_0$ as in the proof of Lemma 2.2.

Define $\Delta(n,m) = n(\beta_j - \beta_i) - m(\beta_l - \beta_j)$, for integers n and m. Then, by (2.30) and (2.31),

$$h(x) = h(x - \Delta(n,m)) + c\Delta(n,m), \tag{2.32}$$

which shows that $|\Delta(n,m)|$ is a period length of $h^*(x)$. However, there may still exist a shorter period. Let $\min\{|\Delta(n,m)|; 1 \le |n| \le N, 1 \le |m| \le N\} =: M_N$. Since $M_N \ge 0$ is at least weakly monotone decreasing for $N = 1, 2, 3, \ldots$, there exists a maximum lower bound, $\inf M_N$; and $\inf M_N \le \Delta(n,m) < \inf M_N + \epsilon$ with arbitrary $\epsilon > 0$ holds for infinitely many N. Suppose that $\inf M_N > 0$, and let, e.g., $\epsilon = 10^{-1}\inf M_N$; moreover, choose n_0, m_0 such that

$$\inf M_N \le \Delta(n_0, m_0) < \inf M_N + \epsilon. \tag{2.33}$$

Let, without loss of generality, $n_0 \ge 1$ (the case $n_0 \le -1$ can be treated analogously): define $\Delta^* = n_0(\beta_j - \beta_i) - |t|\Delta(n_0, m_0)$, where t is the largest integer such that $\Delta^* > 0$; obviously, $t \ge 1$. Since both $\beta_j - \beta_i$ and $\Delta(n_0, m_0)$ are period lengths, so is Δ^*, and

$$h(x) = h(x \mp \Delta^*) \pm c\Delta^*. \tag{2.34}$$

Now define p such that $\Delta^* = p\Delta(n_0, m_0)$. Then it is seen that p satisfies $0 < p < 1$: firstly, $p > 0$, because Δ^* was defined to be > 0; secondly, $p \le 1$, because $p > 1$ would imply that $\Delta^* - \Delta(n_0, m_0) > 0$, which would mean that t is not the largest integer yielding $\Delta^* > 0$. Hence two possibilities remain, $p = 1$ and $0 < p < 1$. The case $p = 1$ can be ruled out, because it would imply

$$\begin{aligned} \Delta^* = n_0(\beta_j - \beta_i) - t\Delta(n_0, m_0) &= \Delta(n_0, m_0), \\ tn_0(\beta_j - \beta_i) - (t+1)m_0(\beta_l - \beta_j) &= 0, \end{aligned}$$

which is clearly impossible, because $(\beta_j - \beta_i)/(\beta_l - \beta_j)$ was assumed to be irrational. Hence $0 < p < 1$. Therefore, from (2.34),

$$h(x) = h(x \mp p\Delta(n_0, m_0)) \pm cp\Delta(n_0, m_0), \tag{2.35}$$

which implies that $p\Delta(n_0, m_0)$ is a period length; but we also have

$$\begin{aligned} h(x) &= h(x - p\Delta(n_0, m_0) + \Delta(n_0, m_0)) + cp\Delta(n_0, m_0) - c\Delta(n_0, m_0), \\ &= h(x + (1-p)\Delta(n_0, m_0)) - c(1-p)\Delta(n_0, m_0), \end{aligned} \tag{2.36}$$

hence $(1-p)\Delta(n_0, m_0)$ is a period length, too. Since $p\Delta(n_0, m_0) + (1-p)\Delta(n_0, m_0) = \Delta(n_0, m_0)$, and $\epsilon = 10^{-1}\inf M_N < 10^{-1}\Delta(n_0, m_0)$, we have $p\Delta(n_0, m_0) < \inf M_N$ or $(1-p)\Delta(n_0, m_0) < \inf M_N$. This is a contradiction to the assumption that $\inf M_N > 0$. Hence, $\inf M_N = 0$, i.e., $M_N \to 0$ as $N \to \infty$, and therefore $h(x)$ is linear, $h(x) = cx + h_0$. □

We can summarize these findings in Lemma 2.3, based on assumption (vi).

(vi) The universe of items consists of a finite number of items, k, with $k \geq 3$, where the ICCs are of the form $f(\theta - \beta_i)$, for $\theta \in \mathbb{R}$ and $i = 1, 2, \ldots, k$, and f is a continuous bijective increasing $\mathbb{R} \rightarrow (0, 1)$ function.

Lemma 2.3 *Let (vi) hold. If all quotients $(\beta_j - \beta_i)/(\beta_q - \beta_p)$* are rational num*bers, then the θ are unique, except for positive linear transformations, at all points $\theta_{in} = \beta_i \pm n\Delta$, for $n = 0, 1, 2, \ldots$ and $i = 1, \ldots, k$, where Δ is the largest real number satisfying $\beta_j - \beta_i = w_{ji}\Delta$ for all pairs (I_i, I_j), with integers w_{ji}. At all points between the θ_{in}, the θ are unique only up to continuous, strictly monotone transformations. If there exist at least 3 items, I_i, I_j, I_l, such that $(\beta_j - \beta_i)/(\beta_l - \beta_j)$ is irrational, the θ-scale is an interval scale.*

(Remark: assumptions (i) and (ii) are contained in (vi) and hence need not be listed explicitly.)

In psychometric literature there also exist other derivations of the RM, based on variations of the sufficiency assumption. One of them, which is due to Andersen (1973a, 1977), deals with consequences of the assumption of a minimal sufficient statistic for ξ that is independent of the item parameters. The assumptions are formalized as follows.

(vii) The ICCs are of the form $g_i(\xi) = g(\xi, \delta_i)$, where the δ_i are scalar item parameters, and g a continuous function $\mathbb{R}^2 \rightarrow$ (0,1), such that, for any fixed $\delta_i \in \mathbb{R}$, $g(\cdot, \delta_i)$ is a bijective increasing $\mathbb{R} \rightarrow (0, 1)$ function, and for any fixed $\xi \in \mathbb{R}$, $g(\xi, \cdot)$ is a bijective decreasing $\mathbb{R} \rightarrow (0, 1)$ function.

(viii) Given that the test consists of items $I_1, \ldots, I_k$, there exists a nontrivial minimal sufficient statistic T for ξ that does not depend on the item parameters $\delta_1, \ldots, \delta_k$.

If (vii) and (viii) are combined with the 'technical' assumption of local independence, (iii), the following theorem can be proved:

Theorem 2.3 *Assumptions (vii), (iii), and (viii) imply that there exist continuous strictly monotone increasing functions ('scale transformations') ϕ: $\xi \rightarrow \theta$ and ψ: $\delta \rightarrow \beta$, both $\mathbb{R} \rightarrow \mathbb{R}$, such that all ICCs attain form (2.1).*

(Remark: assumptions (i) and (ii) are contained in (vii) and hence need not be listed explicitly.)

Proof As Andersen (1973a) observes, any minimal sufficient statistic T which is independent of the δ_i can be constructed by considering the special case $\delta_1 = \delta_2 = \ldots = \delta_k = \delta$, for some fixed value δ. This makes the x_i, $i = 1, \ldots, k$, realizations of identically distributed random variables. Then a result of Bahadur (1954) about the uniqueness of minimal sufficient statistics applies, namely, that such a statistic must be symmetric in its arguments. Therefore, T is some bijective function u of the unweighted raw score R, $T = u(R)$, and hence is equivalent to R.

Now the derivation can proceed as in the proof of Theorem 2.1. □

Another derivation of the RM is due to Pfanzagl (1994). He makes the assumptions that the ICCs are of the additive form (v) for all $\theta, \beta \in \mathbb{R}$, plus the existence of a nontrivial sufficient statistic T for θ that is independent of the item parameters. Since Andersen's (1973a) assumptions are weaker, we shall not present Pfanzagl's proof here.

2.2 Derivation from Composite Transitivity

A derivation similar to that from sufficiency of the raw score R (proof of Theorem 2.1) is due to Roskam and Jansen (1984); they set out from the following deterministic property of the Guttman scale: let $v\mathcal{D}i$ denote the observation that subject S_v solves item I_i (S_v 'dominates' I_i). The Guttman scale is characterized by the rule

$$\text{if}(v\mathcal{D}i \wedge \neg w\mathcal{D}i \wedge w\mathcal{D}j) \Rightarrow v\mathcal{D}j. \tag{2.37}$$

This means, if S_v solves I_i, and S_w does not solve I_i but solves I_j, then S_v solves I_j. This implies a simple *composite ordering* of the S_v and the I_i (Ducamp & Falmagne, 1969).

Let '$i\mathcal{R}j$' denote 'i is right of j' on the latent scale. It then follows that

$$\neg v\mathcal{D}i \wedge v\mathcal{D}j \Rightarrow i\mathcal{R}j, \tag{2.38}$$

irrespective of S_v's precise location between I_i and I_j on the scale. This determines the ordering of the I_i independently of which subject S_v (between I_i and I_j on the common scale) is instrumental for the ordering.

Applying the same reasoning to a probabilistic IRT framework, Roskam and Jansen (1984) introduce the following axiom:

(ix) $P(i\mathcal{R}j) := P(\neg v\mathcal{D}i \wedge v\mathcal{D}j | R_v = 1) = c$,
where R_v is the raw score on items I_i and I_j, and c is a constant independent of v.

Roskam and Jansen (1984) call a probabilistic ordering of the items induced by (ix) a *stochastically consistent ordering.*

The right-hand equality in definition (ix), however, is formally equivalent to the assumption of sufficiency of the raw score R_v for item pair (I_i, I_j), cf. (2.2); hence, the formal derivation of the RM from (ix) can proceed like that from (2.2) for $k = 2$, and need not be repeated here. In order to obtain the RM for test length k, it is necessary to make assumption (ix) for at least $k-1$ pairs of items, e.g., (I_1, I_2), (I_2, I_3), (I_3, I_4), etc. This assumption is conceptually less appealing than the assumption of the sufficiency of the raw score R in a test of length k. Roskam and Jansen (1984) postulated (ix) for all $\binom{k}{2}$ pairs (I_i, I_j). Formally, both assumptions are equivalent, because the Roskam and Jansen assumption implies the RM (2.1) for all item pairs, and hence also for the test of length k;

this entails the sufficiency of R. The sufficiency of R in a test of length k, on the other hand, implies the RM, and this yields (ix) for all item pairs.

Theorem 2.4 *Assumptions (i), (ii), (iii), and (ix) applied to all pairs of items (I_i, I_j), imply that the model is equivalent to an RM with ICCs (2.1).*

2.3 Derivation from Specific Objectivity

Rasch (1967) formulated the postulate of *specific objectivity* (SO) as a general scientific principle: within a given well-defined frame of reference, scientific statements should be *generalizable*. Scientific statements, according to Rasch, refer to comparisons of objects, and any such comparison should be generalizable beyond the particular experimental situation or instrument ('agent') on which it is based. Scientific results satisfying this generalizability requirement were called 'specifically objective comparisons'. Rasch preferred the term 'objective' to 'generalizable' because it conveys the idea that even a subjective or arbitrary choice of an agent should not bias the result. He considered SO a very general epistemic principle, stressing, however, that SO in each case refers to a particular, specific frame of reference: "Two features seem indispensable in scientific statements: they deal with comparisons, and they must be *objective*. To complete these requirements I have to specify the *kind of comparisons* and the *precise meaning of objectivity*. When doing so I do not feel confident that all sorts of what can justifiably be called 'science' are covered, but certainly a large area of science is." (Rasch, 1967, p. 4).

The favorite example used by Rasch (1960, 1967) to illustrate the meaning of the objectivity of comparisons is based on the joint definition and measurement of mass and force in classical mechanics: let the objects of interest be rigid bodies, O_v, $v = 1, 2, \ldots$, and let their masses be M_v. Moreover, let there be some experimental conditions (agents) where forces F_i are applied in turn to each of the masses, such that accelerations A_{vi} become observable. According to the Second Newtonian Axiom (force = mass × acceleration), an observed acceleration is proportional to the force exerted on the object, and inversely proportional to the object's mass,

$$A_{vi} = M_v^{-1} F_i. \tag{2.39}$$

The comparison of any two objects O_v and O_w with respect to their masses M_v and M_w is easily carried out by means of the quotient

$$\frac{A_{vi}}{A_{wi}} = \frac{M_v^{-1} F_i}{M_w^{-1} F_i} = \frac{M_w}{M_v}, \tag{2.40}$$

which is *independent of* F_i. (The quotient (2.40) can alternatively be replaced, upon a logarithmic transformation, by the parameter difference $\ln(A_{vi}/A_{wi}) = \ln M_w - \ln M_v$. This implies that any two masses can be compared without knowledge of, and hence independently of, the force(s) F_i that are applied. In other words, the comparison of any two masses is *generalizable* across the experimental conditions which produce the forces F_i.

It is important to mention that acceleration can be defined independently of the concepts of mass and force, namely, as the second derivative with respect to time of the distance covered; hence, both the *simultaneous definition* of mass and force and the comparison (or *measurement*) of masses (and, similarly, also of forces) can be based on (2.40).

Rasch (1967) undertook to apply basically the same approach to the problem of comparing the abilities of any two subjects, S_v and S_w, based on their reactions to a test item, I_i. He made the following assumptions: let a set of subjects, $S_1, \ldots, S_n$, and a set of test items, $I_1, \ldots, I_k$, be given, such that a response (a random variable R_{vi}) is generated for each pair (S_v, I_i). These reaction variables are assumed to be 'observable', either directly (deterministic case), or indirectly via the registration of their realizations (probabilistic case). Let each subject S_v be fully characterized, as far as his/her reactions to items $I_1, \ldots, I_k$ are concerned, by a latent trait parameter ξ_v; similarly, let each item be fully characterized by a parameter δ_i, and the reaction variable R_{vi}, by a reaction parameter ρ_{vi}. (All these parameters may be vectors, but we restrict the present considerations to scalar parameters.) Furthermore, Rasch demanded that the comparison of any two subjects S_v and S_w with respect to their parameters ξ_v and ξ_w be *always possible* once ρ_{vi} and ρ_{wi} are known, and that the result of the comparison be *always unique*. The existence and uniqueness of the results of such comparisons implies that ρ_{vi} must be some function F ('reaction function') of ξ_v and δ_i,

$$\rho_{vi} = F(\xi_v, \delta_i). \tag{2.41}$$

In psychology, the comparison of any two subjects, S_v and S_w, actually has to be based on realizations of the reaction variables R_{vi} and R_{wi}, rather than on the unobservable parameters ρ_{vi} and ρ_{wi}. Hence, there should exist some function $W(R_{vi}, R_{wi})$ that allows one to carry out the comparison of S_v and S_w. Yet it is quite instructive to begin by looking for a function $U(\rho_{vi}, \rho_{wi})$ that yields a result in terms of ξ_v and ξ_w, because the parameters carry all the pertinent information contained in the reaction variables. So we assume the existence of a function

$$U[F(\xi_v, \delta_i), F(\xi_w, \delta_i)] =: V(\xi_v, \xi_w) \tag{2.42}$$

which, according to the postulate of SO, is independent of δ_i, and hence can be denoted $V(\xi_v, \xi_w)$. Function U is called a *comparator*.

To derive results from (2.42), the assumptions about the parameters and the functions U and V have to be formalized:

(x) *Let the S_v be characterized by one parameter $\xi_v \in \mathbb{R}$ each, the I_i by one parameter $\delta_i \in \mathbb{R}$ each, and let a random variable R_{vi} exist for each pair (S_v, I_i) that is characterized by a reaction parameter $\rho_{vi} \in \mathbb{R}$; let* *the reaction parameters be determined by the respective person and item parameters,* $\rho = F(\xi, \delta)$, *where* F *is a continuous* $\mathbb{R}^2 \to \mathbb{R}$ *function, such that* $z = F(\cdot, y)$, *for any fixed* y, *is an increasing bijective* $\mathbb{R} \to \mathbb{R}$ *function, and* $z = F(x, \cdot)$, *for any fixed* x, *a decreasing bijective* $\mathbb{R} \to \mathbb{R}$ *function.*

(Remark: suppose two subjects S_v and S_w with parameters $\xi_v \neq \xi_w$ had the same reaction variable with respect to some item I_i, implying $\rho_{vi} = \rho_{wi}$; uniqueness of the comparison of S_v and S_w based on I_i would then entail a contradiction to $\xi_v \neq \xi_w$, because the result $\xi_v = \xi_w$ would always be consistent with $\rho_{vi} = \rho_{wi}$. This motivates the monotonicity assumption regarding the function $z = F(\cdot, y)$. The other assumptions have similar justifications.)

Now the assumption of SO is stated as follows.

(xi) *A comparator* $U(\rho_{vi}, \rho_{wi})$, $\mathbb{R}^2 \to \mathbb{R}$, *continuous, strictly monotone increasing in* ρ_{vi} *and decreasing in* ρ_{wi}, *is assumed to exist for all* $\rho_{vi}, \rho_{wi} \in \mathbb{R}$, *satisfying the functional equation*

$$U[F(x, z), F(y, z)] =: V(x, y), \tag{2.43}$$

for all $x, y, z \in \mathbb{R}$, *where* V *is independent of* z.

As will be seen, a strong result can be derived from (x) and (xi). The result is due to Rasch (1968, 1972, 1977); however, one of these articles contains no proofs, and in two of them the proof is barely sketched and uses stronger assumptions than necessary. We shall derive the main result in a more general form employing functional equations theory. Related work is due to Fischer (1987a, 1988), Irtel (1987, 1994), Neuwirth (1988), and Tutz (1989).

Lemma 2.4 *Assumptions (x) and (xi) imply that there exist continuous bijective increasing functions* k, m, u, *and* f, *all* $\mathbb{R} \to \mathbb{R}$, *such that*

$$U(\rho_{vi}, \rho_{wi}) = V(\xi_v, \xi_w) = u[k(\xi_v) - k(\xi_w)] = u(\theta_v - \theta_w),$$
$$F(\xi_v, \delta_i) = f[k(\xi_v) - m(\delta_i)] = f(\theta_v - \beta_i),$$

where $\theta_v := k(\xi_v)$ *and* $\beta_i := m(\delta_i)$.

Proof The strict monotonicity of F in both variables implies that, for any fixed z, $F(y, z) = w$ can be inverted, i.e., there exists a continuous function G such that $G(w, z) = y$, where G is strictly monotone increasing in both variables. Inserting this in (2.43) yields

$$U[F(x, z), w] = V[x, G(w, z)], \tag{2.44}$$

which is a 'generalized functional equation of associativity' (Aczél, 1966, pp. 310–312). Under the continuity and monotonicity assumptions made above, the general continuous solution of (2.44) is of the form

$$\begin{aligned} U(x, y) &= l[a(x) - b(y)], \quad F(x, y) = a^{-1}[k(x) - m(y)], \\ V(x, y) &= l[k(x) - h(y)], \quad G(x, y) = -h^{-1}[-m(x) - b(y)], \end{aligned} \tag{2.45}$$

where a, b, h, k, l, m are continuous bijective increasing $\mathbb{R} \to \mathbb{R}$ functions.

Inserting (2.45) in (2.43) and using the symbol '$\circ$' for the concatenation of functions,

$$l\left[a \circ a^{-1}[k(x) - m(z)] - b \circ a^{-1}[k(y) - m(z)]\right] = l[k(x) - h(y)],$$

$$b \circ a^{-1}[k(y) - m(z)] = h(y) - m(z),$$

and setting $s := k(y)$, so that $y = k^{-1}(s)$, and $t := m(z)$, it is seen that

$$b \circ a^{-1}(s - t) = h \circ k^{-1}(s) - t. \tag{2.46}$$

For $t = 0$, (2.46) implies

$$b \circ a^{-1}(s) = h \circ k^{-1}(s), \tag{2.47}$$

and, similarly, for $s = 0$,

$$b \circ a^{-1}(-t) = h \circ k^{-1}(0) - t,$$

which is equivalent to

$$b \circ a^{-1}(t) = t + c, \tag{2.48}$$

where $c := h \circ k^{-1}(0)$ is a constant. Furthermore, setting $p := a^{-1}(t)$, so that $t = a(p)$, it follows from (2.48) that

$$b(p) = a(p) + c. \tag{2.49}$$

Similarly, (2.47) and (2.48) imply

$$h(q) = k(q) + c. \tag{2.50}$$

Inserting (2.49) and (2.50) in (2.45), with $u(x) := l(x - c)$, yields

$$\begin{aligned} U(x, y) &= l[a(x) - a(y) - c] = u[a(x) - a(y)], \\ F(x, y) &= a^{-1}[k(x) - m(y)], \\ V(x, y) &= l[k(x) - k(y) - c] = u[k(x) - k(y)]. \end{aligned}$$

Reinserting this in (2.43),

$$\begin{aligned} U\left(F(x, z), F(y, z)\right) = u[a(F(x, z)) - a(F(y, z))] = \\ u[a \circ a^{-1}(k(x) - m(z)) - a \circ a^{-1}(k(y) - m(z))] = \\ u[k(x) - k(y)] = V(x, y). \end{aligned} \tag{2.51}$$

Denoting $a^{-1}(x)$ by $f(x)$ for simplicity, the response function F in (2.43) attains its final form

$$F(x, y) = f[k(x) - m(y)], \tag{2.52}$$

and the comparator becomes

$$V(x, y) = u[k(x) - k(y)]. \tag{2.53}$$

Returning to our original notation and defining $k(\xi) =: \theta$ and $m(\delta) =: \beta$, we find that the reaction function can be written as

$$F(\xi_v, \delta_i) = f(\theta_v - \beta_i), \tag{2.54}$$

and the comparator as

$$V(\xi_1, \xi_2) = u(\theta_v - \theta_w), \tag{2.55}$$

which is the desired result. □

This shows that SO, together with continuity and monotonicity, implies a certain 'latent additivity' (Rasch, 1972; or 'subtractivity', in the present notation) of the response function F and the comparator U. Subtractivity of the parameters in F and U is achieved by applying appropriate scale transformations $k(\xi)$ and $m(\delta)$ to the original latent scales, establishing new scales θ and β. Although the original scales ξ and δ are only ordinal by nature of the psychological notion of 'ability' and 'difficulty', the transformations $k(\xi)$ and $m(\delta)$ are highly specific and unique except for linear transformations (see Lemma 2.2 above); that is, the resulting subtractive scales θ and β have interval scale properties.

There is a weakness in the above derivation of latent additivity/subtractivity: SO is defined for each item I_i separately, not for a test of length k. It is conceivable that a comparator $U(\rho_{v1}, \ldots, \rho_{vk}; \rho_{w1}, \ldots, \rho_{wk}) = V(\xi_v, \xi_w)$ might exist, whereas (2.42) does not hold for some or all single items. This question, however, has scarcely been raised in the literature (cf. Fischer, 1987a; Tutz, 1989). Rasch's motivation for dealing with single items was that he demanded the result of the comparison to be independent of the choice of the set of agents (items); any single agent (item) should therefore suffice for the comparison.

The subtractivity of θ and β in the reaction function f, and the analogous subtractivity of θ_v and θ_w in the comparator function U, is very general, without any specificity for IRT. It applies to the case of force and mass in physics (namely, $\ln A_{vi} = \ln F_i - \ln M_v$) in the same way as to ability and item difficulty in psychology. In the latter domain, however, additional specific assumptions can be brought to play, which surprisingly lead to a very concrete result about the family of IRT models compatible with SO: as will be seen, within the usual (dichotomous) IRT framework a family of RMs is the most general class of specifically objective IRT models.

To derive this result, some modification of the assumption of SO is required, and one additional postulate has to be introduced: within IRT, most results about items and/or subjects are based on the *likelihood* of observed responses (dichotomous responses, in our case). It will therefore be posited that the comparator U is a function of an unconditional or conditional likelihood of observed responses. To formalize this assumption, the following notation is needed: let

$A = \{(+\,+), (+\,-), (-\,+), (-\,-)\}$ be the sample space of all possible patterns of responses of two subjects S_v and S_w to one item I_i, and let B and C be subsets of this sample space, $C \subset B \subseteq A$.

(xii) *For any two subjects S_v and S_w, and any item I_i, there exists a comparator U: $(0,1) \times (0,1) \to \mathbb{R}$ of the form $U[g(\xi_v, \delta_i), g(\xi_w, \delta_i)]$, where $g_i(\xi) = g(\xi, \delta_i)$ are ICCs, and where U satisfies the following conditions:*

(a) *$U(x, y)$ is continuous, strictly monotone increasing in x for all y, and strictly monotone decreasing in y for all x;*

(b) *$U[g(\xi_v, \delta_i), g(\xi_w, \delta_i)] =: V(\xi_v, \xi_w)$ is independent of δ_i, for all $\xi_v, \xi_w, \delta_i \in \mathbb{R}$ (i.e., the comparison of S_v and S_w based on any item I_i is specifically objective);*

(c) *Likelihood Principle: U is a function of an unconditional likelihood $P(B)$ or of a conditional likelihood $P(C|B)$.*

Theorem 2.5 *Assumptions (vii), (iii), and (xii) imply that there exist continuous bijective increasing functions k: $\xi \to \theta$ and l: $\delta \to \beta$, both $\mathbb{R} \to \mathbb{R}$, such that the ICCs for all S_v and I_i assume form (2.22). The θ and β are unique except for linear scale transformations with a common multiplicative constant.*

Proof Consider the responses of S_v and S_w to some item I_i. We have to discuss whether there exists some nontrivial likelihood function $P(B)$ with $B \subseteq A$, or some conditional likelihood function $P(C|B)$ with $C \subset B \subseteq A$, satisfying the above assumptions. Rasch (1968) has shown that the two complementary conditional likelihoods $P[(+\,-)|(+\,-) \vee (-\,+)]$ and $P[(-\,+)|(+\,-) \vee (-\,+)]$ are the only answer to this question. The present proof closely follows Rasch (1968), based on an investigation of all possible likelihood functions, requiring a lengthy enumeration of cases. For brevity, however, we shall restrict the present exposition to a few examples.

Unconditional likelihoods $P(B)$ will be discussed first. Let $g(\xi_v, \delta_i)$ be denoted by p_{vi} for simplicity. If B consisted of a single element of set A, e.g., $B = \{(+\,-)\}$, then $P(B) = p_{vi}(1 - p_{wi})$ would have to be of the form $V(\xi_v, \xi_w)$, where V is some function of ξ_v and ξ_w, but is independent of δ_i. Since, however, for $\xi_v = \xi_w$ the left-hand side is $p_{vi} - p_{vi}^2$ and obviously depends on δ_i, a contradiction occurs. In a similar way, all other single elements of A can also be ruled out; and, since triplets of elements of A are complementary to single elements, no triplet is a possible candidate for B either.

Let pairs of elements be considered next, e.g., $B = \{(-\,+), (+\,+)\}$; in this case, $P(B) = (1 - p_{vi})p_{wi} + p_{vi}p_{wi} = p_{wi} = V(\xi_v, \xi_w)$; this obviously contradicts assumptions because the left-hand side is independent of ξ_v. Consider $B = \{(-\,+), (+\,-)\}$ as another example: $P(B) = (1 - p_{vi})p_{wi} + p_{vi}(1 - p_{wi})$, which for $\xi_v = \xi_w$ becomes $2(1 - p_{vi})p_{vi} = 2(p_{vi} - p_{vi}^2) = V(\xi_v, \xi_v)$. This is the same contradiction as in the case discussed in the last paragraph. All other pairs of elements can be eliminated in a similar way.

Now only conditional likelihood functions $P(C|B)$ are left over. Obviously, B cannot be a single element, since C would be the empty set; so B must be a pair or a triplet. There exist 24 possible combinations of triplets B with subsets $C \subset B$, of which, however, many are only permutations of each other. The case $B = \{(-\,-), (-\,+), (+\,+)\}$ with $C = \{(-\,+)\}$ will be considered as an example. We obtain

$$P(C|B) = \frac{(1-p_{vi})p_{wi}}{(1-p_{vi})(1-p_{wi}) + (1-p_{vi})p_{wi} + p_{vi}p_{wi}}. \tag{2.56}$$

For $\xi_v = \xi_w$, (2.56) becomes

$$\frac{p_{vi}(1-p_{vi})}{1-p_{vi}(1-p_{vi})},$$

which should be independent of δ_i. This implies that $p_{vi}(1-p_{vi}) = p_{vi} - p_{vi}^2$ be independent of δ_i, which is the same contradiction as before. All other triplets can be eliminated in a similar manner.

Now only pairs of elements of A remain for set B, of which six exist. Four of them are immediately ruled out, namely, $\{(-\,-), (-\,+)\}$, $\{(-\,-), (+\,-)\}$, $\{(-\,+), (+\,+)\}$, $\{(+\,-), (+\,+)\}$; in all these cases, B would completely determine the response of either S_v or of S_w, such that $P(C|B)$ would equal one of the unconditional probabilities p_{vi}, $1-p_{vi}$, p_{wi}, $1-p_{wi}$, none of which can be independent of δ_i.

Hence, only the two pairs $\{(--), (++)\}$ and $\{(+-), (-+)\}$ remain. In the first of the two, e.g., for $C = \{(++)\}$, we get

$$P(C|B) = \frac{p_{vi}p_{wi}}{(1-p_{vi})(1-p_{wi}) + p_{vi}p_{wi}} = \frac{\dfrac{p_{vi}}{1-p_{vi}}\dfrac{p_{wi}}{1-p_{wi}}}{1+\dfrac{p_{vi}}{1-p_{vi}}\dfrac{p_{wi}}{1-p_{wi}}}. \tag{2.57}$$

Let $\xi_v = \xi_w$; then (2.57) is strictly monotone increasing in p_{vi}, which is not independent of δ_i. So again we meet a contradiction. – The case $C = \{(-\,-)\}$ is analogous.

Hence, only $B = \{(+\,-), (-\,+)\}$ remains. For $C = \{(+\,-)\}$, the conditional likelihood $P(C|B)$ becomes

$$P(C|B) = \frac{p_{vi}(1-p_{wi})}{p_{vi}(1-p_{wi}) + (1-p_{vi})p_{wi}}, \tag{2.58}$$

which is assumed to be independent of δ_i, implying

$$\frac{p_{vi}(1-p_{wi})}{(1-p_{vi})p_{wi}} = \frac{V(\xi_v, \xi_w)}{1-V(\xi_v, \xi_w)}. \tag{2.59}$$

This expression characterizes the comparison of two subjects S_v and S_w on the basis of item I_i and is formally analogous to (2.6) (which, however, refers to the comparison of two items based on the responses of one subject). Hence, here again we may expect the same result as in Theorem 2.1. Let strictly monotone functions $k_l(\xi)$ be introduced for all items I_l as follows:

$$\ln\left(\frac{g_l(\xi)}{1-g_l(\xi)}\right) =: k_l(\xi). \tag{2.60}$$

Using (2.60) for item I_i, (2.59) yields

$$k_i(\xi_v) - k_i(\xi_w) = \ln\left(\frac{V(\xi_v,\xi_w)}{1-V(\xi_v,\xi_w)}\right),$$

which should be independent of I_i. Hence, for an arbitrarily chosen 'standard item' I_0,

$$k_i(\xi_v) - k_i(\xi_w) = k_0(\xi_v) - k_0(\xi_w),$$

$$k_i(\xi_v) - k_0(\xi_v) = k_i(\xi_w) - k_0(\xi_w), \tag{2.61}$$

for all $\xi_v, \xi_w \in \mathbb{R}$. Since the left hand side of (2.61) depends on ξ_v, and the right-hand side on ξ_w, (2.61) must in fact be a constant depending only on items I_i and I_0. The right-hand side of (2.61) may therefore be denoted by $-\beta_{0i}$, giving

$$k_i(\xi_v) = k_0(\xi_v) - \beta_{0i}. \tag{2.62}$$

Inserting (2.62) in (2.60),

$$\ln\left(\frac{g_i(\xi)}{1-g_i(\xi)}\right) = k_0(\xi_v) - \beta_{0i},$$

$$g_i(\xi_v) = \frac{\exp(k_0(\xi_v)-\beta_{0i})}{1+\exp(k_0(\xi_v)-\beta_{0i})}, \tag{2.63}$$

which is (2.1), with $\theta_v := k_0(\xi_v)$ and $\beta_i := \beta_{0i}$, as a special case of (2.22). The uniqueness of the representation up to linear transformations $a\theta+b_1$ and $a\beta+b_2$, $a>0$, and hence also the family (2.22), again follow from Lemma 2.2. □

2.4 Discussion

To give a better overview of the results in this chapter, the theorems and their assumptions are listed in Table 2.1. As this immediately shows, all derivations of RMs (i.e., Theorems 2.1 through 2.5) are based on (i) continuity and strict monotonicity of the ICCs, (ii) definition of limits of the ICCs (exclusion of guessing), and (iii) local stochastic independence. In addition to these, either of the following additional assumptions can be used to derive a family of RMs:

Table 2.1: Overview of the theorems and their assumptions. T2.1 through T2.6 denote the theorems, L2.1 through L2.4 the lemmata, '•' the assumptions employed; '(•)' means that this assumption is needed but is already contained in another one.

	T2.1	L2.1	L2.2	T2.2	L2.3	T2.3	T2.4	L2.4	T2.5
(i) Continuity and strict monotonicity of $g_i(\xi)$, for $\xi \in \mathbb{R}$.	•	•	(•)	(•)	(•)	(•)	•		(•)
(ii) No guessing.	•	•	(•)	(•)	(•)	(•)	•		(•)
(iii) Local stochastic independence.	•	•		•		•	•		•
(iv) Sufficiency of R_v.	•	•		•					
(v) ICCs of the form $f(\theta - \beta)$, for all $\theta, \beta \in \mathbb{R}$, continuity, monotonicity, no guessing.			•	•					
(vi) ICCs of the form $f(\theta - \beta_i)$, for all $\theta \in \mathbb{R}$, $i = 1,\ldots,k$, $k \geq 3$, continuity, monotonicity, no guessing.					•				
(vii) ICCs of the form $g_i = g(\xi, \delta_i)$, for all $\theta, \delta_i \in \mathbb{R}$, continuity, monotonicity, no guessing.						•			•
(viii) Existence of a nontrivial sufficient statistic T for ξ that is independent of the δ_i.						•			
(ix) Stochastically consistent ordering of the items.							•		
(x) Scalar parameterization of persons and items, continuity and monotonicity of the reaction function.								•	
(xi) SO.								•	
(xii) SO plus Likelihood Principle.									•

(iv), sufficiency of the (unweighted) raw score R for person parameter ξ (Theorems 2.1 and 2.2); or,

(vii), scalar parameterization of the items, plus (viii), the existence of a nontrivial sufficient statistic T for ξ that is independent of the item parameters (Theorem 2.3); or

(ix), stochastically consistent ordering of the items (Theorem 2.4); or

(x), SO plus the Likelihood Principle (Theorem 2.5).

Comparing Theorems 2.1 and 2.4, and considering Lemma 2.1, shows that, within the framework of the technical assumptions (i), (ii), and (iii), sufficiency of the raw score and stochastically consistent ordering, postulated for all item pairs, are equivalent assumptions. Summarizing these results, we may say that there exist two essentially different approaches to deriving – and hence justifying – the RM: sufficiency as an important property from point of view of practical IRT applications, and SO as the methodological principle of generalizability of results (comparisons) over the actual testing instrument(s). The preference of one over the other is primarily a matter of taste.

In all cases, the derivations do *not* lead to the RM in its usual form, but rather to a 'family of RMs' (2.22), that is, to a logistic model where all items have the same (unspecified) discrimination parameter. If this family of models holds for a dense universe of items, it may be concluded that the scales for person and item measurement are *interval scales* with the same unit of measurement and with possibly different (arbitrary) zero points. This is in accordance with a general result of Pfanzagl (1971, p. 171) on probabilistic measurement structures, and with Colonius' (1979) and Hamerle's (1982) conclusions regarding Rasch measurement. (The latter author uses the term 'strong latent trait model' for IRT models satisfying (v).) As Lemma 2.3 shows, the result about the scale properties of the RM is slightly weakened if only a finite universe of items is presumed, but without practical consequences for the application of the scales of θ and β.

These conclusions, however, contradict common interpretations of the RM and of its scale properties. There exist two other widespread opinions:

1. Some authors have conjectured that the RM has only rank order scale properties (Roskam, 1983, p. 83; Wottawa, 1980, p. 207). They argue that the psychological concepts of 'ability' and 'item difficulty' are so weak that any monotone scale transformation is admissible at any time. This is, of course, true for the *initial* scales, denoted ξ and δ in the present paper; but whatever scale is chosen for ξ and δ, there always exist monotone scale transformations into new scales θ and β such that θ and β enter subtractively in the argument of the ICCs f_i, and these transformations k: $\xi \rightarrow \theta$ and l: $\delta \rightarrow \beta$ are unique except for linear transformations; this makes the scales of θ and β interval scales.

The argument that instead of θ and β we might still use any strictly monotone functions $\phi(\theta)$ and $\psi(\beta)$, is irrelevant for the present question of scale properties. Consider the case of mass and force: to measure masses, we might also use $\sqrt{M}$, and for forces, $\ln F$, if we were so inclined; but this would not change the validity of the (obviously correct) statement that the scales of M and F underlying (2.40) are ratio scales. In the same sense, scales for θ and β defined on the basis of (v) – or of other equivalent assumptions – are interval scales by virtue of Lemma 2.2.

2. Another rather common interpretation of the RM says that the scales of θ and β, as defined by the ICCs (2.1), are difference scales (i. e., unique except for shifts), which makes the scales defined by the *multiplicative form* of the RM (cf. Section 3.3),

$$P(+|S_v, I_i) = \frac{\xi_v \epsilon_i}{1 + \xi_v \epsilon_i}, \tag{2.64}$$

ratio scales. This statement is formally true only if model (2.1) is assumed to be *given*: (2.1) defines the common discrimination parameter of the items as 1, and hence allows only shifts of the θ and β scales. Taking antilogs in (2.1) yields (2.64), that is, the scales for ξ and ϵ in (2.64) are unique except for multiplication with a constant. But this is of little relevance to the present question of the scale level of the psychological concepts of 'ability' and 'difficulty'. The measurement properties must be determined by empirically testable laws, such as SO (= generalizability over items or subjects). Whether SO is introduced as a postulate, or whether it follows from the assumption of sufficiency of the raw score, is immaterial, because in both cases it is an empirically testable model property. (For tests of fit, see Chapter 5.) Such tests, however, are not sensitive to changes of the common discrimination parameter of all items and hence cannot be used as arguments for specifying $a = 1$ in (2.22).

Nevertheless, Rasch seems to have thought that the multiplicative form of the model (2.64) yields ratio scales of ability and of item difficulty, even if he was not outspoken in this respect. (Terms like 'scale' or 'scale properties' are not traditional concepts in mathematics and statistics, and hence were not used by him.) Rasch (1960, p. 74) explicitly says that comparisons like 'one person is twice as able as another' are empirically meaningful, and elsewhere he similarly writes with respect to item difficulty: "... the standard item having a 'unit of easiness', in multiples of which the degree of easiness of the other items are expressed." (Rasch, 1960, p. 4). Similarly, Rasch's student and co-worker Stene (1968, p. 235) writes "Die Größe ϵ_i/ϵ_j mißt das Verhältnis der Schwierigkeit der beiden Items" ["The magnitude of ϵ_i/ϵ_j measures the ratio of difficulties of the two items"; translated by the present author]. Many others followed Rasch and postulated ratio scale properties for the RM (2.64) (e.g., Fischer, 1974, p. 210; 1983, p. 623; Hamerle, 1982, p. 83).

As will be sufficiently clear from the results presented in this chapter, the present author rejects interpretation 1 and no longer adheres to interpretation 2, because the specific form of the RM in (2.64) can be justified neither by empirically testable nor theoretically motivated assumptions. Interestingly, Rasch (1960, p. 121) also points to exactly this problem; he mentions a class of admissible scale transformations that are equivalent to linear transformations of θ and β, without, however, pursuing the matter further.

All the above conclusions about the measurement properties of the RM rest on the consideration of admissible transformations of the parameters for given probabilities $P(+|\theta, \beta)$. Strictly speaking, this analysis is incomplete, because the probabilities are unobservable; what we do have is observations of Bernoulli variables governed by the $P(+|\theta, \beta)$. It may therefore happen that, in spite of the above results, the parameters are still *not* unique up to positive linear transformations. Fortunately, there exist uniqueness results regarding the maximum likelihood estimators $\hat{\beta}_i$ and $\hat{\theta}_v$ in finite data sets: Fischer (1981) gives necessary and sufficient conditions for the uniqueness of the normalized (discrimination $\alpha = 1$, $\sum_i \beta_i = 0$) unconditional or joint (JML) solution and for the conditional (CML) maximum likelihood solution in the RM (see Chapter 8, Sections 8.4 and 8.5). The essential requirement, namely, 'well-conditionedness' of the data, is obtained almost surely as $n \to \infty$. Hence, uniqueness of the normalized solution will be attained in practically all real-life applications, except in certain cases of structurally incomplete data (cf. Chapters 9 and 10).

The practical methods for measuring person and item properties in specifically objective form rest on the CML approach, where both sets of parameters are separated from each other (for details, see Chapter 3, Section 3.3). Similarly, specifically objective hypothesis tests can be carried out by means of conditional LR tests (see Chapter 5, Section 5.4).

3

Estimation of Item Parameters

Ivo W. Molenaar [1]

ABSTRACT The introduction of this chapter sketches the problem of the estimation of item parameters and the notation in the case of incomplete data. Then the joint, conditional, and marginal maximum likelihood methods are discussed. A final section briefly mentions a few other methods not based on likelihoods.

3.1 Introduction

There are several ways to obtain estimates of item parameters from a data matrix containing the dichotomously scored answers of a sample of n persons to k items. In the early publications on parameter estimation, it was assumed that this matrix is completely observed. It is a major advantage of IRT, however, that its models and methods can easily be generalized to situations where different subsets of the items are presented to certain subgroups of respondents. In many later publications, and also in this chapter, this case of data 'missing by design' is included. If data are missing for other reasons, our derivations are only valid in the case of 'ignorable non-response' (see, e.g., Rubin, 1987), which roughly means that the fact that a certain item was not responded to is not related to the parameters of person and item. Mislevy and Wu (1988) argue convincingly that this will seldom be the case when a subject decides to skip an achievement item after seeing its content. They also discuss models proposed by Lord (1974, 1983b) for achievement items not reached in a time limit test. Also in the case of attitude measurement, ignorability is often not plausible and should be investigated before the methods outlined in this chapter are applied to incomplete data where the incompleteness design was not planned in advance. In order to stay on the safe side, the present text speaks only about items not presented to a person.

If subject S_v was not asked to respond to item I_i, one puts the element b_{vi} of the design matrix $\boldsymbol{B}$ equal to 0 and the corresponding element x_{vi} of the data matrix $\boldsymbol{X}$ equal to some arbitrary value a $(0 < a < 1)$. Thus,

$$x_{vi} = \begin{cases} 1 & \text{if } S_v \text{ responded positively to } I_i, \\ 0 & \text{if } S_v \text{ responded negatively to } I_i, \\ a & \text{if } I_i \text{ was not presented to } S_v, \end{cases}$$

[1] Rijksuniversiteit Groningen, Vakgroep Statistiek en Meettheorie, Grote Kruisstraat 2/I, 9712 TS Groningen; e-mail: W.MOLENAAR@PPSW.RUG.NL

and

$$b_{vi} = \begin{cases} 1 & \text{if } I_i \text{ was presented to } S_v, \\ 0 & \text{if } I_i \text{ was not presented to } S_v. \end{cases}$$

In the case of complete data, $b_{vi} = 1$ for all pairs (v, i) and the formulas can be simplified. It will be assumed that the matrix $\boldsymbol{B}$ is known before the answers are obtained; see above for the more general case of ignorability, and see Glas (1988b) for a discussion of multistage testing in which a part of $\boldsymbol{B}$ depends on the answers to the routing test(s).

Let $\boldsymbol{\beta}$ denote the k-vector of item parameters, and $\boldsymbol{\theta}$ the n-vector of person parameters. For each cell of the data matrix $\boldsymbol{X}$ one has, for $x_{vi} \in \{0, a, 1\}$ and corresponding $b_{vi} \in \{0, 1\}$,

$$P(X_{vi} = x_{vi}|\theta_v, \beta_i) = \frac{\exp[b_{vi}x_{vi}(\theta_v - \beta_i)]}{[1 + \exp(\theta_v - \beta_i)]^{b_{vi}}}. \tag{3.1}$$

In the unobserved cells one has by definition $x_{vi} = a$. By local independence and independence between persons, it follows that, for any choice of zero or one in the observed cells,

$$P(\boldsymbol{X} = \boldsymbol{x}|\boldsymbol{\theta}, \boldsymbol{\beta}) = \prod_{v=1}^{n} \prod_{i=1}^{k} \frac{\exp[b_{vi}x_{vi}(\theta_v - \beta_i)]}{[1 + \exp(\theta_v - \beta_i)]^{b_{vi}}}. \tag{3.2}$$

If all item parameters β_i and all person parameters θ_v were known, this would define a probability distribution on all $n \times k$ matrices with entries from the set $\{0,1\}$ for all real answers, and entries a for all missing observations.

If a test with known item parameters β_i is given to a fresh sample of persons, one can estimate their person parameters θ_v. This is the topic of Chapter 4. Here we deal with the situation in which both θ_v and β_i are unknown and have to be estimated. The prime interest in the present chapter is on estimating the item parameters β_i; the person parameters θ_v act as nuisance parameters.

In this situation, one restriction must be imposed in order to obtain a unique solution. In Chapter 2 the form (2.22) was derived; above it was already assumed that the scale constant a in that formula equals 1, but one must also fix the origin of the scale, which amounts to fixing b in (2.22). It is usual to impose the restriction $\sum_i \beta_i = 0$; this convention is followed in the present chapter unless the contrary is stated. One could also put the population mean of the person parameters equal to 0, or select a fixed item I_i for which the item parameter equals 0. Note that the latter leads to different standard errors of the parameter estimates because they are correlated across items (Verhelst, 1993).

Two basic distinctions between estimation methods underlie the structure of this chapter. The first is whether the person parameters are

- jointly estimated with the item parameters, or
- eliminated by conditioning, or

- integrated out by marginalization.

The second distinction is whether estimation takes place by

- maximum likelihood, or
- some other method.

3.2 Joint Maximum Likelihood (JML)

It follows from (3.2) that the unconditional or joint log-likelihood equals

$$\ln L(\boldsymbol{\theta}, \boldsymbol{\beta}) = \sum_{v=1}^{n}\sum_{i=1}^{k} b_{vi}x_{vi}(\theta_v - \beta_i) - \sum_{v=1}^{n}\sum_{i=1}^{k} b_{vi}\ln(1+\exp(\theta_v - \beta_i)) =$$

$$= \sum_{v=1}^{n} x_{v.}\theta_v \; - \; \sum_{i=1}^{k} x_{.i}\beta_i \; - \; C(\boldsymbol{\theta}, \boldsymbol{\beta}). \tag{3.3}$$

Note that the second double sum, denoted by $C(\boldsymbol{\theta}, \boldsymbol{\beta})$, does not depend on the data. Equation (3.3) shows that one deals with an exponential family in which the column and row sums of observed values,

$$x_{.i} = \sum_{v=1}^{n} b_{vi}x_{vi} \tag{3.4}$$

and

$$r_v = x_{v.} = \sum_{i=1}^{k} b_{vi}x_{vi}, \tag{3.5}$$

are sufficient statistics for the item parameters β_i and the person parameters θ_v, respectively.

By putting the derivatives of (3.3) with respect to each parameter equal to zero, one obtains that the JML estimates satisfy

$$r_v = \sum_{i=1}^{k} \frac{b_{vi}\exp(\theta_v - \beta_i)}{1+\exp(\theta_v - \beta_i)}, \tag{3.6}$$

and

$$x_{.i} = \sum_{v=1}^{n} \frac{b_{vi}\exp(\theta_v - \beta_i)}{1+\exp(\theta_v - \beta_i)}. \tag{3.7}$$

This is a special case of a general result for exponential families, see, e.g., Andersen (1980a): The observed values of the sufficient statistics (left) are equal to their expectations (right).

For the complete data case at first sight there are $n + k$ such equations, but groups of persons or groups of items with the same sufficient statistic lead to identical equations. Person scores can be 0,1,..., k, but zero and perfect scores have a special role (see next paragraph). There are k items, but one of the equations (3.7) is superfluous due to the normalization restriction. So there are for complete data at most $2k - 2$ independent equations to be solved, and a lower number when some person scores have never been observed and/or when some items have equal observed popularity. For incomplete data the situation is more complex, as the same value of the sufficient statistic, combined with a different row or column of the design matrix $\boldsymbol{B}$, leads to a different equation.

A special case arises for zero or perfect sums. When an item is failed by all persons to whom it was presented ($x_{.i} = 0$), the formal solution of (3.7) diverges to $\beta_i = \infty$. For this value one would predict that all future respondents also fail this item. It is more practical to infer that this item is more difficult than the others, but that the present sample offers insufficient information to estimate its location.

Similarly, when an item is passed by all persons to whom it was presented ($x_{.i} = b_{.i}$, or simply $x_{.i} = n$ when there are no missing data), the formal solution diverges to $\beta_i = -\infty$, and one infers that the item is rather easy but there is not enough information to locate it.

Pending further data collection (for more able and less able persons, respectively), such zero and perfect items are omitted from analysis. In most testing situations, the number of subjects n by far exceeds the number of items k, and the items presented to each subject are so chosen that zero and perfect item scores will seldom occur.

The analogous situation for subjects, however, occurs rather frequently. For a subject with no positive answers to any presented item, $r_v = 0$ with formal solution $\theta_v = -\infty$, and for a subject passing all items that were presented, $r_v = b_{v.}$ (thus $r_v = k$ in the complete data case) with formal solution $\theta_v = \infty$. This would suggest that such a subject would fail (pass) all future items from the same domain, so again the practical solution is to remove subjects with zero or perfect scores from the current JML estimation procedure.

Note that there is a theoretical possibility that after this first round of removing persons and items, new items or persons have zero or perfect scores in the remaining data matrix. Fortunately the removal process usually ends after one or two rounds. Similar problems occur in logistic regression or loglinear analysis of contingency tables, where one often adds a small positive constant to zero entries and subtracts the same constant from perfect entries. This trick is not recommended for Rasch analysis.

It remains problematic, however, that for given parameter vectors $\boldsymbol{\beta}$ and $\boldsymbol{\theta}$ there is a positive probability of obtaining zero or perfect scores, for both items and persons. This implies that their JML estimators would have a positive probability of assuming the values $-\infty$ and ∞, and their expectation or higher moments would be undefined.

In the present section it will be assumed that the data collection takes place under circumstances in which the probability of zero or perfect scores is very small, and that all results refer to the truncated distribution obtained by omitting such pathological results.

Even then, however, a finite solution to the estimation equations is not always available (see Fischer, 1974, 261-263, and Fischer, 1981). An exception occurs when the items can be subdivided into two classes such that all persons either have positive (or missing) answers to all items in the first class, or negative (or missing) answers to all items in the second class. Then the items from the second class appear to be 'infinitely more difficult' than those from the first class, and no comparison of item parameters from the two classes is possible. Such data are called *ill-conditioned.* If no such subclasses exist, the data are said to be *well-conditioned.* The latter property is necessary for the existence of a finite, unique normalized JML solution. It is also sufficient, provided that subjects with zero and perfect scores have been removed. Fischer (1981) also presents useful methods for establishing well-conditionedness, both for complete and for incomplete designs. For the complete data case, Pfanzagl (1994) gives asymptotic results indicating that the Fischer conditions are almost always fulfilled for large enough samples of persons.

The JML estimation for the persons and items without zero or perfect scores proceeds by a simple iterative process. Initial values (see Section 3.5) are updated either by alternating the solution of (3.6) and (3.7) (which is easy because each equation contains only one person or item parameter, respectively), or by simultaneously updating all person and item parameter estimates in a routine that maximizes (3.3).

It was already mentioned that there are at most $2k-2$ independent equations for complete data. For a complicated design matrix $\boldsymbol{B}$ the number of equations can be quite large. In this case, as well as for long tests (say $k > 50$) in the complete case, it is recommended to avoid a simultaneous maximization routine that involves the Hessian matrix, such as Newton's algorithm.

The major drawback of JML estimates for the item parameters is that they are inconsistent for $n \to \infty$, k fixed, although consistency does hold for $n \to \infty$, $k \to \infty$, $n/k \to \infty$; see Andersen (1971a, 1973c) and Haberman (1977). One usually has a large sample of subjects and a fixed and limited set of items, in which case the number n of incidental parameters θ_v far exceeds the number k of structural parameters β_i which is mostly viewed as fixed. In this case the JML item parameter estimates are not consistent, and not even asymptotically unbiased. Although for complete data a correction factor $(k-1)/k$ appears to remove most of the bias, most researchers tend to avoid JML estimation. Another reason to do so is that several desirable properties of hypothesis tests, to be discussed later in this book, do not hold when JML estimates are used.

3.3 Conditional Maximum Likelihood (CML)

At the end of the preceding section the problem was raised that for JML estimation the number of nuisance parameters grows with the sample size n, by which the JML estimates of the item parameters are not consistent. By a general property of exponential families, the conditional distribution given the sufficient statistics for the nuisance parameters no longer depends on those parameters. Thus maximization of the conditional log-likelihood, given the person total scores r_v defined in (3.5), leads to improved estimates of the item parameters by a method that is denoted by CML.

In this section the notation $\xi_v = \exp(\theta_v)$ and $\epsilon_i = \exp(-\beta_i)$ is more convenient than the use of θ_v and β_i. In this notation, (3.1) becomes

$$P(X_{vi} = 1) = \frac{\xi_v \epsilon_i}{1 + \xi_v \epsilon_i},$$

$$P(X_{vi} = 0) = \frac{1}{1 + \xi_v \epsilon_i}.$$

In the simple case of only $k = 2$ items, complete data, and one fixed person parameter value ξ_v, one obtains

$$P(X_{v1} = 1, X_{v2} = 0 | X_{v.} = r_v = 1) = \frac{\dfrac{\xi_v \epsilon_1}{(1 + \xi_v \epsilon_1)(1 + \xi_v \epsilon_2)}}{\dfrac{\xi_v \epsilon_1 + \xi_v \epsilon_2}{(1 + \xi_v \epsilon_1)(1 + \xi_v \epsilon_2)}} = \frac{\epsilon_1}{\epsilon_1 + \epsilon_2}, \tag{3.8}$$

which shows that the probability ratio of (1,0) and (0,1) answers is the same for any value ξ_v, and that the person parameter does not occur in the conditional likelihood, see also Section 2.3.

Passing to the general case of k items and n persons, with possibly incomplete data, one collects all multiplicative item parameters into the k-vector $\boldsymbol{\epsilon}$ and all person total scores r_v into the n-vector $\boldsymbol{r}$. Some algebra shows that the conditional likelihood for the data matrix $\boldsymbol{X}$ equals

$$L_C(\boldsymbol{\epsilon} | \boldsymbol{r}) = \prod_v (\prod_{i=1}^{k} \epsilon_i^{x_{vi} b_{vi}} \gamma_{r_v}^{-1}) = (\prod_v \gamma_{r_v})^{-1} \prod_i \epsilon_i^{x_{.i}}, \tag{3.9}$$

where the elementary symmetric functions γ_{r_v} of the variables $\epsilon_i b_{vi}$ are defined as the sum of all products of r_v such variables :

$$\gamma_{r_v}(\epsilon_1 b_{v1}, ..., \epsilon_k b_{vk}) = \sum_{\boldsymbol{y}|r_v} \prod_{i=1}^{k} (\epsilon_i b_{vi})^{y_i}, \tag{3.10}$$

with the summation running across all answer patterns $\boldsymbol{y} = (y_1, ..., y_k)$ with

$$\sum_{i=1}^{k} y_i b_{vi} = r_v.$$

In the complete data case, one simply has

$$\begin{aligned}
\gamma_0 &= 1, \\
\gamma_1 &= \epsilon_1 + \epsilon_2 + \ldots + \epsilon_k, \\
\gamma_2 &= \epsilon_1\epsilon_2 + \epsilon_1\epsilon_3 + \ldots + \epsilon_{k-1}\epsilon_k, \\
&\vdots \\
\gamma_k &= \epsilon_1\epsilon_2\ldots\epsilon_k.
\end{aligned}$$

In the incomplete data case, the ϵ_i are replaced by $b_{vi}\epsilon_i$, so that all products are automatically omitted from the γ-functions for which at least one of the factors corresponds to an unobserved entry.

Maximization of $\ln L_C$ follows by putting all derivatives with respect to ϵ_i equal to zero, for which one uses that

$$\frac{\partial \gamma_{r_v}(b_{v1}\epsilon_1, \ldots, b_{vk}\epsilon_k)}{\partial \epsilon_i} = b_{vi}\gamma^{(i)}_{r_v-1}, \tag{3.11}$$

where $\gamma^{(i)}_{r_v-1}$ denotes the elementary symmetric function of $r_v - 1$ arguments $\epsilon_j b_{vj}$ omitting $\epsilon_i b_{vi}$. Some algebra leads to the following set of CML equations:

$$x_{.i} - \sum_{v=1}^{n} \frac{\epsilon_i b_{vi} \gamma^{(i)}_{r_v-1}}{\gamma_{r_v}} = 0, \tag{3.12}$$

for $i = 1, 2, ..., k$. This again agrees with the general theory of exponential families, because the summands can easily be shown to equal the conditional probability of a positive response to item I_i given a total of r_v positive responses, and thus the observed sufficient statistics are equalized to their conditional expectations, see also Sections 15.2 and 15.3 where CML estimation is discussed in more detail for the polytomous case.

The equations (3.12) must be solved in an iterative process, during which one could meet two technical problems. The first is the evaluation of the elementary symmetric functions; for long tests or tests with widely varying item parameters the result may become inaccurate due to rounding errors. This problem is discussed for example by Fischer (1974, Sec. 14.3), Gustafsson (1977), Jansen (1984), Formann (1986), and Verhelst, Glas and Van der Sluis (1984). The latter authors provide an algorithm that is remarkably efficient in keeping the rounding error within acceptable limits; for an alternative, see Fischer and Ponocny (1994) and Chapters 8, 15, and 19.

The second problem could be slow convergence. This occurs in particular for older software when one item parameter estimate at a time is updated, and hardly for Newton-Raphson and its variants. Fischer (1974, Sec. 14.3) advocates the use of Aitken extrapolation, which is based on the assumption that the differences between item parameter estimates in successive iterations form a decreasing geometric series.

Zwinderman (1991a, Ch. 6) uses simulated data for $n = 100, 1000$ persons and $k = 5, 10, 20, 30$ items to compare several maximization algorithms; he also studies the effect of the choice of initial estimates (see Section 3.5) and the effect of Aitken extrapolation (which is sometimes substantial). On the other hand, most modern computers are fast enough to ignore extrapolation and just run many iterations.

Whereas the software of twenty years ago was not without problems, it appears that its modern counterpart is sufficiently fast and accurate to be used in all standard situations. Only for very long tests, very complicated patterns of incompleteness, or very unusual parameter configurations, one may still meet numerical problems in CML estimation.

Contrary to JML estimates, CML estimates are consistent for $n \to \infty$, k fixed, under mild conditions, see, e.g., Theorems 1 and 2 of Pfanzagl (1994). Because we have an exponential family, the estimated asymptotic covariance matrix of the estimates can be obtained from the matrix of second derivatives with the parameter estimates inserted. For the exact and approximate formula of this conditional information matrix, see Fischer (1974, p. 238–239). For long tests the size of the matrix may preclude fast inversion, in which case one often uses the inverse of the main diagonal of the information matrix. This leads to a slight overestimation of the standard errors. Verhelst (1993) obtains a factor $k/(k-1)$ for the special case of equal parameters; Nap (1994) finds in his simulation study that roughly the same tends to hold for other cases. Of course the restriction mentioned in Section 1 has to be taken into account: at first one parameter estimate is fixed, then the $k-1$ others are estimated with their standard errors and the result is transformed back to k parameter values with sum zero.

For the problems of infinite parameter values, both for zero and perfect item or person totals and for strange configurations of the data matrix $\boldsymbol{X}$, the discussion at the end of the preceding section applies. Fischer (1981) gives necessary and sufficient conditions for the existence of finite CML estimates, both for the complete and the incomplete data case. For the complete case Pfanzagl's (1994) asymptotic results again indicate that such conditions are nearly always fulfilled for large samples.

The CML method maximizes not the full likelihood, but the conditional one given the total scores of the persons. In general such a conditioning implies loss of information. Thus the Cramèr-Rao bound for the asymptotic variance of the estimators would no longer be attained. In our case one might expect that the distribution of the total score also contains information on the item parameters. Andersen (1970, 1973c) has demonstrated the consistency and asymptotic normality of the CML estimates under some weak regularity conditions. This in itself, however, does not show that they are efficient (i.e., attain the Cramèr-Rao bound). A sufficient condition would be that the total score R is ancillary with respect to $\boldsymbol{\beta}$ (that is, its distribution would not depend on $\boldsymbol{\beta}$). This condition can not be shown to hold in the present case.

Andersen (1970, p. 294–296), however, shows for the exponential family that efficiency of the estimates also holds when the dependence of R on $\boldsymbol{\beta}$ has a specific structure, denoted by him as 'weak ancillarity', and later called 'S-ancillarity' by Barndorff-Nielsen (1978). Next, Andersen (1973c) shows that this structure indeed holds for the Rasch model (RM). The conclusion is that the CML estimates are asymptotically efficient, and that the loss of information becomes negligible for $n \to \infty$.

3.4 Marginal Maximum Likelihood (MML)

In this section, the nuisance parameters θ_v will be integrated out, rather than simultaneously estimating them (JML) or conditioning on their sufficient statistics (CML). One may write

$$P(\boldsymbol{X} = \boldsymbol{x}|\boldsymbol{\theta}, \boldsymbol{\beta}) = P(\boldsymbol{X} = \boldsymbol{x}|\boldsymbol{r}, \boldsymbol{\beta})\ P(\boldsymbol{R} = \boldsymbol{r}|\boldsymbol{\theta}, \boldsymbol{\beta}). \tag{3.13}$$

The first factor is maximized in CML. At the end of Section 3.3 it was shown that neglecting the second factor had no adverse asymptotic effects on the quality of the CML estimates. The MML method, however, has other advantages not shared by CML. It leads to finite person parameter estimates even for persons with zero or perfect scores, and such persons are not removed from the estimation process. They are useful for finding the ability distribution, even though they provide no information on the relative position of the item parameters. If the research goal refers to ability distributions, either in the whole group or in meaningful subgroups (e.g., based on gender or race, or of the same persons at different time points), the MML procedure is clearly superior, see also Chapters 4 and 11. On the other hand, MML requires to estimate or postulate a distribution for the latent trait, and if this is wrongly estimated or postulated, the MML estimates may be inferior. Moreover, CML stays closer to the concept of person-free item assessment that was discussed in Chapter 2.

Let $G(\theta)$ denote the cumulative distribution function of the person parameter in the population, and suppose that the person parameters of the observed persons are a random sample from this distribution. As was observed by Glas (1989a, p. 48) it is even sufficient to assume that each person's parameter is a random variable with distribution function G, independent of the other persons' random variables, without reference to a population.

Consider a fixed subject S_v with parameter θ_v. Let $\boldsymbol{x}_v$ be any n-vector with entries a, $0 < a < 1$, for the items for which $b_{vi} = 0$, and arbitrary choices from $\{0,1\}$ for each other entry. Then,

$$P(\boldsymbol{X}_v = \boldsymbol{x}_v|G, \boldsymbol{\beta}) = \int_{-\infty}^{\infty} \prod_{i=1}^{k} \frac{\exp[b_{vi}x_{vi}(\theta - \beta_i)]}{[1 + \exp(\theta - \beta_i)]^{b_{vi}}} dG(\theta). \tag{3.14}$$

The product for $v = 1, \ldots, n$ of these integrals, to be denoted by L_M, then is the marginal likelihood of obtaining the observed data matrix $\boldsymbol{X}$. It is a function of G and of $\boldsymbol{\beta}$.

There are two basic ways of tackling the problem that G is unknown. One is to assume that G belongs to a given parametric family with only a few unknown hyperparameters, say a vector $\boldsymbol{\tau}$, which must then be estimated jointly with $\boldsymbol{\beta}$. In particular, many authors have assumed that θ has a normal distribution with unknown mean and variance. A different parametrization occurs when one postulates an $N(0,1)$ distribution. In this case the scale is fixed by the person parameter distribution, and the indeterminacy discussed in Chapter 2 and in Section 3.1 is resolved without a restriction on the item parameters, but one must estimate the constant a in (2.22).

A second group of authors opt for the non-parametric variant in which G is estimated from the data, see, e.g., Follman (1988) or Lindsay, Clogg and Grego (1991). De Leeuw and Verhelst (1986) have shown, however, that only the first k moments of the distribution function H of $\xi = \exp(\theta)$ can be uniquely estimated. Thus, even when n is large, the limited information per person does not allow to determine the distribution of ξ in great detail. The usual procedure therefore is to choose for G or H a step function with at most entier$((k+2)/2)$ nodes on the θ-axis or ξ-axis, between which a total weight of 1 is divided. Estimation of the node and weight values alternates with estimation of $\boldsymbol{\beta}$ until convergence is reached. Actually, the experience in numerical integration with smooth integrands makes it plausible that a lower number of nodes will often suffice for estimating $\boldsymbol{\beta}$. As is worked out by Lindsay (1991), this means that the IRT model with a continuous distribution for the latent person parameter is replaced by a latent class model in which local independence holds within each class, see Chapter 13.

Space does not permit to work out the MML estimation in the same amount of detail as the JML and CML methods. For the normal distribution case, see the equations for the polytomous case in Section 15.4, or see Andersen and Madsen (1977), Thissen (1982), and Glas (1989a), and for the non-parametric (also called mixture) approach, see Lindsay et al. (1991).

De Leeuw and Verhelst (1986) have shown that CML and non-parametric MML are asymptotically ($n \to \infty$, k fixed) equivalent; see also Follman (1988), Lindsay et al. (1991), and Pfanzagl (1994). This confirms earlier simulation studies for the normal MML case in which the two estimates were very close for data generated according to the model. Both Glas (1989a) and Zwinderman (1991a), however, obtain that normal MML estimates can be grossly biased if the generating person parameter distribution differed from normal.

At the end of Section 3 the asymptotic efficiency of the CML estimates was discussed. One could argue that MML estimates also suffer from information loss, because the parameters describing the ability distribution are estimated from the same data. Engelen (1989, Ch. 5) argues that this information loss will often be small. He presents formulas and simulated values for the item information for a logistic ability distribution and obtains that the MML information is higher than the CML information.

The result by Pfanzagl (1994), finally, shows that the distributional properties of CML and MML estimates must be asymptotically the same, provided that the MML model was correctly specified.

3.5 Other Estimation Methods

In this section, several estimation methods will be mentioned that are not based on some likelihood function. For them, no asymptotic standard errors are available. On the other hand, most of them are much easier to compute, and some produce initial estimates useful for JML, CML, or MML.

The RM for a complete data matrix obviously satisfies

$$\text{logit}\,[P(X_{vi} = 1|\theta_v, \beta_i)] = \theta_v - \beta_i. \tag{3.15}$$

Verhelst and Molenaar (1988) therefore discuss the use of logistic regression methods with iteratively reweighted least squares methods on the fraction of persons with total score r that have a positive answer to item I_i. They show that this is equivalent to the JML method, and thus inconsistent. Their derivation sheds new light, however, on the causes: all persons with total score r have the same sufficient statistic, but not the same true person parameter. In Sec. 4 of the same paper, they derive an efficient and consistent one-step estimator of the item parameters, which in a simulation study compared favorably to CML in terms of accuracy, although not in programming effort.

Fischer (1974, Sec. 14.6 to 14.8) mentions three procedures based on the number m_{ij} of persons that have a positive answer to item I_i and a negative answer to item I_j. Clearly $m_{ii} = 0$ by definition. It follows from (3.8) that the ratio m_{ij}/m_{ji} is an estimate of ϵ_i/ϵ_j. The so-called explicit method multiplies such ratios for all $j \neq i$ and obtains, because $\sum_j \beta_j = 0$ implies $\prod_j \epsilon_j = 1$, that

$$\prod_{j \neq i} \frac{m_{ij}}{m_{ji}} \simeq \epsilon_i^k. \tag{3.16}$$

Taking the k-th root one obtains an estimate of ϵ_i. It is clear, however, that the result is 0 as soon as any $m_{ij} = 0$, and that it is ∞ as soon as any $m_{ji} = 0$ (or undefined when both occur). With small samples and/or widely varying item parameters, a matrix $\boldsymbol{M}$ without any zero off-diagonal entries will be rare.

A second procedure based on the matrix $\boldsymbol{M}$ is called 'symmetrizing'. It minimizes

$$\sum_i \sum_{j<i} \left(\frac{m_{ij}}{\epsilon_i} - \frac{m_{ji}}{\epsilon_j} \right)^2,$$

which can be reduced to an eigenvalue problem for a $k \times k$ matrix, see Fischer (1974, Sec. 14.6).

The third and most promising method based on M is the MINCHI procedure that minimizes

$$\sum_i \sum_{j<i} \frac{(m_{ij}\epsilon_j - m_{ji}\epsilon_i)^2}{(m_{ij}+m_{ji})\epsilon_i\epsilon_j}.$$

Pairs (i,j) for which $m_{ij} + m_{ji} = 0$ are of course omitted here. The equations that result by putting all partial derivatives equal to zero are given by Fischer (1974, Sec. 14.8) as

$$\epsilon_h^{-2} = \frac{\sum_i y_{ih}^2 \epsilon_i^{-1}}{\sum_j y_{hj}^2 \epsilon_j}, \tag{3.17}$$

where

$$y_{ij} = \frac{m_{ij}}{\sqrt{m_{ij}+m_{ji}}}. \tag{3.18}$$

See also Zwinderman (1991a, Sec. 2.2.4). Convergence is rapid, and in simulations fairly good agreement with CML estimates is found. The name MINCHI is used rather than 'Minimum Chi-Quadrat' as chosen by Fischer (1974), because it is correct that each summand has asymptotically a χ^2-distribution with df $= 1$, but the summands are not independent and hence the sum has no χ^2-distribution.

Because the ratio m_{ij}/m_{ji} is a consistent estimate of ϵ_i/ϵ_j, it is clear that all three methods will be consistent. Their asymptotic efficiency is an open question, however, because little is known about their standard errors. They share the advantage that incomplete data are easily accomodated (only persons to whom both items I_i and I_j have been presented are counted in the calculation of m_{ij} and m_{ji}).

Eggen and Van der Linden (1986) and Engelen (1989) discuss a method for item parameter estimation that is based on viewing the answers of each person to two items I_i and I_j as a the result of a pairwise comparison experiment that may result in a tie. Their method only considers a subset of item pairs such that there is no logical dependence among the results, but in other respects it comes close to the three methods just discussed.

The methods from this section may be used in order to obtain initial estimates for the JML, CML, or MML methods discussed in the earlier sections. Although good initial estimates will be beneficial when the iterative method is slow per iteration step or has to be repeated very often due to slow convergence, see Nap (1994), it is generally found for modern software that the choice of initial estimates has little effect when a powerful iteration algorithm is used on a powerful computer.

Zwinderman (1991a) has compared three sets of initial values for the item parameters ϵ_i:

- put $\epsilon_i = \dfrac{x_{.i}}{n - x_{.i}}$;

- put $\epsilon_i = 1$ for all i;
- use the proposal of Gustafsson (1979) :

$$\ln(\epsilon_i) = \frac{x_{.i} - \dfrac{x_{..}}{k}}{\displaystyle\sum_{r=1}^{k-1} n_r \frac{r(k-r)}{k(k-1)}},$$

where n_r denotes the number of persons with total score r. He obtains a slight superiority for the Gustafsson initial values in terms of the number of iterations required for the same accuracy criterion.

Swaminathan and Gifford (1982) assume that the item parameters can be viewed as independent drawings from an exchangeable prior distribution characterized by a few hyperparameters. They then use as item parameter estimates the joint mode of the posterior distribution of the item parameters given the data. For their method one has to specify or estimate the prior distribution. In the view of the present author, it will be rare that enough knowledge is available to validly do so while at the same time the information would be the same for each item parameter. With a large sample of persons, on the other hand, the precise choice of the prior will be relatively unimportant, and it is an advantage of the Bayesian method that also zero and perfect items can be accomodated.

Summarizing this chapter, the conclusion must be that there exists a wealth of methods for estimating unknown item parameters. In many cases, however, the asymptotic standard errors are unknown. Most simulation studies appear to indicate that the quality difference for finite samples is not spectacular, but the results do not always agree between studies. The fast and simple methods have lost some of their appeal now that computing power is so easily available. Unless there are clear reasons for a different decision, the present author would recommend to use CML estimates.

4

On Person Parameter Estimation in the Dichotomous Rasch Model

Herbert Hoijtink and Anne Boomsma[1]

ABSTRACT An overview is given of person parameter estimation in the Rasch model. In Section 4.2 some notation is introduced. Section 4.3 presents four types of estimators: the maximum likelihood, the Bayes modal, the weighted maximum likelihood, and the Bayes expected a posteriori estimator. In Section 4.4 a simulation study is presented in which properties of the estimators are evaluated. Section 4.5 covers randomized confidence intervals for person parameters. In Section 4.6 some sample statistics are mentioned that were computed using estimates of θ. Finally, a short discussion of the estimators is given in Section 4.7.

4.1 Introduction

At first sight person parameter estimation in the dichotomous Rasch model (RM) seems to be an easy and well defined problem: given the responses $\boldsymbol{X} = \boldsymbol{x}$ of a person to a set of k items, estimate the person parameter θ. At closer look a number of difficulties become apparent.

Statistical theory about the asymptotic behavior (bias, efficiency, consistency, normality) of estimators based on independently and identically distributed random variables is standard and readily applicable (Lehmann, 1983, Ch. 5 and 6; Huber, 1981, pp. 43–54 and 68–72; Bickel & Freedman, 1981; Miller, 1974). However, as far as the estimation of θ is concerned, the sample size equals the number of items, k, a person has responded to. In most applications this number is relatively small, and it is therefore questionable whether asymptotic results can be used to evaluate the behavior of various estimators. The only finite sample result known to the authors is due to Klauer (1991a), who discusses randomized confidence intervals for θ (see Section 4.5).

Since the number of items, k, usually is small, the amount of information available for the estimation of θ is also small. To tackle this problem, Klinkenberg (1992) proposes to use an estimator that incorporates information about the relation between θ and manifest person characteristics (e.g., gender, age).

Furthermore, since a person's responses are not identically distributed (the distribution of each response depends on the difficulty of the respective item), even asymptotic results are not easily obtained. Important work in this area has been done by Lord (1983a), Klauer (1990), and Samejima (1993) for the max-

[1]Rijksuniversiteit Groningen, Vakgroep Statistiek en Meettheorie, Grote Kruisstraat 2/1, 9712 TS Groningen, The Netherlands; e-mail: H.J.A.HOYTINK@PPSW.RUG.NL

imum likelihood estimator, and by Warm (1989) and Chang and Stout (1993) for the Bayes modal estimator.

For coping with the difficulty that the distribution of each response depends on an unknown item parameter, it is common practice to substitute estimates of the item parameters and pretend that they are known. Tsutakawa and Soltys (1988) and Tsutakawa and Johnson (1990) tackle this problem more seriously using a version of the *Bayes expected a posteriori* estimator of θ (Meredith & Kearns, 1973) which incorporates uncertainty about the item difficulties.

Another difficulty is given by model specification errors. The RM is not fully adequate for responses when, for example, subjects copy some responses from their neighbours, or are careless. Several estimators were developed to deal with response patterns containing such aberrant responses: Mislevy and Bock (1982) use a so-called M-estimator (Huber, 1981, pp. 43 ff.), that is, a robust type of maximum likelihood estimator; Wainer and Wright (1980) experiment with a jackknifed (Efron, 1982; Boomsma, 1991) estimator; and, Liou and Yu (1991) use a bootstrapped (Efron, 1982; Boomsma, 1991) estimator. For none of these estimators theoretical solutions to the difficulties mentioned above exist. However, the respective authors invariably expect that the finite sample behavior (bias, standard error, level α confidence intervals) of these estimators is better than the finite sample behavior of the well-known and often used maximum likelihood estimator. Since the asymptotic and finite sample properties of these estimators still have to be explored, they will not be discussed here.

Stout (1990) and Junker (1991) discuss a different model specification error: violations of the unidimensionality assumption. Junker proves that under mild violations of the unidimensionality assumption for IRT models in general, the maximum likelihood estimator and the Bayes modal and expected a posteriori estimators are still consistent estimators of θ. However, the usual formulation of the error variance no longer holds, and the asymptotic distribution is not necessarily normal.

4.2 Notation

Before estimators of θ can be discussed, some notation has to be introduced. Note that here and in the sequel, all sums and products will run over $i = 1, ..., k$, or $r = 0, ..., k$. According to the RM the distribution of response X, which can take on the values 0 and 1, conditional on the person parameter θ and item difficulty β, is governed by the probabilities

$$P_i = P(X_i = 1|\theta, \beta_i) = \frac{\exp(\theta - \beta_i)}{1 + \exp(\theta - \beta_i)}. \tag{4.1}$$

The distribution of response vector $\boldsymbol{X}$, conditional on the person parameter θ and the item difficulties $\boldsymbol{\beta}$, and assuming local stochastic independence, is

$$P(\boldsymbol{X} = \boldsymbol{x}|\theta, \boldsymbol{\beta}) = \prod_i P_i^{x_i}(1 - P_i)^{(1-x_i)} = B(\theta)\exp(r\theta)h(\boldsymbol{x}), \tag{4.2}$$

where

$$B(\theta) = \prod_i [1 + \exp(\theta - \beta_i)]^{-1} \tag{4.3}$$

(see also Chapter 3). The observed raw score is $r = \sum_i x_i$, and

$$h(\boldsymbol{x}) = \exp\left(-\sum_i \beta_i X_i\right). \tag{4.4}$$

The distribution of raw score R, conditional on the person parameter θ and the item difficulties $\boldsymbol{\beta}$, can be written as

$$P(R = r|\theta, \boldsymbol{\beta}) = B(\theta)\exp(r\theta) \sum_{\boldsymbol{x}|R=r} h(\boldsymbol{x}), \tag{4.5}$$

and the Fisher information of $\boldsymbol{X}$ with respect to θ is

$$\mathcal{I}(\theta) = -E\left(\frac{\partial^2 \log P(\boldsymbol{X} = \boldsymbol{x}|\theta)}{\partial^2 \theta}\right) = \sum_i P_i(1 - P_i). \tag{4.6}$$

The notation

$$\mathcal{J}(\theta) = \sum_i \frac{\partial^2 P_i}{\partial \theta^2} \tag{4.7}$$

will be used in the sequel. The following abbreviations will be used to denote the estimators to be discussed: maximum likelihood estimator (MLE); Bayes modal estimator (BME); weighted likelihood estimator (WLE); and Bayes expected a posteriori estimator (EAP).

4.3 Estimators

4.3.1 The Maximum Likelihood Estimator

Assuming the item parameters to be known, the MLE of θ is obtained by solving

$$\frac{\partial \log P(\boldsymbol{X} = \boldsymbol{x}|\theta, \boldsymbol{\beta})}{\partial \theta} = 0 \tag{4.8}$$

for θ, i.e., by finding the value of θ for which $r = \sum_i P_i$. Note that the MLE of θ is unique since the sum of the probabilities is strictly increasing in θ.

Lord (1983a) derived the variance and bias of the MLE of θ, which is correct to the order specified:

$$\text{Var(MLE)} = \frac{1}{\mathcal{I}(\text{MLE})} + o(k^{-1}), \tag{4.9}$$

and

$$\begin{aligned} \text{Bias(MLE)} &= E(\text{MLE}) - \theta \\ &= \frac{-\mathcal{J}(\text{MLE})}{2\mathcal{I}^2(\text{MLE})}, \end{aligned} \tag{4.10}$$

where $\mathcal{I}$(MLE) is defined by (4.6) with MLE substituted for θ. Note that Bias(MLE) is $O(k^{-1})$. Roughly speaking, both $o(k^{-1})$ and $O(k^{-1})$ imply that the corresponding quantities become negligible if $k \to \infty$ (see Serfling, 1980, pp. 1–2). But as will be seen in Section 4.4, for small k the formulas only give an approximation of the desired parameters.

Klauer (1990) proves that the MLE is asymptotically normally distributed; more specifically,

$$(\text{MLE} - \theta)\sqrt{\mathcal{I}(\text{MLE})} \xrightarrow{L} N(0,1). \tag{4.11}$$

This result can be used to construct level-α confidence intervals that are asymptotically correct:

$$\text{MLE} - \frac{z_{\alpha/2}}{\sqrt{\mathcal{I}(\text{MLE})}} < \theta < \text{MLE} + \frac{z_{\alpha/2}}{\sqrt{\mathcal{I}(\text{MLE})}}. \tag{4.12}$$

The MLE does not exist for response patterns with zero or perfect raw score (it tends to minus or plus infinity, respectively). The probability of zero and perfect raw scores, however, tends to zero for $k \to \infty$ for all finite θ. Interesting in this context is a result by Warm (1989) who provides 'rational bounds' for the MLE. Warm uses the term rational, because outside these bounds the bias of the estimate increases faster than the estimate itself. For response patterns with a raw score of zero the MLE tends to minus infinity, i.e., the lower rational bound provides a better estimate. For response patterns with a perfect raw score, the MLE tends to plus infinity, i.e., the upper rational bound provides a better estimate.

4.3.2 The Bayes Modal Estimator and the Weighted Likelihood Estimator

Assuming that the item parameters are known, and making an appropriate choice for the density function $g(\theta)$ of θ (either via the use of a prior density function or via estimation of the parameters of $g(\theta)$ from an appropriate sample, see Chapter 3), a Bayes modal estimate (BME) of θ is obtained maximizing the posterior density function of θ, conditional on $\boldsymbol{X}$ and $\boldsymbol{\beta}$, with respect to θ:

$$P(\theta|\boldsymbol{X},\boldsymbol{\beta}) = \frac{B(\theta)\exp(\theta r)g(\theta)}{\int B(\theta)\exp(\theta r)g(\theta)d\theta} \propto B(\theta)\exp(\theta r)g(\theta). \tag{4.13}$$

(Here and in the sequel all integrals are to be taken over the whole range of the integration variable.) Chang and Stout (1993) prove that the BME of θ has the same asymptotic distribution (4.11) as the maximum likelihood estimate.

The bias of the BME, correct to the order specified, can be found in Warm (1989):

$$\text{Bias(BME)} = \text{Bias(MLE)} + \frac{\frac{\partial \log g(\theta)}{\partial\theta}|_{\theta=\text{BME}}}{\mathcal{I}(\text{BME})}. \tag{4.14}$$

Note that Bias(BME) is $O(k^{-1})$. The error variance is given by

$$\text{Var(BME)} = \left[\mathcal{I}(\text{BME}) - E\left(\frac{\partial^2 \log g(\theta)}{\partial^2\theta}|_{\theta=\text{BME}}\right)\right]^{-1} + o(k^{-1}). \tag{4.15}$$

Warm suggests to take $g(\theta)$ such that

$$\text{Bias(MLE)} = -\frac{1}{\mathcal{I}(\theta)}\frac{\partial \log g(\theta)}{\partial\theta}, \tag{4.16}$$

in order to obtain an estimator with the same asymptotic distribution as the maximum likelihood estimate, but with a smaller bias. For the RM, $g(\theta)$ equals the Jeffreys (uninformative) prior (Lehmann, 1983, p. 241),

$$g(\theta) = \sqrt{\mathcal{I}(\theta)}. \tag{4.17}$$

Substituting this prior for g(θ) in (4.13), Warm obtains the weighted likelihood estimator (WLE) with

$$\text{Bias(WLE)} = o(k^{-1}), \tag{4.18}$$

and asymptotic error variance

$$\text{Var(WLE)} = \frac{1}{\mathcal{I}^2(\text{WLE})}\frac{\mathcal{I}(\text{WLE}) + \mathcal{J}^2(\text{WLE})}{4\mathcal{I}^2(\text{WLE})} + o(k^{-1}). \tag{4.19}$$

A nice feature of both the BME and the WLE is their existence for response patterns with zero or perfect raw score. However, whereas the former needs prior knowledge of $g(\theta)$, the latter does not, since for the WLE $g(\theta)$ is a function of the item parameters.

4.3.3 Bayes Expected A Posteriori Estimator

The expected a posteriori estimator (EAP) is obtained taking the expectation of the posterior density function of θ, conditional on $\boldsymbol{X}$ and $\boldsymbol{\beta}$:

$$\text{EAP} = E(\theta|\boldsymbol{X},\boldsymbol{\beta}) = \frac{\int \theta B(\theta)\exp(\theta r)g(\theta)d\theta}{\int B(\theta)\exp(\theta r)g(\theta)d\theta} = \int \theta f(\theta|r)d\theta. \tag{4.20}$$

The variance of the posterior density function of θ, conditional on $\boldsymbol{X}$ and $\boldsymbol{\beta}$, is given by

$$\widehat{\text{Var}}(\text{EAP}) = \int (\theta - E(\theta|\boldsymbol{X}))^2 f(\theta|r)d\theta. \tag{4.21}$$

Note the similarity between the middle part of (4.20) and the second term of (4.13). In (4.13) the value of θ maximizing $P(\theta|\boldsymbol{X},\boldsymbol{\beta})$ is computed, in (4.20) $E(\theta|\boldsymbol{X},\boldsymbol{\beta})$ is computed. A nice feature of the EAP estimator is its existence for patterns with a zero or perfect raw score. However, so far no theoretical results with respect to asymptotic or finite sample behavior of the EAP estimator and its bias are available.

One could replace θ in (4.20) by a linear combination of centralized manifest predictors $\boldsymbol{Y}$ plus an error ϵ, and integrate with respect to the density function of the error term,

$$E(\theta|\boldsymbol{X},\boldsymbol{\beta},\boldsymbol{\gamma}) = \int (\boldsymbol{\gamma}'\boldsymbol{Y} + \epsilon)f(\boldsymbol{\gamma}'\boldsymbol{Y} + \epsilon|r)d\epsilon, \tag{4.22}$$

where the elements of $\boldsymbol{\gamma}$ are known linear regression coefficients. Note that the definition of $f(\cdot)$ is the same as in (4.20). The relation between the manifest predictors $\boldsymbol{Y}$ and θ (as expressed by the regression coefficients $\boldsymbol{\gamma}$) has to be established before (4.22) can be computed, i.e., the regression coefficients are assumed to be known (see Zwinderman, 1991b, for procedures to estimate $\boldsymbol{\gamma}$). Klinkenberg (1992) notes that the variance of estimator (4.22) is expected to be substantially smaller than the variance of estimator (4.20), since two sources of information ($\boldsymbol{X}$ and $\boldsymbol{Y}$) are used for the estimation of θ. He also notes the danger of a large bias if the difference between the linear combination of manifest predictors $\boldsymbol{Y}$ and θ is large.

Tsutakawa and Johnson (1990) generalize the EAP estimator such that uncertainty about the item parameters can be taken into consideration when the parameter θ has to be estimated. Their basic idea is to compute the posterior mean and variance of θ, conditional both upon a person's response pattern $\boldsymbol{X}$ and upon the data matrix $\boldsymbol{U}$ used for estimating the item parameters:

$$E(\theta|\boldsymbol{X},\boldsymbol{U}) = \int E(\theta|\boldsymbol{X},\boldsymbol{\beta})L(\boldsymbol{\beta}|\boldsymbol{U})d\boldsymbol{\beta}, \tag{4.23}$$

and

TABLE 4.1. Estimates, asymptotic bias, and asymptotic error variance for the WLE, MLE, BME, and EAP, for raw scores 0 to 5. The item parameters are −1.5, −0.75, 0, 0.75, 1.5, and $g(\theta) = N(0,1)$.

	WLE			MLE			BME			EAP		
R	$\hat{\theta}$	Var	Bias	$\hat{\theta}$	Var	Bias	$\hat{\theta}$	Var	Bias	$\hat{\theta}$	Var	Bias
0	−2.95	4.40	.00				−1.30	.55		−1.34	.57	–
1	−1.45	1.37	.00	−1.70	1.45	−.32	−.76	.52	.79	−.79	.54	–
2	−.46	1.04	.00	−.51	1.05	−.06	−.25	.51	.20	−.26	.52	–
3	.46	1.04	.00	.51	1.05	.06	.25	.51	−.20	.26	.52	–
4	1.45	1.37	.00	1.70	1.45	.32	.76	.52	−.79	.79	.52	–
5	2.95	4.40	.00				1.30	.55		1.34	.57	–

$$\widehat{\text{Var}}(\text{EAP}|\boldsymbol{X}, \boldsymbol{U}) = \int E[(\theta - E(\theta|\boldsymbol{X}, \boldsymbol{U}))^2|\boldsymbol{X}, \boldsymbol{\beta}]L(\boldsymbol{\beta}|\boldsymbol{U})d\boldsymbol{\beta}, \tag{4.24}$$

where $L(\boldsymbol{\beta}|\boldsymbol{U})$ can be interpreted as the likelihood of $\boldsymbol{\beta}$ conditional on $\boldsymbol{U}$, or asymptotically ($n \to \infty$) as a multivariate normal variable with mean vector $\hat{\boldsymbol{\beta}}$ and covariance matrix $[\mathcal{I}(\hat{\boldsymbol{\beta}})]^{-1}$.

Tsutakawa and Johnson (1990) find for the three-parameter logistic model that EAP estimates and error variances change substantially if uncertainty about the item parameters is taken into account. However, for the two-parameter logistic model Tsutakawa and Soltys (1988) found only small changes. Generalizing this trend, it may be that taking uncertainty about the item parameter estimates into account is not all that important in the RM.

4.3.4 Examples

To illustrate some features of the estimators discussed, the WLE, MLE, BME, EAP, and their asymptotic variances and biases were computed for a five items test with raw scores 0 to 5. The hypothetical item parameters where chosen to be −1.5, −0.75, 0, 0.75, 1.5, with $g(\theta)$ distributed $N(0,1)$. The result, displayed in Table 1, can be used to illustrate the theoretical properties discussed above. The EAP estimator was computed by applying an eight-step Gauss-Hermite quadrature to the $N(0,1)$ density function of θ.

The MLE is the only estimator that yields infinite estimates for response patterns with zero or perfect raw score. Since Bias(BME) is a function of the MLE, its bias estimate does not exist for response patterns with zero or perfect raw score. Finally, the asymptotic bias of the EAP is not yet known.

Comparing the estimates and error variances of the BME and the EAP, it can be seen that the Bayes modal and a posteriori estimators hardly differ from each other.

A nice feature of the WLE is its asymptotic unbiasedness. The bias of the BME and MLE differ in direction; the BME regresses towards the mean of the density function of θ (in our example, the absolute value of the estimates is smaller than the absolute value of the corresponding WLE), the MLE is biased outwards (the absolute value of the estimates is larger than the absolute value of the corresponding WLE).

A final observation concerns the variances. Since both the BME and the EAP use information about the density function of θ in addition to the information contained in the item responses, their error variance is substantially smaller than the error variance obtained for the WLE and MLE. Note also that the error variance of the WLE is slightly smaller than the error variance of the MLE. The reason is that (4.9) and (4.19) are slightly different for small k.

4.4 Finite Sample Properties of Estimators of θ

Two criteria can be used to evaluate the finite sample behavior of estimators. The first criterion concerns the empirical size of the bias of the estimates,

$$\text{Bias}(\hat{\theta}) = V^{-1} \sum_{v=1}^{V} \hat{\theta}_v - \theta, \tag{4.25}$$

and the error variance of the estimates,

$$\text{Var}(\hat{\theta}) = V^{-1} \sum_{v=1}^{V} \left(\hat{\theta}_v - V^{-1} \sum_{v=1}^{V} \hat{\theta}_v \right)^2, \tag{4.26}$$

where $\hat{\theta}$ may be MLE, BME, WLE, or any other estimator. Notice that the empirical bias and error variance are closely related to the classical definitions of bias and error variance. The main difference is that the expectations appearing in the latter are replaced by summations over $v = 1, ..., V$ replicated estimates of a fixed value of θ in (4.25) and (4.26). The smaller the empirical values of these statistics, the better the estimator.

The second criterion is the accuracy with which these quantities can be estimated for unknown θ, based on the observed raw score r. Estimates may be obtained using formulas that are asymptotically correct, e. g., (4.9), (4.10), (4.14), (4.15), (4.18), and (4.19).

Figures 4.1 through 4.4 present the main results of a simulation study in which the asymptotic and empirical bias and error variance are compared for the MLE, WLE and BME. At each of the θ-values $0.00, 0.25, \ldots, 2.75, 3.00$, a set of 1000 response vectors was simulated for sets of 5 and 15 items each. For each response vector, the MLE, WLE, BME, and their asymptotic bias and error variance were estimated. Subsequently, at each value of θ, the empirical bias and error variance were computed, using (4.25) and (4.26), respectively (with the sums taken across the elements of the simulated sample of size $V = 1000$). Furthermore, at each of the θ-values, the expected values of the asymptotic bias and error variance of the MLE, WLE, and BME were computed for the simulated sample. Note that, especially for large values of θ, the results for the MLE and the asymptotic bias of the BME are somewhat distorted. Since the MLE does not yield finite estimates for response vectors with zero or perfect raw score, it was assigned the arbitrary values of -5 and 5, respectively. Note that the vertical scale differs for each of the Figures 4.1 through 4.4.

In each of the Figures 4.1 through 4.4, three pairs of lines (two for each estimator) are displayed. One line corresponds to an empirical property of an estimator (the bias in Figures 4.1 and 4.2, the error variance in Figures 4.3 and 4.4), the other line corresponds to the asymptotic property. From Figures 4.1 and 4.2 it can be seen that the WLE has the smallest (along a substantial range almost zero) empirical bias across the whole range of θ; to obtain the results for negative θ, the picture has to be mirrored around the point $(0, 0)$. Furthermore, the asymptotic approximation of the empirical bias is best for the WLE. Going from 5 to 15 items, the size of the empirical bias decreases, and the difference between asymptotic and empirical bias decreases for all estimators (compare Figures 4.1 and 4.2). Observe also that the differences between the three estimators become smaller when k becomes larger. This is not surprising: from Warm (1989), Klauer (1990), and Chang and Stout (1993) it may be inferred that all three estimators are asymptotically ($k \to \infty$) equivalent. Finally, note that for all estimators the empirical values of the bias are large along a substantial range of the θ-continuum in the five-item set, and substantial for the MLE and BME in the fifteen-item set. As a consequence, it is important to have good approximations of the empirical bias if the item sets are small (one could use bias-corrected estimators for the BME and the MLE). However, as can be seen in Figures 4.1 and 4.2, the asymptotic approximation of the empirical bias is only accurate for the WLE with fifteen items.

Figures 4.3 and 4.4 show that the BME has the smallest variance across the whole range of θ; for obtaining the results for negative θ, the picture has to be mirrored in the Y-axis. Furthermore, the agreement between empirical and asymptotic error variance is best for the BME.

Going from 5 to 15 items, the size of the empirical error variance decreases for all estimators. For the fifteen-item set it can be observed that across a large range of the θ-continuum the agreement between empirical and asymptotic error variance is almost perfect for the WLE and the MLE.

Due to the asymptotic equivalence of the three estimators, the similarity between the empirical variances will increase with k, the size of the item set. Even with only fifteen items, the MLE and the WLE are already very similar. Note that the empirical error variances resulting from the simulation with fifteen items are still substantial, i.e., there still is a lot of uncertainty about a person's location after it has been estimated from the responses to (only) fifteen items.

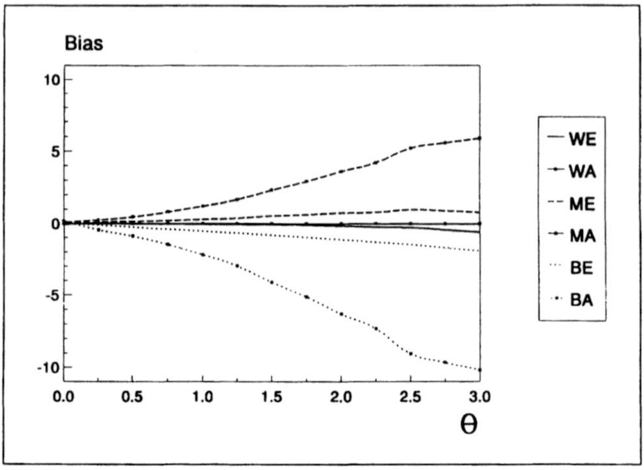

FIGURE 4.1. Empirical (E) versus asymptotic (A) bias for the WLE (W), MLE (M), and BME (B), for a set of 5 items with parameters −1.5, −0.75, 0, 0.75, 1.5, where $g(\theta)$ is the N(0,1).

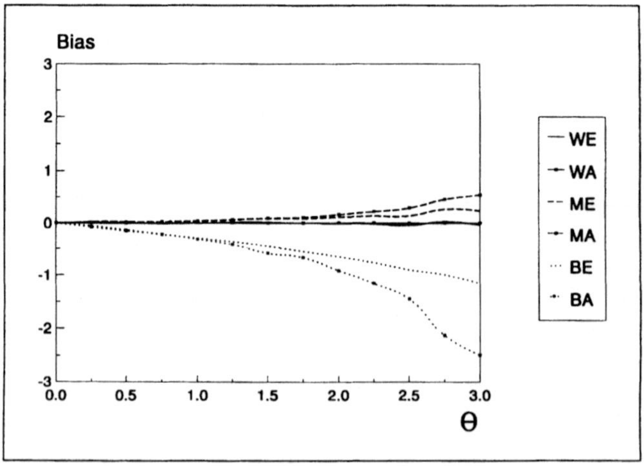

FIGURE 4.2. Empirical (E) versus asymptotic (A) bias for the WLE (W), MLE (M), and BME (B), for a set of 15 items with parameters −2.5, −2.14, …, 2.14, 2.5, where $g(\theta)$ is the N(0,1).

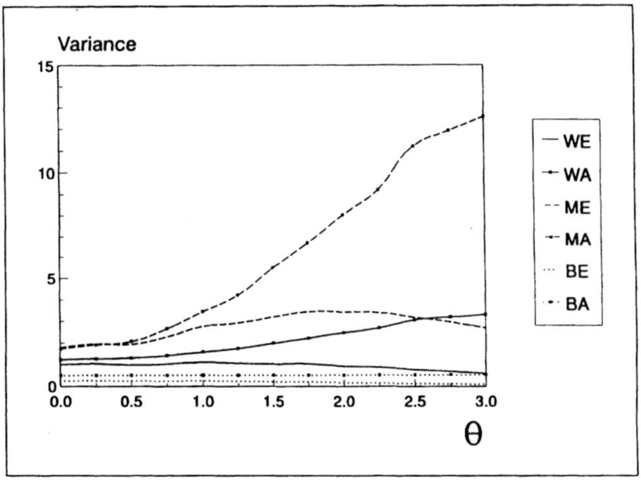

FIGURE 4.3. Empirical (E) versus asymptotic (A) error variance for the WLE (W), MLE (M), and BME (B), for a set of 5 items with parameters −1.5, −0.75, 0, 0.75, 1.5, where $g(\theta)$ is the N(0,1).

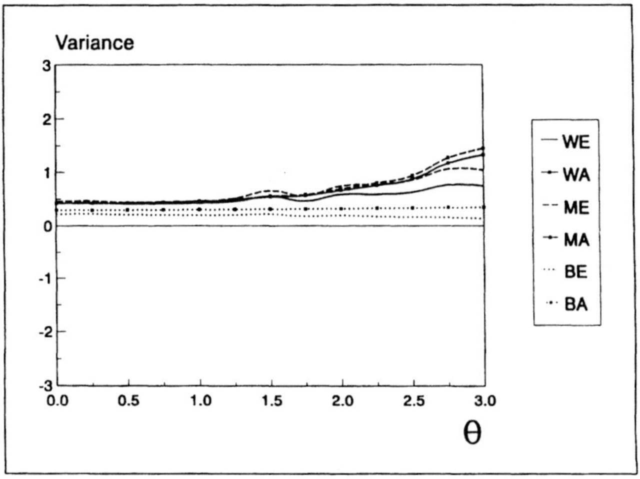

FIGURE 4.4. Empirical (E) versus asymptotic (A) error variance for the WLE (W), MLE (M), and BME (B), for a set of 15 items with parameters −2.5, −2.14, ..., 2.14, 2.5, where $g(\theta)$ is the N(0,1).

Although the three estimators are asymptotically equivalent, their properties differ in finite samples. With respect to the empirical bias, the WLE is superior; with respect to the empirical error variance, the BME. Computation of the mean squared error would not be decisive in this respect. Superiority will depend on the configuration of the item parameters and the θ-level.

There are other studies that evaluate the finite sample properties of estimators of θ. However, these studies almost never make a distinction between empirical and asymptotic properties, and are usually not aimed at the RM. The most interesting one is presented in Warm (1989), others are found in the papers mentioned in the introduction.

4.5 Optimal (Non-)Randomized Confidence Intervals

Confidence intervals are usually constructed assuming that the estimates are normally distributed and that an estimate of the variance of that distribution is available. Asymptotically, with test length k going to infinity, the MLE, WLE and BME were proved to have a normal distribution (Warm, 1989; Klauer, 1990; Chang & Stout, 1993). However, the practical relevance of this result is questionable since the test length often is small. The simulation study discussed in Section 4.4 indicated that the distributions of the MLE and WLE are non-normal, even for $k = 25$ (not reported). The distribution of the BME appeared to be normal for $k = 15$ and $k = 25$. This is probably explained by the normal prior used for $g(\theta)$ in the simulations. As a consequence, the finite sample distribution of some estimates of θ may be non-normal, and the asymptotic approximation of their variance may be poor.

Klauer (1991a) derives smallest exact confidence intervals for the ability parameter of the RM guaranteeing a given coverage probability $1 - \alpha$ of θ. His optimal randomized confidence interval for θ does not depend on asymptotic properties. For practical purposes, Klauer also presents a related smallest non-randomized confidence interval for the ability parameter. Both types of optimal intervals will be treated in some detail now.

Optimal exact confidence intervals for θ can be obtained from uniformly most powerful (UMP) unbiased tests (Lehmann, 1986, pp. 135–140) for the hypothesis

$$H_0 : \theta = \theta_0 \text{ vs. } H_1 : \theta \neq \theta_0, \tag{4.27}$$

with test size α, where the true value is denoted by θ, and θ_0 is any value along the θ-continuum. Since a UMP unbiased test for evaluating H_0 versus H_1 cannot be achieved on the basis of the discretely distributed test score R, it is necessary to modify the discrete test score in order to obtain optimal (UMP unbiased) tests. Following a standard procedure, Klauer (1991a) suggests to add a random term $U = u$, drawn from a uniform distribution on the interval $[0, 1)$, to the observed score $R = r$, yielding the randomized test score $w = r + u$.

The randomized test variable $W = R + U$, where R and U are stochastically independent, is a continuous variable. Since the distribution of W belongs to the exponential family of distributions that is parameterized by an ability parameter θ, it can be shown (see Witting, 1978, pp. 98 ff.) that a UMP unbiased test statistic $\phi(w)$ exists for deciding between H_0 and H_1. This test statistic is given by

$$\phi(w) = \begin{cases} 0 & \text{if } c_1(\theta) \leq w \leq c_2(\theta), \quad \text{i.e., do not reject } H_0, \\ 1 & \text{otherwise,} \qquad\qquad\qquad \text{i.e., reject } H_0, \end{cases} \tag{4.28}$$

where the cut-off scores $c_1(\theta)$ and $c_2(\theta)$, which are strictly increasing functions of the unknown ability parameter θ, are solutions of the set of two equations,

$$E(\phi(W)|\theta) = P(\phi(W) = 1|\theta) = \alpha \tag{4.29}$$

and

$$E(\phi(W)R|\theta) = \alpha E(R|\theta). \tag{4.30}$$

The cut-off scores $c_1(\theta)$ and $c_2(\theta)$ of the UMP unbiased test are computed by numerical methods (see Klauer, 1991a, pp. 538–540). Once they have been calculated, an optimal confidence interval, denoted by $C_R(w)$, can be obtained. Here, an optimal confidence interval C^* is one that overlaps wrong ability values with uniformly smallest probability, for all θ and θ_0, $\theta \neq \theta_0$, and covers the right ability values uniformly with a probability of at least $1 - \alpha$; see Witting (1978, p. 40). Such intervals have also been called uniformly most accurate confidence limits. The interval $c_1(\theta) \leq w \leq c_2(\theta)$ constitutes the non-rejection region of the test (4.27) for a given value of θ. Instead, what is needed is an optimal interval of size $1 - \alpha$ for the unknown parameter θ. For each randomized score w the optimal confidence interval of size $1 - \alpha$ is determined by the range of the set of parameters θ for which $\phi(w) = 0$.

In order to calculate $C_R(w)$, inverse functions $c_1^{-1}(w)$ and $c_2^{-1}(w)$ of the cut-off scores are used, which are computed by bisection methods. Finally, this yields the *optimal randomized confidence interval* for θ,

$$C_R(w) \;=\; [\theta_l(w),\, \theta_u(w)], \tag{4.31}$$

where the lower and upper confidence limits $\theta_l(w)$ and $\theta_u(w)$ depend upon w, and obviously also on α and the number of items k (see Klauer, 1991a, p. 540).

Two remarks can be made. First, the randomized confidence interval (4.31) has peculiar properties around the ends of the randomized score scale: for $0 \leq w \leq \alpha$, both lower and upper bound are $-\infty$; for $\alpha < w \leq 2-\alpha$, the lower bound is $-\infty$; for $k-1+\alpha \leq w <$ k+1-α, the upper bound is ∞; and for $k+1-\alpha \leq w < k+1$, both lower and upper bound are ∞.

Secondly, Klauer (1991a, p. 540) notes the following disadvantage of the randomized confidence interval: two examinees with the same observed score are assigned different randomized scores, and hence are assigned slightly different confidence intervals for the ability parameter underlying their performance, "a statistical subtlety that may not be acceptable for all practitioners in spite of the optimum properties it entails".

His solution to this problem is the so-called Clopper-Pearson exact confidence interval, denoted by $C_{NR}(w)$, which is the smallest non-randomized interval containing the (optimal) randomized interval. It can be derived easily from (4.31) by replacing the lower and upper bounds, $\theta_l(w)$ and $\theta_u(w)$, by the values obtained at the closest integer values, $r = [w]$ and $r = [w] + 1$, respectively. Hence, the *non-randomized Clopper-Pearson confidence interval* for the ability parameter θ is defined as

$$C_{NR}(r) = [\theta_l(r), \theta_u(r+1)] = [\theta_l([w]), \theta_u([w]+1)], \tag{4.32}$$

which is only a function of the non-randomized score $r = [w]$, α, and the number of items k.

The reader is referred to Klauer (1991a) for further details. From a numerical comparison of several interval estimation methods Klauer (1991a, pp. 546–547) concludes that the Clopper-Pearson interval $C_{NR}(r)$ yields a "substantially more accurate confidence estimation than asymptotic intervals", like the traditional one, using normality assumptions and Fisher information $\mathcal{I}(\theta)$ for a test of k items (Birnbaum, 1968). The use of these intervals is also of practical importance, because it allows the estimation of confidence intervals for examinees with zero and perfect raw scores, besides the guarantee that the coverage probabilities are at least $1 - \alpha$ for all test sizes and all item parameters that may arise. Klauer also shows that a further "substantial gain of the confidence estimation is achieved through the use of the optimal interval", $C_R(w)$. For this reason the latter should be recommended, despite a minor computational burden, for which a computer program has been made available.

4.6 Statistics Based on Estimates of θ

It is common practice to estimate some basic parameters, like the expectation and variance of θ across all persons in the population, directly by the sample mean and sample variance, respectively, of the estimates $\hat{\theta}$ of θ. Lord (1983a) and Hoijtink and Boomsma (1991) show that such results may be severely biased (especially when small item sets are used) if bias and error variance of the estimates $\hat{\theta}$ are not taken into account.

Solutions that are asymptotically correct are easily obtained using

$$\theta = \hat{\theta} - e, \tag{4.33}$$

where e denotes the error in the estimation. The expectation of θ then becomes

$$\begin{aligned} E(\theta) = E(\hat{\theta} - e) = E(\hat{\theta}) - E(E(e|\theta)) &= E(\hat{\theta}) - E(\text{Bias}(\hat{\theta})) \\ &\approx E(\hat{\theta}) - E(\widehat{\text{Bias}}(\hat{\theta})). \end{aligned} \tag{4.34}$$

The variance of θ becomes

$$\begin{aligned}\sigma^2(\theta) &= \sigma^2(\hat{\theta} - e) \\ &= \sigma^2(\hat{\theta}) - E(\mathrm{Var}(\hat{\theta})) - 2\mathrm{Cov}(\hat{\theta}, \mathrm{Bias}(\hat{\theta})) + \sigma^2(\mathrm{Bias}(\hat{\theta})) \\ &\approx \sigma^2(\hat{\theta}) - E(\widehat{\mathrm{Var}}(\hat{\theta})) - 2\mathrm{Cov}(\hat{\theta}, \widehat{\mathrm{Bias}}(\hat{\theta})) + \sigma^2(\widehat{\mathrm{Bias}}(\hat{\theta})).\end{aligned} \tag{4.35}$$

See Hoijtink and Boomsma (1991) for more detailed derivations. Three remarks with respect to (4.34) and (4.35) can be made. First, note that they apply in principle to the MLE, WLE, and BME. A disadvantage of the MLE, which also affects the BME via its bias, is that it does not exist for response patterns with zero or perfect raw score, which poses a problem when such response patterns are actually observed in a sample. Second, note that (4.34) and (4.35) simplify substantially for the WLE. Since the WLE is asymptotically unbiased, all terms containing estimates of the bias disappear. Third, note that the size of the bias and error variance of an estimator are relatively unimportant when statistics are based on estimates of θ. The quality of the approximation is determined completely by the accuracy by which the expectations of the asymptotic bias and error variance approximate the empirical bias and error variance, respectively.

Another statistic which is frequently used and involves estimates of θ is the index of subject separation, a reliability coefficient at the θ-level (Gustafsson, 1979):

$$\rho = \frac{\sigma^2(\theta)}{\sigma^2(\hat{\theta})} = \frac{\sigma^2(\hat{\theta} - e)}{\sigma^2(\hat{\theta})}. \tag{4.36}$$

This index of subject separation compares the variance of θ (say, true score), with the variance of the estimates of θ (say, observed score). Whenever the first covers a large portion of the latter (say, $\rho > .30$), the measurement instrument used offers at least some discrimination among the persons in the sample. Much higher values of ρ are of course desirable; they are often found unless either the test is very short or the ability range in the sample is very narrow.

4.7 Discussion

There exist many estimators for the person parameter in the RM. This chapter focusses on the MLE, WLE, BME, and to some lesser extent, the EAP estimator. Although the asymptotic behavior of the MLE, WLE, and BME estimators is identical, it turns out that their finite sample behavior shows marked differences. The WLE has the smallest bias, the BME the smallest error variance, and the MLE shows the worst performance in both respects. The most important drawback of the MLE, however, is the fact that it does not exist for response patterns with zero or perfect raw score.

Unfortunately, neither for the WLE, BME, MLE, and EAP, nor for the estimators developed to deal with model specification errors (see Section 4.1) finite sample results are available. The simulation study in Section 4.4 illustrates that use of asymptotic properties to estimate empirical quantities should not be recommended when the number of items is small. The only finite sample results known to the authors are the randomized confidence intervals of Klauer (1991a).

A general recommendation would be to estimate person parameters only if the number of items is at least ten. For smaller item sets both the bias and the error variance will be very large and one cannot even rely on asymptotic results to estimate these quantities. Two possible solutions exist for this problem: either the derivation of finite sample results (this however does not change the fact that it is hard to determine where a person is located from the responses to only a few items); or incorporation of extra information with respect to θ in the estimator, using a person's score on a number of manifest predictors (Klinkenberg, 1992).

Finally, in Section 4.6, it was illustrated how the mean and variance of θ should be estimated if only estimates of θ are available. It turned out that valid inferences are only possible if the bias and error variance of the estimates are taken into consideration.

5

Testing the Rasch Model

Cees A. W. Glas and Norman D. Verhelst[1]

ABSTRACT In this chapter, it is shown that the problem of evaluating model fit can be solved within the framework of the general multinomial model, and it is shown how tests for this framework can be adapted to the Rasch model. Four types of tests are considered: generalized Pearson tests, likelihood ratio tests, Wald tests, and likelihood ratio tests. The statistics presented not only support the purpose of a global overall model test, but also provide information with respect to specific model violations, such as violation of sufficiency of the sum score, strictly monotone increasing and parallel item response functions, unidimensionality, and differential item functioning.

5.1 Introduction

One of the aims of this chapter is to create some order in the plethora of testing procedures for the Rasch model (RM). To present an overview of the various approaches to testing model fit, a taxonomy will be used that is based on three aspects: the assumptions and properties of the model which are tested, the type of statistic on which the test is based, and the mathematical sophistication of the procedure, particularly, the extent to which the distribution of the statistic is known. These three aspects will first be discussed in some detail.

The first aspect of the taxonomy of model tests concerns the assumptions and properties of the model to be tested. Although model tests are constructed to have power against specific alternatives, it is not really possible to completely separate the assumptions in order to test them one at a time. Some confounding will always occur. Given this restriction, the model tests can be ordered as follows.

First, there are tests that focus on the assumptions of sufficiency of the sum score and of strictly monotone increasing and parallel item response functions. To this class belong the Martin-Löf (1973) T-test, the Van den Wollenberg (1982) Q_1-test, the Glas (1988a, 1989) R_1-test, the Molenaar (1983) U_i-test, and the S_i- and M-tests (Verhelst & Eggen, 1989; Verhelst, Glas, & Verstralen, 1994). An important property of the Rasch model is that, under very mild regularity assumptions (see Pfanzagl, 1994), consistent item parameter estimates can be obtained from a sample of any subgroup of the population where the model holds. So item parameter estimates obtained using different samples from different subgroups (say, gender or ethnic subgroups) of the population should, apart from

[1]National Institute for Educational Measurement, P.O.-Box 1034, 6801 MG Arnhem, The Netherlands; fax: (+31)85-521356

random fluctuations, be equal. Not only the estimates obtained from subgroups formed on the basis of background variables should be approximately equal, also estimates obtained at different score levels should, within chance limits, be equal. Andersen's likelihood ratio test (Andersen, 1973b), the Fischer-Scheiblechner test (Fischer & Scheiblechner, 1970; Fischer, 1974), and the Wald test proposed in this chapter are constructed to be sensitive to violation of this property.

The assumptions of unidimensionality of the parameter space and of local stochastic independence are the focus of the Martin-Löf (1973, 1974) likelihood ratio test, the Van den Wollenberg (1982) Q_2-test, and the Glas (1988a, 1989) R_2-test.

The additional assumptions with respect to the distribution of ability, which are made in MML estimation, are the focus of the R_0-test by Glas and Verhelst (1989).

Finally, tests for the RM can also be based on explicit extensions of the RM, such as the class of log-linear models by Kelderman (1984, 1989), the OPLM presented in Chapter 12, and the RM with a multivariate ability distribution by Glas (1989, 1992).

The second aspect of the taxonomy of model tests concerns the type of statistic used. The classification related to this aspect generally follows the usual classification in discrete statistical models. So tests can be based on Pearson-type statistics, that is, statistics involving differences between observed frequencies and their expected values, likelihood ratio statistics, and Wald statistics. In addition, in Section 5.5 the application of Lagrange multiplier statistics to the RM will be discussed. To the practitioner, the first aspect of the taxonomy, that is, the assumptions that are tested, will be much more important than the second aspect, that is, the type of statistic used. However, from a theoretical point of view and for reasons of presentation, the second aspect has gained priority in this chapter.

The third aspect of the taxonomy of model tests concerns the mathematical sophistication of the procedure, especially the extent to which the (asymptotic) distribution of the statistic is known. In most instances, the tests for model fit can be arranged into two classes: a class of statistics where the (asymptotic) distribution is known, and a class of statistics which can be viewed as approximations of the statistics in the first class. It must be stressed that it is perfectly feasible to construct a test statistic which does have a known asymptotic distribution, but which is completely uninformative with respect to model violations and has little power. So the requirement of having a test focussed on specific alternatives precedes the mathematical sophistication of the approach. But from the viewpoint of interpretation of the outcome, test statistics where the (asymptotic) distribution is known, are far preferable to statistics where this is not the case. Also generalizations of the model, such as the OPLM, and specializations of the model such as the LLTM (Chapter 8 of this book), are fostered by a firm statistical framework.

5.2 Pearson-Type Tests

In this section a detailed overview of Pearson-type test statistics for the RM will be given. In Subsection 5.2.1 a theoretical framework for these statistics will be sketched. In the following sections the actual statistics fitting this framework and their approximations will be discussed.

5.2.1 A Class of Generalized Pearson Tests

Item response models can be viewed as multinomial models with the set of all possible response patterns as observational categories (Cressie & Holland, 1983). Therefore, the problem of evaluating model fit in IRT models is solved within the framework of the general multinomial model. Consider a multinomial model with M mutually exclusive outcomes occurring with probability $\pi_1(\boldsymbol{\phi}), ..., \pi_M(\boldsymbol{\phi})$, $\boldsymbol{\phi}$ denoting a q-vector of model parameters. When testing IRT models, the sample space is the set of all possible response patterns $\boldsymbol{x}$. In the sequel, it will be assumed that all persons have responded to the same set of k items, unless otherwise stated. So, if a test consists of k items, the number of possible response patterns M equals 2^k. In the approach of this chapter, it will be irrelevant whether one takes the traditional view that the source of randomness is situated within the subject, or that one adopts the view that response patterns are randomly sampled from some distribution of response patterns and that the 'stochastic subject' may not exist (for a discussion, see Holland, 1990, and Chapters 1 and 6 of this book). The essential aspect of the approach of this chapter will be that observed response patterns are described by a multinomial distribution parameterized via an IRT model.

Let n independent observations be drawn, where the M outcomes occur with sample proportions $p_1, .., p_M$, and let $\hat{\pi}_m$ denote $\pi_m(\boldsymbol{\phi})$ evaluated at a BAN estimate (Best Asymptotically Normal estimate, for instance a maximum likelihood estimate or a minimum chi-square estimate) of $\boldsymbol{\phi}$. It is a well-known result of asymptotic theory that under very mild regularity conditions (see Birch, 1964), for $n \to \infty$,

$$X^2 = n \sum_{m=1}^{M} \frac{(p_m - \hat{\pi}_m)^2}{\hat{\pi}_m} \tag{5.1}$$

is asymptotically chi-square-distributed with $M-q-1$ degrees of freedom. In the sequel, the regularity conditions are assumed to be satisfied, and in applications they will be satisfied; for a discussion of the regularity conditions, see Bishop, Fienberg, and Holland (1975, pp. 509–511).

Notice that in cases where the test is applied to an IRT model, the sum runs over all $M = 2^k$ possible response patterns $\boldsymbol{x}$. Even for tests of moderate length, the number of possible response patterns will be very large, which leads to two drawbacks. Firstly, the number of frequencies on which the statistic is based will be too large to be informative with respect to model violations. Secondly, the expected frequencies will be very small and thus the claims concerning the asymptotic distribution of the statistic will be jeopardized. Of course, the above asymptotic result remains true, but the required sample size n needed to obtain acceptable expected frequencies will become enormous. Therefore, Glas and Verhelst (1989) introduced a class of asymptotically χ^2-distributed test statistics which evades these problems. To define this class of statistics, let the M-vectors $\boldsymbol{p}$ and $\boldsymbol{\pi}$ be $(p_1, ..., p_M)'$ and $(\pi_1, ..., \pi_M)'$, respectively, and let $\hat{\boldsymbol{\pi}}$ be the vector $\boldsymbol{\pi}$ evaluated using a BAN estimate of $\boldsymbol{\phi}$. Moreover, let $\boldsymbol{D}_\pi$ be the diagonal $M \times M$ matrix of the elements of $\boldsymbol{\pi}$, and let a vector of deviates $\boldsymbol{b}$ be defined as

$$\boldsymbol{b} = n^{1/2} (\boldsymbol{p} - \hat{\boldsymbol{\pi}}). \tag{5.2}$$

It is immediately seen that, with these definitions, (5.1) can be rewritten as

$$X^2 = \boldsymbol{b}' \hat{\boldsymbol{D}}_\pi^{-1} \boldsymbol{b}. \tag{5.3}$$

The tests in the class to be presented will be based on a vector of G linear combinations $\boldsymbol{d} = \boldsymbol{U}'\boldsymbol{b}$, where the $M \times G$ matrix $\boldsymbol{U}$ is chosen such that $G \ll M$ and the linear combinations may show specific model violations, based on expected probabilities that are sufficiently large for applying asymptotic theory. In the sequel, $\boldsymbol{U}$ will be called the matrix of contrasts. Consider the statistic

$$Q = Q(\boldsymbol{U}) = \boldsymbol{b}'\boldsymbol{U}\left(\boldsymbol{U}'\hat{\boldsymbol{D}}_\pi^{-1}\boldsymbol{U}\right)^{-}\boldsymbol{U}'\boldsymbol{b} = \boldsymbol{d}'\boldsymbol{W}^{-}\boldsymbol{d}\,, \tag{5.4}$$

where $\left(\boldsymbol{U}'\hat{\boldsymbol{D}}_\pi\boldsymbol{U}\right)^{-}$ and $\boldsymbol{W}^{-}$ stand for the generalized inverse of $\left(\boldsymbol{U}'\hat{\boldsymbol{D}}_\pi\boldsymbol{U}\right)$ and $\boldsymbol{W}$, respectively. In what follows, the vector $\boldsymbol{d}$ and the matrix $\boldsymbol{W}$ will be called the vector of deviates and the matrix of weights, respectively.

Glas and Verhelst (1989) have derived sufficient conditions for Q to be asymptotically χ^2-distributed with degrees of freedom equal to rank$(\boldsymbol{U}'\boldsymbol{D}_\pi\boldsymbol{U})$ $-q-1$. Since the proof is beyond the scope of the present chapter, only the result will be presented here; for the proof, the interested reader is referred to Glas and Verhelst (1989). A sufficient condition for (5.4) to be asymptotically χ^2-distributed can be given as follows. Let $\boldsymbol{A}$ be an $M \times q$ matrix $((a_{mj}))$ defined by

$$a_{mj} = \pi_m^{-1/2} \frac{\partial \pi_m}{\partial \phi_j}, \tag{5.5}$$

such that $\boldsymbol{A} = \boldsymbol{D}_\pi^{-1/2} \partial\boldsymbol{\pi}/\partial\boldsymbol{\phi}$. Furthermore, for an arbitrary matrix $\boldsymbol{P}$, let $\mathcal{M}(\boldsymbol{P})$ be the linear manifold of the columns of $\boldsymbol{P}$, that is, the collection of all vectors that can be written as a linear combination of the columns of $\boldsymbol{P}$. Then $Q(\boldsymbol{U})$ is asymptotically χ^2-distributed if the following two conditions are satisfied:

(A) The columns of $\boldsymbol{A}$, defined by (5.5), belong to $\mathcal{M}\left(\boldsymbol{D}_\pi^{1/2}\boldsymbol{U}\right)$;

(B) There exists a vector of constants $\boldsymbol{c}$ such that $\boldsymbol{U}\boldsymbol{c} = \mathbf{1}$, where $\mathbf{1}$ is the M-vector with all elements equal to 1.

Before applying these conditions to the construction of specific tests, it should be shown how the RM fits in the general framework of the multinomial model. Using MML offers no problems: the probability of a response pattern $\boldsymbol{x}$ is clearly a function of the item parameters and the parameters of the distribution of θ, so $\boldsymbol{\phi}$ is the concatenation of the item parameters $\boldsymbol{\beta}$ and the population parameters $\boldsymbol{\lambda}$. In the case of CML estimation, the situation is somewhat more complicated. Let n_r be the number of respondents attaining sum score r, and $\pi(\boldsymbol{x}|r)$ the probability of observing response pattern $\boldsymbol{x}$ given r. Then the set of all possible response patterns given sum score r has a multinomial distribution with parameters n_r and $\pi(\boldsymbol{x}|r)$ for all $\boldsymbol{x}$ compatible with r. The number of respondents with score r is treated as a given constant. This model will be enhanced with a saturated multinomial model which perfectly describes the empirical score distribution. Letting ω_r denote the parameters of this distribution, the conditional model can be expanded into a multinomial model with probabilities $\pi(\boldsymbol{x}) = \omega_r \pi(\boldsymbol{x}|r)$. Of course, to apply conditions (A) and (B), the parameters ω_r are to be considered as model parameters. A more formal treatment of this approach can be found in Haberman (1974).

In general, one could imagine two strategies for the construction of the matrices $\boldsymbol{U}$ satisfying the conditions (A) and (B). As a first strategy, one could adopt a matrix of contrasts $\boldsymbol{U}$ which meets one's requirements, and try to prove that the conditions are satisfied. Several examples of this strategy will be given below. As a second strategy, which will also be illustrated below, one could try to develop a procedure which guarantees that the conditions are automatically satisfied. Such a procedure is easy to construct in the case of exponential family models. In general its rationale is as follows: consider a matrix $\boldsymbol{U}$ that can be partitioned $\boldsymbol{U} = (\boldsymbol{T}|\boldsymbol{Y})$. Let $\boldsymbol{Y}$ be the matrix which, via $\boldsymbol{Y}'\boldsymbol{b}$, produces the observed and expected frequencies one is interested in. Now, let $\boldsymbol{T}$ be a matrix for which the two conditions are satisfied, that is, there exists a vector $\boldsymbol{c}$ such that $\boldsymbol{T}\boldsymbol{c} = \boldsymbol{1}$, and the columns of $\boldsymbol{A}$ belong to $\mathcal{M}(\boldsymbol{D}_\pi^{1/2}\boldsymbol{T})$. Then these conditions are obviously also satisfied by $\boldsymbol{U} = (\boldsymbol{T}|\boldsymbol{Y})$. One could object that in this way the contrasts contained in $\boldsymbol{T}$ are added to the contrasts contained in $\boldsymbol{Y}$, while one is only interested in the contrasts $\boldsymbol{Y}$. However, in exponential family models, $\boldsymbol{T}$ can be chosen in such a way that $\boldsymbol{T}'\boldsymbol{b} = \boldsymbol{0}$, which can be loosely qualified as: 'the contribution of the contrasts contained in $\boldsymbol{T}$ to the test statistic $Q(\boldsymbol{U})$ is zero'. These ideas will now be formalized.

A model belongs to the exponential family if the likelihood function, given an observation $\boldsymbol{x}$, can be written as

$$L(\boldsymbol{\phi};\boldsymbol{x}) = c(\boldsymbol{x})\frac{\exp[\boldsymbol{\phi}'\boldsymbol{t}(\boldsymbol{x})]}{a(\boldsymbol{\phi})}, \tag{5.6}$$

where $\boldsymbol{t}(\boldsymbol{x})$ is a vector of functions of $\boldsymbol{x}$, and $c(\cdot)$ and $a(\cdot)$ are functions only of $\boldsymbol{x}$ and $\boldsymbol{\phi}$, respectively. The likelihood of the RM, enhanced with a saturated

TABLE 5.1. Matrix **T** for 3 items

$\boldsymbol{x}$			$\boldsymbol{T}_1$			$\boldsymbol{T}_2$			
x_1	x_2	x_3	$t(\boldsymbol{x})_1$	$t(\boldsymbol{x})_2$	$t(\boldsymbol{x})_3$	δ_{0r}	δ_{1r}	δ_{2r}	δ_{3r}
0	0	0	0	0	0	1	0	0	0
1	0	0	1	0	0	0	1	0	0
0	1	0	0	1	0	0	1	0	0
0	0	1	0	0	1	0	1	0	0
1	1	0	1	1	0	0	0	1	0
1	0	1	1	0	1	0	0	1	0
0	1	1	0	1	1	0	0	1	0
1	1	1	1	1	1	0	0	0	1

multinomial model for the score distribution, can be written as

$$L(\boldsymbol{\beta}, \boldsymbol{\omega}; \boldsymbol{x}) = \frac{\exp\left(\sum_{i=1}^{k} x_i \beta_i + \sum_{j=0}^{k} \delta_{jr} \ln \omega_j\right)}{\gamma_r(\boldsymbol{\epsilon})}, \tag{5.7}$$

where δ_{jr} is the Kronecker delta, taking the value 1 if $j = r$, and zero otherwise; and $\gamma_r(\boldsymbol{\epsilon})$ is an elementary symmetric function as defined in Chapter 3. By comparing (5.6) and (5.7) it can easily be verified that (5.7) defines an exponential family. Notice that the restriction $\sum_{r=0}^{k} \omega_r = 1$ implies that there are k free parameters in the saturated model for the frequency distribution of the respondents' sum scores. Since also the item parameters need a restriction to produce a unique solution of the likelihood equations, the total number of free parameters q is equal to $2k-1$. The indeterminacy can be resolved, for instance, by subjecting (5.7) to the reparametrization $\beta_i^* = \beta_i - \beta_1$ for $i = 1, ..., k$, and $\omega_r^* = \omega_r - \omega_0$ for $r = 0, ..., k$, such that β_1^* and ω_0^* are fixed. However, it proves convenient to first consider the overparameterized model (5.7); the step to the identified model will be commented upon later.

Let $\boldsymbol{T}$ be defined as an $M \times (q+2)$ matrix which, for the M different patterns, has the sufficient statistics $\boldsymbol{t}(\boldsymbol{x})'$, defined in (5.6), as rows. The rows are in an arbitrary but fixed order. The matrix $\boldsymbol{T}$ will be partitioned as $(\boldsymbol{T}_1|\boldsymbol{T}_2)$. The matrix $\boldsymbol{T}_1$ has k columns, each column corresponding to an item parameter, the matrix $\boldsymbol{T}_2$ has $k+1$ columns, each column corresponding to a score, including the zero score. An example of the $\boldsymbol{T}$-matrix for 3 items is given in Table 5.1. For convenience, the 8 possible response patterns are given in the first 3 columns of Table 5.1. Using (5.7), the reader can verify that, for any response pattern $\boldsymbol{x}$, $\boldsymbol{T}_1$ has an equivalent row. Moreover, it is easy to see that the columns of $\boldsymbol{T}_2$ are nothing else than indicator vectors of the scores.

In exponential family models, the following three lemmas hold.

Lemma 5.1

$$\frac{\partial \boldsymbol{\pi}}{\partial \boldsymbol{\phi}'} = \boldsymbol{D}_\pi \boldsymbol{T} - \boldsymbol{\pi}\boldsymbol{\pi}'\boldsymbol{T} .$$

Proof Since the probabilities $\pi_{\boldsymbol{x}}$ sum to one, the factor $a(\boldsymbol{\phi})$ in (5.6) can be written as $\sum_{\boldsymbol{x}} c(\boldsymbol{x}) \exp[\boldsymbol{\phi}'\boldsymbol{t}(\boldsymbol{x})]$. Let $\boldsymbol{y}$ be some response pattern. Then $\partial \pi_{\boldsymbol{y}}/\partial \phi_j = t_j(\boldsymbol{y})\pi_{\boldsymbol{y}} - \pi_{\boldsymbol{y}} \sum_{\boldsymbol{x}} t_j(\boldsymbol{x})\pi_{\boldsymbol{x}}$, for $j = 1, ..., q$, and the result follows. □

Lemma 5.2 *In exponential family models, it holds that* $\boldsymbol{T}'\boldsymbol{b} = \boldsymbol{0}$, *where* $\boldsymbol{b}$ *is defined by (5.2) and evaluated using ML-estimates.*

Proof The log-likelihood function can be written in the form $\ln L(\boldsymbol{\phi}; \boldsymbol{X}) = \sum_{\boldsymbol{x}} p_{\boldsymbol{x}} \ln \pi_{\boldsymbol{x}} + c$, where c does not depend on the parameters $\boldsymbol{\phi}$. Taking partial derivatives gives $\partial \ln L(\boldsymbol{\phi}; X)/\partial \boldsymbol{\phi}' = \boldsymbol{p}'\boldsymbol{D}_\pi{}^{-1}(\partial \boldsymbol{\pi}/\partial \boldsymbol{\phi}') = \boldsymbol{p}'\boldsymbol{D}_\pi^{-1}(\boldsymbol{D}_\pi \boldsymbol{T} - \boldsymbol{\pi}\boldsymbol{\pi}'\boldsymbol{T}) = \boldsymbol{p}'\boldsymbol{T} - \boldsymbol{p}'\boldsymbol{D}_\pi^{-1}\boldsymbol{\pi}\boldsymbol{\pi}'\boldsymbol{T}$. But $\boldsymbol{D}_\pi^{-1}\boldsymbol{\pi} = \boldsymbol{1}$, and since $\boldsymbol{p}'\boldsymbol{1} = 1$, the likelihood equations can be written as $(\boldsymbol{p} - \boldsymbol{\pi})' \boldsymbol{T} = \boldsymbol{0}'$. □

Before giving the third lemma, notice that condition (B) is satisfied because summing the column vectors of $\boldsymbol{T}_2$ gives the unit vector.

Lemma 5.3 *If condition (B) holds, the columns of* $\boldsymbol{A}$, *defined by (5.5) and Lemma 5.1, belong to* $\mathcal{M}(\boldsymbol{D}_\pi^{1/2}\boldsymbol{T})$.

Proof Notice that $\boldsymbol{A} = \boldsymbol{D}_\pi^{-1/2}(\boldsymbol{D}_\pi \boldsymbol{T} - \boldsymbol{\pi}\boldsymbol{\pi}'\boldsymbol{T}) = \boldsymbol{D}_\pi^{1/2}\boldsymbol{T} - \boldsymbol{\pi}^{1/2}\boldsymbol{\pi}'\boldsymbol{T}$. Clearly, $\boldsymbol{D}_\pi^{1/2}\boldsymbol{T}$ belongs to $\mathcal{M}(\boldsymbol{D}_\pi^{1/2}\boldsymbol{T})$. From condition (B) it follows that there exists a vector of constants $\boldsymbol{c}$ such that $\boldsymbol{T}\boldsymbol{c} = \boldsymbol{1}$, so $\boldsymbol{\pi}^{1/2}$ is in $\mathcal{M}(\boldsymbol{D}_\pi^{1/2}\boldsymbol{T})$. □

For an arbitrary contrast matrix $\boldsymbol{Y}$, define $\boldsymbol{U} = (\boldsymbol{T}|\boldsymbol{Y})$ and $\boldsymbol{d}' = (\boldsymbol{d}_0'|\boldsymbol{d}_1') = (\boldsymbol{b}'\boldsymbol{T}|\boldsymbol{b}'\boldsymbol{Y})$. Using Lemma 5.2, $Q(\boldsymbol{U})$ can be written as

$$\begin{aligned} Q(\boldsymbol{U}) &= \begin{pmatrix} \boldsymbol{d}_0' & \boldsymbol{d}_1' \end{pmatrix} \begin{pmatrix} \boldsymbol{T}'\hat{\boldsymbol{D}}_\pi \boldsymbol{T} & \boldsymbol{T}'\hat{\boldsymbol{D}}_\pi \boldsymbol{Y} \\ \boldsymbol{Y}'\hat{\boldsymbol{D}}_\pi \boldsymbol{T} & \boldsymbol{Y}'\hat{\boldsymbol{D}}_\pi \boldsymbol{Y} \end{pmatrix}^- \begin{pmatrix} \boldsymbol{d}_0 \\ \boldsymbol{d}_1 \end{pmatrix} \\ &= \boldsymbol{d}_1' \left(\boldsymbol{Y}'\hat{\boldsymbol{D}}_\pi \boldsymbol{Y} - \boldsymbol{Y}'\hat{\boldsymbol{D}}_\pi \boldsymbol{T} \left(\boldsymbol{T}'\hat{\boldsymbol{D}}_\pi \boldsymbol{T} \right)^- \boldsymbol{T}'\hat{\boldsymbol{D}}_\pi \boldsymbol{Y} \right)^- \boldsymbol{d}_1 \\ &= \boldsymbol{d}_1' \boldsymbol{W}_1^- \boldsymbol{d}_1. \end{aligned} \tag{5.8}$$

From (5.8) it can be seen how the matrix $\boldsymbol{T}$ influences $Q(\boldsymbol{U})$: Although it has no contribution to the vector of deviates, it acts as a kind of correction on the matrix of the quadratic form. Generally speaking, the reason why this correction has to be carried out lies in the restrictions on the vector of deviates. Although $\boldsymbol{b}$ has M elements, they can not all vary freely in the M-dimensional cube with edges $(0, 1)$, because their sum is zero and there are q restrictions imposed by the likelihood equations. So the matrix of the quadratic form reflects the fact that the parameters are estimated from the data.

This section will be concluded with presenting an alternative, but completely equivalent way of constructing $\boldsymbol{U}$ that uses the identified version of the model.

In this case let $\boldsymbol{U} = (\boldsymbol{T}^*|\boldsymbol{Y})$. Suppose the model is identified as was suggested above, that is, by fixing β_1^* and ω_0^* to zero. The associated columns of $\boldsymbol{T}$ will be removed and $\boldsymbol{T}^*$ will contain $\boldsymbol{T}_1$ and $\boldsymbol{T}_2$ minus their first columns. To fulfill condition (B), a last column consisting entirely of ones must be added to $\boldsymbol{T}^*$. The resulting matrix will be called $\boldsymbol{T}^{**}$. The fact that also in this strategy condition (A) is satisfied follows from the observation that $\mathcal{M}(\boldsymbol{T}^{**}) = \mathcal{M}(\boldsymbol{T})$. The next three sections will be devoted to test statistics which belong to the general class defined in this section.

5.2.2 THE MARTIN-LÖF TEST, THE R_1-TEST, THE Q_1-TEST

The tests of this section are developed to have power against violation of the axiom of monotone increasing, parallel item response functions. First, three tests for the framework of CML estimation will be presented: the Martin-Löf test, the R_{1c}-test, and the Q_1-test. Next, one of these tests, the R_{1c}-test, will be generalized to the framework of MML estimation as the R_{1m}-test.

For computing these statistics, the group of persons taking a test is partitioned into subgroups in the following manner. The score range 1 trough $k-1$ is partitioned into G disjoint score regions and the group of persons is divided into G subgroups according to their score level. At the end of this section, it will be briefly discussed how the statistics to be presented can be generalized to incomplete designs, that is, to designs where different subgroups of persons respond to different sets of items. There it will also be discussed how the statistics can be applied in the situation where one wants to divide the persons into subgroups using their value on some external variable, such as sex or age.

To define the Martin-Löf T-statistic and the R_{1c}- and Q_1-statistics, let the stochastic variable M_{1gi}, with realization m_{1gi}, be the count of the number of persons belonging to score region g and giving a positive response to item I_i, and let $E(M_{1gi}|\hat{\omega}, \hat{\boldsymbol{\beta}})$ be its CML expected value, that is, its expected value given the frequency distribution of the respondents' sum scores and the CML estimates of the item parameters. The tests are based on the differences

$$d_{1gi}^* = m_{1gi} - E(M_{1gi}|\hat{\omega}, \hat{\boldsymbol{\beta}}). \tag{5.9}$$

Since the sum score is a sufficient statistic for ability, checking M_{1gi} against its expected values across various score levels may reveal differences between the observed proportions and expected probabilities of producing positive responses at various ability levels. If, for instance, the observed number of positive responses is too small at low score levels and too large at high score levels, the item response function is steeper than predicted by the RM. Of course, too steep a slope may also mean that there are one or more items that have too weak a slope, one should keep in mind that the slope is evaluated using an RM that fits the 'average' item of the test. To interpret the magnitude of the differences d_{1gi}^*, so-called scaled deviates can be computed by dividing the differences (5.9) by their estimated standard deviation, and thus producing a standardized binomial variable z_{1gi}. Squaring and summing results in a global index of model fit. However, for deriving the distribution of such an index, the dependency among the (scaled) deviates has to be accounted for. This can be done within the framework of the generalized Pearson statistics introduced in Section 5.2.1. Consider the statistic given by

$$R_{1c} = \sum_{g} \boldsymbol{d}'_{1g} \boldsymbol{W}^{-}_{1g} \boldsymbol{d}_{1g}, \tag{5.10}$$

where $\boldsymbol{d}_{1g}$ is the vector of elements d_{1gi}, with $d_{1gi} = d^*_{1gi}/\sqrt{n}$, for all items in the test, and $\boldsymbol{W}_{1g}$ is the matrix of weights $\boldsymbol{U}'\hat{\boldsymbol{D}}_{\pi}\boldsymbol{U}$. It turns out that this matrix is equal to the covariance matrix of $\boldsymbol{d}_{1g}$, and the laborious multiplication $\boldsymbol{U}'\hat{\boldsymbol{D}}_{\pi}\boldsymbol{U}$ needs not actually be performed when computing the statistic. For a more detailed expression for $\boldsymbol{W}_{1g}$, one is referred to Glas (1988a, 1989), where it is also shown that R_{1c} has an asymptotic χ^2-distribution with df $= (G-1)(k-1)$. A sketch of the proof will be given below. Glas (1981) has shown that the R_{1c}-statistic is equivalent to the Martin-Löf (1973) T-statistic, meaning that computation of these two statistics yields exactly the same result, regardless of their different weight matrices. The motivation for transforming the Martin-Löf T-statistic to the R_{1c}-statistic is that the latter fits the framework of generalized Pearson statistics, which makes its generalization to a wide variety of applications feasible, as will be shown below. Another statistic based on the deviates (5.9) is the Van den Wollenberg (1982) Q_1-statistic. It is defined by

$$Q_1 = \frac{k-1}{k} \sum_{i=1}^{k} \sum_{g=1}^{G} z^2_{1gi}, \tag{5.11}$$

where z_{1gi} is a scaled deviate, computed by dividing d_{1gi} by its estimated standard deviation. Simulation studies (Van den Wollenberg, 1982) support the conjecture that the distribution of Q_1 is approximately χ^2-distributed with df $= (G-1)(k-1)$. The Q_1 statistic can be viewed as an approximation of R_{1c} where the matrix of weights is replaced with a diagonal matrix consisting of the diagonal elements of the complete covariance matrix.

An example of the $\boldsymbol{U}$ matrix of the R_{1c}- statistic is given in Table 5.2. It concerns the same example of 3 items that was used for the $\boldsymbol{T}$ matrix in Table 5.1. Moreover, it is assumed that two score levels are used for computing the statistic, so $G = 2$; the first level relates to respondents obtaining a score one, the second level relates to respondents scoring two. For convenience, the rows of the $\boldsymbol{U}$ matrix in Table 5.2 are numbered and the last column contains a symbolic representation of the probabilities of the response patterns. To verify that $\boldsymbol{U}'(\boldsymbol{p} - \boldsymbol{\pi})$ results in a vector of deviates $\boldsymbol{d}$ compatible with the definition of (5.9), proceed as follows: consider the inner product of u_{21} with the vector $\boldsymbol{\pi}$; this inner product is equivalent to $\pi(110) + \pi(101)$, the result is the probability

TABLE 5.2. The **U**-matrix

	U_0		U_1			U_2		U_3	
nr	$u_{0.}$	u_{11}	u_{12}	u_{13}	u_{21}	u_{22}	u_{33}	$u_{3.}$	π
1	1	0	0	0	0	0	0	0	$\pi(000)$
2	0	1	0	0	0	0	0	0	$\pi(100)$
3	0	0	1	0	0	0	0	0	$\pi(010)$
4	0	0	0	1	0	0	0	0	$\pi(001)$
5	0	0	0	0	1	1	0	0	$\pi(110)$
6	0	0	0	0	1	0	1	0	$\pi(101)$
7	0	0	0	0	0	1	1	0	$\pi(011)$
8	0	0	0	0	0	0	0	1	$\pi(111)$

of scoring two and getting item 1 correct. Using Table 5.2 the reader may further verify that $\boldsymbol{U}'\hat{\boldsymbol{D}}_\pi\boldsymbol{U}$ is of block-diagonal form, hence the summation of quadratic forms $\boldsymbol{d}'_{1g}\boldsymbol{W}^-_{1g}\boldsymbol{d}_{1g}$ in (5.10).

Next, a sketch of the proof of the asymptotic distribution of R_{1c} will be given. Checking that the columns of $\boldsymbol{T}$ belong to $\mathcal{M}(\boldsymbol{U})$ proceeds as follows: the first column of $\boldsymbol{T}_1$ can be formed by adding the columns labeled u_{11}, u_{21} and $u_{3.}$, of $\boldsymbol{U}_1$,$\boldsymbol{U}_2$ and $\boldsymbol{U}_3$, respectively. The other columns of $\boldsymbol{T}_1$ are constructed in a similar way. The column δ_{0r} of Table 5.1 is the same as $u_{0.}$ in Table 5.2. In the same manner, δ_{3r} is equivalent to u_3. Finally, the columns δ_{1r} and δ_{2r} are constructed summing the elements of u_{1j}, $j = 1, ..., 3$, or u_{2j}, $j = 1, ..., 3$, and dividing by r. This completes the sketch of the proof.

The R_{1c}-test has an analogue R_{1m}, which can be used in the framework of MML estimation. Before turning to this statistic, another one for the MML framework will be discussed. Although the effects of misfitting items and incorrect assumptions about the ability distribution can hardly be separated, it is most practical to start an evaluation of the model fit by testing the appropriateness of the specified distribution: incorrect modeling of the ability distribution has an impact on the evaluation of the model fit for all items, whereas a hopefully small number of misfitting items may have little effect on the evaluation of the model for the ability distribution. Since the person's sum score is a sufficient statistic for the ability parameter, the statistic is based on evaluating the difference between the observed and MML expected score distribution (the score distribution given the MML estimates of the item and population parameters); that is, the test is based on the differences

$$d^*_{0r} \;=\; n_r - E(N_r|\hat{\boldsymbol{\beta}}, \hat{\boldsymbol{\lambda}}), \tag{5.12}$$

for $r = 0, ..., k$, where the stochastic variable N_r represents the number of persons obtaining score r, n_r is its realization, and $\hat{\boldsymbol{\beta}}$ and $\hat{\boldsymbol{\lambda}}$ are the MML estimates of the item parameters and population parameters, respectively. These differences are combined into the overall statistic

$$R_0 = \boldsymbol{d}_0'\boldsymbol{W}_0^-\boldsymbol{d}_0, \tag{5.13}$$

where the vector $\boldsymbol{d}_0$ has elements d_{0r}, with $d_{0r} = d^*_{0r}/\sqrt{n}$, for $r = 0, ..., k$, and $\boldsymbol{W}_0$ is a matrix of weights to be discussed below. In general, R_0 has an asymptotic χ^2-distribution with df $= k - \dim(\boldsymbol{\lambda})$ (Glas, 1989; Glas & Verhelst, 1989). If a normal ability distribution is assumed, R_0 has an asymptotic χ^2-distribution with df $= k - 2$.

A detailed proof of the asymptotic distribution of R_0 will not be given here, the reader is referred to the references above and to Chapter 18 of this book. However, some remarks are in order, which may lead to a better understanding of the other statistics for the MML framework. Although the derivation of the computational formulae for the generalized Pearson goodness of fit tests is complicated, conceptually the construction of the tests is fairly simple: The columns of the matrix $\boldsymbol{U} = [\boldsymbol{T}|\boldsymbol{Y}]$ define linear combinations of the deviations $(\boldsymbol{p} - \hat{\boldsymbol{\pi}})$. Within the framework of quadratic forms $Q(\boldsymbol{U})$ defined by (5.4), the test-statistic R_0 can be defined as $R_0 = Q(\boldsymbol{T})$, so in this case matrix $\boldsymbol{Y}$ is omitted. In exponential families, $(\boldsymbol{p} - \hat{\boldsymbol{\pi}})'\boldsymbol{T} = \boldsymbol{0}$. However, the enhanced RM, that is, the RM extended with a structural model for the distribution of ability, is no exponential family, and the relation $(\boldsymbol{p} - \hat{\boldsymbol{\pi}})'\boldsymbol{T} = \boldsymbol{0}$ does not hold. Therefore, R_0 is not identical to zero. Within the CML framework, the measurement model was enhanced with a saturated multinomial model for the prediction of the score distribution. In the MML framework, the theoretical distribution of the scores is a function of both the item parameters and the parameters of the ability distribution, which have to be estimated from the data, and the observed frequency distribution of the respondents' sum scores will in general not match the predicted frequency distribution. It is precisely this deviation which is measured by the R_0-statistic. Hence, the R_0-test will be especially sensitive for deviations from the assumed latent ability distribution. Since thus far there exists no result that $\boldsymbol{T}$ may be further reduced without violating conditions (A) and (B), it is necessary to evaluate

$$R_0 = n(\boldsymbol{p} - \hat{\boldsymbol{\pi}})'[\boldsymbol{T}(\boldsymbol{T}'\hat{\boldsymbol{D}}_\pi\boldsymbol{T})^-\boldsymbol{T}']^-(\boldsymbol{p} - \hat{\boldsymbol{\pi}}). \tag{5.14}$$

The matrix of weights of (5.13) is of order $k + 1$, and the matrix of weights of (5.14) is of order $2k + 1$, so (5.14) is not strictly equivalent to (5.13). However, from the above remark on the sensitivity of R_0 to badly fitting score distributions, one might conjecture that, analogously to Lemma 5.2, the vector $(\boldsymbol{p} - \hat{\boldsymbol{\pi}})'\boldsymbol{T}_1$ will be identically zero, since $\boldsymbol{T}_1$ in a sense represents the measurement model, not the structural model. Fortunately, numerous numerical examples (Glas, 1989) corroborated this conjecture; unfortunately, a mathematical proof for the conjecture is still lacking. If the conjecture is valid, this implies that the matrix of weights reduces to $[\boldsymbol{T}_2'\hat{\boldsymbol{D}}_\pi\boldsymbol{T}_2 - \boldsymbol{T}_2'\hat{\boldsymbol{D}}_\pi\boldsymbol{T}_1(\boldsymbol{T}_1'\hat{\boldsymbol{D}}_\pi\boldsymbol{T}_1)^-\boldsymbol{T}_1'\hat{\boldsymbol{D}}_\pi\boldsymbol{T}_2]$. To keep things conceptually simple, the simplification resulting from the conjecture has been carried through in (5.13).

The analogue of the R_{1c}-test for the MML framework, the R_{1m}-test, has a rationale that is analogous to the test for the CML framework. If the sample is partitioned on the basis of sum scores, it must be assumed that the number of misfitting items and the seriousness of the misfit are such that the person's sum scores are not invalidated as estimates of the person's ability level. So although the RM does not strictly hold, the sum score must still be a more or less adequate measure of the ability level of a respondent. If this is the case, the difference between the observed and expected value of M_{1gi} may show differences in discrimination between the items. However, in a CML framework no assumptions concerning the ability distribution need be made, whereas in an MML framework this assumption will always interfere with item fit. A statistic for testing the assumptions concerning the ability distribution was discussed above.

As with R_{1c}, the sample of persons taking a test is divided into G subgroups according to their score level; these subgroups will be indexed $g = 1, ..., G$. Let d_{1gi} be the difference between the number of positive responses on item I_i in subgroup g and its MML expected value, that is,

$$d^*_{1gi} = m_{1gi} - E(M_{1gi}|\hat{\boldsymbol{\beta}}, \hat{\boldsymbol{\lambda}}). \tag{5.15}$$

For diagnostic purposes they are transformed into standardized binomial variables, i.e., scaled deviates. For a global test statistic a vector $\boldsymbol{d}_{1g}$ is defined. Apart from the differences d_{1gi}, $d_{1gi} = d^*_{1gi}/\sqrt{n}$, the vector $\boldsymbol{d}_{1g}$ will also include differences d_{0r}, $d_{0r} = d^*_{0r}/\sqrt{n}$, and d^*_{0r} as in (5.12) for the scores included in group g. Finally, for establishing the asymptotic distribution of the statistic, an expression for the zero and perfect score must be included,

$$c_0 = \frac{[n_0 - E(N_0|\hat{\boldsymbol{\beta}}, \hat{\boldsymbol{\lambda}})]^2}{E(N_0|\hat{\boldsymbol{\beta}}, \hat{\boldsymbol{\lambda}})} \tag{5.16}$$

and

$$c_k = \frac{[n_k - E(N_k|\hat{\boldsymbol{\beta}}, \hat{\boldsymbol{\lambda}})]^2}{E(N_k|\hat{\boldsymbol{\beta}}, \hat{\boldsymbol{\lambda}})}. \tag{5.17}$$

Using these definitions,

$$R_{1m} = \left(c_0 + \sum_g \boldsymbol{d}'_{1g} \boldsymbol{W}^-_{1g} \boldsymbol{d}_{1g} + c_k\right), \tag{5.18}$$

with $\boldsymbol{W}_{1g}$ a suitable matrix of weights, has an asymptotic χ^2-distribution with df $= G(k-1) + 1 - \dim(\boldsymbol{\lambda})$ (Glas, 1988a, 1989).

This section will be concluded by a brief discussion on how these statistics can be generalized to incomplete designs, that is, to designs where different subgroups of persons respond to different sets of items. Furthermore, it will be discussed how a division of the sample of persons into subgroups using their value on some external variable fits into this framework. In Chapter 3, the concept of incomplete data was presented, using the item administration variables b_{vi}. In the present chapter, this general concept will be supplemented with some additional terminology. Let $\boldsymbol{b}_v$ be the item administration vector of person S_v, defined by $\boldsymbol{b}_v = (b_{v1}, ..., b_{vi}, ..., b_{vk})'$. In the sequel, it will become clear that it must be possible to split up the complete sample of respondents into subgroups b, $b = 1, ..., B$, in which all respondents have the same item administration vector $\boldsymbol{b}_b$. The item administration vectors $\boldsymbol{b}_b$, $b = 1, ..., B$, will be called booklets. So the item administration design consists of B booklets, and all persons have responded to one of the booklets. This is completely equivalent to the definition of Chapter 3. However, it will become clear that every booklet must be administered to a substantial number of persons. An item administration design may be either linked or non-linked. A definition of a linked design is given as follows. Let $\boldsymbol{b}_a$ and $\boldsymbol{b}_b$ be any two booklets in the design. A design is linked if all booklets $\boldsymbol{b}_a$ and $\boldsymbol{b}_b$ in the design are linked, that is, if there exists a sequence of booklets $\boldsymbol{b}_a, \boldsymbol{b}_{h1}, \boldsymbol{b}_{h2}, ..., \boldsymbol{b}_b$, such that any two adjacent booklets in the sequence have common items, or, in the case of MML estimation, are assumed to be administered to samples from the same distribution of ability. Notice that the sequence may just consist of $\boldsymbol{b}_a$ and $\boldsymbol{b}_b$. A non-linked design is a design in which there are at least two booklets which are not linked. Assumptions with respect to ability distributions do not play any part in CML estimation. So CML estimation is only possible if the design is linked by common items. Linking by common distributions only applies to MML estimation.

Generalization of the R_{1c}-, Q_1-, R_{1m}- and R_0- statistics essentially boils down to estimating the parameters using all the data in the (linked) incomplete design, for all booklets computing the statistic in the same manner as defined above for a complete design, and summing the statistics obtained in the booklets to an overall statistic. For instance, for the R_{1c}-statistic, for every booklet the sample of respondents must be partitioned into a number of subgroups on the basis of their total scores, and the statistic will be based on the difference between the observed and expected number of correct responses on item I_i, in a subgroup g of a booklet b, given the CML estimates of the item parameters. The overall statistic R_{1c} is computed as $R_{1c} = \sum_b R_{1c}^{(b)}$, where $R_{1c}^{(b)}$ is identical to the statistic for the case of a design with one booklet as defined by (5.10). The generalization of Q_1 and R_{1m} proceeds in the same manner. Also the generalization of R_0 is essentially the same, that is, the test will be based on the difference between the observed score frequencies in the booklets and expected score frequencies in the booklets given the MML estimates of the item and population parameters. Again, the overall statistic R_0 is computed as $R_0 = \sum_b R_0^{(b)}$, where $R_0^{(b)}$ is identical to the statistic for the case of a design with one booklet as defined by (5.14). The theoretical justification of this approach can be found in Glas (1988a,

1989) and Glas and Verhelst (1989), where also the degrees of freedom of the various statistics are summarized. Now it is also clear why every booklet must be administered to a substantial number of persons, for only in this case the counts and their expected values will be sufficiently large to justify application of the asymptotic results.

The notion of a design made up of booklets can also be used for introducing background variables into the model testing procedure. Suppose the background variable of interest is gender. In that case two booklets can be defined which consist of the same items, one of which is assumed to be administered to the boys, and one to the girls. Differences in the number of correct responses to items in the score level groups of the two gender groups may reflect a model violation known as item bias or differential item functioning (DIF). For instance, the differences in observed and expected numbers of correct responses can be used for detecting items that are more difficult for boys than for girls, or vice versa. Also differences in slope between the groups can be detected this way. It must be stressed that in case MML is used, different ability distributions for the two groups must be introduced, for item bias does not concern the hypothesis that the probability of a correct response as such differs between groups, but it concerns the hypothesis that response probabilities differ between groups conditionally on ability level. A detailed example of testing for DIF, using generalizations of the tests proposed here to polytomous items, is given in Chapter 18.

5.2.3 Item-Oriented Tests

In this section, three test statistics will be presented which are based on the same kind of deviates as the tests of Section 5.2.2, but explicitly focus on specific items. Above it was mentioned that the scaled deviates z_{1gi} could be used as a diagnostic tool for evaluating variations in item discrimination. Elaborating on this approach, for the context of CML estimation, Molenaar (1983) suggested a statistic where the sign of the outcome indicates whether the discrimination of the item is too high or too low. Let c_1 and c_2 be cutoff-points dividing the score range into a low, middle, and high region. Only the low and high regions are considered in the test. Since the scaled deviates z_{1gi} are standardized binomial variables, the statistic

$$U_i = \frac{\sum_{g=1}^{c_1} z_{1gi} - \sum_{g=c_2}^{k-1} z_{1gi}}{(c_1 + k - c_2)^{1/2}} \tag{5.19}$$

has approximately a standard normal distribution. The designation 'approximately' expresses that the scaled deviates z_{1gi} are dependent, both by their sampling scheme and their reliance on the parameter estimates. The boundaries c_1 and c_2 are usually chosen in such a way that both summations include 25% of the persons of the sample. Notice that, if the item response function is steeper than the model permits, U_i is negative, and if the reverse is true, U_i is positive

TABLE 5.3. The matrix $\mathbf{U} = (\mathbf{T}_1|\mathbf{T}_2|\mathbf{Y})$ used for the M-tests for item I_2 (same example as in Tables 5.1 and 5.2)

T_1			T_2				Y
x_1	x_2	x_3	δ_{0r}	δ_{1r}	δ_{2r}	δ_{3r}	y_1
0	0	0	1	0	0	0	0
1	0	0	0	1	0	0	0
0	1	0	0	1	0	0	1
0	0	1	0	1	0	0	0
1	1	0	0	0	1	0	-1
1	0	1	0	0	1	0	0
0	1	1	0	0	1	0	-1
1	1	1	0	0	0	1	0

(Molenaar, 1983).

This statistic has also been brought into the framework of generalized Pearson tests (Verhelst & Eggen, 1989; Verhelst, Glas, & Verstralen, 1994). Squaring the deviates d_{1gi} underlying (5.19) and scaling them by a variance-covariance matrix yields in a statistic with an asymptotic χ^2-distribution with df $= 1$. An example of the $\boldsymbol{U}$ matrix for constructing this generalized Pearson test is shown in Table 5.3. The example concerns the matrix $\boldsymbol{Y}$ for the contrast for item I_2 between score group 1 and score groups 2 and 3.

Further details of this so-called M-statistic, including a proof of its asymptotic distribution, can be found in Chapter 12.

The last test of this section is the S_i-test (Verhelst & Eggen, 1989; Verhelst, Glas, & Verstralen, 1994). Like the M-tests, the S_i-test is defined at item level; it is based on the differences between the observed and expected proportion of responses in homogeneous score groups, and it has power against differences in discrimination between the items. The statistic is based on the same deviates d_{1gi} that were defined in (5.9) for the R_{1c}-statistic. So the score range is partitioned into equivalence classes $g = 1, ..., G$ and, for every score level G, the difference between the observed and expected number of correct responses to an item is computed. For $g = 1, ..., G$, let the vector $\boldsymbol{d}_i$ have elements d_{1gi}. Then the statistic

$$S_i = \boldsymbol{d}_i' \boldsymbol{W}_i^- \boldsymbol{d}_i \tag{5.20}$$

has an asymptotic χ^2-distribution with df $= G - 1$. The statistic is based on the $\boldsymbol{U}$ matrix of Table 5.4. Notice that the two-column vectors of the $\boldsymbol{Y}$ matrix sum to the column vector of x_2 minus the column vector under δ_{3r}. Therefore, the statistic based on the contrast defined by this $\boldsymbol{Y}$ matrix will have df $= 1$.

Like the R_{1c}-, Q_1-, R_{1m}- and R_0- statistics, the S_i-statistic can also be generalized to incomplete designs. Also in this case it is possible to compute the statistic

TABLE 5.4. The matrix $\mathbf{U} = (\mathbf{T}_1|\mathbf{T}_2|\mathbf{Y})$ used for the S_i-tests for item I_2 (same example used as in Tables 5.1 and 5.2)

T_1			T_2				Y	
x_1	x_2	x_3	δ_{0r}	δ_{1r}	δ_{2r}	δ_{3r}	y_1	y_2
0	0	0	1	0	0	0	0	0
1	0	0	0	1	0	0	0	0
0	1	0	0	1	0	0	1	0
0	0	1	0	1	0	0	0	0
1	1	0	0	0	1	0	0	1
1	0	1	0	0	1	0	0	0
0	1	1	0	0	1	0	0	1
1	1	1	0	0	0	1	0	0

for all booklets in the design in the same manner as defined above, and sum the statistics obtained in the booklets to an overall statistic. However, Verhelst and Eggen (1989) have developed a different approach. Above, the subgroups were formed within booklets. Here, G subgroups are specified and respondents are assigned to one subgroup each on the basis of their total scores, taking the booklet responded to into account in such a way that homogeneous ability groups are formed. So, in this approach the overall statistic is not made up of a summation of statistics over booklets, but the observed and expected frequencies within the subgroups are summed across booklets in such a way that the subgroups are homogeneous with respect to ability level. As with the R_{1c}-, Q_1-, and R_{1m}-statistics, also background variables can be introduced into the model testing procedure by defining different booklets for respondents with different values on that variables.

5.2.4 The R_2- and Q_2-Test

The statistics of the previous section can be used for testing the assumption of parallel item response functions. Another assumption underlying the RM is unidimensionality. Suppose unidimensionality is violated. If the person's position on one latent trait is fixed, the assumption of local stochastic independence requires that the association between the items vanishes. In the case of more than one dimension, however, the person's position in the latent space is not sufficiently described by one unidimensional ability parameter and, as a consequence, the association between the responses to the items given this one ability parameter will not vanish. Therefore, tests for unidimensionality are based on the association between the items.

Let M_{2ij} be the number of persons attaining a score greater than one, who get both item I_i and item I_j correct. Furthermore, $E(M_{2ij}|\hat{\boldsymbol{\omega}}, \hat{\boldsymbol{\beta}})$ is its CML expected value, and m_{2ij} its realization. Then the difference

$$d^*_{2ij} = m_{2ij} - E(M_{2ij}|\hat{\omega}, \hat{\boldsymbol{\beta}}) \tag{5.21}$$

may show violations of the dimensionality axiom. Consider the quadratic form

$$\boldsymbol{d}_2' \boldsymbol{W}_2^{-1} \boldsymbol{d}_2, \tag{5.22}$$

where $\boldsymbol{d}_2$ is a vector of the differences $d^*_{2ij}/\sqrt{n}$ of all pairs of items, and $\boldsymbol{W}_2^{-1}$ the inverse of its estimated covariance matrix. To facilitate the derivation of the distribution of the statistic presented below, also persons answering only one item correct have to be taken into account. Let $\boldsymbol{d}_1$ be a vector which has as elements, for all items, the differences between the numbers of persons giving a correct response on the item and attaining a score one, and its CML expected value, both divided by the square root of n. Moreover, let $\boldsymbol{W}_1^{-1}$ be its estimated covariance matrix. Then

$$R_{2c} = \boldsymbol{d}_2' \boldsymbol{W}_2^{-1} \boldsymbol{d}_2 + \boldsymbol{d}_1' \boldsymbol{W}_1^{-1} \boldsymbol{d}_1, \tag{5.23}$$

has an asymptotic χ^2-distribution with df $= k(k-1)/2$. Again this statistic belongs to the class of generalized Pearson statistics; for a proof of the assertion concerning its asymptotic distribution, see Glas (1988a, 1989).

Notice that the order of $\boldsymbol{W}_2$ is equal to the number of pairs of items in the test. For instance, if the test contains 15 items, the order of $\boldsymbol{W}_2$ is 105. Computing the statistic is only feasible as long as $\boldsymbol{W}_2$ can be inverted. Not limited by the number of items in a test is the Q_2-statistic by Van den Wollenberg (1982). As with many of the tests above, the sample of persons taking a test is partitioned into G subgroups indexed $g = 1, ..., G$ (both score levels and external variables may be used). The differences (5.21) between the observed and expected number of persons who answer both item I_i and item I_j correctly are split up further; so let M_{2gij} be the number of persons in subgroup g who answer both item I_i and item I_j correctly. The difference between its realization and its CML expected value is

$$d_{2gij} = m_{2gij} - E(M_{2gij}|\hat{\omega}, \hat{\boldsymbol{\beta}}). \tag{5.24}$$

Dividing these differences by their estimated standard deviations yields the standardized variables z_{2gij}, which can be combined to form the statistic

$$Q_2 = \frac{k-3}{k-1} \sum_{i=1}^{k-1} \sum_{j=i+1}^{k} \sum_{g=1}^{G} z^2_{2gij}. \tag{5.25}$$

To get a good approximation of a χ^2-distribution, the item parameters need to be estimated in each subgroup, and for that case simulation studies showed that Q_2 approaches a χ^2-distribution with df $= Gk(k-3)/2$ (Van den Wollenberg, 1982).

The assumption of unidimensionality can be tested also in the MML framework by evaluating the observed and expected association between items. The difference between the observed and MML expected value of M_{2ij} being denoted

$$d^*_{2ij} = m_{2ij} - E(M_{2ij}|\hat{\boldsymbol{\beta}}, \hat{\boldsymbol{\lambda}}), \tag{5.26}$$

the core of the statistic is

$$\boldsymbol{d}_2' \boldsymbol{W}_2^{-1} \boldsymbol{d}_2, \tag{5.27}$$

where $\boldsymbol{d}_2$ is a vector of the differences d_{2ij}, with $d_{2ij} = d^*_{2ij}/\sqrt{n}$, of all pairs of items I_i and I_j, and $\boldsymbol{W}_2^{-1}$ the inverse of a matrix of weights. To create a statistic where the asymptotic distribution can be derived, the persons attaining a score zero, a score one, and a perfect score, must be taken into account also. Let $\boldsymbol{d}_1$ be a vector which has as elements, for all items, the differences between the numbers of persons giving a positive response on the item and attaining a score one, and its MML expected value, both divided by the square root of n; and denote its covariance matrix by $\boldsymbol{W}_1^{-1}$. For the zero and perfect score, c_0 and c_k defined by (5.16) and (5.17) must be added. Then,

$$R_{2m} = c_0 + \boldsymbol{d}_2' \boldsymbol{W}_2^{-1} \boldsymbol{d}_2 + \boldsymbol{d}_1' \boldsymbol{W}_1^{-1} \boldsymbol{d}_1 + c_k \tag{5.28}$$

has an asymptotic χ^2-distribution with df $= (k(k+3)/2-1) - k - \dim(\boldsymbol{\lambda})$ (Glas, 1988a, 1989).

5.3 Likelihood Ratio Tests

The well-known general principle of the likelihood ratio (LR) test can be summarized as follows. Let $L_0(\boldsymbol{\phi}_0)$ be the likelihood function of a model which is a special case of some more general model with likelihood function $L_1(\boldsymbol{\phi}_1)$, where $L_1(\boldsymbol{\phi}_1)$ is believed to be true. It is assumed that $L_0(\boldsymbol{\phi}_0)$ and $L_1(\boldsymbol{\phi}_1)$ are functions of s_0 and s_1 parameters, respectively, and that $s_0 < s_1$. The validity of the special or restricted model can be tested against the more general alternative using the statistic $LR = -2\ln[L_0(\hat{\boldsymbol{\phi}}_0/L_1(\hat{\boldsymbol{\phi}}_1)]$, where the likelihood function of the restricted model and that of the general model are evaluated at their maxima, that is, they are evaluated using ML estimates. Under certain mild regularity assumptions, the LR statistic has an asymptotic χ^2-distribution with df $= s_1 - s_0$ (see, for instance, Lehmann, 1986).

In this section, first two tests for the RM based on this principle will be discussed: Andersen's LR test and the Martin-Löf LR test. Both apply to the CML context. Next, other applications of the LR principle to testing the RM will be discussed, such as the possibilities of applying the LR principle in the MML context.

Andersen's LR test is based on the outstanding feature of CML estimation that under very mild regularity assumptions (see Pfanzagl, 1994) the item parameters can be consistently estimated in any (not necessarily random) sample from a population where the model applies. Therefore, a testing procedure can be based on an evaluation of the differences between the CML estimates of the item parameters in different subgroups, which again can be formed on the basis of score levels or on the basis of external criteria. If items fit the model, the parameter estimates are equal, apart from random fluctuations. The parameter estimates of non-fitting items will generally not be equal. For these items the pattern of the parameter estimates can give an indication of the cause of misfit. For example, if the parameter estimate is too low in the low score group, the low score group gives more correct responses than expected; there might be much guessing on that item.

The test statistic is defined as follows. Assume that $L_C(\hat{\boldsymbol{\beta}}; \boldsymbol{X})$ is the conditional likelihood function evaluated using the CML estimates of the item parameters obtained using all data, say $\boldsymbol{X}$. The persons are divided into subgroups, let $\boldsymbol{X}_g$ be the data of subgroup g, and let $L_C(\hat{\boldsymbol{\beta}}_g; \boldsymbol{X}_g)$ be the conditional likelihood function evaluated using the CML estimates of the item parameters obtained in subgroup g. If items are present where the parameters cannot be estimated in each subgroup, CML estimates are obtained by restricting these parameters to be the same for all subgroups. Using these definitions, the likelihood-ratio statistic

$$LR = 2\left(\sum_{g=1}^{G} \ln L_C(\hat{\boldsymbol{\beta}}_g; \boldsymbol{X}_g) - \ln L_C(\hat{\boldsymbol{\beta}}; \boldsymbol{X})\right) \tag{5.29}$$

has an asymptotic χ^2-distribution. The degrees of freedom are equal to the number of parameters estimated in the subgroups minus the number of parameters estimated in the total data set. With respect to the handling of items that cannot be estimated in all subgroups, this statistic is an adaptation of Andersen's LR statistic (Andersen, 1973b). Comparing Andersen's LR test with the R_{1c}- and Q_1-test, it comes as no surprise that all these tests have power against similar model violations. For instance, if the subgroups are score regions, all three tests are sensitive to differences in discrimination between the items, and their results are usually not much different. In the present case, background variables can again be introduced to study item bias.

The second application of the LR principle is a test proposed by Martin-Löf (1973). Since the statistic is devised to test whether two sets of items form a Rasch scale, it can be viewed as a test of the dimensionality axiom. Let the items be partitioned into two subsets of k_1 and k_2 items, respectively. Let $\boldsymbol{r} = (r_1, r_2)'$, $r_1 = 0, ..., k_1$ and $r_2 = 0, ..., k_2$, be score patterns on the two subtests, and $n_{\boldsymbol{r}}$ the number of persons obtaining score pattern $\boldsymbol{r}$. As above, r is the respondents' total sum score, so $r = r_1 + r_2$, and n_r is the number of persons attaining this score. The statistic is defined as

$$LR = 2\Big(\sum_r n_r \ln\left(\frac{n_r}{N}\right) - \sum_r n_r \ln\left(\frac{n_r}{N}\right) - \ln L_C + \ln L_C^{(1)} + \ln L_C^{(2)}\Big), \quad (5.30)$$

where L_C, $L_C^{(1)}$, and $L_C^{(2)}$ are the likelihood functions evaluated using CML estimates in the complete test, the first subtest, and the second subtest, respectively. If the items form one Rasch scale, (5.30) has an asymptotic χ^2-distribution with $\mathrm{df} = k_1k_2 - 1$.

Superficially, the last three terms of (5.30) seem to represent a proper LR statistic; however, adding the first two sums is necessary because L_C is conditional on the frequency distribution of sum scores on the entire test of k items, while $L_C^{(1)}$ and $L_C^{(2)}$ are conditional on the frequency distributions of the subtests.

The two tests above are just two examples of application of the LR principle to the RM. In the framework of CML estimation, LR tests are also used to evaluate the validity of an RM with linear restrictions on the item parameters, a so-called LLTM (see Chapter 8), against the unrestricted RM. Also nested LLTM's can be compared with each other (Fischer, 1974). Kelderman (1984) takes the approach even further, in the sense that parameters are added to the RM to construct so-called log-linear RMs. The validity of models belonging to this class are also compared using LR statistics evaluated using CML estimates.

In the framework of MML estimation, the counterparts of the tests given above are not always feasible. For instance, Andersen's LR test may require estimation in score regions. However, consistent estimation of the population parameters $\boldsymbol{\lambda}$ requires a random sample from the ability distribution. As a consequence, splitting up the resulting frequency distribution to obtain estimates in subgroups will not produce consistent estimates. If external variables are used for splitting up the original sample, various counterparts of the Andersen LR test can be constructed. Suppose the external variable is gender, and one wants to test the hypothesis that the RM holds for both sexes. Then one may start the testing procedure with specifying a general model where both groups have different item parameters and different population parameters and a restricted model with equal item parameters in both gender groups and unequal ability distributions. If the hypothesis of equal item parameters is supported, one can go further and test the hypothesis of equal ability distributions using an LR statistic.

The MML counterpart of the Martin-Löf LR statistic has some complications. Probably, the latent variables relating to subtests which do not form one Rasch scale will still be associated. In the framework of normal latent ability distributions, this amounts to postulating a multivariate normal distribution of latent ability. How to estimate and test this model is discussed in Glas (1989, 1992). Testing the validity of the unidimensional RM against this compound RM can be done using the LR statistic

$$LR = 2[\ln L(\hat{\boldsymbol{\beta}}^{(1)}, \hat{\boldsymbol{\beta}}^{(2)}, ..., \hat{\boldsymbol{\mu}}, \hat{\boldsymbol{\Sigma}}) - \ln L(\hat{\boldsymbol{\beta}}, \hat{\mu}, \hat{\sigma})], \quad (5.31)$$

with $\hat{\boldsymbol{\beta}}^{(1)}, \hat{\boldsymbol{\beta}}^{(2)}, \ldots$ the item parameter estimates of the subtests, $\hat{\boldsymbol{\mu}}$ and $\hat{\boldsymbol{\Sigma}}$ those of the parameters of the multivariate normal ability distribution, and $\hat{\mu}$ and $\hat{\sigma}$ those of the parameters of the univariate normal ability distribution.

This section is concluded with a well-known, but often disregarded word of caution. When applying LR tests, it must always be kept in mind that, if the general model of the alternative hypothesis does not fit, a non-significant outcome of the LR test cannot be considered as support for the restricted model of the null hypothesis. The most one can conclude in these cases is that both models are equally improbable. Closely related to this problem is the problem of the power of the test. Applying Andersen's LR test using an external variable poorly related to ability, say, the respondent's body length, results in a test with little power, and a non-significant result offers no support for the hypothesis that the RM fits the data.

5.4 Wald-type Tests

The Wald test has much in common with the LR test. One starts with defining a general model, and tests whether certain restrictions hold. As with many of the tests considered above, most applications of the Wald test to the RM focus on meaningful subgroups of the sample of respondents. In this section, only the case of two subgroups will be considered, but the generalization to more subgroups is straightforward. Let the model parameters for the g-th subgroup be denoted $\boldsymbol{\phi}_g = (\phi_{g1}, \ldots, \phi_{gm})'$, $g = 1, 2$, and m be the number of free parameters, which is assumed constant over subgroups (although, strictly speaking, this is not necessary). In the application discussed here, $m = k - 1$ if CML is used, and $m = k + 1$ in the case of MML with a normal ability distribution. It is important that the same normalization is used in both subgroups. Assume that the parameter of the k-th item is fixed to zero and that the parameters $\phi_{g1}, \ldots, \phi_{g,k-1}$ refer to the difficulty parameters of the items 1 to $k - 1$, respectively.

Estimating the parameters in the two subgroups separately amounts to estimating $2m$ parameters. This means that the k items in the first group are considered as possibly different from the k items in the other group. In a CML framework, the hypothesis of two different sets of item parameters for the two subgroups can be translated into a design with two groups taking two different sets of items. However, the two sets of items are not linked, and only $2(k - 1)$ parameters are free to vary. In an MML framework the same design is considered, and it will be assumed that the two means and variances are free to vary. Finally, let $\boldsymbol{\phi}' = (\boldsymbol{\phi}_1', \boldsymbol{\phi}_2')$. The null hypothesis can be understood as the requirement that q functions $h_j(\boldsymbol{\phi})$, $j = 1, .., q$, are zero. If only equality of item parameters is assumed, then $q = k - 1$, and the q restrictions are

$$h_j(\boldsymbol{\phi}) \;=\; \phi_{1j} - \phi_{2j} \;=\; 0, \qquad j = 1, .., q. \tag{5.32}$$

Let $\boldsymbol{h}(\boldsymbol{\phi})' = (h_1(\boldsymbol{\phi}), ..., h_q(\boldsymbol{\phi}))$, and let $\boldsymbol{\Sigma}_g$, $g = 1, 2$, be the variance-covariance matrix of the ML estimator of $\boldsymbol{\phi}_g$. Since the responses of the two subgroups are independent, it follows that the variance-covariance matrix of the ML estimator (which may be either the CML or the MML estimator) of $\boldsymbol{\phi}$ is

$$\boldsymbol{\Sigma} = \begin{pmatrix} \boldsymbol{\Sigma}_1 & 0 \\ 0 & \boldsymbol{\Sigma}_2 \end{pmatrix}. \tag{5.33}$$

The Wald test statistic is given by the quadratic form

$$W = \boldsymbol{h}'(\hat{\boldsymbol{\phi}})[\boldsymbol{T}'(\hat{\boldsymbol{\phi}})\boldsymbol{\Sigma}\boldsymbol{T}(\hat{\boldsymbol{\phi}})]^{-1}\boldsymbol{h}(\hat{\boldsymbol{\phi}}), \tag{5.34}$$

where $T(\boldsymbol{\phi})$ is a $2m \times q$ matrix $[\, t_{gj}]$ defined by

$$t_{gj} = \frac{\partial h_j(\boldsymbol{\phi})}{\partial \phi_g}. \tag{5.35}$$

In the case of such simple restrictions as (5.32), it is easily verified that (5.34) simplifies to

$$W = (\hat{\boldsymbol{\phi}}_1^* - \hat{\boldsymbol{\phi}}_2^*)'[\boldsymbol{\Sigma}_1^* + \boldsymbol{\Sigma}_2^*]^{-1}(\hat{\boldsymbol{\phi}}_1^* - \hat{\boldsymbol{\phi}}_2^*), \tag{5.36}$$

where an asterisk denotes that (pairs of) parameters not involved in any restriction are deleted from all the vectors and matrices. Wald (1943) has proved the now classical result that, under mild regularity assumptions, W is asymptotically χ^2-distributed with df $= q$. In general, the number of degrees of freedom equals the number of restrictions. In the particular case discussed above, the number of degrees of freedom is $k - 1$.

In interpreting the differences between groups of the item parameter estimates, one should be careful with respect to the origin and unit chosen for the scale in the two subgroups, because comparing two estimates which are on different scales is pointless. The problems associated with the origin, however, are a little bit more subtle. In interpretative terms, one is used to the item as the unit of interpretation. However, with k binary items, there are only $k - 1$ free parameters, so there is always some degree of arbitrariness in assigning specific values to the item parameter estimates. Influencing the value of the W-statistic by this arbitrary assignment should be avoided. In other words, one should choose a restriction, which, unlike (5.32), is independent of the chosen normalization in both subgroups. To see how a normalization may influence the W-statistic, imagine a test of ten items, all having the same parameter value (zero, say) in one population, while in the other population item I_1 has a parameter value of -1.0, and the other nine have parameters equal to zero. Estimating these parameters in two large samples will yield approximately these values as estimates, or more precisely, the mutual differences between any two parameter estimates in the first sample will be approximately zero, as will be the case for the items I_2 to I_{10} in sample 2, while the difference with the estimate of item I_1 will be approximately one. If, for both groups, a normalization is chosen which fixes the parameter of item I_1 at a given constant, the differences $\hat{\beta}_{1i} - \hat{\beta}_{2i}$, $i > 1$, will be approximately 1.0, leading to the conclusion (with sufficiently large samples) that nine items are biased. Moreover the power of the Wald test is the same for all items, so if DIF is detected for one item, it will be detected with high probability for all items. Although this conclusion cannot be disputed because one cannot formally distinguish between one item being biased and the remaining ones being biased, the interpretation is not elegant and may be difficult in more complicated cases. Moreover, taking another item as reference will lead to the correct conclusion that only item I_1 is biased.

Therefore, it might be better to use a restriction which is independent of the chosen normalization. Such restrictions are obtained if they are functions of the differences between a particular item parameter estimate and the estimates of the other item parameters. Such a restriction is

$$h_{1i} = \sum_{j \neq i} (\hat{\beta}_{1i} - \hat{\beta}_{1j}) - \sum_{j \neq i} (\hat{\beta}_{2i} - \hat{\beta}_{2j}) = 0, \quad (i = 1, .., k), \tag{5.37}$$

where the first subscript of the $\hat{\beta}$ refers to the sample, and the subscript j takes the values 1 to k, excluding i. However, since the terms in both sums can be positive or negative, cancellation can occur. To avoid this and at the same time to preserve differentiability of the restrictions, the differences can be squared, yielding as a second set of restrictions

$$h_{2i} = \sum_{j \neq i} (\hat{\beta}_{1i} - \hat{\beta}_{1j})^2 - \sum_{j \neq i} (\hat{\beta}_{2i} - \hat{\beta}_{2j})^2 = 0, \quad (i = 1, .., k). \tag{5.38}$$

If in both cases the scale is chosen such that the sum of the parameters is zero, (5.37) and (5.38) reduce to

$$h_{1i} = 2(\hat{\beta}_{1i} - \hat{\beta}_{2i}) = 0, \tag{5.39}$$

and

$$h_{2i} = (k\beta_{1i}^2 + \sum_j \hat{\beta}_{1j}^2) - (k\hat{\beta}_{2i}^2 + \sum_j \hat{\beta}_{2j}^2) = 0, \tag{5.40}$$

for $i = 1, ..., k$. As is clear from (5.40), the test will be sensitive to differences in the variance of the item parameters as well as to differences in the square of a particular item parameter estimate.

First, (5.39) will be substituted into the general expression for the Wald statistic. If $\boldsymbol{\Sigma}$ denotes the variance-covariance matrix of a solution normalized to a zero sum in both groups, (5.34) reduces to

$$W_{1i} = \frac{(\hat{\beta}_{1i} - \hat{\beta}_{2i})^2}{\sigma_{1i}^2 + \sigma_{2i}^2}, \tag{5.41}$$

where σ_{1i} and σ_{2i} denote the i-th diagonal element of $\boldsymbol{\Sigma}_1$ and $\boldsymbol{\Sigma}_2$, respectively. This gives a test for DIF of item I_i, because W_{1i} is asymptotically χ^2-distributed with df $= 1$. This statistic is closely related to a statistic proposed by Fischer and Scheiblechner (1970, see also Fischer, 1974, p. 297). The main difference between the two is that for the latter only the diagonal of the information matrix is used for computing the variances in the denominator of (5.41). Therefore, the Fischer and Scheiblechner statistic can be viewed as an approximation of W_{1i}. Taking the square root, and giving it the sign of the difference between the two parameter estimates, yields a directional test statistic which is asymptotically normally distributed. It should be noted that the W_{1i}-statistics are not independent of each other, nor are they completely specific. To refer to the example given above, if one item is biased, all W_{1i}-tests will yield significant results if the samples are large enough. However, what matters is that W_{1i} will give a significant result with higher probability if only item I_i is biased, than if only a single other item is biased.

Substituting (5.40) in the general expression for the Wald statistic gives

$$W_{2i} = \frac{h_{2i}^2}{\boldsymbol{t}_1'\boldsymbol{\Sigma}_1\boldsymbol{t}_1 + \boldsymbol{t}_2'\boldsymbol{\Sigma}_2\boldsymbol{t}_2}, \tag{5.42}$$

where $\boldsymbol{t}_g$, $g = 1, 2$, is a k-dimensional vector defined by

$$t_{gj} = 2(1 + \delta_{ij}k)\beta_{gj}, \tag{5.43}$$

and δ_{ij} is the Kronecker symbol.

5.5 Lagrange Multiplier Tests

It is well-known that likelihood ratio (LR), Wald, and Lagrange multiplier (LM) tests are closely related; in fact, in some instances they are identical (Buse, 1982). It is therefore natural to search for applications of the LM test in the framework of IRT in general, and applications to the Rasch model in particular. The idea behind the LM statistic (Aitchison & Silvey, 1958), and the equivalent efficient score statistic (Rao, 1948), can be summarized as follows. Let $\boldsymbol{\phi}$ be the vector of parameters of some general model, and let $\boldsymbol{\phi}_0$ be the parameters of a special case of the general model. So, as with the LR and Wald tests, there is a general and a restricted model. In the present case it will be assumed that the restricted model is derived from the general model by setting one or more parameters of the general model to fixed constants. Hence, $\boldsymbol{\phi}_0$ is partitioned, $\boldsymbol{\phi}_0' = (\boldsymbol{\phi}_{01}', \boldsymbol{\phi}_{02}') = (\boldsymbol{\phi}_{01}', \boldsymbol{c}')$, where $\boldsymbol{c}$ is a vector of constants. Next, let $\boldsymbol{b}_{\boldsymbol{\phi}}(\boldsymbol{\phi}^*)$ be the partial derivative of the log-likelihood function of the general model evaluated at $\boldsymbol{\phi}^*$, that is, $\partial \ln L(\boldsymbol{\phi})/\partial\boldsymbol{\phi}$ evaluated at $\boldsymbol{\phi}^*$. This partial derivative vector gives the change in the log-likelihood for local changes in $\boldsymbol{\phi}^*$. Furthermore, $\boldsymbol{I}_{\boldsymbol{\phi},\boldsymbol{\phi}}(\boldsymbol{\phi}^*, \boldsymbol{\phi}^*)$ is defined as $-\partial^2 \ln L(\boldsymbol{\phi}, \boldsymbol{\phi})/(\partial\boldsymbol{\phi}\partial\boldsymbol{\phi}')$ evaluated at $\boldsymbol{\phi}^*$. The statistic is defined as

$$\text{LM} = \boldsymbol{b}(\hat{\boldsymbol{\phi}}_0')\boldsymbol{I}_{\phi,\phi}(\hat{\boldsymbol{\phi}}_0, \hat{\boldsymbol{\phi}}_0)^{-1}\boldsymbol{b}(\hat{\boldsymbol{\phi}}_0). \tag{5.44}$$

The nice aspect of (5.44) is that it is evaluated using the ML estimates of the parameters of the restricted model $\hat{\boldsymbol{\phi}}_0$. Apart from the computational convenience, this also gives the motivation for the test. The unrestricted elements of $\hat{\boldsymbol{\phi}}_0$, that is, $\hat{\boldsymbol{\phi}}_{01}$, have a partial derivative equal to zero, because their values originate from solving the likelihood equations. The magnitude of the elements of $\boldsymbol{b}_{\phi}(\hat{\boldsymbol{\phi}}_0)$ corresponding with fixed parameters, the elements of $\boldsymbol{b}_{\phi}(\boldsymbol{c})$, determine the value of the statistic. The statistic itself has an asymptotic χ^2-distribution with degrees of freedom equal to the number of fixed parameters. Since $\boldsymbol{b}_{\phi}(\hat{\boldsymbol{\phi}}_{01}) = 0$, the LM statistic can also be computed as

$$\text{LM}(\boldsymbol{c}) \;=\; \boldsymbol{b}_{\phi}(\boldsymbol{c}')\boldsymbol{W}^{-1}\boldsymbol{b}_{\phi}(\boldsymbol{c}), \tag{5.45}$$

with

$$\boldsymbol{W} \;=\; \boldsymbol{I}_{\phi_2,\phi_2}(\boldsymbol{c},\boldsymbol{c}) - \boldsymbol{I}_{\phi_2,\phi_1}(\boldsymbol{c},\hat{\boldsymbol{\phi}}_{01})\boldsymbol{I}_{\phi_1,\phi_1}(\hat{\boldsymbol{\phi}}_{01},\hat{\boldsymbol{\phi}}_{01})^{-1}\boldsymbol{I}_{\phi_1,\phi_2}(\hat{\boldsymbol{\phi}}_{01},\boldsymbol{c}), \tag{5.46}$$

where $\boldsymbol{I}_{\phi_2,\phi_2}(\boldsymbol{c},\boldsymbol{c})$ is defined as $-\partial^2 \ln L(\boldsymbol{\phi}_2, \boldsymbol{\phi}_2)/(\partial\boldsymbol{\phi}_2\partial\boldsymbol{\phi}_2')$ evaluated at $\boldsymbol{c}$, etc. Notice that, if the restrictions are tested one at a time, $\boldsymbol{W}$ is a number, which takes little effort to invert, and $\boldsymbol{I}_{\phi_1,\phi_1}(\hat{\boldsymbol{\phi}}_{01},\hat{\boldsymbol{\phi}}_{01})$ needs to be inverted only once.

In the chapter on the OPLM (Chapter 12), the LM approach will be used to build item oriented tests in the context of MML estimation. In the present chapter, an application to the testing the validity of linear restrictions on the item parameters in the RM will be briefly mentioned. In the LLTM (Chapter 8), the k item parameters $\boldsymbol{\beta}$ are a linear function of m so-called basic parameters $\boldsymbol{\eta}$, that is, $\boldsymbol{\beta} \;=\; \boldsymbol{Q}\boldsymbol{\eta}$. The matrix $\boldsymbol{Q}$ has rows $\boldsymbol{q}_{i*}$, columns $\boldsymbol{q}_{*j}$, and elements q_{ij}, for $i = 1,...,k$ and $j = 1,...,m$. The elements of $\boldsymbol{Q}$ are fixed constants. If they had to be estimated, the likelihood function would not be an exponential family, in fact, the expression for the model would contain multiplications of parameters, complicating the estimation process further. Still, the practitioner often wants to evaluate the validity of the restrictions imposed. This can be accomplished using the LM($\boldsymbol{c}$)-statistic defined in (5.45). The statistic can be computed from the perspective of the items and the elementary parameters, and for the separate elements of $\boldsymbol{Q}$. This results in the statistics LM($\boldsymbol{q}_{i*}$), LM($\boldsymbol{q}_{*j}$) and LM(q_{ij}), for $i = 1,...,k$ and $j = 1,...,m$. Using these statistics of item fit to the LLTM, the appropriateness of the defined basic parameters and the appropriateness of the magnitude of the elements of $\boldsymbol{Q}$ can be evaluated.

5.6 Conclusion

A first conclusion that can be drawn is that the testing machinery for the RM is extensive and that the properties of the model are such that these procedures are statistically well-founded, that is, the testing procedures can be based on statistics with a known asymptotic distribution. A question that immediately comes to mind is which procedures to chose. First of all, it must be mentioned that computational considerations are nowadays hardly important, because the performance of computers is impressive and still increasing. Furthermore, software is readily available, most of the procedures described above are implemented in RSP (Glas & Ellis, 1993) and OPLM (Verhelst, Glas, & Verstralen, 1994). So the scientific way may, after all, be to choose 'statistics all' and to give the alternative hypothesis that the RM does not hold as much chance as possible. However, just as important as having a global test statistic is the information which the statistic provides with respect to causes of a possible lack of model fit. When it comes to the diagnostic function of a testing procedure, practitioners generally have their personal preferences. One practitioner may prefer to look at the differences between observed and expected frequencies and use Pearson-type statistics, while another may be more attracted to differences in parameter estimates and use likelihood ratio and Wald statistics. Diagnostic analysis based on these statistics can, of course, be supported by various graphic displays, but this is beyond the scope of the present chapter. In fact, the emphasis of this chapter has been on the statistics on which testing the RM can be based, and little has been said about the actual process of item analysis. It must be stressed that item analysis is an iterative process. Computation of the statistics involves the estimates of the parameters of all items, and as a result the effects of the misfit of the items become entangled. It may well be that an item that was removed from a scale can be added again when other misfitting items are deleted.

If the model fails, two directions can be followed depending on the practitioners' motives. If the motive is to construct a Rasch-homogeneous measurement instrument, items must be removed. If the motive is fitting a model to the data, alternative models must be chosen by relaxing the assumptions of the RM. If the assumptions of sufficiency of the sum score and of strictly monotone increasing and parallel item response functions do not hold, the OPLM may be the model of choice. In Chapter 12 it will be shown that also for this model CML estimation is feasible, and the testing procedures for the RM presented here can easily be adapted to the OPLM. If the assumption of parallel item response functions fails, one can also take recourse to the two- and the three-parameter logistic model (Birnbaum, 1968). However, for these two models, only MML estimation is feasible (see Bock & Aitkin, 1981) and the theoretical advantages of CML estimation are not supported here. Moreover, the lack of non-trivial sufficient statistics for the parameters prevents the generalization of most of the testing procedures described above; for instance, Pearson-type statistics and Andersen's likelihood ratio statistic cannot be used.

If the assumptions of unidimensionality and local stochastic independence are violated, alternative models are available. The models of Kelderman (1984) and Jannarone (1986) do not require the assumption of local independence, while preserving the possibility of CML estimation. The role of local independence is modified in the dynamic test model by Verhelst and Glas (1993; see Chapter 10), while unidimensionality is explicitly lacking in the RM with a multivariate ability distribution by Glas (1989, 1992). However, these last two models only allow for MML estimation. Unidimensionality is also explicitly absent in the LLRA-type models (Chapter 9) where the possibility of CML estimation is preserved.

6

The Assessment of Person Fit

Karl Christoph Klauer[1]

ABSTRACT Person fit is discussed from the perspective of testing whether or not an individual's test behavior is consistent with the Rasch model. A general method is considered for deriving person fit indices and person fit tests that are powerful for testing against a prespecified type of deviation from the Rasch model. Power analyses are performed, and previous approaches are subsumed under the general framework where possible.

6.1 Introduction

The Rasch model (RM) assumes local independence, unidimensionality, and a specific shape of the item characteristic curve. The item characteristic curve for item I_i is given by

$$f_i(\theta) = \frac{\exp(\theta - \beta_i)}{1 + \exp(\theta - \beta_i)}, \tag{6.1}$$

where β_i is the item difficulty parameter. Equation (6.1) can be interpreted in at least two ways (Holland, 1990). As explained by Molenaar in Chapter 1, according to the *random sampling* rationale (Birnbaum, 1968), θ defines strata or subpopulations of the population of examinees, with the same ability. The meaning of $f_i(\theta)$ is the proportion of people in the stratum labeled θ who will answer item I_i correctly if tested. Given the random sampling view, it does not make sense to estimate an individual's ability. The latent variable θ is little more than a variable over which one integrates to determine the so-called manifest probabilities (Cressie & Holland, 1983), and (6.1), taken together with local independence, is little more than a useful tool for specifying Cressie and Holland's (1983) *manifest* RM (Holland, 1990). The goodness-of-fit of this manifest RM can be determined by means of traditional tests of the RM (Andersen, 1973b; Gustafsson, 1980b; Molenaar, 1983; Kelderman, 1984; Glas & Verhelst, see Chapter 5).

Most authors subscribe to an alternative view of (6.1), the *stochastic subject* rationale (Holland, 1990). An individual's responses are assumed to be governed by a stochastic mechanism that is characterized by an ability parameter θ and by parameters that describe the characteristics of the items (for example, Lord & Novick, 1968, p. 30). Thus, (6.1) specifies the probability that a subject S_v with person parameter $\theta_v = \theta$ answers item I_i correctly. The stochastic subject

[1]Universität Heidelberg, Psychologisches Institut, Hauptstraße 47–51, 69117 Heidelberg, FR Germany; e-mail: KLAUER@URZ.UNI-HEIDELBERG.DE

rationale underlies many applications of the RM such as estimation and inference on person parameters (Klauer, 1990, 1991a; Junker, 1991; Hoijtink & Boomsma, Chapter 4), mastery testing, and adaptive testing (Hambleton & Swaminathan, 1985), among others. The stochastic subject rationale views the RM as a model for given individuals and thus requires goodness-of-fit tests at the level of the *conditional* probabilities (Trabin & Weiss, 1983). The objective of this chapter is to derive means of testing the constraints that the RM imposes on the conditional probabilities, using single response vectors. Such tests will be called *person tests*.

If an individual answers in accordance with the RM, a person fit assessment may nevertheless wrongly indicate lack of fit and thus, an error of the first kind can occur in testing person fit. The probability of an error of this kind should not be a function of the individual's ability, so that the person test is independent of the ability under the null hypothesis. A person test statistic is said to be 'standardized' (Levine & Drasgow, 1988) if its conditional distribution, given ability, does not vary as a function of ability for individuals whose responses follow the RM. Most of the traditional person fit indices (e.g., Wright, 1977; Rudner, 1983; Trabin & Weiss, 1983; Tatsuoka, 1984; Levine & Drasgow, 1983, 1988) fail to meet this basic requirement (Molenaar & Hoijtink, 1990). If a given individual could be tested many times using the same or statistically equivalent items, a standardized person test could be derived on the basis of traditional multinomial theory for contingency tables (Klauer, 1988). Such a test would have test power to detect every possible deviation from the RM. In practice, however, only one response vector can be obtained for a given individual. A person test based on this one response vector can reasonably be expected to exhibit only limited sensitivity for a small subset of possible deviations from the RM. The strategy proposed in this chapter capitalizes on this observation, starting out from the specification of simple alternatives to the RM. A person test is then derived for each alternative, focussing on maximum sensitivity for detecting deviations in the direction of the specified alternative. The alternatives considered are two-parameter exponential families of distributions (Barndorff-Nielsen, 1978) that involve a second person-specific parameter η_v over and above the ability parameter θ_v.

The chapter is organized as follows: Section 2 presents three alternatives that model different types of deviations. Section 3 discusses the statistical theory for deriving person tests against these alternatives, and Section 4 compares the test power of each person test for all three alternatives considered in this chapter. Section 5 addresses the question of how many of these person tests should be conducted and proposes a rational method for selecting person tests in a given testing context.

6.2 Alternative models

The alternatives are two-dimensional generalizations of the RM that contain a second person-specific parameter η_v in addition to the ability parameter θ_v. The parameter η describes size and direction of deviations from the original RM. For a certain value η_0 of η, in general for $\eta_0 = 0$, the generalized model becomes the RM. In all cases, the alternative model defines a two-parameter exponential family of distributions with so-called canonical parameters θ and η and canonical statistics R and T, where R is the raw score variable, and T is a statistic that depends on the particular alternative considered. Thus, using suitable functions μ and h, the alternative models can be represented as follows:

$$P(\boldsymbol{X} = \boldsymbol{x} \mid \theta, \eta) = \mu(\theta, \eta) h(\boldsymbol{x}) \exp[\eta T(\boldsymbol{x}) + \theta R(\boldsymbol{x})], \tag{6.2}$$

where $\boldsymbol{x}= (x_1, \ldots, x_k)$ is a response vector and $\boldsymbol{X}$ is the response vector variable. In the following, three alternative models are introduced.

6.2.1 Example 1: Non-Invariant Abilities

The RM implies that an individual's ability is invariant over subtests of the total test. Thus, if a given test is split into two subtests, A_1 and A_2, the individual's true abilities, θ_{v1} and θ_{v2}, describing the individual's ability for subtests A_1 and A_2, respectively, should be identical: $\theta_{v1} = \theta_{v2}$. In the sequel, the person index v is dropped for simplicity. The first alternative model allows for lack of invariance of the ability parameter. According to this model, the individual with abilities θ_j answers according to the RM for each subtest A_j, $j = 1, 2$, and thus,

$$P(\boldsymbol{X}_j = \boldsymbol{x}_j \mid \theta_j) = \mu_j(\theta_j) h_j(\boldsymbol{x}_j) \exp[\theta_j R_j(\boldsymbol{x}_j)], \qquad j = 1, 2,$$

where

R_j is the raw score variable for subtest A_j,
$\boldsymbol{x}_j$ is a response pattern for subtest A_j,
$h_j(\boldsymbol{x}_j) = \exp\left(-\sum_{i:i \in A_j} x_i \beta_i\right)$, and
$\mu_j(\theta_j) = \prod_{i:i \in A_j} [1 + \exp(\theta_j - \beta_i)]^{-1}$,
A_j is the set of indices i of items of the j−th subtest.

The joint distribution is, however, described by a two-dimensional generalization of the RM ($\boldsymbol{x} = (\boldsymbol{x}_1, \boldsymbol{x}_2)$) :

$$P(\boldsymbol{X} = \boldsymbol{x} \mid \theta_1, \theta_2) = \mu_1(\theta_1) \mu(\theta_2) h_1(\boldsymbol{x}_1) h_2(\boldsymbol{x}_2) \exp[\theta_1 R_1(\boldsymbol{x}_1) + \theta_2 R_2(\boldsymbol{x}_2)].$$

Setting $\eta = \theta_1 - \theta_2$, $\theta = \theta_2$, $\mu(\theta, \eta) = \mu_1(\theta_1)\mu_2(\theta_2)$, and $h(\boldsymbol{x}) = h_1(\boldsymbol{x}_1) h_2(\boldsymbol{x}_2)$, simple manipulations yield the equivalent representation

$$P(\boldsymbol{X} = \boldsymbol{x} \mid \theta, \eta) = \mu(\theta, \eta) h(\boldsymbol{x}) \exp[\eta R_1(\boldsymbol{x}) + \theta R(\boldsymbol{x})], \tag{6.3}$$

that is, a two-parameter exponential family of distributions with canonical parameters η and θ as well as canonical statistics R_1 and R. The RM is obtained for $\eta = 0$. The person-specific parameter η describes size and direction of deviations from invariance of the ability parameter. As discussed by Trabin and Weiss (1983), Klauer and Rettig (1989), and Klauer (1991b), response tendencies such as carelessness, guessing, multidimensionality, and others cause specific violations of the invariance hypothesis for suitable subdivisions of the total test.

6.2.2 Example 2: Person-Specific Item Discrimination

The following alternative model is helpful for understanding a person fit test recently proposed by Molenaar and Hoijtink (1990). Using a second person-specific parameter η, it is defined as follows:

$$P(\boldsymbol{X} = \boldsymbol{x} \mid \theta, \eta) = \mu(\theta, \eta) \exp\Big(\sum_{i=1}^{k} x_i \eta(\theta - \beta_i)\Big), \tag{6.4}$$

where

$$\mu(\theta, \eta) = \prod_{i=1}^{k} [1 + \exp(\eta(\theta - \beta_i))]^{-1}.$$

An examinee whose responses follow (6.4) basically still responds in accordance with an RM; the responses obey local independence, the ICCs are logistic functions with equal item discriminations and without guessing parameters. The parameter η, however, regulates the overall level of the *item discrimination* operating for the examinee.

For $\eta = 1$, the original RM is obtained. For $0 < \eta < 1$, the level of item discrimination is lowered and thus, the actual variance in item difficulty decreases relative to the original RM. For $\eta = 0$, the items are equally difficult, and for $\eta < 0$, the difficulty order of items is reversed.

For $\eta > 1$, on the other hand, the overall level of item discrimination is higher than that of the RM so that the effective variance in item difficulty is increased. For persons with large η–values, the items even approach a Guttman scale, in which the examinee solves all items below his or her ability but none above that ability.

Setting $M(\boldsymbol{x}) = -\sum_{i=1}^{k} \beta_i x_i$, and absorbing η into the ability parameter, $\theta := \eta\theta$, the model is seen to represent a two-parameter exponential family with canonical parameters θ and η, and canonical statistics R and M:

$$P(\boldsymbol{X} = \boldsymbol{x} \mid \theta, \eta) = \mu(\theta, \eta) \exp[\eta M(\boldsymbol{x}) + \theta R(\mathrm{x})]. \tag{6.5}$$

6.2.3 Example 3: Violations of Local Independence

Jannarone (1986) discusses a class of models that represent specific violations of the local independence assumption. One such model is the following:

$$P(\boldsymbol{X} = \boldsymbol{x} \mid \theta, \eta) = \mu(\theta, \eta) h(\boldsymbol{x}) \exp\Big(\eta \sum_{i=1}^{k-1} X_i(\boldsymbol{x}) X_{i+1}(\boldsymbol{x}) + \theta R(\boldsymbol{x})\Big), \tag{6.6}$$

where $\boldsymbol{X} = (X_1, \ldots, X_k)$ and $\mu(\theta, \eta)$ is a normalizing factor so that these values sum to one. Again, a two-parameter exponential family with canonical parameters θ and η is defined. The canonical statistics are R and $\sum_{i=1}^{k-1} X_i X_{i+1}$. The case $\eta = 0$ corresponds to the RM. For a person S_v with $\eta_v > 0$, giving a correct response to item I_i facilitates responding correctly to the subsequent item I_{i+1}. For $\eta < 0$, on the other hand, solving item I_i correctly lowers the probability of responding correctly to I_{i+1}. The first case occurs when the solution of an item provides new insights that are useful for solving a subsequent item. The second case arises when the solution process is very exhausting.

6.3 Statistical Theory for Optimal Person Tests

6.3.1 Definitions

The alternative models represent two-parameter exponential families containing the RM for a certain value η_0 of the second person-specific parameter η. Testing for consistency with the RM corresponds to the testing problem with null hypothesis H_0: $\eta = \eta_0$ and alternative hypothesis H_1: $\eta \neq \eta_0$. Since in general there is no *a priori* information about the direction of the deviation, the one-sided testing problems (for example, H_1: $\eta < \eta_0$) will often be less interesting although they can be treated analogously.

A *randomized test* for evaluating H_0 versus H_1 is a function ϕ that assigns to response vectors $\boldsymbol{x}$ probabilities of rejecting H_0. In particular, if $\phi(\boldsymbol{x}) = 0$, H_0 is accepted; in the case of $\phi(\boldsymbol{x}) = 1$, on the other hand, H_0 is rejected. In between, $0 < \phi(\boldsymbol{x}) = c < 1$, the decision hinges on a suitable additional random experiment such that H_0 is rejected with probability c. For example, a value u of a random variable is drawn that follows a uniform distribution on $[0, 1)$ (that is, $0 \leq u < 1$). If $u \leq c$, H_0 is rejected; otherwise, it is accepted.

A test ϕ is a *test of level* α, if, for $\eta = \eta_0$ and all θ,

$$P(\text{reject } H_0 \mid \theta, \eta_0) = E(\phi \mid \theta, \eta_0) \leq \alpha,$$

where E denotes expected values. Thus, the probability of falsely rejecting H_0 does not exceed α. A test of level α is called *unbiased* if, for all θ and $\eta \neq \eta_0$, the probability of correctly rejecting H_0 is at least α,

$$P(\text{reject } H_0 \mid \theta, \eta) = E(\phi \mid \theta, \eta) \geq \alpha.$$

An unbiased test of level α is a *uniformly most powerful* (UMP) unbiased test if, for all θ, $\eta \neq \eta_0$,

$$P(\text{reject } H_0 \mid \theta, \eta) = E(\phi \mid \theta, \eta) = \sup,$$

where the supremum is taken over all alternative unbiased tests of level α, separately for each pair (θ, η) (Witting, 1978, Section 2).

6.3.2 Uniformly most powerful unbiased tests

As described above, the alternative models form two-parameter exponential families of distributions with canonical statistics R and T. Whereas in each case, R is the raw score variable, T is a statistic that depends on the particular alternative considered. For evaluating the testing problem H_0 versus H_1, a UMP unbiased test can be derived using well-known results (Lehmann, 1986; Witting, 1978) for exponential families of distributions.

According to these results, a UMP unbiased test ϕ^* for evaluating H_0: $\eta = \eta_0$ versus H_1: $\eta \neq \eta_0$ exists. It uses T as test statistic, and H_0 is accepted if the statistic falls into an acceptance region that is spanned by a lower and an upper cut-off score, c_1 and c_2, respectively. The cut-off scores are a function of the raw score, so that the optimal test ϕ^* may be considered a score-conditional test. At the boundaries of the acceptance region, that is, for $T = c_1$ or $T = c_2$, additional random experiments are required to come to a decision on consistency with H_0. Formally, the test is given by

$$\phi^*(\boldsymbol{x}) = \begin{cases} 0, & \text{if } c_1(r) < T(\boldsymbol{x}) < c_2(r), \\ \gamma_1(r), & \text{if } T(\boldsymbol{x}) = c_1(r), \\ \gamma_2(r), & \text{if } T(\boldsymbol{x}) = c_2(r), \\ 1, & \text{otherwise,} \end{cases}$$

where r is the raw score associated with response vector $\boldsymbol{x}$, and $c_1(r), c_2(r)$, and $\gamma_1(r), \gamma_2(r)$, with $0 \leq \gamma_i(r) < 1$, are solutions of the following equations:

$$\begin{aligned} E(\phi^*(\boldsymbol{X}) \mid R = r, \eta = \eta_0) &= \alpha, \\ E(T(\boldsymbol{X})\phi^*(\boldsymbol{X}) \mid R = r, \eta = \eta_0) &= \alpha E(T(\boldsymbol{X}) \mid R = r, \eta = \eta_0). \end{aligned} \tag{6.7}$$

It is not difficult to solve (6.7) for the cut-off scores c_1, c_2, and γ_1, γ_2. Since the raw score variable R is a sufficient statistic for the ability parameter θ, the distribution of T, given R, is not a function of θ. Let $t_1 < t_2 < \ldots < t_m$ be the m different values that T can assume, and define $f(t \mid r)$ by

$$f(t \mid r) = P(T = t \mid R = r, \eta = \eta_0).$$

With the abbreviations,

$$\begin{aligned}
F_1(c_1, \gamma_1|r) &= \textstyle\sum_{j:\, t_j<c_1} f(t_j|r) + \gamma_1 f(c_1|r), \\
F_2(c_2, \gamma_2|r) &= \textstyle\sum_{j:\, t_j>c_2} f(t_j|r) + \gamma_2 f(c_2|r), \\
G_1(c_1, \gamma_1|r) &= \textstyle\sum_{j:\, t_j<c_1} t_j f(t_j|r) + \gamma_1 c_1 f(c_1|r), \\
G_2(c_2, \gamma_2|r) &= \textstyle\sum_{j:\, t_j>c_2} t_j f(t_j|r) + \gamma_2 c_2 f(c_2|r), \\
\tau(r) &= \textstyle\sum_{j=1}^{m} t_j f(t_j|r),
\end{aligned}$$

(6.7) can be written explicitly as

$$F_1(c_1, \gamma_1|r) + F_2(c_2, \gamma_2|r) = \alpha, \tag{6.8}$$

$$G_1(c_1, \gamma_1|r) + G_2(c_2, \gamma_2|r) = \alpha\tau(r). \tag{6.9}$$

The equations have to be solved for the desired $c_1(r)$, $\gamma_1(r)$, and $c_2(r)$, $\gamma_2(r)$. Note that the cut-off scores $c_1(r)$ and $c_2(r)$ will be found among the possible values of T, and that $0 \leq \gamma_1(r) < 1$ and $0 \leq \gamma_2(r) < 1$. For given values c_1 and γ_1, (6.8) can be solved for c_2 and γ_2: c_2 is the t_j with the smallest j for which the left-hand side of (6.8) is not larger than α ($\gamma_2 = 0$). Given c_2, the equation can be solved for γ_2 algebraically. Thus, c_2, γ_2, and $G(c_1, \gamma_1 \mid r) = G_1(c_1, \gamma_1 \mid r) + G(c_2(c_1, \gamma_1), \gamma_2(c_1, \gamma_1) \mid r)$ can be expressed as functions of c_1 and γ_1. The desired $c_1(r)$ of the solution of (6.8) and (6.9) can now be determined as the unique t_j for which $G(t_j, 1 \mid r) \leq \alpha\tau(r) \leq G(t_j, 0 \mid r)$; $c_2(r)$ can be determined anlogously, and given $c_1(r)$ and $c_2(r)$, (6.8) and (6.9) constitute a system of linear equations in γ_1 and γ_2 that can be solved directly for the desired $\gamma_1(r)$ and $\gamma_2(r)$, completing the solution of (6.8) and (6.9).

The computational demand of this algorithm is primarily a function of m, the number of values that T can assume. For the alternative of Example 1, T is the raw score of the first subtest of k_1 items, so that $m = k_1 + 1$. Thus, T assumes only a few different values, and the algorithm can be implemented easily. For the alternative of Example 2, on the other hand, T will in general assume a different value for each response vector, so that computations will become highly demanding as the test length increases. Useful computational short-cuts and simplifications for this case are discussed by Molenaar and Hoijtink (1990) and Liou and Chang (1992).

In practice, it is often inconvenient to work with randomized tests, and a certain loss of test power is sometimes acceptable to obtain a non-randomized test. The non-randomized test that approximates the optimal randomized test most closely is obtained by also accepting H_0 when T assumes a boundary value c_1 or c_2 of the acceptance region. The non-randomized test is still a test of level α but is less powerful than the optimal randomized test. Furthermore, using methods proposed by Klauer (1991b), a standardized person test statistic X^2 for assessing the size of deviations in the direction of the specified alternative can be derived. When several persons are tested, it is also possible to aggregate the individual deviations, using X^2, to perform a χ^2-distributed overall test for the entire group of examinees. Since with several persons tested, a few significant results are expected by chance alone, it is useful to apply the overall test to evaluate first whether the overall pattern of deviations is consistent with a chance interpretation under the RM.

6.4 Power Analyses

In the previous section, person tests were derived for detecting deviations in the direction of prespecified two-dimensional generalizations of the RM. For example, tests against the first alternative discussed in Section 2 are tests for invariance of the ability parameter. This example actually defines a large class of person tests corresponding to the different splits of the total test into subtests that can be used. As argued by Klauer and Rettig (1990) and Klauer (1991b), different types of split, based on item difficulty, item content, item format, or presentation order, among others, reveal different kinds of deviant responses. In addition, the methods discussed above are also useful in analyzing a given person test. For example, using these techniques, it is not difficult to see that the person test proposed by Molenaar and Hoijtink (1990), based on the statistic $M(\boldsymbol{x}) = -\sum_{i=1}^{k} \beta_i x_i$, is the (closest non-randomized approximation of the) UMP test for the one-sided alternative H_1: $\eta < \eta_0 = 1$, given the variable discrimination model of Example 2. The one-sided UMP test rejects response patterns with small values of the statistic M. The test is given by (cf. Witting, 1978)

$$\phi^*(\boldsymbol{x}) = \begin{cases} 0, & \text{if } M(\boldsymbol{x}) > c(r), \\ \gamma(r), & \text{if } M(\boldsymbol{x}) = c(r), \\ 1, & \text{otherwise,} \end{cases}$$

where r is the raw score associated with response vector $\boldsymbol{x}$, and $c(r)$, and $\gamma(r)$, $0 \leq \gamma(r) < 1$, are solutions of the equation:

$$E(\phi^*(\boldsymbol{X}) \mid R = r, \eta = \eta_0) = \alpha.$$

The non-randomized test is obtained by also accepting H_0 if $M(\boldsymbol{x}) = c(r)$, and is identical with the person test proposed by Molenaar and Hoijtink (1990). Thus, that test is most sensitive for deviations that lead to decreased variance in the item difficulties or downright reversals in the order of the item difficulties.

Each person test tests for deviations in the direction of a prespecified alternative. As shown by Klauer (1991b), the test power for detecting deviations in the order of $|\eta| = 2$ logits can be satisfactory for tests of moderate lengths already. Unfortunately, little information is obtained about possible deviations in directions *different* from the prespecified alternative. Although additional person tests could be performed on the same response vector in order to test for these different kinds of deviations, the problem of multiple testing (accumulation of the α−risk) strictly limits the number of person tests that can be performed on a given person's responses.

For illustrating the specificity of each person test for the alternative that it is constructed to detect, power analyses were made for several person tests and the alternative models described above. In the following, ϕ_1, ϕ_2, and ϕ_3 denote the two-sided tests against the following alternatives:

- ϕ_1: Test against non-invariant abilities (Example 1 of Section 2), where the test is split into two subtests of equal size according to the order of presentation,
- ϕ_2: test against variable discrimination (Example 2 of Section 2), and
- ϕ_3: test against local dependence (Example 3 of Section 2).

The power analyses are based on a test of 10 items with difficulty parameters $\beta_1, \ldots, \beta_{10}$ given by -1.5, 0.05, 1.0, -1.0, 1.8, -2.0, -0.5, 0.5, 2.3, -0.05, respectively. For each test and alternative model, the test power was computed for deviations η between -5 and 5. The significance level was set at $\alpha = 10\%$. The Figures 1, 2, and 3 show the results for an examinee with ability parameter $\theta_v = 0$. Similar results are obtained for other abilities. For computing the test power for a given test and alternative model, all possible response vectors are generated and subjected to the test. If the test rejects a response vector as inconsistent with the RM, its probability under the alternative model with person-specific parameter $\theta = 0$ and η is computed and added to the test power. At the boundaries of the test, where an additional random experiment has to be performed, the probability of the response vector is added only in proportion to the appropriate probability $\gamma_1(r)$ or $\gamma_2(r)$ with which the response vector is rejected as a result of the additional random experiment.

Figures 1, 2, and 3 show the test power of each test for detecting deviations in the direction of the model with non-invariant abilities, of the model with variable discrimination, and of the model with violations of local independence, respectively. Thus, ϕ_x is the optimal two-sided test for the deviations considered in Figure x for $x = 1, \ldots, 3$.

As can be seen, ϕ_1 is most powerful against the alternative with non-invariant abilities (Figure 1), as expected. The remaining person tests, ϕ_2 and ϕ_3, do not exhibit strong sensitivity for detecting deviations of this kind. The test power of ϕ_2 even falls below the α−level. Analogous results are obtained for the other alternatives (Figures 2 and 3): each test is most powerful in detecting the kind of deviation for which it was constructed, but does not exhibit satisfactory test power for the other kinds of alternatives. In the case of the alternative with local dependence (Figure 3), even the optimal test ϕ_3 has little test power for deviations with $\eta > 0$. Note that, around $\eta = 1$, ϕ_1 has slightly higher test power than ϕ_3. This is possible because ϕ_3 is optimal in the set of two-sided, unbiased tests, and ϕ_1 is not unbiased. Slightly higher test power is achieved by the optimal *one-sided* test whose test power by necessity also exceeds that of ϕ_1

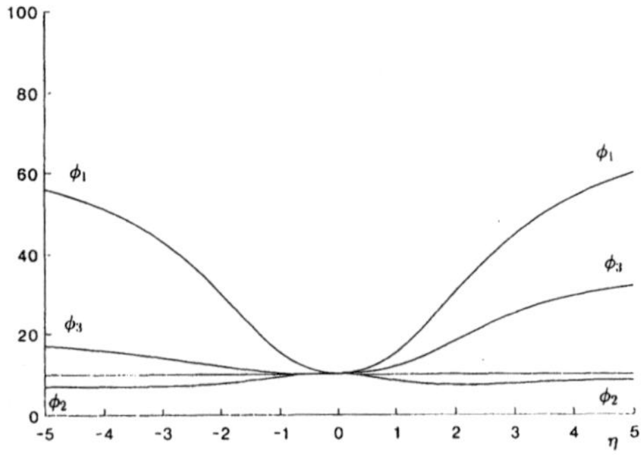

FIGURE 6.1. Test power (%) against local dependence

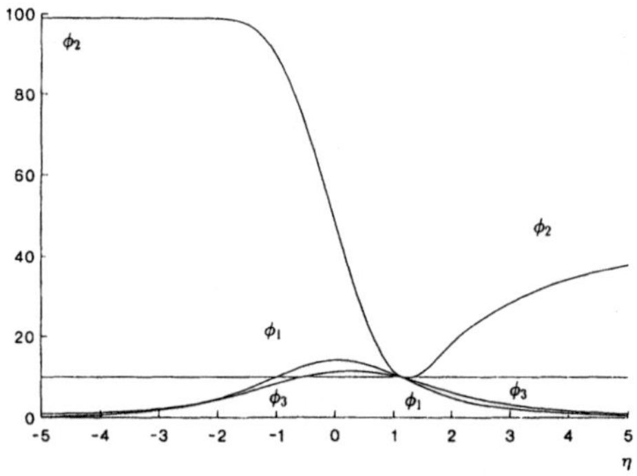

FIGURE 6.2. Test power (%) against variable discrimination

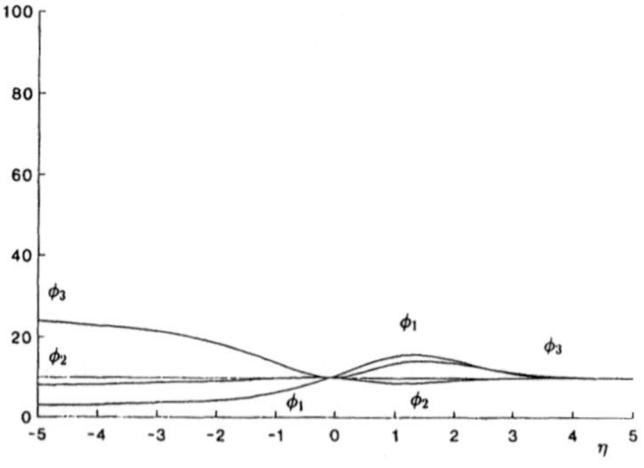

FIGURE 6.3. Test power (%) against non-invariant abilities

for $\eta > 0$.

In the case of large positive η, there is strong local dependence: one correct response will strongly facilitate subsequent correct responses, so that response patterns with only a few false responses at the very beginning followed by mostly correct responses will prevail. Such response patterns are also consistent with the RM since they can arise under that model with high probability, given an individual with high ability. This explains the very small test power for detecting deviations from local independence with large positive η. Thus, there are deviations from the RM that cannot be detected on the basis of non-replicated observations even when the optimal test is used.

Molenaar (personal communication) points out that although the unconditional test power is very small in the case of local dependence with large positive η, the score-conditional test power may be larger, given raw scores in the middle of the score scale. The unconditional test power reported above is the weighted average of the score-conditional test powers. The weights are given by the probabilities with which the examinee produces the various raw scores. The conditional test power will usually be larger than the unconditional one given raw scores in the middle of the score scale, and it will be smaller given extreme raw scores such as $k-2$ and $k-1$. As explained above, the poor unconditional test power for detecting local dependence with large positive η is mostly due to the fact that individuals with this type of deviation are very likely to produce response vectors with extreme raw scores, where the conditional test power is poor.

The conclusions from these analyses are straightforward: a particular person test circumscribes a small class of detectable deviations and leaves out a large class of similarly plausible alternatives that cannot be evaluated. This is also true if an explicit alternative has not been specified for a given person test. In fact, using the theory described here, the relevant class of alternative models can be identified *a posteriori* as exemplified for the person test proposed by Molenaar and Hoijtink (1990).

6.5 A Rational Strategy for Evaluating Person Fit

As discussed in the introduction, many applications of the RM and of more general latent trait models imply the *stochastic subject view* (Holland, 1990). The stochastic subject view conceives the RM as a model for individuals and requires goodness-of-fit tests at the level of individuals (Trabin & Weiss, 1983). Traditional goodness-of-fit tests (see Chapter 5), operating on the level of population marginals, are well suited to the *random sampling view* of the RM (Holland, 1990), but are irrelevant for assessing goodness-of-fit under the stochastic subject view (Klauer, 1988, 1990).

As most chapters of this volume, the present chapter adopts the stochastic subject rationale. A method for deriving person tests is considered. These person tests meet a number of important requirements:

- they are *standardized* in the sense that the decisions on consistency with the RM do not depend on the individual's ability under the null hypothesis;
- they make explicit which deviations they are sensitive against, and are most powerful tests for detecting these deviations;
- they can be transformed into standardized person fit indices that assess the *size* of the deviation in the direction of the specified alternative (Klauer, 1991b);
- individual deviations can be aggregated to perform an overall group test for consistency with the RM.

As in Levine and Drasgow's (1988) *optimal appropriateness* program, it is proposed to make the alternative underlying a given person fit index explicit and to determine an optimal test for the resulting testing problem. As with Levine and Drasgow (1988), use of optimal tests suggests investigations into the question of whether it is possible in principle to detect a given kind of deviation from the RM on the basis of single response vectors. For example, it has been seen that even the optimal test ϕ_3 for detecting violations of local independence has hardly any test power for $\eta > 0$ (Figure 3).

In contrast to the *optimal appropriateness* approach, single response vectors are not used to test constraints at the level of Cressie and Holland's (1983) manifest probabilities. Instead, the emphasis is on testing hypotheses that the RM imposes on a given individual's test behavior. As discussed above, this difference between the two approaches reflects the difference between the random sampling and the stochastic subject view of the RM (Holland, 1990).

Person fit investigation has been presented in the form of formal significance tests, based on the response vector of one person, of the null hypothesis against a specific class of alternatives. The null hypothesis is the RM with known item parameters. The decision-theoretic approach meshes well with the view frequently taken in the literature: preliminary classification of all persons in the sample into "probably Rasch" and "probably aberrant" (for example, Molenaar & Hoijtink, 1990), into "deviant" and "non-deviant", and so forth. On the other hand, the present chapter has not explicitly emphasized possible sources of aberrance frequently considered in the literature, like cheating, sleeping, alignment error, failure on a subdomain. The idea is, of course, to determine, in each case, how these specific sources translate into specific alternative models (for example, Trabin & Weiss, 1983) for which a person test is then constructed through the use of the methods discussed here. In some cases, work on person fit has *implicitly* defined a specific alternative by using a simulated Rasch data matrix to which a group of deviant response vectors is added (for example, Rudner, 1983). Deviant response vectors are constructed by one of a variety of procedures. For example, thirty percent of the responses x_i, $i = 1, \ldots, k$, in a response vector generated under the RM are replaced by wrong responses, $x_i = 0$, or by correct responses $x_i = 1$, or by the opposite responses $y_i = 1 - x_i$. The particular procedure used implicitly defines a specific alternative model generating the deviant response vectors although often this is not realized. Person fit indices are then compared with regard to their ability to discriminate between consistent and deviant response vectors. The work summarized in this chapter has demonstrated that the relative superiority of one person fit index over another is largely a function of the particular alternative considered, so that results will not generalize to other kinds of alternatives. What is more, it is frequently not clear why the particular alternative defined implicitly by a given simulation is worth considering at all.

If a given individual could be tested many times using the same or statistically equivalent items, a standardized person test could be derived that exhibits satisfactory test power for detecting every possible deviation from the RM (Klauer, 1988). In practice, only one set of responses can be obtained from each individual, and each person test based on this one response vector is sensitive for only a small subset of possible deviations. The present work demonstrated that each standardized person test is most powerful against a certain alternative that takes the form of a two-dimensional generalization of the RM. Furthermore, as shown above, each person test in general exhibits strongly reduced test power against other alternatives.

Thus, performing a given person test implies a bias for detecting the test-specific deviations and a bias against detecting other kinds of deviations. Although further person tests designed to detect other kinds of deviations could also be performed, the problems associated with multiple testing strictly limit the number of person tests that can reasonably be performed for a given individual.

On the other hand, a given test is usually constructed with a specific testing purpose in mind. For example, the test may have been constructed for institutional selection purposes, or for individual diagnosis using adaptive testing strategies, and so forth. One possibility of escaping the dilemma sketched above is to focus on the specific purpose and to investigate first, by means of robustness studies, what types of deviations are most damaging for the purpose in question and against what types of deviations the procedures employed are robust. For example, if the goal is to use the test in adaptive testing, violations of invariance of ability are highly damaging: adaptive testing is based on the assumption that ability is invariant over subtests of the total test, so that a different subtest can be administered to each individual. Thus, person tests for invariance of the ability parameter would probably be used to guard against lack of invariance. On the other hand, a robustness study might reveal that the adaptive ability estimation is not so sensitive to violations of local independence (Junker, 1991), so that a person test for local independence could be omitted.

In sum, as a first step of a rational strategy to assess person fit, a robustness study is conducted with the objective to identify the kinds of deviations that are most detrimental for the intended application. In the second step, these deviations are modeled by means of a few two-parameter exponential families as exemplified in Section 2. Using the statistical theory outlined in Section 3, optimal person tests are finally derived for the critical alternatives and used to evaluate observed response patterns.

The approach proposed in this paper poses a number of problems. For example, computing the exact significance levels for a given person test can be very costly (Molenaar & Hoijtink, 1990; Liou & Chang, 1992), although this is not always the case. Furthermore, even the optimal test may exhibit poor test power in absolute terms, as can be seen in Figures 1, 2, and 3 for a test of ten items. Test power usually increases strongly with increasing test length, and in an example analyzed by Klauer (1991b) assumed acceptable values between 80% and 90% for tests of 45 items, under favorable conditions. Thus, for assessing person fit, it is important to avail of data sets of reasonable sizes.

7

Test Construction from Item Banks

Ellen Timminga[1] and Jos J. Adema[2]

ABSTRACT This chapter discusses the topic of computerized test construction from item banks if a dichotomous Rasch model holds for all items of the bank. It is shown how the task of selecting items optimally can be formulated as a 0-1 linear programming problem. Next, integer linear programming is introduced as an easier way to near-optimal item selection. Then, the techniques of 0-1 and integer linear programming are discussed. Finally, an overview of the literature in this field and a discussion of various approaches to the construction of parallel tests is given.

7.1 Introduction

Computerized test construction assumes the existence of a large pool of pre-calibrated items, a so-called item bank, which stores psychometric and other characteristics of items. Here we assume that all items in an item bank are scored dichotomously and perfectly fit a Rasch model (RM). Moreover, the difficulty parameters of the items are treated as known true values. In practice, however, there will exist no banks of items perfectly fitting a RM, and difficulty parameters will always be estimates rather than true values. Thus, the importance of selecting the test with the best psychometric qualities should not be exaggerated. Satisfying the practical constraints is the major goal.

Two approaches to computerized test construction can be distinguished:

1. the construction of complete tests, and

2. adaptive testing.

The construction of complete tests, which is the topic of this chapter, means to select, on each occasion, a test from the item bank meeting certain requirements; this test is then given to one or more examinees. In adaptive testing, examinees are presented with individualized tests. One item is selected at a time on the basis of the currently available estimate of the examinee's ability (e.g., Lord, 1980; Weiss, 1982). The ability estimate is updated after the administration of each item, and the next optimal item is selected. A criterion is needed for deciding when to stop the process; such criteria often focus on accuracy of ability

[1]University of Groningen, Grote Rozenstraat 15, 9712 TG Groningen;
e-mail: E.TIMMINGA@PPSW.RUG.NL
[2]PTT-Research, P. O. Box 421, 2260 AK Leidschendam, The Netherlands;
e-mail: ADEMA@PTTRNL.NL

estimation. This chapter, however, does not deal with the problem of adaptive testing.

Good tests meet requirements that are specified by the test constructor. Meeting these requirements, however, is problematic for at least two reasons: one is the impossibility of examining all possible tests in order to find the one that best fits the test constructor's requirements. Even for a relatively small item bank of 200 items, say, in theory $2^{200} - 1$ tests have to be compared. The other is related to the practical requirements that tests have to meet. The problem is that several requirements have to be taken into account simultaneously, which often is almost impossible.

The main concern in this chapter is to optimize the psychometric qualities of the test to be selected. It will also be indicated how to consider other kinds of requirements. The psychometric characteristics used are item difficulty parameters, and item and test information functions.

It is well-known that the maximum likelihood estimator of θ asymptotically has the smallest variance attainable. This minimum variance is $[\mathcal{I}(\theta)]^{-1}$, where $\mathcal{I}(\theta)$ is called the test information. Thus, an information function indicates the maximum accuracy with which an examinee's ability parameter can be estimated over the total ability range. The general formula of the test information function for the RM is (see also (4.6)):

$$\mathcal{I}(\theta) = \sum_{i=1}^{k} P_i(\theta)[1 - P_i(\theta)], \tag{7.1}$$

where k is the number of test items.

Observe that (7.1) consists of independent and additive contributions of items. This is a result of the local independence assumption that is made for maximum likelihood estimation (see Chapters 1 and 4). Thus, the information provided by a test is simply the sum of the information functions of the items in the test:

$$\mathcal{I}(\theta) = \sum_{i=1}^{k} \mathcal{I}_i(\theta), \tag{7.2}$$

where $\mathcal{I}_i(\theta)$ denotes the information function of item I_i, which can also be formulated as follows:

$$\mathcal{I}_i(\theta) = P_i(\theta)[1 - P_i(\theta)] = [1 + \exp(-(\theta - \beta_i))]^{-1}. \tag{7.3}$$

For all items fitting the same RM, the information functions are identically bell-shaped functions that only differ in their location. Expression (7.3) clearly shows that the maximum of the information function is obtained at ability level $\theta = \beta_i$, and that this maximum amount is equal to 0.25. A thorough discussion of information functions can, for instance, be found in Birnbaum (1968), Lord (1980), and Hambleton and Swaminathan (1985).

The property of additivity is used by Birnbaum (1968) in his procedure for test construction, which can be summarized as follows:

1. indicate the shape of the desired test information function, called the target test information function; and

2. select the items for which the summed information functions fill the area under this target.

A description of this procedure is also given by Lord (1980) and Hambleton and Swaminathan (1985). It is not obvious how to implement this procedure systematically in practice, the test construction methods discussed here originate from Birnbaum's ideas, and employ mathematical programming methods for test construction from RM-based item banks. Mathematical programming techniques are optimization techniques that belong to the field of Operations Research, also called Management Science (cf., Daellenbach, George, & McNickle, 1978; Nemhauser & Wolsey, 1988; Rao, 1985). The paper of Feuerman and Weiss (1973) discusses the use of mathematical programming techniques for test construction, but does not consider the use of IRT nor the selection of tests from item banks. Yen (1983) was the first to propose the use of linear programming (LP) for IRT-based test construction from item banks, but she did not formulate a specific model. Theunissen (1985) formulated an LP-model that minimizes test length. Since then the application of LP-techniques to test construction problems has been considered thoroughly by several researchers. This research shows that test construction problems can indeed be formulated as LP-models. Two main advantages of this approach are: the possibility to consider all kinds of psychometric and other practical test requirements, and the availability of algorithms and heuristics for solving such problems.

7.2 A 0-1 Linear Programming Approach

In LP, an objective function is maximized or minimized subject to a number of constraints. Both the objective function and the constraints are linear functions of the unknown decision variables, which are allowed only to take real values greater than or equal to zero. An LP-model is called a 0-1 LP-model if the decision variables may only take the values 0 or 1. In most test construction applications, the decision variables z_i are defined as follows:

$$z_i = \begin{cases} 1, & \text{item } I_i \text{ is selected for the test;} \\ 0, & \text{item } I_i \text{ is not selected for the test.} \end{cases}$$

From the point of view of an RM, the item difficulty parameters are of major importance with respect to measurement of the qualities of a test. In theory, test item selection can be exclusively based on these parameters. In this chapter, however, both selection on the basis of difficulty parameters and information functions is discussed, and it is shown when the use of information functions is to be preferred.

The remainder of this section discusses several test construction approaches using 0-1 LP-techniques. Possible objective functions and several types of constraints are described in turn. Finally, some considerations are given for putting together an LP-model consisting of an objective function and some constraints.

7.2.1 Objective Functions

This section presents several objective functions that can be used for RM-based test construction. The special case of mastery test selection will be discussed first, and more general functions will be outlined later.

Mastery Tests

Mastery tests are defined in this chapter as tests with one cut-off point θ_f on the ability scale. The optimal test consists of those items that have difficulty parameters closest to the cut-off point, or with maximum amounts of information at this point, leading to the same test, of course. Item selection is simple in this case and can easily be done by hand, or by using certain options in a database package. If more realistic constraints have to be considered, however, it is no longer easy to select the items by hand. It then becomes impractical to keep an eye on satisfying all constraints, and hence LP-techniques are needed.

There are two possible objective functions, namely the following:

1. Maximize the information at θ_f for the items in the test, or

2. minimize the deviation between the cut-off point θ_f and the β_i of the items selected for the test.

The first objective function is formulated as

$$\text{maximize} \sum_{i=1}^{n} \mathcal{I}_i(\theta_f) z_i, \tag{7.4}$$

where $\mathcal{I}_i$ denotes the information function for item I_i, and n the number of items in the item bank. The second objective function is

$$\text{minimize} \sum_{i=1}^{n} |\theta_f - \beta_i| z_i. \tag{7.5}$$

Observe that the $|\theta_f - \beta_i|$ are constants in the model, they do not cause the objective function to be non-linear.

Other Tests

In cases of non-mastery tests the test constructor is interested in more than one point ($f = 1, \dots, f^*$) on the ability scale. For example, suppose a classification test has to be constructed with cut-off points $\theta_1 = -2$ and $\theta_2 = 2$. In this case, the item selection can be done without LP as long as additional test requirements do not render this impossible, using item difficulty values to select the best items. However, the use of difficulty values for selecting the items will be problematic if several ability points have to be regarded, each requiring different levels of measurement accuracy. In this case, the use of information functions is more desirable.

Four types of objectives are distinguished:

1. exact target information values are specified that should be reached as nearly as possible;

2. maximum information is required at all ability points, i.e., the ability points are equally important;

3. relative target information values are considered, implying that only the 'relative shape' (not the exact values) of the target information function is specified at some ability points;

4. test length is minimized while the obtained test information values are required to be greater or equal to the target values.

Objective functions for these four possibilities are discussed below. In the case of LP, target information values are specified at some ability points that can be freely chosen by the test constructor. It is not necessary to specify a complete information function. Test information values at nearby ability points will hardly differ, because information functions of Rasch items are bell-shaped smooth functions.

There are several possible objective functions for *realizing exact target information values.* As it will never be possible to reach the target information values exactly, the test constructor has to decide what type of deviation from the target should be minimized. Three possibilities are presented below.

Minimize the sum of positive deviations from the target using the following objective function and corresponding set of constraints:

$$\text{minimize} \sum_{i=1}^{n} \sum_{f=1}^{f^*} \mathcal{I}_i(\theta_f) z_i, \tag{7.6}$$

subject to

$$\sum_{i=1}^{n} \mathcal{I}_i(\theta_f) z_i \geq T(\theta_f), \quad f = 1, ..., f^*. \tag{7.7}$$

The constraints in (7.7) indicate that the target values $\mathrm{T}(\theta_f)$ should at least be reached, thus the obtained test information values can be minimized through objective function (7.6). A disadvantage of this objective function is the possibility that the target is just reached at one point, while much more information than required is obtained at another point.

The following objective function and corresponding constraints minimize the sum of absolute deviations from the target:

$$\text{minimize} \sum_{f}^{f^*} (y_f + u_f), \tag{7.8}$$

subject to

$$\sum_{i=1}^{n} \mathcal{I}_i(\theta_f) z_i - y_f + u_f = T(\theta_f), \quad f = 1, ..., f^*, \tag{7.9}$$

$$y_f, u_f \geq 0, \quad f = 1, ..., f^*. \tag{7.10}$$

Two sets of decision variables y_f and u_f ($f = 1, ..., f^*$) are introduced to indicate positive and negative deviations from the target values. A positive deviation is represented by y_f (≥ 0), and a negative deviation by u_f (≥ 0). Observe that, because the sum of the y_f and u_f is minimized and both are greater than or equal to zero, it always holds that one of them is zero, and that the other is equal to the deviation from the target.

The third objective function minimizes the largest deviation from the target. Let y, $y \geq 0$, be the decision variable that indicates the largest deviation. Then the objective function and corresponding constraints are formulated as:

$$\text{minimize } y, \tag{7.11}$$

subject to

$$\sum_{i=1}^{n} \mathcal{I}_i(\theta_f) z_i - T(\theta_f) \leq y, \quad f = 1, ..., f^*, \tag{7.12}$$

$$T(\theta_f) - \sum_{i=1}^{n} \mathcal{I}_i(\theta_f) z_i \leq y, \quad f = 1, ..., f^*. \tag{7.13}$$

Maximizing information at several ability points actually requires an objective function like (7.4) for each ability point. However, as only one objective function at a time can be considered in LP, they have to be re-formulated. Let the decision variable y, $y \geq 0$, indicate the minimum amount of information obtained at the ability points. Then, by maximizing this variable, the information values at all points can be maximized. Note that in this case the number of items in the test should be restricted one way or another. In LP-formulation the objective function and its constraints are:

$$\text{maximize } y, \tag{7.14}$$

subject to

$$\sum_{i=1}^{n} \mathcal{I}_i(\theta_f) z_i \geq y, \quad f = 1, ..., f^*. \tag{7.15}$$

The third type of psychometric requirement considers the specification of a so-called *relative target information function.* This can be interpreted as specifying the shape rather than the exact height of the target information function at certain ability points. Actually this is much easier for a test constructor than specifying exact target information values if he/she has no reference with respect to these values. Van der Linden and Boekkooi-Timminga (1989) proposed this approach. They formulated a so-called maximin model that can be used for maximizing information values, while the relative information values should at least be reached. Constants g_f are introduced in the model to indicate the relative information values desired at all ability points θ_f. The objective function and constraints formulated below have the effect of blowing up the 'relative' information function, so that the information values are maximized. The decision variable $y, y \geq 0$, indicates the extent to which the information is maximized. The objective function and corresponding constraints are:

$$\text{maximize } y, \tag{7.16}$$

subject to

$$\sum_{i=1}^{n} \mathcal{I}_i(\theta_f) z_i \geq g_f y, \quad f = 1, ..., f^*. \tag{7.17}$$

Note the similarity between (7.14)–(7.15) and (7.16)–(7.17).

The fourth objective function *minimizes test length,* while the obtained test information is higher than the target values specified. This objective function and its constraints are:

$$\text{minimize } \sum_{i=1}^{n} z_i, \tag{7.18}$$

subject to

$$\sum_{i=1}^{n} \mathcal{I}_i(\theta_f) z_i \geq T(\theta_f), \quad f = 1, ..., f^*. \tag{7.19}$$

If it is desired to minimize test length, one should be aware not to fix the number of items in the test via practical constraints. Syntactically this would be correct, but the objective function would loose its role for the model; it would be redundant, and another objective function would be preferable in this case.

7.2.2 Practical Constraints

In practice, test constructors always consider additional characteristics of items. For example, administration time is taken into account, and/or test composition is considered to obtain the required face validity of the test. As a matter of fact, such test requirements are considered to be of major importance. Another reason for the consideration of such requirements is that the local independence assumption will often be violated more or less. The researcher therefore will tend to prevent certain groups of items from entering the same test. The general formulations of several test construction constraints are given below.

An LP-model results if an objective function is combined with a set of constraints. This should be done with care: even if there are no syntactical errors in the model, there is always the possibility that no feasible solution exists (infeasibility). This occurs, for example, if it is required to select 10 multiple choice items with difficulty parameters between -1 and 1 while there are only 6 such items in the item bank. Furthermore, one should always take care that the constraints are not contradictory. A disadvantage of LP is that no reasons for infeasibility are given; the only message given is 'infeasible'.

Another possible outcome is 'unbounded solution'. A solution is unbounded if the optimum solution is to select all items in the item bank. This can occur, for example, if the maximin model (7.16)–(7.17) is applied without limiting the number of items in the test.

Numbers of Items

These constraints provide the test constructor with the possibility to control the length of a test as a whole, but also the number of items in the test from certain subsets of the item bank. Let J_m be a subset of items in the bank, and k_m the number of items to be selected from this subset, then the general constraint is formulated as follows:

$$\sum_{I_i \in J_m} z_i = k_m. \tag{7.20}$$

Observe that, if J_m includes the total item bank, k_m defines the required test length. In (7.20) an equality sign is included, however, inequality signs can also be used if one wants to specify a range of acceptable numbers. Several constraints of this type can be included in a test construction model . It can also be used for handling certain types of inter-item-dependencies, for example, if at most one item should be selected from a certain subset of dependent items.

It is also possible to specify the number of items to be selected from subsets in proportion to one another. For example, the requirement that twice as many items should be selected from subset J_1 as from subset J_2, is formulated as follows:

$$\frac{1}{2}\sum_{I_i \in J_1} z_i = \sum_{I_i \in J_2} z_i. \tag{7.21}$$

Constraints of this type are often used in combination with the objective function that minimizes test length.

Sums
It is possible to limit the sum of certain characteristics of the items in the test, for example, specifying target information values like in (7.7), or setting a bound to the maximum test administration time. The general constraint is

$$\sum_{I_i \in J_m} t_i z_i \leq b_m. \tag{7.22}$$

For example, let t_i be the item administration time of item I_i, and J_m the subset of multiple-choice items, then (7.22) states that the sum of the administration times of the multiple-choice items in the test should maximally be b_m. In such constraints, equality signs should be used with great care because infeasibility of the model may be the result. Instead, upper and lower bounds can be put on the required sum if it has to be close to a certain value.

Means
If it is required to limit the mean of a certain characteristic of the items in the test, the above constraint has to be adapted slightly. An example of such a constraint is the following: mean item difficulty of the test items should be within the range -1 to 1. Again subsets of items can be taken into account. The general formulation for the mean is

$$\frac{1}{k_m}\sum_{I_i \in J_m} a_i z_i \leq b_m. \tag{7.23}$$

In (7.23) it is assumed that the number of items to be selected from the subset J_m, denoted k_m, is known. Thus a constraint like (7.20) should be added to the model to fix k_m. Again, equality signs should be used with great care.

All or None
In some practical situations selecting one item entails selecting other items as well. This may be the case if a certain subset of items is related to the same auxiliary material, for example, a reading text. Thus, all or none of the items from subset J_m should be selected for the test. This is formulated as

$$\sum_{I_i \in J_m} z_i = |J_m| z_{i_m}, \tag{7.24}$$

where $|J_m|$ indicates the number of items in subset J_m, and z_{i_m} is the decision variable of a random item from subset J_m. For example, if a subset consists of items $I_1, \ldots, I_5$, (7.24) can be represented as:

$$\sum_{i=1}^{5} z_i = 5z_1. \tag{7.25}$$

Include
A test constructor may require that certain items should enter the test. There are two possibilities to treat this type of constraint. One is to fix the decision variables of the items to be selected to one (e.g., $z_{15} = 1$, if item I_{15} should enter the test), so that they definitely will be selected for the test. Another more laborious approach is to exclude these items from the item selection process, and next to adapt the objective function and constraints such that the characteristics of the pre-selected items are taken into account. For example, if the test should consist of 40 items, and 5 items are fixed to be selected, then the constraint is reformulated to require the selection of 35 items. The second approach is preferable if the number of pre-selected items is large compared to the number of those that still have to be selected. However, implementing this approach may often be too time consuming, so that the first approach is preferred.

Exclude
Excluding items can be done in two ways. The items do not enter the test construction model, and thus are not considered for selection, or their decision variables are set to zero. The first approach is to be preferred especially if large numbers of items are excluded from being selected. Ignoring the items will reduce the size of the model, and thus also the memory space and computation time needed.

7.3 An Integer Linear Programming Approach

An LP-model is called an integer LP-model if the decision variables are allowed to take integer values only. Thus, 0-1 LP is a special case of integer LP. Integer LP test construction models assume that an item bank is divided into a number of mutually exclusive subsets of items (clusters J_c, c = 1, ..., c^*) with approximately equal information functions. Because item information functions are fully determined by the difficulty values of the items, it is easy to determine a small set of clusters for the RM on the basis of these values. For other IRT models it is much more complicated to determine clusters, and more clusters will be needed because the information functions differ more.

For RMs an integer LP-approach can be interesting if none or only a few practical constraints are added to the psychometric requirements. The advantage of the integer LP-approach is that it reduces the number of decision variables in the model dramatically, which implies that it can be solved much faster. A discussion on the computational complexity of 0-1 and integer LP problems is given in the next section of this chapter. Here we outline the main idea at the base of the integer LP-method.

Items in a cluster are viewed to be interchangeable, and, for example, mean item difficulty of the items in a cluster is used for computing the representative information function for all items in the cluster. The problem is now to decide how many items from each cluster should be selected for the test. This requires a re-definition of the decision variables, which are: z_c is the number of items selected for the test from cluster J_c, where $0 \leq z_c \leq u_c$, and u_c is the number of items available in cluster J_c.

The objective functions and constraints that have been described in the section titled 'objective functions' can easily be reformulated for integer LP by replacing the index i by c and adjusting the upper summation limit accordingly. For example, (7.4) is now reformulated as

$$\text{maximize} \sum_{c=1}^{c^*} \mathcal{I}_c(\theta_f) z_c, \tag{7.26}$$

where $\mathcal{I}_c(\theta_f)$ is the information of an item in cluster c at ability level θ_f. It is not that simple, however, to reformulate the constraints in the section 'practical constraints' using the z_c. This is due to those decision variables referring to subsets of items for which *only* the psychometric characteristics are 'equal'; all other characteristics of the items in the same subset may differ from item to item. Thus, in order to consider practical constraints with respect to another characteristic, say, item format, the existing subsets have to be split into subsets of items that are 'equal' with respect to both psychometric *and* format characteristics. As the number of constraints on different characteristics increases, a large number of subsets consisting of small numbers of items are required. Then the number of subsets will approach the number of items in the bank and, consequently, the advantage of reducing the number of decision variables will be lost.

The main advantage of the integer LP-method, applied without additional constraints, is that it is extremely fast, while the quality of tests is comparable to the quality of tests constructed using 0-1 LP-methods. Furthermore, the construction of parallel tests is very easy and takes even less time than the construction of a single test. Suppose four parallel tests of 40 items have to be constructed. Then, for example, the following cluster-based maximin model can be used:

$$\text{maximize } y, \tag{7.27}$$

subject to

$$\sum_{c=1}^{c^*} \mathcal{I}_c(\theta_f) z_c \geq g_f y, \quad f = 1, ..., f^*, \tag{7.28}$$

$$\sum_{c=1}^{c^*} z_c = 40, \tag{7.29}$$

$$0 \leq z_c \leq \frac{u_c}{4}, \quad c = 1, ..., c^*, \tag{7.30}$$

$$y \geq 0. \tag{7.31}$$

The numbers of items to be selected from each of the clusters for each test are the result of solving this problem. For each test, these numbers of items are then selected randomly from the clusters.

7.4 Solving 0-1 and Integer LP-Models

This section discusses how to solve 0-1 and integer LP-models. One of the most popular procedures for solving 0-1 and integer LP-problems is outlined. It is argued that solving practical test construction problems *optimally* is not realistic in general because of the computational complexity. Several heuristic procedures have therefore been derived from the optimal procedure; they produce solutions close the optimum, yet consume realistic amounts of computer time.

A test construction example is given in which the computation times and objective values of several heuristics are compared with the optimal solution.

It is important to note that optimal solutions to standard LP-problems ($z_i \geq 0$) can be obtained very fast by using the Simplex algorithm, for example, which is described in many textbooks on mathematical programming. Solving 0-1 and integer LP-problems is computationally much more complex, because of the restriction that the decision variables may only take integer values. Except for some problems that have a special structure, it is generally believed that no fast algorithms exist for these problems. Unfortunately, realistic test construction problems do not have such a special structure.

A popular procedure for solving 0-1 and integer LP-problems optimally is as follows: (a) solve the relaxed 0-1 or relaxed integer LP-problem, which is the LP-version of the problem, using the Simplex algorithm; this means taking $0 \leq z_i \leq 1$ instead of $z_i \in \{0,1\}$ for 0-1 LP-problems, and (b) apply a 'branch-and-bound' technique until the best integer solution is obtained. Solving the relaxed problem in (a) takes a small amount of time. The solution obtained will include a few fractional decision variable values most of the time. If all decision variables are integer–valued, the optimal solution has been found, and step (b) can be skipped. The objective function value obtained in step (a) is an upper bound for that of the integer solution. This is a useful feature that provides the opportunity to evaluate solutions, when heuristics have been used in step (b).

Step (b) consists of a tree-search. The relaxed model in step (a) is the root of the tree. There are several possible ways to search the tree. At every deeper node an extra constraint is added to the model, for example, fixing the decision variable of the item with the largest fractional value to 1. The deeper the nodes,

TABLE 7.1. Information function values for the example with 6 items.

Item	$I_i(\theta_1)$	$I_i(\theta_2)$
1	0.051	0.116
2	0.113	0.199
3	0.154	0.167
4	0.248	0.105
5	0.192	0.041
6	0.092	0.174

the more constrained the problem. The relaxed problem of the model is solved at every node under consideration. After a solution has been obtained, it is checked whether the objective function value is better than that of the best integer solution found so far. Backtracking to a previous node is required if the objective function value of the new solution is not better or if the solution is integer. In the latter case the solution is compared with the best integer solution found so far; the best one is saved. The tree search continues until all branches have been checked. The obtained solution is the best possible solution. Although not all possible nodes of the tree have to be checked, an extremely large number of nodes have to be considered in most cases. This number depends on the number of decision variables in the model. Because 0-1 and integer LP-problems need step (b), they are much harder to solve than standard LP-problems requiring only step (a).

An example of the branch-and-bound method is given below for the maximin model (7.14)–(7.15) with 6 items. A more realistic problem is not described, because the tree for such a problem would be too large and the ideas underlying the branch-and-bound method can be shown just as clearly by a small problem.

The model to be solved contains objective function (7.14), constraints (7.15), and a constraint implying that three items should be selected: $\sum_{i=1}^{n} z_i = 3$. The values of $I_i(\theta_f)$, $f = 1, 2$, are given in Table 7.1.

First the relaxed model with $0 \leq z_i \leq 1$ instead of $z_i \in \{0, 1\}$ was solved by means of the Simplex algorithm. The objective function value for this model is 0.4845. This is Node 1 in the branch-and-bound tree in Figure 7.1.

The tree is not known before the branch-and-bound method ends. Here the tree is given in advance, so we can refer to it. In this implementation of the branch-and-bound method the variable with a non-integer value closest to 0 or 1 was chosen as branching variable. Two new nodes of the tree were generated by setting the branching variable equal to 0 or 1. In the solution of the relaxed model, variable z_4 turned out to be the branching variable, so that two new models (nodes) were created: one with $z_4 = 1$ (Node 2) and one with $z_4 = 0$

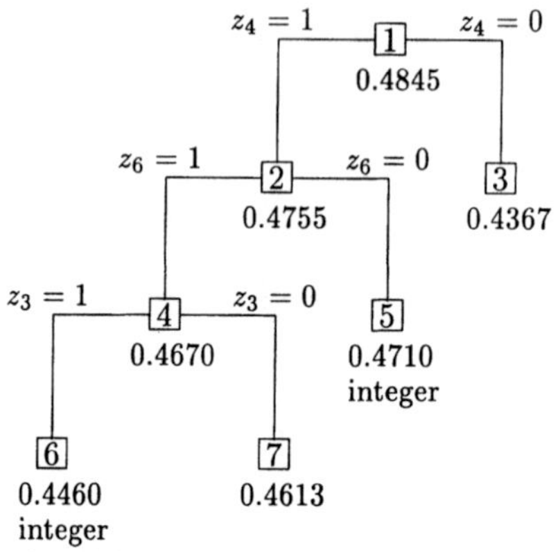

FIGURE 7.1. Branch-and-bound tree for the example with 6 items. Nodes are denoted by squares containing its number.

(Node 3). The model in Node 2 was solved first. Its objective function value was 0.4755. In this branch-and-bound method, a depth-first search was implemented, so that the search continued by generating two new models by including $z_6 = 1$ (Node 4) and $z_6 = 0$ (Node 5) in the model corresponding to Node 2. Next the model of Node 4 was solved. The objective function value was 0.4670, and not all variables in the solution had the value 0 or 1. Variable z_3 was selected as the branching variable, and the Nodes 6 and 7 were created. The solution of the model corresponding to Node 6 was integer valued, the objective function value was 0.4460. Generating new nodes starting from Node 6 did not make sense, because the corresponding models would be more restrictive, so that the objective function value would be worse or the solution would not be feasible. Therefore, the search should be continued from another node. The choice was between Nodes 3, 5, and 7. Node 3 was chosen because its mother node (Node 1) had the highest objective function. The objective function value corresponding to this node, 0.4367, was smaller than the objective function value of the best solution found so far (Node 6). This implied that no better solution could be found in this branch, so the search continued with Node 5. Again an integer solution was found ($z_1 = 0; z_2 = 1; z_3 = 1; z_4 = 1; z_5 = 0; z_6 = 0$), with objective function 0.4710, higher than the best solution found so far. Now only Node 7 remained. Its objective function value was smaller than that of Node 5. Since all branches had been searched, the solution of the model in Node 5 was the best, implying that items 2, 3, and 4 should be selected for the test.

For a 0-1 or integer LP-problem with more than 20 decision variables (items in the item bank), say, computation times for the branch-and-bound procedure

TABLE 7.2. Computation times and objective function y for the example with $k = 20$, for four computational methods

Computational method	CPU-time	y	Deviation from y Simplex
Simplex	49 seconds	3.5505	–
Branch-and-bound	23.5 hours	3.5186	(0.90%)
First integer	56 seconds	3.4149	(3.82%)
Adema heuristic	60 seconds	3.4986	(1.46%)

described are too high. A simple heuristic that often gives good results involves rounding the fractional decision variables obtained in step (a) upwards or to the nearest integer. Other heuristics have also been proposed, such as optimal rounding, and the 'Adema–heuristic' that fixes some decision–variable values after obtaining the relaxed solution and stops the branch-and-bound process if an acceptable solution has been obtained. The simple rounding heuristic turns out to give very good results for the RM in most cases, the optimal 0-1 or integer solution is equal to or very close to this rounded solution.

Computation times for solving the relaxed problems in step (a) mostly are higher for the RM than for other IRT models. The explanation for this is that the information functions of Rasch items are less different than those of items fitting other models. Thus, finding the best combination of items may take more time. It depends particularly on the LP-model to be solved whether step (b) is carried out faster for the RM than for other IRT models.

7.4.1 An Example

From an item bank of 200 items, a test was constructed using different computational methods. Computation times and objective function values were compared. The computational methods used were: the Simplex method for obtaining the relaxed solution; the branch-and-bound method for obtaining the optimal solution; the first integer solution; and the 'Adema-heuristic'. An MS-DOS computer (25 MHz) with hard disk and mathematical coprocessor was used to select the tests.

The requirements for the test were the following: (1) test length $k = 20$ items; (2) 10 multiple choice items; (3) all or none of the items from a certain subset of five items. As an objective function, (7.11) – (7.12) was used, and θ_1, θ_2, θ_3 $= -2, 0, 2$, respectively.

The results in Table 7.2 clearly illustrate the computational complexity of 0-1 linear programming problems. To complete the total branch-and-bound process and thus to obtain the optimal solution took 23.5 hours, while very good solutions could be found within a minute by ending the search process earlier.

Several packages for solving (integer) linear programming problems exist for PC and mainframe computers. An overview article that compares several packages is that by Sharda (1992). Most program libraries include, for instance, the Simplex algorithm, which can be used for programming specific applications. There exist several general purpose packages for solving (integer) LP-problems. Packages for a PC that are easy to use cannot handle problems consisting of more than app. 1000 variables and app. 1000 constraints. Some packages require the user to program a matrix generator that generates the input matrix for every LP-problem to be solved. Others use a complex control language that is difficult to learn. This is the reason that efforts were made to develop specialized test construction packages. So far one package has been developed for test construction using linear programming that is named OTD (Verschoor, 1991). Another package called CONTEST is under development (Boekkooi-Timminga & Sun, 1992) .

7.5 Discussion

This chapter deals with the construction of tests from banks of items that fit the same RM. It describes how 0-1 and integer linear programming methods can be used for selecting tests from item banks. Some references to papers on 0-1 LP-models for test construction are: Adema (1990a, 1990b, 1992a, 1992b); Adema, Boekkooi-Timminga, and Gademann (1992); Adema, Boekkooi-Timminga, and Van der Linden (1991); Adema and Van der Linden (1989); Armstrong, Jones, and Wu (1992); Baker, Cohen, and Barmish (1988); Boekkooi-Timminga (1987, 1989, 1990a); Theunissen (1985, 1986); Van der Linden and Boekkooi-Timminga (1988, 1989). More details and experimental results on the integer LP-approach can be found in Boekkooi-Timminga (1990b). De Gruijter (1990) also applied integer LP.

Papers that discuss heuristics for solving 0-1 and integer LP-problems are, for example: Adema (1988; 1992b); Adema, Boekkooi-Timminga and Van der Linden (1991); Armstrong and Jones (1992); Armstrong, Jones, and Wu (1992); Stocking and Swanson (1993); Swanson and Stocking (1993); Van der Linden and Boekkooi-Timminga (1989). Computation times for Rasch item banks and others are compared by Boekkooi-Timminga (1990a). Furthermore, Boekkooi-Timminga (1990b) compares objective function values obtained applying several heuristics. Computational complexity of 0-1 and integer LP-problems is discussed, for example, by Lenstra and Rinnooy Kan (1979), and Papadimitriou and Steiglitz (1982). A readable book that addresses a wide variety of issues for model building in mathematical programming is Williams (1985). The infeasibility problem has recently been considered by Timminga and Adema (in press), who propose a procedure for detecting some types of problematic constraints and mention some possible adaptations for such constraints.

Several researchers have addressed the problem of constructing all kinds of special types of tests (e.g., Adema, 1990a; Boekkooi-Timminga, 1987; Theunissen, 1986). Adema and Van der Linden (1989) describe a test construction approach considering IRT-based item banks using item parameters from classical test theory. Especially the construction of weakly parallel tests has received much attention (e.g., Adema, 1992a; Armstrong & Jones, 1992; Armstrong, Jones, & Wu 1992; Boekkooi-Timminga, 1990a, 1990b). Weakly parallel tests are defined to be tests with approximately equal information functions (Samejima, 1977). Van der Linden and Boekkooi-Timminga (1988) proposed a computerized version of Gulliksen's matched random subtests method using 0-1 LP for the development of parallel tests using classical test theory; in fact, the method for parallel test construction proposed by Armstrong et al. (1992) is an improved version of this method. The authors succeeded in formulating part of the parallel test construction problem as a standard assignment problem. This is a clear advantage because such problems can be solved very fast.

The LP-approaches to parallel test construction by Adema (1992a), Armstrong and Jones (1992), and Armstrong et al. (1992) are very efficient. However, they have the drawback of having difficulties in considering additional constraints. For the RM, parallel test construction using the cluster based method is also very fast, but has the same drawback. On the other hand, standard algorithms can be used for solving this problem easily, which is not the case for the methods proposed by Adema, Armstrong and Jones, and Armstrong et al. They require specialized computer programs that may involve some programming efforts.

Armstrong and Jones', and Armstrong et al.'s method match the parallel tests on item level. The other LP-approaches mentioned match the tests as a whole. Another difference between these methods and the other methods is that the former do not use information functions. Instead, a matrix of difference measures between the items is used. The proposed difference measures are functions of the item parameters. Differences between information functions of items can be used in their method, but it would take much time to compute the complete difference matrix for all items in the bank. Boekkooi-Timminga (1990a) discusses several methods for parallel test construction that can consider all kinds of additional constraints. However, computation time may be a problem; the extent to which this is the case depends on the method used.

Some other approaches to parallel test construction from a non-LP point of view are described by Ackerman (1989), and Luecht and Hirsch (1992). Their methods work well, but they also have the problem of adding extra constraints. Furthermore, it is difficult to see whether a constructed test is close to optimal, which can easily be assessed if LP-methods are used.

Part II
Extensions of the Dichotomous Rasch Model

8

The Linear Logistic Test Model

Gerhard H. Fischer[1]

ABSTRACT The linear logistic test model is an extension of the Rasch model where certain linear constraints are imposed on the item parameters. Its development originated from the idea of modeling item difficulty as a function of the difficulty of cognitive operations involved. The chapter presents the model, discusses its equivalence with a Bradley-Terry-Luce model with linear constraints on the object parameters, and deals with parameter estimation, uniqueness, and hypothesis testing within a conditional maximum likelihood framework. Many typical applications are mentioned.

8.1 Introduction

This chapter deals with an extension of the Rasch model (RM) that has many useful applications: the 'linear logistic test model' (LLTM), which assumes that the item parameters β_i of the RM (1.2) can be decomposed into a weighted sum of 'basic parameters', α_l. The idea motivating this extension was that, in some tests, item difficulty can be conceived as a function of certain cognitive operations involved in the solution process (Scheiblechner, 1972), each of which has a difficulty parameter α_l. In other cases, item difficulty can similarly be modeled as a function of experimental or other conditions of the test situation or administration, of training or therapy or communication that changes the difficulty of the items. (What actually changes is θ_v, of course, but formally θ_v can be assumed constant and the change described as a change of β_i; Fischer, 1972a, 1976, 1977a, 1983a, 1987a, 1989; see Chapter 9). All these cases can be handled successfully within the same formal framework where item difficulty is described as an additive function of basic parameters α_l. To see what the additivity in terms of basic parameters means, imagine an item I_i that requires two cognitive steps or operations, with associated difficulty parameters α_1 and α_2, and an item I_j requiring the same two operations plus a third one with parameter α_3. If the item parameters are assumed to be of the additive form $\beta_i = \alpha_1 + \alpha_2 + c$ and $\beta_j = \alpha_1 + \alpha_2 + \alpha_3 + c$, then $\beta_j - \beta_i = \alpha_3$, that is, the third operation alone accounts for the *difficulty difference* of the items. This is an example of the core assumption of the LLTM, namely, that differences between item parameters are due to cognitive operations involved in (or experimental conditions effective for) one, but not involved in (or effective for) another item. Examples of applications are found in Section 8.7.

[1]Department of Psychology, University of Vienna, Liebiggasse 5, A–1010 Vienna, Austria; e-mail: GH.FISCHER@UNIVIE.AC.AT

The LLTM originated from the following considerations: the RM defines the reaction parameter ρ_{vi} (cf. Chapter 2) as a difference of one person parameter and one item parameter,

$$\rho_{vi} = \ln \frac{f_i(\theta_v)}{1 - f_i(\theta_v)} = \theta_v - \beta_i. \tag{8.1}$$

Clearly, (8.1) can be extended in the form

$$\rho_{vi} = \theta_v - \beta_i + \lambda_l + \mu_m + \nu_n + \ldots + \omega_o, \tag{8.2}$$

comprising any number of additional factors $L, M, N, \ldots, O$ with respective parameters $\lambda_l, \mu_m, \nu_n, \ldots, \omega_o$, which influence the reaction probability. If the factor levels of all factors are combined systematically (like the person factor S and the item factor I in the usual RM), the model is called a 'complete multifactorial RM'. Multifactorial RMs have been suggested by Rasch (1965), Micko (1969, 1970), Falmagne (cited in Micko, 1970), and have been developed explicitly by Scheiblechner (1971) and Kempf (1972). The latter two authors showed that the parameters of each factor can be separated from – and thus estimated independently of – all other factors, analogous to the separation of the item factor I and the person factor S in the RM. To estimate the parameters λ_k of factor L, for instance, all one has to do is to consider L, in a purely formal sense, as a 'virtual item factor', I^*, and all combinations $S \times I \times M \times N \times \ldots$ as levels of a 'virtual person factor', S^*, of an RM; if the data are rearranged in a data matrix such that the rows correspond to virtual persons, and the columns to virtual items, it is immediately possible to estimate the λ_k by means of the usual CML method (see Chapter 3). This amounts to defining all combinations of factor levels $S_v \times I_i \times M_j \times N_k \times \ldots$ as virtual persons, as if the responses of S_v under each of the combinations of levels $I_i \times M_j \times N_k \times \ldots$ stemmed from different (namely, virtual) persons S^*_w.

One separate normalization condition, however, will then be needed for making the parameters within each factor, $I, L, M, N, \ldots$, unique. These multifactorial RMs are special cases of the LLTM, which does not require completeness of the factor combinations.

Formally, the LLTM is defined as an RM subject to the side conditions

$$\beta_i = \sum_{l=1}^{p} w_{il}\alpha_l + c, \quad \text{for } i = 1, \ldots, k, \tag{8.3}$$

where

β_i is the difficulty parameter of item I_i in the RM,

α_l, $l = 1, \ldots, p$, are the 'basic' parameters of the LLTM,

w_{il} are given weights of the basic parameters α_l, and

c is the usual normalization constant.

The role of c in (8.3) can be explained as follows: the LLTM describes differences of item difficulty parameters as $\beta_j - \beta_i = \sum_l (w_{jl} - w_{il})\alpha_l$, for all pairs (I_j, I_i). This implies $\beta_j - \sum_l w_{jl}\alpha_l = \beta_i - \sum_l w_{il}\alpha_l =: c$, for all (I_j, I_i), which is equation (8.3). Whatever (admissible) shift of the item parameters β_i is carried out, e.g., $\beta_i \to \beta_i + d$, this shift is immediately compensated by the same shift $c \to c + d$ of the normalization constant in (8.3), whereas the structural basic parameters α_l are not affected.

8.2 Estimation of the α_l

Three methods have been proposed for estimating the item parameters β_i under the constraints (8.3). They are completely parallel to the most frequently used estimation methods in the RM: conditional (CML), marginal (MML), and unconditional or joint (JML) maximum likelihood. The CML method is easy to apply, reasonably fast and accurate, and has the major advantage that the asymptotic behavior of the estimators is known. Moreover, conditional likelihood ratio tests are easily made after parameter estimation, without any additional computations. As regards MML, two cases have to be distinguished: the parametric MML approach, on the one hand, is not consistent if the true distribution of person parameters θ differs from the hypothesized distribution (cf. Pfanzagl, 1994); since we consider it unlikely that a researcher has reliable evidence of the true latent distribution, and since the robustness under misspecification of the latent distribution is unknown, we refrain from treating the parametric MML method here (cf. Chapter 11). The semi-parametric MML approach (Follmann, 1988), on the other hand, is asymptotically equivalent to the CML method if the number of support points is at least $(k+1)/2$ (De Leeuw & Verhelst, 1986; Pfanzagl, 1994; see also Chapter 13); hence this method will not be required if a CML solution is available. Therefore we shall not describe this approach either. Finally, the disadvantages of the JML method are the same as those mentioned for the RM in Chapter 3; they are the reasons why JML for the LLTM has hardly ever been considered (see Fischer & Formann, 1982a).

To facilitate the derivation of the CML equations, a particular normalization of the item parameters will be adopted: observe that (8.3) implies $\beta_i^* := \beta_i - c = \sum_l w_{il}\alpha_l$. That is, there exists *one* particular normalization of the item parameters such that c in (8.3) vanishes. The conditional likelihood function being independent of the normalization, however, we may presume this particular normalization without loss of generality as far as CML estimation is concerned. For simplicity, we even drop the asterisks, thus writing

$$\beta_i = \sum_{l=1}^{p} w_{il}\alpha_l, \tag{8.4}$$

but keeping in mind that the β_i in (8.4) cannot be subjected to any other normalization. By this convention, a formally somewhat more complicated treatment of the conditional likelihood function (cf. Fischer, 1983a) can be avoided.

Let $\boldsymbol{X}$ be a complete data matrix, $x_{.i}$ and r_v the item and person marginal sums, respectively, $\epsilon_i = \exp(-\beta_i)$ item easiness parameters, and $\gamma_r(\epsilon_1, \dots, \epsilon_k)$ the elementary symmetric function of order r of $\epsilon_1, \dots, \epsilon_k$ (see Chapter 3). The conditional likelihood function L_C of the data $\boldsymbol{X}$, given the person marginal sums r_v for $v = 1, \dots, n$, is the same as (3.9), namely,

$$L_C = P(\boldsymbol{X}|\boldsymbol{\epsilon}, \boldsymbol{r}) = \prod_v \frac{\prod_i \epsilon_i^{x_{vi}}}{\gamma_{r_v}} = \frac{\exp(-\sum_i x_{.i}\beta_i)}{\prod_r \gamma_r^{n_r}}, \tag{8.5}$$

where n_r is the number of persons with raw score r. Since (8.5) holds for whatever normalization of the β_i, it also holds for the particular normalization underlying (8.4). Therefore, we may insert (8.4) in (8.5) obtaining

$$\begin{aligned} L_C &= \frac{\exp[-\sum_i x_{.i} \sum_l w_{il}\alpha_l]}{\prod_r \gamma_r^{n_r}} \\ &= \frac{\exp[-\sum_l t_l \alpha_l]}{\prod_r \gamma_r^{n_r}}, \end{aligned} \tag{8.6}$$

where

$$t_l = \sum_i x_{.i} w_{il}, \tag{8.7}$$

for $l = 1, \dots, p$.

Notice that (8.4) can be rewritten in matrix form as

$$\boldsymbol{\beta} = \boldsymbol{W\alpha}, \tag{8.8}$$

where $\boldsymbol{\beta}$ and $\boldsymbol{\alpha}$ are column vectors of parameters, and $\boldsymbol{W}$ the weight matrix. Now the derivative of $\ln L_C$ with respect to $\boldsymbol{\alpha}$ is obtained by means of the chain rule for differentiation of a scalar function with respect to a vector (cf. Rao, 1965, p. 71) as

$$\frac{\partial \ln L_C}{\partial \boldsymbol{\alpha}} = \left[\frac{\partial \boldsymbol{\beta}}{\partial \boldsymbol{\alpha}}\right]' \frac{\partial \ln L_C}{\partial \boldsymbol{\beta}} = \boldsymbol{W}' \frac{\partial \ln L_C}{\partial \boldsymbol{\beta}}. \tag{8.9}$$

Since we know from Chapter 3 that the elements of the vector $\partial \ln L_C / \partial \boldsymbol{\beta}$ are (3.12), setting all $b_{vi} = 1$ we immediately get the p CML equations as

$$\sum_{i=1}^{k} w_{il} \left[x_{.i} - \sum_{v=1}^{n} \frac{\epsilon_i \gamma_{r_v - 1}^{(i)}}{\gamma_{r_v}} \right] = 0, \tag{8.10}$$

for $l = 1, \dots, p$. These equations (for complete data sets) are very similar to the CML equations of the RM, except that the differences between the item marginal sums $x_{.i}$ and their conditional expectations – right-hand terms within the brackets in (8.10) – are no longer zero; rather, p weighted sums of differences, each of them with weights w_{il} for $i = 1, \ldots, k$, are set to zero. As will be shown in Section 8.4, under weak conditions on both $\boldsymbol{W}$ and $\boldsymbol{X}$ the CML equations possess a unique solution $\hat{\boldsymbol{\alpha}}$.

It should be remarked that the functions γ_r and $\gamma_{r-1}^{(i)}$ depend on the actual normalization, but the expressions $\epsilon_i \gamma_{r-1}^{(i)}/\gamma_r$ in (8.10) are independent thereof; that is, the CML equations also apply under decomposition (8.3) with whatever normalization the researcher may have chosen.

Sometimes it is preferable to write the CML equations (8.10) in the form

$$\sum_i w_{il} \left[x_{.i} - \sum_r n_r \frac{\epsilon_i \gamma_{r-1}^{(i)}}{\gamma_r} \right] = 0, \tag{8.11}$$

for $l = 1, \ldots, p$, allowing one to reduce the computational labor by grouping persons with the same raw score r into score groups of sizes n_r.

Suppose, as a special case, that the weights in (8.3) are $w_{il} = \delta_{il}$, where δ_{il} is the Kronecker symbol. This means that there is an identity of item and basic parameters, $\beta_i = \alpha_i$, $i = 1, \ldots, k$, so that the LLTM becomes the RM. The CML equations (8.10) then become those of the RM, namely (3.12). We conclude that in formal respects the *LLTM is more general than the RM,* that is, the latter is a special case of the former. As we already know from Chapter 3, the equations (3.12) do not possess a unique solution unless the β_i are normalized by some restriction like $\beta_1 = 0$ or $\sum_i \beta_i = 0$. This result will later be shown also to be a special case of a more general uniqueness theorem for the LLTM.

To solve equations (8.10) or (8.11) with respect to the CML estimates $\hat{\alpha}_l$, computation of the combinatorial functions γ_r and $\gamma_{r-1}^{(i)}$ and a suitable iterative algorithm are required to improve provisional estimates. The γ_r and $\gamma_{r-1}^{(i)}$ can be computed using recurrence relations based on simple algebraic (or combinatorial) properties of these functions. A basic property of γ_r is that

$$\gamma_r(\epsilon_1, \ldots, \epsilon_k) = \sum_{l=l_1}^{l_2} \gamma_{r-l}(\epsilon_1, \ldots, \epsilon_{k'}) \gamma_l(\epsilon_{k'+1}, \ldots, \epsilon_k), \tag{8.12}$$

with $l_1 = \max(0, r-k')$ and $l_2 = \min(k-k', r)$. Formula (8.12) is obvious because the left-hand side, γ_r, is composed of products of the ϵ_i corresponding to patterns of responses $\boldsymbol{x}$ with $\sum_i x_i = r$; the right-hand side of (8.12) constructs these patterns from partial patterns having raw score $r-l$ in subtest $J_1 = \{I_1, \ldots, I_{k'}\}$ combined with partial patterns having raw score l in subtest $J_2 = \{I_{k'+1}, \ldots, I_k\}$. The limits of the summation, l_1 and l_2, are easily derived from the obvious restrictions $l \geq 0$, $l \geq r-k'$, $l \leq k-k'$, $l \leq r$.

The recurrence relation (8.12) can be used for an efficient computation of the γ_r (Verhelst & Veldhuijzen, 1991), but this algorithm is complicated and so will not be described here. (Other possibilities are discussed by Jansen, 1984; Verhelst, Glas, & Van der Sluis, 1984; Formann, 1986). A useful special case of (8.12), however, is $J_1 = \{I_1, \ldots, I_{t-1}\}$, $J_2 = \{I_t\}$,

$$\gamma_r(\epsilon_1, \ldots, \epsilon_t) = \gamma_r(\epsilon_1, \ldots, \epsilon_{t-1}) + \gamma_{r-1}(\epsilon_1, \ldots, \epsilon_{t-1})\epsilon_t, \tag{8.13}$$

for $0 \leq r \leq t$, $t = 1, \ldots, k$, with $\gamma_0 = 1$ and $\gamma_r = 0$ for $r < 0$ or $r > t$. From (8.13) one obtains in turn

$$\gamma_r(\epsilon_1, \epsilon_2) = \gamma_r(\epsilon_1) + \gamma_{r-1}(\epsilon_1)\epsilon_2$$
$$(\text{with } \gamma_0(\epsilon_1) = 1, \ \ \gamma_1(\epsilon_1) = \epsilon_1, \ \ \gamma_r(\epsilon_1) = 0 \text{ for } r > 1),$$
$$\gamma_r(\epsilon_1, \epsilon_2, \epsilon_3) = \gamma_r(\epsilon_1, \epsilon_2) + \gamma_{r-1}(\epsilon_1, \epsilon_2)\epsilon_3,$$
$$\text{etc., up to } \gamma_r(\epsilon_1, \epsilon_2, \ldots, \epsilon_k).$$

The partial derivatives $\gamma_{r-1}^{(i)}$ can be obtained by means of the same recursion, simply deleting one item I_i at a time.

This method was first proposed by Andersen (1972; see also Fischer, 1974, p. 230) and was later called the 'summation method' (Gustafsson, 1980), stressing that it involves only summations of positive terms, which makes the method numerically stable. While it is reasonably fast for the γ_r, it is not very efficient for the $\gamma_{r-1}^{(i)}$, because the whole recursion has to be carried out separately for each $i = 1, \ldots, k$.

Fischer and Ponocny (1994) therefore recommend using the summation method for the γ_r only and combining it with a modified version of the 'difference method' (the latter was used by Fischer and Allerup, 1968; see also Fischer, 1974, pp. 243–244) for the $\gamma_{r-1}^{(i)}$. From the definition of γ_r and $\gamma_{r-1}^{(i)}$ it immediately follows that

$$\gamma_r = \epsilon_i \gamma_{r-1}^{(i)} + \gamma_r^{(i)}, \tag{8.14}$$

$$\gamma_r^{(i)} = \gamma_r - \epsilon_i \gamma_{r-1}^{(i)}, \tag{8.15}$$

which is a recurrence relation for the $\gamma_r^{(i)}$, $r = 0, \ldots, k-1$. From (8.15) one obtains

$$\begin{aligned} \gamma_0^{(i)} &= 1, \\ \gamma_1^{(i)} &= \gamma_1 - \epsilon_i, \\ \gamma_2^{(i)} &= \gamma_2 - \epsilon_i \gamma_1^{(i)}, \\ &\text{etc., for all } \gamma_r^{(i)}. \end{aligned}$$

A problem, however, is the numerical accuracy of the difference method as described so far: those $\gamma_r^{(i)}$ where γ_r and $\epsilon_i \gamma_{r-1}^{(i)}$ are of approximately the same magnitude, are very prone to numerical error, because computing the difference on the right-hand side of (8.15) in these cases entails a rapid loss of significant digits. Therefore it is necessary to stop the described recursion at some suitably chosen r^* and to use essentially the same recursion 'top-down' as follows:

$$\begin{aligned}
\gamma_k^{(i)} &= 0, \\
\gamma_{k-1}^{(i)} &= \prod_{l \neq i} \epsilon_l, \\
\gamma_{k-2}^{(i)} &= \epsilon_i^{-1}(\gamma_{k-1} - \gamma_{k-1}^{(i)}), \\
\gamma_{k-3}^{(i)} &= \epsilon_i^{-1}(\gamma_{k-2} - \gamma_{k-2}^{(i)}), \\
&\vdots \qquad\qquad \vdots
\end{aligned}$$

for k, $k-1, \ldots, r^*$. This leads to two separate values for $\gamma_{r^*}^{(i)}$, which can be used for a check of the accuracy. If the two values do not correspond sufficiently well, it is recommended to re-compute all $\gamma_r^{(i)}$, $r = 1, \ldots, k-1$, for this value of i, by means of the summation method.

This combination of the summation and difference methods guarantees high numerical precision and acceptable computing times. (A similar generalized recursion for the functions γ_r and $\gamma_{r-1}^{(i)}$ in polytomous RMs will be described in Chapter 19; see also Fischer and Ponocny, 1994).

As an iterative method for solving the nonlinear systems (8.10) or (8.11), the Broyden-Fletcher-Goldfarb-Shanno (BFGS) procedure (a Quasi-Newton-method, cf. Churchhouse, 1981) is very efficient. Its advantage is that the second-order partial derivatives of $\ln L_C$ that are required, e.g., for the Newton-Raphson method, need not be computed, but the inverse of the asymptotic information matrix is still obtained as a result of the iteration procedure. The same method is also recommended for the polytomous RMs, see Chapter 19, or Fischer and Ponocny, 1994. The BFGS method requires, however, that the LLTM is in 'minimal form', see Section 8.4.

The second-order partial derivatives could easily be obtained also from (8.9), applying again the rule for differentiation of a scalar function of a vector, as

$$\frac{\partial^2 \ln L_C}{\partial \boldsymbol{\alpha} \partial \boldsymbol{\alpha}'} = \boldsymbol{W}' \frac{\partial^2 \ln L_C}{\partial \boldsymbol{\beta} \partial \boldsymbol{\beta}'} \boldsymbol{W}, \tag{8.16}$$

see also Fischer, 1983a, p. 7.

Finally, for the sake of completeness we also give the JML equations:

$$\sum_{i=1}^{k} w_{il} \left[x_{.i} - \sum_{v=1}^{n} \frac{\exp(\theta_v - \beta_i)}{1 + \exp(\theta_v - \beta_i)} \right] = 0, \quad \text{for } l = 1, \ldots, p, \tag{8.17}$$

$$r_v - \sum_{i=1}^{k} \frac{\exp(\theta_v - \beta_i)}{1 + \exp(\theta_v - \beta_i)} = 0, \quad \text{for } v = 1, \ldots, n, \tag{8.18}$$

where the basic parameters α_l are implicitly contained in the β_i, see (8.3). The estimation equations (8.17), by virtue of (8.9), follow immediately from those of the RM, (3.7), and equations (8.18) are identical to (3.6), with all $b_{vi} = 1$.

Note that the JML equations for the item parameters of the RM are a special case of (8.17) obtained by setting $w_{il} = \delta_{il}$ (where δ_{il} is the Kronecker Delta), that is, where item and basic parameters are identical. The numerical solution of (8.17) and (8.18) can be computed as in the case of the RM, see Chapter 3.

8.3 Incomplete data

As will be seen especially in Chapter 9, the LLTM has many interesting applications to structurally incomplete designs. We therefore have to extend the estimation equations to cases of incomplete data. Let the following notation be used again as in Chapter 3:

$\boldsymbol{B} = ((b_{vi}))$ is an $n \times k$ design matrix with $b_{vi} = 1$ if item I_i is presented to S_v, and $b_{vi} = 0$ otherwise. We shall assume $\sum_i b_{vi} \geq 2$ for all S_v, i.e., persons have responded to at least two items. (Response patterns with $\sum_i b_{vi} = 1$ have a conditional likelihood $= 1$ and do not contribute to the CML estimation; they may therefore be discarded.)

$x_{vi} = a$, with $0 < a < 1$, is defined for $b_{vi} = 0$;

$r_v := \sum_i x_{vi} b_{vi}$ is the raw score of S_v;

$x_{.i} := \sum_v x_{vi} b_{vi}$ is the item marginal sum of I_i;

γ_{r_v} and $\gamma^{(i)}_{r_v - 1}$ are now redefined as elementary symmetric functions of the variables $\epsilon_1 b_{v1}, \ldots, \epsilon_k b_{vk}$, cf. (3.10) in Chapter 3.[2]

Using this notation, we get, in direct analogy to (8.5), the conditional likelihood function of the (possibly incomplete) data $\boldsymbol{X}$, given the design matrix $\boldsymbol{B}$ and the persons' raw scores r_v, as

$$L_C = P(\boldsymbol{X}|\boldsymbol{\epsilon}, \boldsymbol{r}; \boldsymbol{B}) = \prod_v \frac{\prod_i (\epsilon_i b_{vi})^{x_{vi} b_{vi}}}{\gamma_{r_v}}$$

$$= \frac{\exp(-\sum_i x_{.i}\beta_i)}{\prod_v \gamma_{r_v}} = \frac{\exp[-\sum_l t_l \alpha_l]}{\prod_v \gamma_{r_v}}, \tag{8.19}$$

with

$$t_l = \sum_v \sum_i x_{vi} b_{vi} w_{il}. \tag{8.20}$$

[2]For formal exactness, however, the notation should be 'γ_{v,r_v}', because the function γ depends on v via the b_{vi}, not only on S_v's raw score r_v.

For consistency, we define that $(\epsilon_i b_{vi})^{x_{vi}b_{vi}} = 0^0 = 1$ for $b_{vi} = 0$, that is, when no response has been observed. As in Section 8.2 above, the CML estimation equations follow from taking partial derivatives of (8.19), using (3.11), which yields equations (8.21),

$$\sum_{i=1}^{k} w_{il} \left[x_{.i} - \sum_{v=1}^{n} \frac{\epsilon_i b_{vi} \gamma_{r_v-1}^{(i)}}{\gamma_{r_v}} \right] = 0, \tag{8.21}$$

for $l = 1, \ldots, p$; they have a structure similar to (8.10), however, under the present more general definition of the item marginal sums $x_{.i}$, raw scores r_v, and combinatorial functions γ_{r_v} and $\gamma_{r_v-1}^{(i)}$. For practical purposes it will again be advisable to group persons, first by design vectors, yielding groups G_g, and then by raw scores within groups G_g, such that the CML equations become

$$\sum_{i} w_{il} \left[x_{.i} - \sum_{g} \sum_{r} n_{gr} \frac{\epsilon_i \gamma_{g,r-1}^{(i)}}{\gamma_{gr}} \right] = 0, \tag{8.22}$$

for $l = 1, \ldots, p$, with n_{gr} for the number of persons with raw score r within group G_g, and γ_{gr} and $\gamma_{g,r-1}^{(i)}$ the respective elementary symmetric functions defined for the subset of items given to group G_g, however, with the additional definition $\gamma_{g,r-1}^{(i)} = 0$ if $b_{vi} = 0$ for the $S_v \in G_g$, that is, if item I_i was not presented to the testees of group G_g.

These equations, by the way, are general enough even to cope with situations where the structures of the items, that is, the $(w_{i1}, \ldots, w_{ip})$, depend on the person group, such as arise in repeated measurement designs; this complication can be dealt with by introducing the concept of 'virtual items' and by applying (8.22) to these rather than to the real items (cf. Chapter 9).

The CML equations (8.10) were derived by Fischer (1973), computer programs were first published by Fischer and Formann (1972) and Fischer (1974). A generalization to polytomous items is also found in Fischer (1974, 1977b). The incomplete data case was developed in Fischer (1983a). A more general polytomous 'linear rating scale model' (LRSM), which comprises the LLTM as a special case, and respective computer software were published more recently by Fischer and Parzer (1991a, 1991b), and another generalization, the 'linear partial credit model' (LPCM), by Fischer and Ponocny (1994); see Chapter 19.

8.4 Existence and Uniqueness of the CML Solution

For many areas of application of the LLTM, especially in structurally incomplete designs, it is important to ascertain whether a CML solution $\hat{\boldsymbol{\alpha}}$ exists and whether it is unique. A trivial necessary condition for the uniqueness of $\hat{\boldsymbol{\alpha}}$ is that a given vector of item parameters β_i, inserted in (8.3), can be decomposed uniquely into the α_l and c, i.e., that the system $\boldsymbol{\beta} = \boldsymbol{W}\boldsymbol{\alpha} + c\mathbf{1}$, where $\mathbf{1}$ is a column vector of ones, has a unique solution $(\boldsymbol{\alpha}, c)$. Clearly, this holds if and only if the rank of the matrix

$$\boldsymbol{W}^+ = \begin{pmatrix} w_{11} & \dots & w_{1p} & 1 \\ w_{21} & \dots & w_{2p} & 1 \\ \dots & \dots & \dots & \dots \\ w_{k1} & \dots & w_{kp} & 1 \end{pmatrix}$$

is maximal, that is, $p+1$. Denoting the rank of $\boldsymbol{W}^+$ by $r(\boldsymbol{W}^+)$, we state that $r(\boldsymbol{W}^+) = p+1$ is necessary for the uniqueness of $\hat{\boldsymbol{\alpha}}$. If this requirement is violated, the LLTM contains too many basic parameters, so that some of them can be eliminated (the exponential family is not in 'canonical' or 'minimal' form, cf. Barndorff-Nielsen, 1978; Andersen, 1980a). We shall say that the LLTM is in minimal form if $r(\boldsymbol{W}^+) = p+1$.

A more complex necessary and sufficient condition for the existence and uniqueness of a CML solution $\hat{\boldsymbol{\alpha}}$ has been given by Fischer (1983a). It is based on a characterization of the relevant structural properties of the data $\boldsymbol{X}$ by means of a certain directed graph. Important special cases of this general result employ the concepts of 'well-conditioned' vs. 'ill-conditioned' data $\boldsymbol{X}$. To define these, we have to introduce some notation first:

Let C be a directed graph (digraph) with vertices V_i assigned 1–1 to the items I_i, such that a directed line (diline) $\overrightarrow{V_iV_j}$ exists iff some person S_v has given a right response to I_i and a wrong response to I_j. Let furthermore C_l, $l = 1, \dots, u$, be the strong components of C. (Remark: the set of strong components corresponds to the partition of the vertices into the largest subsets C_l such that any vertex within C_l can be reached via a directed path from any vertex in C_l; the C_l are therefore called strongly connected or 'strong'. Mutual reachability does not hold, on the other hand, for vertices belonging to different components, C_l and C_m, say, even if a diline $\overrightarrow{V_iV_j}$ exists for some $V_i \in C_l$ and some $V_j \in C_m$. Powerful algorithms for determining the strong components of a digraph exist in graph theory, see, e.g., Christofides, 1975).

Based on the digraph C of $\boldsymbol{X}$, we give the following definitions:

Definition 8.1 *The data $\boldsymbol{X}$ are said to be well-conditioned iff digraph C of $\boldsymbol{X}$ is strongly connected (i.e., has only one strong component $C_l = C$).*

Definition 8.2 *The data $\boldsymbol{X}$ are said to be ill-conditioned iff $\boldsymbol{X}$ is not well-conditioned (i.e., C has at least two strong components).*

Digraph C and its property of connectedness or unconnectedness allow us to formulate useful results on the existence and uniqueness of a CML solution of (8.10) and (8.21). Note that these definitions apply equally well to complete and incomplete data, so that the uniqueness results can be formulated for both cases.

Theorem 8.1 *Let the strong components of digraph C be $C_1, \ldots, C_u$. A unique CML solution $\hat{\boldsymbol{\alpha}}$ for the LLTM exists iff the ('alternative') linear system of equations (8.23)–(8.24),*

$$\sum_{l=1}^{p} w_{il} y_l = d_t, \quad \text{for all } V_i \in C_t, \text{ and } t = 1, \ldots, u, \tag{8.23}$$

where the d_t are unknown constants satisfying the inequality constraints

$$d_s \leq d_t \quad \text{if a diline } \overrightarrow{V_a V_b} \text{ exists for some } V_a \in C_s \text{ and some } V_b \in C_t, \tag{8.24}$$

has only the trivial solution $\boldsymbol{y} = \boldsymbol{0}$.

(Remark: Theorem 8.1 does not refer to a particular normalization of the β_i because the α_l are independent of such a normalization.)

The system (8.23)–(8.24) will be called *alternative system* (AS), because either the CML system, (8.10) or (8.21), has a unique finite solution and the AS has only the trivial null-solution $\boldsymbol{y} = \boldsymbol{0}$, or the CML equations do not possess a unique finite solution and the AS has some non-null solution $\boldsymbol{y} \neq \boldsymbol{0}$. Hence, in order to determine the uniqueness of the CML solution, it suffices to look at the set of solutions of the linear AS.

A direct, self-contained proof of Theorem 8.1 is quite lengthy, see Fischer (1983a). A slightly shorter proof, using a recent more general result of Jacobsen (1989) for discrete exponential distributions, is given by Fischer and Tanzer (1994), and is also to be found in Appendix A of the present chapter.

Still, in many typical applications the AS may be difficult to handle; however, there exist special cases of great practical relevance. In particular, it is interesting to see what happens if the data are well-conditioned, which is what we may expect with most realistic empirical data, or if there are exactly two strong components.

Suppose the data are well-conditioned; this implies that there is only one strong component, $C_1 = C$. Then the AS reduces to a system of linear equations,

$$\sum_{l} w_{il} y_l = d, \quad \text{for all } V_i, \tag{8.25}$$

which possesses only the trivial solution $\boldsymbol{y} = \boldsymbol{0}$ iff the matrix $\boldsymbol{W}^+$ is of full column rank, $r(\boldsymbol{W}^+) = p + 1$. This allows us to formulate the following result:

Corollary 8.1 *If the data $\boldsymbol{X}$ are well-conditioned, a unique CML solution $\hat{\boldsymbol{\alpha}}$ of the LLTM exists iff matrix $\boldsymbol{W}^+$ has full column rank, $r(\boldsymbol{W}^+) = p + 1$.*

Next we consider the special case of $u = 2$ strong components. Let, without loss of generality, vertices (items) $V_1, \ldots, V_t$ belong to C_1, and $V_{t+1}, \ldots, V_k$ to

C_2. Then the AS is the following:

$$\sum_l w_{il} y_l = d_1, \quad \text{for } i = 1, \dots, t, \text{ and}$$
$$\sum_l w_{il} y_l = d_2, \quad \text{for } i = t+1, \dots, k.$$

This AS has only the trivial solution $\boldsymbol{y} = \boldsymbol{0}$ iff the matrix $\boldsymbol{W}^{++}$,

$$\boldsymbol{W}^{++} = \begin{pmatrix} w_{11} & \dots & w_{1p} & 1 & 0 \\ \dots & \dots & \dots & \dots & \dots \\ w_{t1} & \dots & w_{tp} & 1 & 0 \\ w_{t+1,1} & \dots & w_{t+1,p} & 0 & 1 \\ \dots & \dots & \dots & \dots & \dots \\ w_{k1} & \dots & w_{kp} & 0 & 1 \end{pmatrix}, \tag{8.26}$$

has full column rank $r(\boldsymbol{W}^{++}) = p+2$. Note that the inequality restriction (8.24), $d_1 \leq d_2$, which holds if there exists a diline $\overrightarrow{V_i V_j}$ for some $V_i \in C_1$ and some $V_j \in C_2$, has now become irrelevant: if it should happen that, for a given solution $\boldsymbol{y} \neq \boldsymbol{0}$, $d_1 \geq d_2$ holds contrary to the existence of $\overrightarrow{V_i V_j}$, it suffices to replace the solution by $-\boldsymbol{y}$, which is also a solution and satisfies the required inequality restriction. Hence we have the following result:

Corollary 8.2 *If the data* $\boldsymbol{X}$ *are ill-conditioned with exactly* $u = 2$ *strong components,* $\{V_1, \dots, V_t\} = C_1$, $\{V_{t+1}, \dots, V_k\} = C_2$, *then a unique finite CML solution* $\hat{\boldsymbol{\alpha}}$ *of the LLTM exists iff matrix* $\boldsymbol{W}^{++}$ *has full column rank,* $r(\boldsymbol{W}^{++}) = p+2$.

It is an advantage of these results (Theorem 8.1, Corollaries 8.1 and 8.2) that no distinction between complete and incomplete data sets is necessary; the definition of each element c_{ij} of the adjacency matrix of digraph C rests only on the existence/nonexistence of persons S_v with $x_{vi} = 1$ and $x_{vj} = 0$.

One important question associated with the estimates $\hat{\alpha}_l$ is their asymptotic behavior as $n \to \infty$. It is an attractive feature of the CML approach that strong asymptotic results are available: under regularity conditions, $\hat{\boldsymbol{\alpha}}$ is asymptotically multivariate normal around the true $\boldsymbol{\alpha}$ with $N(\boldsymbol{\alpha}, \boldsymbol{\mathcal{I}}^{-1})$, where $\boldsymbol{\mathcal{I}}$ is the information matrix at $\hat{\boldsymbol{\alpha}}$. The regularity conditions essentially require (a) that the matrix $\boldsymbol{W}^{+}$ be of full column rank $p+1$, which implies that the item parameters β_j can be decomposed uniquely into the α_l, so that $\boldsymbol{\mathcal{I}}$ is nonsingular; and (b) that equation (8.27),

$$\sum_{v=1}^{\infty} \exp(-|\theta_v|) = \infty, \tag{8.27}$$

is satisfied for the infinite sequence of person parameters θ_v, $v = 1, 2, \dots$, (Pfanzagl, 1994). Condition (b) implies that the numbers n_r of persons with raw score

r, for $r = 1, \ldots, k-1$, tend to ∞ as $n \to \infty$ (see also Chapter 3). Thus, we may say that the CML estimator $\hat{\boldsymbol{\alpha}}$ will be consistent and asymptotically normal around $\boldsymbol{\alpha}$ for $n \to \infty$ under realistic assumptions.

The RM is one more special case of interest: let $\boldsymbol{W}$ be the identity matrix of order $k \times k$, that is, $p = k$, such that each basic parameter α_i becomes identical to one item parameter β_i; then the LLTM becomes an RM. Since now the last column of $\boldsymbol{W}^+$ is the sum of the columns of $\boldsymbol{W}$, $r(\boldsymbol{W}^+) = k = p < p+1$, implying that the CML estimates $\hat{\beta}_i = \hat{\alpha}_i$ cannot be unique, whatever the data $\boldsymbol{X}$. This is, however, what one would expect, because the $\hat{\beta}_i$ of an RM are not unique unless they are normalized. Setting $\beta_k = 0$, for instance, thus eliminating α_k, would mean $w_{ii} = 1$ for $i = 1, \ldots, k-1$, and $w_{ij} = 0$ otherwise, leading to the deletion of column k of $\boldsymbol{W}$. Then, the new $\boldsymbol{W}^+$ is of rank $r(\boldsymbol{W}^+) = k$, which equals $p+1$, where p now is the (reduced) number of parameters. Therefore, according to Corollary 8.1, the $\hat{\beta}_i = \hat{\alpha}_i$ are unique if the data are well-conditioned.

If, however, the data are ill-conditioned with strong components C_t, $t = 1, \ldots, u$, $u \geq 2$, the AS (8.23)–(8.24) is

$$
\begin{aligned}
y_i &= d_t \quad \text{for all } V_i \in C_t,\ i = 1, \ldots, k-1;\ t = 1, \ldots, u; \\
0 &= d_l \quad \text{for that one } C_l \text{ for which } V_k \in C_l; \\
d_s &\leq d_t \quad \text{if there exists some } \overrightarrow{V_a V_b} \text{ with } V_a \in C_s \text{ and } V_b \in C_t, \text{ for all } s \neq t.
\end{aligned}
$$

This system of equations always has a solution $\boldsymbol{y} \neq \boldsymbol{0}$: it suffices to choose any values $d_t \neq 0$, $t = 1, \ldots, u$, sufficing the above inequality restrictions, except for index l, where $d_l = 0$; and then to define the y_i accordingly. This immediately gives a solution $\boldsymbol{y} \neq \boldsymbol{0}$. Hence, if the data are ill-conditioned, no unique CML solution exists for the RM.

Corollary 8.3 *The normalized CML estimate $\hat{\boldsymbol{\beta}}$ of the RM is unique iff the data are well-conditioned.*

This result is due to Fischer (1981). Again we see that results about the RM are special cases of more general theorems for the LLTM.

8.5 LLTM and LBTL

The 'Bradley-Terry-Luce Model' (BTL; Zermelo, 1929; Bradley & Terry, 1952; Ford, 1957; Luce, 1959) has a long tradition in psychometrics and mathematical psychology. It explains the outcome of the comparison of each two objects O_i and O_j from a set $\{O_1, \ldots, O_o\}$ by means of the probabilities $P(O_i \succ O_j)$, where '$O_i \succ O_j$' means that in a particular comparison O_i is preferred to O_j, or 'wins' over O_j. All comparisons are assumed to be stochastically independent. A direct relationship between the RM and the BTL is obvious: the RM can be considered as an incomplete BTL for the comparisons of persons with items, where either the person solves the item ('wins' over the item), or the item resists solution ('wins' over the person). This redefinition of the RM is quite useful for obtaining insights into the nature of the RM (Fischer, 1981). Similarly, the LLTM can be viewed as a BTL with linear restrictions imposed on the BTL parameters, which we denote as the 'Linear Bradley-Terry-Luce Model' (LBTL).

The BTL is defined by the equations

$$P(O_i \succ O_j) = \frac{1}{1+\exp(\tau_j - \tau_i)}, \tag{8.28}$$

for all ordered pairs (O_i, O_j). Each τ_i is a parameter characterizing one object (more precisely, its 'value' or 'power', depending on the nature of the objects and of the comparisons). Since model (8.28) is overparameterized, the τ_i have to be normalized by some condition like $\tau_1 = 0$ or $\sum_i \tau_i = 0$.

The data for estimating the τ_i is a matrix $\boldsymbol{N}$ of frequencies n_{ij} of preferences '$O_i \succ O_j$'. The τ_i can easily be estimated by ML (cf. David, 1963). The ML equations have a unique normalized solution iff, for each partition of the objects into two subsets J_1 and J_2, some object $O_i \in J_1$ has been preferred at least once to some object $O_j \in J_2$. This necessary and sufficient condition is equivalent to strong connectedness of a certain digraph G of the data, defined as follows: Assign vertices V_i 1–1 to objects O_i, and insert a diline $\overrightarrow{O_iO_j}$ iff O_i has at least once been preferred to O_j.

The BTL was actually first formulated by Zermelo (1929); remarkably, he both developed the ML estimation method and derived the above uniqueness result. The latter was independently rediscovered by Ford (1957). Its formulation in terms of connectedness of digraph G is found in Harary et al. (1965). We call the condition of G being strong the 'Zermelo-Ford Condition'.

We define the LBTL as follows: in addition to (8.28), let the object parameters satisfy the linear restrictions

$$\tau_i = \sum_{l=1}^{q} w_{il}\alpha_l + c, \tag{8.29}$$

for $i = 1, \ldots, o$ (objects), $l = 1, \ldots, q$ (basic parameters), with $q < o$. An extended weight matrix $\boldsymbol{W}^+ = (\boldsymbol{W}, \boldsymbol{1})$ is again defined as in the LLTM and is assumed to have rank $q+1$, such that the α_l can be determined uniquely if the τ_i are known.

There is a complete equivalence of the LLTM with an LBTL that allows us to deduce interesting insights in both models. Firstly, the unconditional likelihood of any RM (and therefore also of any LLTM) can be considered a likelihood function of a BTL (or an LBTL, respectively). What is less obvious is that the likelihood function of any BTL can also be interpreted as the conditional likelihood of a particular incomplete RM with $k_v = 2$ items per person, where $k_v = \sum_i b_{vi}$, with $b_{vi} = b_{vj} = 1$, and $b_{vl} = 0$ otherwise. The unconditional probability (8.28) is then the same as the conditional probability that person S_v, given raw score $r_v = 1$, solves I_i, but does not solve I_j, see (3.8) in Chapter 3. (A more detailed and formal treatment of this equivalence is given by Tanzer, 1984). This implies that the *unconditional likelihood function of any RM* (or LLTM) can be considered a *conditional likelihood function of a particular incomplete RM* (or LLTM). From this equivalence relation, a result on the uniqueness of an (unconditional or joint) JML solution can be derived for any LLTM. First, however, we will formulate the uniqueness result for the LBTL.

Let digraph G be defined as before. We state the following theorem:

Theorem 8.2 *Let the strong components of digraph G be G_t, $t = 1, \ldots, u$. A unique ML solution $\hat{\boldsymbol{\alpha}}$ for the LBTL exists iff the ('alternative') linear system (8.30)–(8.31),*

$$\sum_{l=1}^{q} w_{il} y_l = d_t, \quad \text{for all } V_i \in G_t, \text{ and } t = 1, \ldots, u, \tag{8.30}$$

where the d_t are unknown constants satisfying the inequality constraints

$$d_s \leq d_t \quad \text{if a diline } \overrightarrow{V_a V_b} \text{ exists for some } V_a \in G_s \text{ and some } V_b \in G_t, \tag{8.31}$$

has only the trivial solution $\boldsymbol{y} = \mathbf{0}$.

Proof The proof of Theorem 8.2 is found in Fischer and Tanzer (1994).

Again an especially interesting case occurs if graph G is strongly connected (Zermelo-Ford Condition). It implies that $d_t = d$ for all t and that no inequality constraints (8.31) exist; in that case, system (8.30) has only the trivial null-solution $\boldsymbol{y} = \mathbf{0}$ iff $\boldsymbol{W}^+$ has full column rank, $r(\boldsymbol{W}^+) = q + 1$. This immediately yields the following result:

Corollary 8.4 *The ML solution $\hat{\boldsymbol{\alpha}} = (\hat{\alpha}_1, \ldots, \hat{\alpha}_q)$ of an LBTL, where $\boldsymbol{W}^+$ has full column rank, $r(\boldsymbol{W}^+) = q + 1$, is unique iff the data $\boldsymbol{N}$ satisfy the Zermelo-Ford Condition.*

The LBTL as an extension of the BTL seems not to have been considered in the psychometric literature, except, e.g., for unpublished papers by Fischer (1983c) and Tanzer (1984). In statistics, however, Bradley and El-Helbawy (1976) studied the BTL under linear restrictions of the form

$$\boldsymbol{B\tau} = \mathbf{0}, \tag{8.32}$$

where $\boldsymbol{B}$ is a matrix of s zero-sum orthonormal contrast vectors (see also Bradley, 1984). El-Helbawy and Bradley (1977) proved that, if (8.32) holds, the Zermelo-Ford Condition is *sufficient* for the uniqueness of the normalized ML solution.

This is a special case of our Corollary 8.4. El-Helbawy and Bradley (1977) furthermore showed that hypotheses of the form (8.32) can be tested by means of asymptotic LR tests.

Theorem 8.2 can be used for obtaining a nontrivial result about the joint maximum likelihood (JML) solution in the LLTM . The unconditional likelihood of the data $\boldsymbol{X}$ under an LLTM can be considered the likelihood function of a structurally incomplete LBTL for the comparison of persons with items, where both the p basic parameters of the LLTM, α_l, and the $k-1$ person parameters, θ_v, become basic parameters of the LBTL (i.e., $q = p+k-1$). Let the incomplete data of this LBTL be $\boldsymbol{N}$. A unique ML solution of the LBTL exists iff the AS (8.30)–(8.31) has only the trivial null-solution. As can be shown, the AS (8.30)–(8.31), defined for the data $\boldsymbol{N}$, is equivalent to AS (8.23)–(8.24) defined for the data $\boldsymbol{X}$ of the corresponding LLTM, plus the restrictions $1 \le r_v \le k_v - 1$ (see Fischer & Tanzer, 1994). Hence we have the following result:

Theorem 8.3 *Let the strong components of digraph C be $C_1, \ldots, C_u$. A unique normalized JML solution $(\hat{\boldsymbol{\alpha}}, \hat{\boldsymbol{\theta}})$ for the LLTM exists iff*

(a) *the linear system (8.33)–(8.34),*

$$\sum_{l=1}^{p} w_{il} y_l = d_t, \qquad \text{for all } V_i \in C_t, \text{ and } t = 1, \ldots, u, \tag{8.33}$$

where the d_t are unknown constants satisfying the inequality constraints

$$d_s \le d_t \qquad \text{if a diline } \overrightarrow{V_a V_b} \text{ exists for some } V_a \in C_s \text{ and } V_b \in C_t, \tag{8.34}$$

has only the trivial null-solution $\boldsymbol{y} = \boldsymbol{0}$, and

(b) *$1 \le r_v \le k_v - 1$ holds for all S_v.*

(Remark: the word 'normalized' refers to the item and person parameters, not the basic parameters. In contrast to the CML case, see Theorem 8.1, the JML solution has to be normalized, because otherwise the θ_v could never be unique.)

Proof of Theorem 8.3 The result follows from the structural equivalence of LLTM and LBTL. A complete formal proof using Jacobsen's (1989) theorem (see Lemma 8.1 in the Appendix to this chapter) is given by Fischer and Tanzer (1994, Theorem 3).

Assuming once more that the digraph C of the LLTM is strongly connected, Theorem 8.3 immediately allows one to deduce the following result:

Corollary 8.5 *If the data $\boldsymbol{X}$ of an LLTM are well-conditioned, a unique normalized JML solution $(\hat{\boldsymbol{\alpha}}, \hat{\boldsymbol{\theta}})$ exists iff*

(a) *$r(\boldsymbol{W}^{+}) = p + 1$, and*

(b) $1 \le r_v \le k_v - 1$ *holds for all* S_v.

Proof The proof is obvious because the AS (8.33)–(8.34) reduces to the homogeneous linear system (8.25), as in the derivation of Corollary 8.1. □

Finally, we can now deduce one more result for the RM:

Corollary 8.6 *The normalized JML estimate* $(\hat{\boldsymbol{\beta}}, \hat{\boldsymbol{\theta}})$ *of the RM is unique iff*

(a) the data are well-conditioned, and

(b) all raw scores satisfy $1 \le r_v \le k_v - 1$.

Proof The RM under the normalization $\beta_k = 0$ is an LLTM with weights $w_{il} = \delta_{il}$, $i = 1, \ldots, k$ and $l = 1, \ldots, k-1$, where δ_{il} denotes the Kronecker symbol. Hence, $p = k - 1$ and $r(\boldsymbol{W}^+) = k = p + 1$. By virtue of Corollary 8.5, we conclude that, if (a) the data are well-conditioned and (b) $1 \le r_v \le k_v - 1$, the (normalized) JML solution is unique. This proves the sufficiency of (a) and (b). On the other hand, it is obvious that no finite solution exists, or that an existing solution is not unique, whenever either (a) or (b) are violated. This proves the necessity. □

8.6 Testing the Fit of an LLTM

As was shown in Section 8.2, as a *formal model* the LLTM is more general than the RM, because likelihood function, estimation equations, and uniqueness conditions for the RM are special cases of those for the LLTM. With respect to empirical applications, however, the LLTM requires that (a) the RM fits the data, and (b) the item parameters satisfy the linear equations (8.3). The LLTM is therefore *empirically* more restrictive than the RM, i.e., the RM is more general than the LLTM (in the sense of a broader applicability).

Testing the fit of an LLTM requires two independent steps:

(i) testing the fit of the RM, and

(ii) testing the restrictions (8.3).

How (i) can be done has been described in Chapter 5 and need not be discussed here again. As regards (ii), a first check can be made by drawing points in $\mathbb{R}^2$ with abscissae $\hat{\beta}_i^{(R)}$ (item parameter estimates based on the RM) and ordinates $\hat{\beta}_i^{(\mathrm{LLTM})}$ (item parameter estimates in the LLTM subject to (8.3)); if the LLTM fits, the points should scatter around the 45° line through the origin. This graphical method, however, has heuristic merits rather than as a test of fit in the strict sense.

Within the CML framework, a powerful test can be carried out as follows: compute the conditional likelihood function $L_C(\boldsymbol{X}|\boldsymbol{r}, \hat{\boldsymbol{\beta}}) =: L_1$ defined by (8.5) under the RM as the H_1, and $L_C(\boldsymbol{X}|\boldsymbol{r}, \hat{\boldsymbol{\alpha}}, \boldsymbol{W}) =: L_0$ defined by (8.6) as the H_0; then, the likelihood ratio statistic,

$$-2\ln\left(\frac{L_0}{L_1}\right) \overset{\text{as.}}{\sim} \chi^2, \quad \text{with } df = k - p - 1, \tag{8.35}$$

yields an asymptotic test of the LLTM against the empirically more general alternative of an RM.

The justification of this procedure is as follows: since the LLTM restricts the item parameters β_i by means of (8.3), testing the H_0: LLTM against the H_1: RM is a test of a composite hypothesis about the item parameters β_i. Under regularity conditions, such a likelihood ratio statistic is asymptotically χ^2-distributed with df = number of parameters under H_1 minus number of parameters under H_0. The regularity conditions are that (a) the number of parameters remains fixed while $n \to \infty$, and (b) $n_r \to \infty$ for all r. Condition (a) is clearly satisfied due to the CML approach chosen, and (b) is satisfied if (8.27) holds for the sequence of person parameters θ_v.

In applications, these LR tests mostly turn out significant and hence lead to the rejection of the LLTM. Then the question becomes relevant whether the LLTM is still useful even in the presence of a significant LR test: if the formal model of item difficulty specified by an LLTM for a given item universe is at least approximately true, it should be possible to *predict item difficulty for new items of the same universe in new samples of persons.* (Such a prediction refers only to the *relative* item difficulty, $\beta_i - \beta_l$, however, irrespective of the chosen normalization.) Nährer (1977, 1980) showed in an empirical study that such predictions were possible in three item domains. This type of model control is quite useful for validating the cognitive theory underlying a given LLTM and can serve as an alternative to the strict LR test of fit.

Other alternatives are the Lagrange multiplier test allowing one to assess the appropriateness of single elements of $\boldsymbol{W}$, outlined in Chapter 5, or quadratic assignment techniques (cf. Hubert & Schultz, 1976; Hubert, Golledge, Costanzo, Gale, & Halperin, 1984) as recently employed by Medina-Díaz (1993).

8.7 Typical Applications and Some Problems Involved

All early applications of the LLTM focussed on explaining item difficulty in terms of the difficulty of the cognitive operations involved in solving items. Scheiblechner (1972) was the first to apply regression analysis to the Rasch item parameter estimates $\hat{\beta}_i^{(R)}$ obtained in an analysis of the cognitive complexity of logical propositions presented to the persons in graphical form. (Since operations in formal logic are isomorphic to set operations, any logical proposition can be depicted graphically as a sequence of operations applied to sets of points, i.e., as Venn diagrams). Scheiblechner's main result was that in his items difficulty could quite well be explained by means of just three operations ('Negation', 'Disjunction', 'Considering Asymmetry'); however, no rigorous test of fit of the LLTM was available at that time. Fischer (1973) incorporated the linear structure (8.3) in the RM, developed the CML method for the LLTM, and analyzed the difficulty of tasks from calculus in terms of the required rules of differentiation. In the area of science didactics, Spada (1976; see also Spada & Kluwe, 1980; Spada & May, 1982) studied tasks from elementary mechanics, and Kubinger (1979) considered tasks posed to students in a statistics course. Other investigations dealt with reasoning-test items (to name a few: Embretson, 1985; Hornke & Habon, 1986; Gittler & Wild, 1988; Hornke & Rettig, 1988; Smith, Kramer, & Kubiak, 1992). All such studies are limited, of course, to item domains where the researcher is equipped with good hypotheses about the set of relevant cognitive operations for each and every item. Misspecification of the matrix $\boldsymbol{W}$, that is, of the linear system (8.3), obviously leads to systematic errors in the estimates $\hat{\alpha}_l$ and $\hat{\beta}_i$ and may seriously affect the fit of the model.

No systematic studies on the effects of misspecification have been made so far. One recent paper by Baker (1993) is based on some simulations indicating that a small percentage of errors in $\boldsymbol{W}$ is not damaging, whereas a higher percentage is. Although this paper does shed some light on the misspecification problem, its results are not generalizable because (a) the ratio of the number of item parameters to the number of basic parameters, $(k-1)$:p, was kept constant, and (b) no distinction was made between effects on the estimates $\hat{\alpha}_j$ and on the fit of the model. Varying the ratio $(k-1)$:p and studying both the distortion of the $\hat{\alpha}_l$ and effects on the fit seems essential for the following reasons: suppose that $k-1=p$; in that case, the $\hat{\alpha}_l$ are trivial transformations of the $\hat{\beta}_i$, and whatever the 'misspecification', it has no effect on the fit as measured by the LR test. (The LR statistic would trivially be zero.) If $k-1$ is slightly greater than p, a misspecification of even one element of $\boldsymbol{W}$ may affect some $\hat{\alpha}_l$ and the fit statistic severely. If $k-1$ is much greater than p, the system of equations (8.3) is largely overdetermined, and the $\hat{\alpha}_l$ might be rather robust against misspecification of a few elements of $\boldsymbol{W}$; however, the misspecification may still affect the fit for at least some items. Hence, the investigation of the consequences of misspecification of $\boldsymbol{W}$ is rather complex.

Problems entailed by misspecification and lack of fit of the LLTM became apparent even in the early applications; in particular, applications of the LLTM to existing tests did not seem entirely adequate because traditionally tests had not been constructed with a well-defined and limited set of cognitive operations in mind. Test development based on a prespecified set of operations yielding a well-defined universe of items, however, seems to be a promising road to *test construction* in the literal sense of the word.

Formann (1973) constructed a nonverbal matrices test similar to Raven's Progressive Matrices (published later in revised form as 'Wiener Matrizentest',

WMT, by Formann & Piswanger, 1979; see also Fischer & Pendl, 1980, or Fischer & Formann, 1982a). Three factors or facets were defined determining the cognitive operations; one factor (Rule) had the levels Continuation, Variation, and Superimposition; another (Direction) had the levels Horizontal, Vertical, and Horizontal plus Vertical, defining the direction for the application of the relevant rule; the third factor (Graphic Component) characterized the graphical elements of the matrices to which the rule should be applied and to which therefore the persons should pay attention, with the levels Form, Pattern, Number, and Array. These three factors defined a universe of items, of which a sample of 42 was subjected to empirical analysis. Fitting the LLTM was only partially successful, but the LLTM was still found to be an effective tool for improving test construction and analysis. A more recent example demonstrating the usefulness of the LLTM in test construction is that of Gittler (1991, 1992, see also Gittler & Wild, 1988) who developed a new cubes test of space perception; this is one of the very few studies reporting a good fit of the LLTM.

One novel aspect of test theory germane to LLTM analyses is the detection of practice effects *within* testing sessions. Traditional (classical or IRT) methods clearly cannot reveal practice effects within testing occasions, because such effects are invariably confounded with item difficulty. The LLTM, however, is capable, as was shown above, of describing and *predicting* item difficulty on the basis of a formalized theory of the cognitive complexity of items. If items tend to become easier the later they appear in the test, a systematic deviation of the $\hat{\beta}_i^{(\text{LLTM})}$, as compared with the $\hat{\beta}_i^{(R)}$ estimates, should be observable. This deviation can again be modeled within the LLTM, for instance, by introducing a practice effect parameter η and extending (8.3) in the form

$$\beta_i = \sum_l w_{il}\alpha_l + (i-1)\eta + c, \tag{8.36}$$

that is, assuming that solving or trying to solve an item always entails a constant practice effect η. (In (8.36) it is assumed for convenience that the items are presented in the order $I_1, I_2, \ldots, I_k$.) The occurrence of such practice effects was demonstrated, for instance, in the work of Spada (1976), who also considered more complicated models of learning.

These models are limited to learning that occurs independently of the person's response to an item. From learning theory, however, it is well-known that most learning processes are response-contingent; actually solving an item will generally produce a different (i.e., larger) practice effect than trying an item unsuccessfully. At first sight, it seems a simple matter to generalize (8.36) to response contingent learning, leading to a 'dynamic RM' (Fischer, 1972b),

$$\begin{aligned} \beta^*_{v1} &= \beta_1 = 0, \\ \beta^*_{vi} &= \beta_i + \eta\sum_{l=1}^{i-1} x_{vl} + c, \quad \text{for } i = 2, 3, \ldots, k. \end{aligned} \tag{8.37}$$

This model, however, cannot be estimated via CML, because once the coefficients of the basic parameters in the right-hand side of (8.37) are given, the whole response pattern is determined and hence contains no information on the parameters β_i and η. (This elegant argument is due to Verhelst and Glas, 1993). It is interesting to see how one single additional parameter η transforms the RM into a model that is unidentifiable within the CML framework.

Verhelst and Glas (1993; see also Chapter 10) show that this dynamic RM *is* estimable under the MML approach (which, however, extends the model by means of a distributional assumption). Yet there is one more argument that induces us to remain careful: Fischer (1982), using the result of Theorem 8.1 above, showed that the conditional likelihood function (8.6) – which in this case is no longer a proper likelihood function, see the argument of Verhelst and Glas – cannot have a finite maximum, whatever the sample of data $\boldsymbol{X}$. This implies, by virtue of Theorem 8.3, that the unconditional likelihood function has no finite maximum either, which means that the indeterminacy of the dynamic model is not just a consequence of the CML approach, but is a deeper problem of the model structure. If the MML approach *does* yield unique estimates of the β_i and η, the question arises to what extent these are a result of the additional distributional assumptions.

Returning to the more 'classical' applications of the LLTM, it is worth mentioning that the LLTM has added a new facet to cross-cultural research: the comparison of the availability of cognitive functions in different cultures. Piswanger (1975; see also Fischer & Formann, 1982a) gave the WMT matrices test to 2485 Austrian high-school students aged 14 to 18 and to 159 Togolese and 200 Nigerian students of the same ages. The comparison of the basic parameters of all cognitive operations involved (resulting from combinations of factors $\times$ levels) revealed a striking and interpretable difference with regard to the factor Direction: for the Austrian testees it was easiest to apply a rule in the Horizontal Direction (more concretely: from left to right, because the item format requires that the testees fill in the right-hand bottom of the matrices); Horizontal plus Vertical was more difficult, and Vertical slightly more so. For the Nigerians, these differences were less marked, and for the Togolese all directions appeared to be almost equally difficult. These results can be interpreted as the effect of culture: while Austrian children are embedded in a society where reading and writing is exclusively from left to right, the older generations of the two African populations (that is, the parents of the testees) still mostly used Arabic writing (i.e., from right to left). This obviously influenced even the way young students in these cultures look at, and/or process, optical stimuli. Trying to process a matrix item from right to left, however, makes it more difficult due to item format. A recent study investigated the effect of variation of item format in the Persian and Austrian cultures (Mohammadzadeh-Koucheri, 1993). – Some other similar LLTM studies on item bias or cross-cultural differences are due to Whitely and Schneider (1981), Mislevy (1981), and Van de Vijver (1988).

The LLTM requires not only homogeneity of the items in the sense of the RM, but also homogeneity in regard to the cognitive strategies employed by the persons. Heterogeneity in terms of strategies of the persons may be one major cause for lack of fit of the LLTM. (Note, however, that this heterogeneity is at the same time a contra-indication to applying the RM.) Some researchers have therefore sought for ways of circumventing this problem: Mislevy (1988) incorporated a random variation of item parameters β_i with the same generic structure $\boldsymbol{w}_i = (w_{i1}, \ldots, w_{ip})$ by assuming a multivariate normal density $N(\boldsymbol{\alpha}'\boldsymbol{W}', \phi^2\boldsymbol{I})$. This Bayesian approach – whilst interesting in its own right – clearly leaves the Rasch framework, and will therefore not be discussed here. Van Maanen, Been, and Sijtsma (1989), having seen that the LLTM did not fit their balance scale items, split the sample of persons (by means of a cluster method) into four strategy-homogeneous subsamples, and found fit of the LLTM for at least one of the groups. This approach, if systematized, should lead to a mixture distribution model as developed by Rost (1990a); this possibility, as far as the RM is concerned, is explored in Chapter 14.

Other interesting uses of the LLTM are applications to structurally incomplete data. An example of such data is found in connection with cognitive development described by the 'Saltus Model' of Wilson (1989). Fischer (1992) observes that the Saltus Model, in the form as actually applied by Wilson, is an LLTM, so that the identification and estimation problems can be solved by means of the results for the LLTM.

Another extension of the LLTM – or, rather, a reinterpretation – arises in longitudinal designs. As will be shown, applications to repeated measurement designs even allow one to drop the unidimensionality assumption of the RM; such multidimensional models for change will be considered in Chapter 9.

Appendix: Proof of Theorem 8.1

The following proof closely follows Fischer and Tanzer (1994).

Jacobsen (1989) studies likelihood functions of the form

$$L(\boldsymbol{x}_v|\boldsymbol{\alpha}) = \Big(1 + \sum_{\boldsymbol{z}_h \in D_v} \exp[\langle \boldsymbol{t}(\boldsymbol{z}_h) - \boldsymbol{t}(\boldsymbol{x}_v), \boldsymbol{\alpha}\rangle + g(\boldsymbol{z}_h) - g(\boldsymbol{x}_v)]\Big)^{-1}, \tag{8.38}$$

where

$\langle\cdot,\cdot\rangle$ denotes the inner vector product,

$\boldsymbol{\alpha} = (\alpha_1, \ldots, \alpha_p)$ is a vector of parameters,

$\boldsymbol{x}_v$ is a given vector-valued observation, for $v = 1, \ldots, n$ (i.e., the response vector of person S_v with raw score r_v, in our case),

D_v denotes the discrete sample space corresponding to observation $\boldsymbol{x}_v$, excluding the observed x_v (i.e., the set of response vectors $\boldsymbol{z}_v \neq \boldsymbol{x}_v$ with the same design vector $(b_{v1}, \ldots, b_{vk})$ and with the same raw score r_v, in our case),

$\boldsymbol{z}_h = (z_{h1}, \ldots, z_{hk})$ is any point in D_v,

$\boldsymbol{t} = (t_1, \ldots, t_p)$ is a vector of minimal sufficient statistics corresponding to $\boldsymbol{\alpha}$,

g: $D_v \rightarrow \mathbb{R}$ is some real-valued function defined on the sample space D_v.

Jacobsen's (1989) result is the following:

Lemma 8.1 *For a given set of observations $\boldsymbol{x}_v$, $v = 1, \ldots, n$, a unique maximum of the likelihood function $\prod_v L(\boldsymbol{x}_v|\boldsymbol{\alpha})$ exists iff the linear system (8.39),*

$$\langle \boldsymbol{t}(\boldsymbol{z}_h) - \boldsymbol{t}(\boldsymbol{x}_v), \boldsymbol{\mu} \rangle \geq 0, \quad \text{for all } \boldsymbol{z}_h \in D_v \text{ and } v = 1, \ldots, n, \tag{8.39}$$

has only the trivial null-solution $\boldsymbol{\mu} = \boldsymbol{0}$.

We shall apply condition (8.39) to the conditional likelihood (8.19). Notice that the sufficient statistic corresponding to parameter α_l, as defined in (8.38) and computed from a single data point $\boldsymbol{x}_v$ as a special case of (8.20), is $t_l(\boldsymbol{x}_v) = -\sum_{i=1}^{k} x_{vi} b_{vi} w_{il}$. Hence we obtain

$$-\sum_i (z_{hi} - x_{vi}) b_{vi} \sum_l w_{il} \mu_l \geq 0, \tag{8.40}$$

for any $\boldsymbol{z}_h \in D_v$ and all $v = 1, \ldots, n$. This is Jacobsen's (alternative) system (8.39) applied to the LLTM.

Without loss of generality let C have u strong components, $C_1, \ldots, C_u$, and the AS (8.23)–(8.24) have a nontrivial solution $\boldsymbol{y} \neq \boldsymbol{0}$; moreover, let the strong components be ordered such that the d_t in (8.23) satisfy $d_1 \geq d_2 \geq \ldots \geq d_u$, and let the items be ordered such that items $I_1, \ldots, I_{k_1}$ belong to C_1, items $I_{k_1+1}, \ldots, I_{k_2}$ to C_2, $\ldots$, $I_{k_{t-1}+1}, \ldots, I_{k_t}$ to C_t, etc. This implies that no diline $\overrightarrow{V_a V_b}$ exists for $V_a \in C_s$, $V_b \in C_t$, and $s < t$: no person can have a right response to some item I_a belonging to C_s and a wrong response to some item I_b belonging to C_t, for $s < t$. Hence the structure of the data must be as in Table 8.1 (see also Fischer, 1983a). The main diagonal of the super-matrix in Table 8.1 consists of item response matrices; all elements above the main diagonal are matrices comprising only ones or missing responses (the latter denoted by asterisks), and all elements below the main diagonal similarly are matrices consisting of zeros or missing responses. This ordering of strong components, items, and persons automatically defines person groups $M_1, \ldots, M_u$ as shown in Table 8.1.

We shall show that $\boldsymbol{\mu} = -\boldsymbol{y}$ is a solution of (8.40). Inserting $-\boldsymbol{y}$ for $\boldsymbol{\mu}$ in the left-hand side of the inequalities (8.40) yields, for all $S_v \in M_t$,

$$\sum_{j=1}^{u} \sum_{i=k_{j-1}+1}^{k_j} (z_{hi} - x_{vi}) b_{vi} d_j = \sum_{j=1}^{u} \Delta_{vj}(\boldsymbol{z}_h) d_j, \tag{8.41}$$

TABLE 8.1. The structure of ill-conditioned data with u strong components C_l; asterisks denote missing observations, 0/1 item responses.

C_1	.....	C_t	C_{t+1}	.....	C_u	
$1 .... k_1$	.....	$k_{t-1}+1 ... k_t$	$k_t+1...k_{t+1}$	.....	$k_{u-1}+1 ... k$	
	.....	$1\,1\,1\,1\,1\,1*1$	$1*11*11$	.....	$*11111*1$	
$0/1/*$	.....	....	....	.....	....	M_1
	.....	$*11*11*1$	$*111111$	.....	$1**111*$	
$0*00000$	.....	$11*11111$	$111111*$	.....	11111111	
....	.....	....	....	.....	....	M_2
$00000**$	.....	$111111*1$	$1111*11$	.....	$1*111111$	
....	.....	....	....	.....	....	
....	.....	....	....	.....	....	$\vdots$
$000000*$	.....		11111111	.....	$111*111*$	
....	.....	$0/1/*$	....	.....	....	M_t
$00*000*$	.....		11111111	.....	11111111	
....	.....	....	....	.....	....	
....	.....	....	....	.....	....	
....	.....	....	....	.....	....	$\vdots$
....	.....	....	....	.....	....	
....	.....	....	....	.....	....	
....	.....	....	....	.....	....	
$000*000$	.....	$0000*0*0$	00000000	.....		
....	.....	....	....	.....	$0/1/*$	M_u
$0000*00$	.....	00000000	$0*00000*$	.....		

where $k_0 = 0$, $k_u = k$, and $\Delta_{vj}(z_h)$ denotes $\sum_{i=k_{j-1}+1}^{k_j}(z_{hi} - x_{vi})b_{vi}$ for short. From this definition of Δ_{vj} it follows that $\sum_{j=1}^{u} \Delta_{vj}(z_h) = 0$ for any $S_v \in M_t$ and all $z_h \in D_v$. Hence,

$$\sum_{j=1}^{t} \Delta_{vj}(z_h) = - \sum_{j=t+1}^{u} \Delta_{vj}(z_h) \geq 0, \qquad (8.42)$$

for any fixed t with $1 \leq t \leq u$; note that the inequality in (8.42) follows from the definition of M_t in Table 8.1. The assumption $d_1 \geq d_2 \geq \ldots \geq d_u$ together with (8.42) implies that the left hand side of (8.40), with $-y$ inserted for μ, is

$$\sum_{j=1}^{u} \Delta_{vj}(z_h)d_j \quad \geq \quad d_t \sum_{j=1}^{t} \Delta_{vj}(z_h) + d_{t+1} \sum_{j=t+1}^{u} \Delta_{vj}(z_h) =$$

$$= (d_t - d_{t+1}) \sum_{j=1}^{t} \Delta_{vj}(\boldsymbol{z}_h) \geq 0. \tag{8.43}$$

Hence, the vector $-\boldsymbol{y}$ satisfies the Jacobsen system (8.40) too. This means that, if there exists a solution $\boldsymbol{y} \neq \boldsymbol{0}$ to the alternative system (8.23)–(8.24), the Jacobsen system (8.40) has the solution $\boldsymbol{\mu} = -\boldsymbol{y} \neq \boldsymbol{0}$.

The reverse is even easier to prove. Suppose that a solution $\boldsymbol{\mu} \neq \boldsymbol{0}$ of (8.40) exists. Consider any two vertices (items) within the same strong component C_t, V_a and V_b, which are connected by a path $\overrightarrow{V_aV_g}, \overrightarrow{V_gV_f}, \ldots, \overrightarrow{V_pV_b}$, say, whereby a diline $\overrightarrow{V_cV_d}$ corresponds to some person S_v with $x_{vc} = 1$ and $x_{vd} = 0$. Amongst the response vectors $\boldsymbol{z}_h$ with the same raw score r_v, there exists one vector $\boldsymbol{z}_j$ which is identical to $\boldsymbol{x}_v$ except for the elements $z_{jc} = 0$ and $z_{jd} = 1$. Hence, $\overrightarrow{V_cV_d}$ together with (8.40) implies

$$\sum_l w_{cl}\mu_l \geq \sum_l w_{dl}\mu_l,$$

so that the path $\overrightarrow{V_aV_g}, \overrightarrow{V_gV_f}, \ldots, \overrightarrow{V_pV_b}$ implies

$$\sum_l w_{al}\mu_l \geq \sum_l w_{gl}\mu_l \geq \sum_l w_{fl}\mu_l \geq \ldots \geq \sum_l w_{bl}\mu_l. \tag{8.44}$$

But since V_a and V_b belong to the same strong component C_t, it can similarly be concluded that

$$\sum_l w_{al}\mu_l \leq \ldots \leq \sum_l w_{bl}\mu_l, \tag{8.45}$$

so that (8.44) and (8.45) give

$$\sum_l w_{al}\mu_l = \sum_l w_{bl}\mu_l =: d_t, \tag{8.46}$$

for all vertices $V_a, V_b \in C_t$. All equations (8.46) taken together are equivalent to (8.23). – The inequalities (8.24) follow from similar arguments. □

9

Linear Logistic Models for Change

Gerhard H. Fischer[1]

ABSTRACT The linear logistic test model (LLTM) described in Chapter 8 is reinterpretable as a model of change and for measuring treatment effects. Chapter 9 deals with this reinterpretation. Moreover, the LLTM for change can be reparameterized in such a way that – quite unlike the RM – change in multidimensional item domains can be analyzed and monitored. This model is called the 'linear logistic model with relaxed assumptions' (LLRA) because it dispenses with the unidimensionality requirement of the RM. The present chapter discusses properties of these models of change, gives estimation equations, uniqueness results, and hypothesis tests, and mentions typical applications.

9.1 The LLTM for Measuring Change in Unidimensional Traits

In the early seventies, in applications of the linear logistic test model (LLTM) the emphasis was placed on explaining item difficulty in terms of cognitive operations involved. Such applications were restricted, however, to test materials where sound hypotheses on the cognitive structure of the tasks were available; moreover, it was often difficult to fit the LLTM to real data sets. Yet, another, broader class of possible applications emerged in the measurement of change (Fischer, 1972a, 1974, 1976, 1977a,b, 1983a,b, 1989; Fischer & Formann, 1982b). These were based on the central idea that one item, given to the same person S_v at two different time points, T_1, T_2, can be considered as a pair of 'virtual items' (V-items): within an IRT framework of Rasch-type models, any change of θ_v occurring between the testing occasions can be described without loss of generality as a change of the item parameters, instead of describing change in terms of the person parameter. Therefore, one *real* item I_i generates two V-items, denoted, e.g., I_l^* and I_j^*, with associated V-item parameters β_l^* and β_j^*. If the amount of change induced by a treatment B between time points T_1 and T_2 is η, constant over all persons S_v, the pair of V-items generated by real item I_i with parameter β_i is characterized by the V-item parameters $\beta_l^* = \beta_i$ (for T_1) and $\beta_j^* = \beta_i + \eta$ (for T_2). Hence, the V-item parameters can be conceived as weighted sums of (real) item parameters β_i and treatment effect(s) η. The linearly additive composition of the β^*-parameters as functions of item parameters β and treatment effects η makes the model an LLTM.

---

[1]Department of Psychology, University of Vienna, Liebiggasse 5, A-1010 Vienna, Austria; e-mail: GH.FISCHER@UNIVIE.AC.AT

This research was supported in part by the Fonds zur Förderung der Wissenschaftlichen Forschung, Vienna, under Grant No. P10118-HIS.

A general methodology for these kinds of LLTM applications has been presented especially in Fischer (1989), comprising designs with any number of multidimensional item sets (i.e., subdomains of items), any number of time points, and possibly different subsets of items selected for different occasions. An empirical example, employing a more restricted design with two time points and with the same test presented twice, is given by Gittler (1992, 1994): he presented the 3DW (a unidimensional cubes test fitting the Rasch model; cf. Gittler, 1991) to high-school students aged 16 (time T_1), and again at age 18 (time T_2). Between T_1 and T_2, one sample of students (Treatment Group) took courses in mechanical geometry – presumed to enhance the development of spatial ability – while another sample (Control Group) chose some course unrelated to space perception instead (e.g., biology). If it is true that studying mechanical geometry enhances the development of spatial ability, the difference of the item parameters of the same items I_i between T_1 and T_2 should be larger in the Treatment Group than in the Control Group. By means of the LLTM this can be formalized as follows: (a) suppose that the test consists of items $I_1, \dots, I_k$, (b) let the items be unidimensional (Rasch-homogeneous), and (c) assume a trend effect τ (that represents development occurring independently of the treatment of interest plus possible effects of taking the same test twice), a gender effect γ (effect of being female rather than male), a treatment effect η, and a treatment $\times$ gender interaction ρ. Then, the LLTM for change is defined by the matrix $\boldsymbol{W}$ given in Table 9.1.

Note that it is a requirement for any LLTM that the matrix $\boldsymbol{W}$ be given *a priori*, or more precisely, should not depend on the item scores $\boldsymbol{X}$. This would trivially hold true if the students were assigned to the Treatment vs. Control Group as in a laboratory experiment. This obviously is not the case here; students choose one of the courses according to their individual interests, which probably correlate with their spatial ability parameters θ_v. Such a dependence, however, does not *per se* contradict the LLTM assumptions, because it is one major asset of the RM and LLTM that group differences in θ have no systematic influence on the estimates of item or basic parameters (to be more precise: do not preclude consistent estimation of these parameters). So, in the present design, even if there is a covariation of preferences for the mechanical geometry course on the one hand, and scores on the spatial ability test on the other, the amount of change can be estimated and compared between groups as long as the item responses depend on the course preferences only via the θ_v. Although there is no proof that this holds true, such an assumption has some plausibility.

It is readily seen from Table 9.1 that the V-item parameters β_l^*, $l = 1, \dots, 5k$, are linearly additive functions (in fact, simple sums) of $p = k+4$ basic parameters $\beta_1, \dots, \beta_k$, τ, γ, η, ρ. (Note that now the real item parameters β_i of the RM become basic parameters of an LLTM for $5k$ V-items.) The design is *structurally incomplete* because each person, depending on his/her group membership and gender, responds to a subset of the V-items only: in the Control Group, males

TABLE 9.1. Weight matrix **W** for the $5k$ V-item parameters as functions of the basic parameters, for Gittler's (1992, 1994) study; all elements $\neq 1$ are 0. After setting $\beta_1 = 0$ and thus excluding the first column of **W** and $\mathbf{W}^+$, the reduced matrix $\mathbf{W}_R^+$ has rank $r(\mathbf{W}_R^+) = k - 1 + 4 + 1 = p + 1$.

Basic Parameters

V-items	β_1	β_2	$\cdots$	β_{k-1}	β_k	τ	γ	η	ρ		
β_1^*	1									1	Block 1: T_1, All Subjects
β_2^*		1								1	
$\vdots$			$\ddots$							$\vdots$	
β_{k-1}^*				1						1	
β_k^*					1					1	
β_{k+1}^*	1					1				1	Block 2: T_2, Control Group, Males
β_{k+2}^*		1				1				1	
$\vdots$			$\ddots$			$\vdots$				$\vdots$	
β_{2k-1}^*				1		1				1	
β_{2k}^*					1	1				1	
β_{2k+1}^*	1					1	1			1	Block 3: T_2, Control Group, Females
β_{2k+2}^*		1				1	1			1	
$\vdots$			$\ddots$			$\vdots$	$\vdots$			$\vdots$	
β_{3k-1}^*				1		1	1			1	
β_{3k}^*					1	1	1			1	
β_{3k+1}^*	1					1		1		1	Block 4: T_2, Treatment Group, Males
β_{3k+2}^*		1				1		1		1	
$\vdots$			$\ddots$			$\vdots$		$\vdots$		$\vdots$	
β_{4k-1}^*				1		1		1		1	
β_{4k}^*					1	1		1		1	
β_{4k+1}^*	1					1	1	1	1	1	Block 5: T_2, Treatment Group, Females
β_{4k-2}^*		1				1	1	1	1	1	
$\vdots$			$\ddots$			$\vdots$	$\vdots$	$\vdots$	$\vdots$	$\vdots$	
$\vdots$				1		1	1	1	1	1	
β_{5k}^*					1	1	1	1	1	1	

W

$\mathbf{W}^+$

$\mathbf{W}_R^+$

respond to Blocks 1 and 2, and females to Blocks 1 and 3; in the Treatment Group, males respond to Blocks 1 and 4, and females to Blocks 1 and 5.

The application of Corollary 8.1 (Chapter 8) to the weight matrix $\boldsymbol{W}$ in Table 9.1 shows that the $k+4$ basic parameters are not identifiable because (as the reader may wish to verify) $r(\boldsymbol{W}^+) = k+4 = p$, whereas a rank of $k+5 = p+1$ is necessary for uniqueness. This is no disaster, though, because uniqueness of the CML solution can immediately be attained – if the data are well-conditioned – by normalizing the real item parameters β_i by setting $\beta_1^* = \beta_1 = 0$. This is equivalent to deleting the first column of $\boldsymbol{W}^+$, yielding a reduced matrix $\boldsymbol{W}_R^+$; this implies a reduction of the number of basic parameters to $p = (k-1)+4 = k+3$, so that $p+1 = k+4$, while $r(\boldsymbol{W}_R^+)$ is still $k+4$.

The LLTM not only allows one to estimate the parameters via the CML approach (for the estimation equations, see Chapter 8), but also to test hypotheses on the effect parameters by means of conditional LR tests: let the H_0 be $\gamma = \eta = \rho = 0$, that is, hypothesize that there exist no causes of change other than ('natural') development, here represented by τ. Then, the LLTM can be estimated with the last three columns of $\boldsymbol{W}$ set to zero, yielding the conditional likelihood L_0, which is compared to L_1 (under the full model that is assumed to be true) by means of the statistic

$$-2(\ln L_0 - \ln L_1) \overset{\text{as.}}{\sim} \chi^2,$$

with $df = (k-1)+4-((k-1)+1) = 3$. (On conditional LR tests in general, see Andersen, 1973a,b,c, and Chapter 5; on LR tests in the LLTM, see Chapter 8.)

This approach to estimation and to testing hypotheses about the 4 effect parameters, τ, γ, η, ρ, could, in principle, be impaired by the presence of the k item parameters β_i which enter the conditional likelihood. The presence of the former may bias the results about the latter. Two ways to circumvent this problem suggest themselves:[2]

(a) If data of a – hopefully much larger – previous calibration study are available, this sample could be added to the Block-1-data, because the calibration sample responded to V-items $I_1^*, \ldots, I_k^*$ (identical to real items $I_1, \ldots, I_k$, see Table 9.1). These additional data would help to stabilize the $\hat{\beta}_i$ and thereby increase the precision of the effect parameter estimates.

(b) The β_i may be eliminated altogether by conditioning out both the β_i and the θ_v, so that the conditional likelihood becomes a function of the 4 effect parameters only. This approach is nearer to Rasch's (1967, 1968, 1972, 1977) methodological postulate of *specific objectivity*, that is, the demand that the result should depend only on the parameters of current interest. (On specific

[2]It is an interesting observation that the bias induced by the β_i in this design is actually minimal; all three estimation methods give practically the same results about the effect parameters.

objectivity, see also Chapter 2.) This leads to the models described in Section 9.2.

9.2 The LLRA for Two Time Points

In order to eliminate the item parameters β_i along with the θ_v, it is convenient to introduce new parameters $\theta_v - \beta_i =: \theta_{vi}$. We shall treat these θ_{vi} as *independent* trait parameters without any assumptions about the values they take, and whether or not they are actually independent or somehow mutually dependent. (This can be done without loss of generality because the θ_{vi} are conditioned out of the likelihood, so their mutual dependence or independence becomes irrelevant.) Each person S_v is now characterized by a vector $\boldsymbol{\theta}_v = (\theta_{v1}, ..., \theta_{vk})$. Hence, replacing $\theta_v - \beta_i$ by θ_{vi} is a *generalization* of the model rather than a reparameterization. For one person S_v and a given pair of items (I_i, I_l), we now may have $\theta_{vi} > \theta_{vl}$, but for another S_w, we may have $\theta_{wi} < \theta_{wl}$. This means that the items no longer need to be unidimensional (but may still be so). So we arrive at a class of more general IRT models for change where the typical unidimensionality restriction of IRT models is abandoned. The model discussed in this section has therefore been called the 'linear logistic model with relaxed assumptions' (LLRA; Fischer, 1977a).

For two time points, T_1 and T_2, let

$$P(X_{vi1} = 1|S_v, I_i, T_1) = \frac{\exp(\theta_{vi})}{1 + \exp(\theta_{vi})}, \tag{9.1}$$

where θ_{vi} is S_v's position on that ability or trait which accounts for S_v's success or failure on item I_i, at time point T_1; similarly, let

$$P(X_{vi2} = 1|S_v, I_i, T_2) = \frac{\exp(\theta'_{vi})}{1 + \exp(\theta'_{vi})} = \frac{\exp(\theta_{vi} + \delta_v)}{1 + \exp(\theta_{vi} + \delta_v)}, \tag{9.2}$$

where $\delta_v = \theta'_{vi} - \theta_{vi}$, for $i = 1, \ldots, k$, denotes the amount of change of S_v between T_1 and T_2. Parameter δ_v is assumed to depend on S_v only via the individual treatment (combination), and will, for the moment, be assumed to be constant across items. Moreover, the model decomposes δ_v into treatment effects, treatment interactions, and a trend parameter,

$$\delta_v = \sum_{j=1}^{m} q_{vj}\eta_j + \tau + \sum_{j<l} q_{vj}q_{vl}\rho_{jl}, \tag{9.3}$$

where

q_{vj} is the (given) dosage of treatment B_j applied to S_v between T_1 and T_2,

η_j is the effect of one dosage unit of B_j,

TABLE 9.2. Repeated response matrices $\mathbf{X}_1$ and $\mathbf{X}_2$ of a sample of n persons to k items (or symptoms). The entries are x_{vit} for $v = 1, \dots, n$ persons, $i = 1, \dots, k$ items, and $t = 1, 2$ time points.

Real Persons	Time Point T_1 Matrix $\mathbf{X}_1$				Time Point T_2 Matrix $\mathbf{X}_2$			
S_1	x_{111}	x_{121}	......	x_{1k1}	x_{112}	x_{122}	......	x_{1k2}
S_2	x_{211}	x_{221}	......	x_{2k1}	x_{212}	x_{222}	......	x_{2k2}
$\vdots$	...	...	......	...	...	...	......	...
S_n	x_{n11}	x_{n21}	......	x_{nk1}	x_{n12}	x_{n22}	......	x_{nk2}

τ is a 'trend' effect, i.e., the combined effect of all causes of change that are unrelated to the treatment(s), and

ρ_{jl} is the interaction effect of $B_j \times B_l$.

Note what the *multidimensional* parameterization of individual differences in this model means: to each and every item, one separate latent dimension D_i is attached. Change is modeled as a migration of a person along all dimensions by the same amount in the same direction. Even so, the multidimensionality of the traits is of great advantage in many domains of application: suppose, for instance, that the items are clinical symptoms, such as sleep disorders, dizziness, lack of energy, impairment of concentration, disorders of thinking, anxiety and other emotional disturbances, lack of social contacts, etc. As is well-known from clinical psychology, patients may be differently prone to display any of these symptoms, that is, they may exhibit individual response patterns even if they are afflicted by similar disorders of their neuro-transmitter systems, and react positively to the same medical treatment. Therefore, it does make sense to assign different latent traits to each of the items (i.e., proneness to particular symptoms), yet at the same time to characterize the treatments by uniform effect parameters η_j (and possibly interaction parameters ρ_{jl}). It should be borne in mind, however, that no assumptions are made, nor are such needed, about the mutual dependence/independence of these traits: they may be completely independent, or statistically or functionally related, or even identical.

For a formal treatment of the LLRA, note that (9.3) is linearly additive in all basic parameters η_j, τ, ρ_{jl}, so that (9.3) can be rewritten for simplicity as

$$\delta_v = \sum_{j=1}^{m} q_{vj}\eta_j, \tag{9.4}$$

where η is now a generic symbol for all effect parameters, and where the weights q_{vj} and the number of effect parameters, m, are redefined appropriately. It will be assumed throughout that the dosages q_{vj} are given constants and the effects

TABLE 9.3. Rearrangement of the data of Table 9.2 as *one* incomplete data matrix $\mathbf{X}^*$, for $2k$ virtual persons and items; '–' denotes missing observations.

	Time Points					
Virtual	T_1		T_2			
Persons	V-item		V-items			
	I_0^*	I_1^*	I_2^*	...		I_n^*
S_{11}^*	x_{111}	x_{112}	–	–	–	–
S_{12}^*	x_{121}	x_{122}	–	–	–	–
$\vdots$	$\vdots$	$\vdots$	$\vdots$	$\vdots$	$\vdots$	$\vdots$
S_{1k}^*	x_{1k1}	x_{1k2}	–	–	–	–
S_{21}^*	x_{211}	–	x_{212}	–	–	–
S_{22}^*	x_{221}	–	x_{222}	–	–	–
$\vdots$	$\vdots$	$\vdots$	$\vdots$	$\vdots$	$\vdots$	$\vdots$
S_{2k}^*	x_{2k1}	–	x_{2k2}	–	–	–
$\vdots$	$\vdots$	$\vdots$	$\vdots$	$\vdots$	$\vdots$	$\vdots$
S_{n1}^*	x_{n11}	–	–	–	–	x_{n12}
S_{n2}^*	x_{n21}	–	–	–	–	x_{n22}
$\vdots$	$\vdots$	$\vdots$	$\vdots$	$\vdots$	$\vdots$	$\vdots$
S_{nk}^*	x_{nk1}	–	–	–	–	x_{nk2}

η_j are finite. The observations consist of two corresponding data matrices $\boldsymbol{X}_1$ (for T_1) and $\boldsymbol{X}_2$ (for T_2), see Table 9.2.

It is easy to show that the LLRA, (9.1), (9.2), and (9.4), is equivalent to an LLTM with two items per person. To see this, consider again the concept of 'virtual (V-) items' and, similarly, of 'virtual (V-)persons': due to the local independence assumption plus the multidimensional parameterization, the two responses of S_v to one item I_i, on the one hand, and the two responses of S_v to another item I_l, on the other, could as well stem from different persons S_{vi}^* and S_{vl}^*, with parameters θ_{vi} and θ_{vl}. Except for the treatment effect parameters, there is nothing at all in the model that relates these two pairs of responses to each other. Hence, the data matrices $\boldsymbol{X}_1$ and $\boldsymbol{X}_2$ can be rearranged as in Table 9.3, yielding one incomplete data matrix $\boldsymbol{X}^*$, where each V-person S_{vi}^* responds to the same V-item I_0^* (with parameter $\beta_0^* = 0$, see probability (9.1)) and to just one other V-item I_v^* (with parameter $\beta_v^* = \delta_v = \sum_{j=1}^m q_{vj}\eta_j$, see (9.2) and (9.4)). The number of virtual items is equal to the number of persons plus 1, $n+1$ (if each person receives a different combination of dosages $\times$ treatments), or otherwise equal to the number of treatment groups plus 1. Person S_1 is now replaced by k V-persons $S_{11}^*, \ldots, S_{1k}^*$, S_2 by $S_{21}^*, \ldots, S_{2k}^*$, etc., and S_n by $S_{n1}^* \ldots, S_{nk}^*$, with corresponding parameters θ_{vi}. This reinterpretation of the data in $\boldsymbol{X}^*$ shows that the LLRA is formally an LLTM with incomplete data, namely, responses to two V-items per V-person.

This model was first suggested in slightly different form – for survey data in opinion research rather than repeated measurement (panel) designs – by Fischer (1972, 1974, 1977b); the requirement that the same persons be observed or tested twice can be dropped if homogeneous strata of the population are defined by means of socio-ecomomic and/or socio-cultural variables, assuming that all persons within the same stratum are characterized by the same latent trait (e.g., attitude) parameters. This can be justified, at least as an approximation, for the area of opinion or market research. In the present chapter, however, we presume that we can test all persons twice, so that individuals are the equivalent of strata.

It is obvious that the LLRA can be generalized to longitudinal designs with more than two time points, $T_1, \ldots, T_t, \ldots, T_s$. One real item I_i then generates a set of V-items I^*_{it}, $t = 1, \ldots, s$, with response probabilities

$$P(X_{vit} = 1|S_v, I_i, T_t) = \frac{\exp(\theta_{vi} + \sum_j q_{vtj}\eta_j)}{1 + \exp(\theta_{vi} + \sum_j q_{vtj}\eta_j)}. \tag{9.5}$$

The q_{vtj} are now the dosages of the treatments received up to time point T_t, i.e., all $q_{v1j} = 0$ for $t = 1$. If the trend effect is $\tau = \eta_m$, say, then $q_{vtm} = T_t - T_1$ for all S_v and all T_t.

The V-items generated by real item I_i, i.e., $I^*_{i1}, \ldots, I^*_{is}$, are now items of an LLTM with test length s. To different real items I_i there correspond V-persons with different parameters θ_{vi} and possibly different raw scores $r_{vi} = \sum_t x_{vit}$. The model can be estimated within the CML framework by conditioning on the r_{vi}. This kind of LLTM application, and some still more general designs, have been described by Fischer (1989); see also Embretson (1991). In what follows, we only discuss designs with two time points.

9.3 CML Estimation in the LLRA

The formal equivalence of the LLRA with an LLTM allows us to apply all results given in Chapter 8 directly to the LLRA. In particular, we begin by specializing the CML estimation equations (8.11) to the present test length $k = 2$,

$$\sum_{i=1}^{2} w_{il} \left[x_{.i} - n_1 \frac{\gamma_0^{(i)} \epsilon_i}{\gamma_1} - n_2 \frac{\gamma_1^{(i)} \epsilon_i}{\gamma_2} \right] = 0,$$

for $l = 1, \ldots, m$ (effect parameters). Since $\gamma_0^{(i)} = 1$ and $\gamma_1^{(i)} \epsilon_i = \gamma_2$, we get

$$\sum_{i=1}^{2} w_{il} \left[x_{.i} - n_1 \frac{\epsilon_i}{\gamma_1} - n_2 \right] = 0.$$

Let n_{10} and n_{01} be the numbers of persons with response patterns (1,0) and (0,1), respectively, then $x_{.1} - n_2 = n_{10}$, $x_{.2} - n_2 = n_{01}$, $n_1 = n_{01} + n_{10}$, and the CML equations become

$$w_{1l}\left[n_{10} - (n_{10}+n_{01})\frac{\epsilon_1}{\epsilon_1+\epsilon_2}\right] + w_{2l}\left[n_{01} - (n_{10}+n_{01})\frac{\epsilon_2}{\epsilon_1+\epsilon_2}\right] = 0. \quad (9.6)$$

Now apply this to all pairs of V-items of the LLRA as defined by Table 9.3. Each V-person S^*_{vi} responds to $k^*_v = 2$ V-items with parameters $\epsilon_1 = \exp(0)$ and $\epsilon_2 = \exp(\delta_v) = \exp(\sum_j q_{vj}\eta_j)$, respectively, and the weights are $w_{1l} = 0$ and $w_{2l} = q_{vl}$. Thus, the m estimation equations are[3]

$$\sum_v q_{vl} \sum_i (x_{vi1} - x_{vi2})^2 \left[x_{vi2} - \frac{\exp(\sum_j q_{vj}\eta_j)}{1+\exp(\sum_j q_{vj}\eta_j)}\right] = 0, \quad (9.7)$$

for $l = 1, \ldots, m$, where the term $(x_{vi1} - x_{vi2})^2$ is instrumental for selecting the pairs of responses with $x_{vi1} + x_{vi2} = 1$ and deleting all pairs with $x_{vi1} + x_{vi2} = 0$ or $= 2$. (Alternatively, these equations can be derived by explicitly writing down the conditional log-likelihood function $\ln L_C$,

$$\ln L_C = \sum_v \sum_i (x_{vi1} - x_{vi2})^2 \left[x_{vi2} \sum_j q_{vj}\eta_j - \ln\Big(1 + \exp\big(\sum_j q_{vj}\eta_j\big)\Big)\right], \quad (9.8)$$

differentiating with respect to each η_l, and setting the partial derivatives to zero. The left-hand terms in (9.7) are these partial derivatives.)

The estimation equations are remarkably simple because $k^*_{vi} = 2$ for all S^*_{vi}: it is readily seen that (9.7) are the well-known unconditional estimation equations for a logit model where the dependent dichotomous responses are the response patterns (1,0) vs. (0,1), for all combinations of (real) persons $S_v \times$ (real) items I_i. Equations (9.7) are solved conveniently by means of the Newton-Raphson method. The required second-order partial derivatives (the negative elements of the information matrix $\boldsymbol{I}$) are easily obtained by differentiating the left-hand side of (9.7), with $l = a$, with respect to η_b, which yields

$$\frac{\partial^2 \ln L_C}{\partial\eta_a \partial\eta_b} = -\sum_v q_{va}q_{vb} \sum_i (x_{vi1} - x_{vi2})^2 \frac{\exp(\sum_j q_{vj}\eta_j)}{\left[1+\exp(\sum_j q_{vj}\eta_j)\right]^2}, \quad (9.9)$$

for $a, b = 1, \ldots, m$. For the Newton-Raphson procedure, these derivatives are evaluated in each step at the last (approximate) estimate of $\boldsymbol{\eta}$. Very little computing time per iteration and just a few iterations are usually needed to obtain the CML solution $\hat{\boldsymbol{\eta}}$. Note that no test length, however large, can create problems, because technically each V-person S^*_{vi} responds only to $k^*_{vi} = 2$ V-items.

---

[3]The CML equations were given in Fischer (1977a, 1983b) in different, but equivalent, form.

The existence and uniqueness questions regarding the solution $\hat{\boldsymbol{\eta}}$ of (9.7) can be answered by applying Theorem 8.1. For the present situation of $k_{vi}^* = 2$ items, the following special case obtains: let

M_1 be the set of pairs of V-items (I_0^*, I_l^*) (i.e., treatment combinations) where some response sequence (0,1) and some (1,0) have been observed,

M_2 the set of pairs of V-items (I_0^*, I_l^*) (i.e., treatment combinations) where some response sequence (0,1), but no (1,0) have been observed, and

M_3 the set of pairs of V-items (I_0^*, I_l^*) (i.e., treatment combinations) where some response sequence (1,0), but no (0,1) have been observed.

Theorem 9.1 *The CML equations (9.7) of the LLRA have a unique finite solution $\hat{\boldsymbol{\eta}}$ iff the ('alternative') system of linear inequalities (9.10),*

$$\begin{aligned} \textstyle\sum_j q_{aj} y_j &= 0, \quad \text{for all } (I_0^*, I_a^*) \in M_1, \\ \textstyle\sum_j q_{bj} y_j &\geq 0, \quad \text{for all } (I_0^*, I_b^*) \in M_2, \\ \textstyle\sum_j q_{cj} y_j &\leq 0, \quad \text{for all } (I_0^*, I_c^*) \in M_3, \end{aligned} \tag{9.10}$$

possesses only the null-solution $\boldsymbol{y} = \boldsymbol{0}$.

The equivalent result for the logit model was first given by Haberman (1977), see also Schumacher (1980), or Jacobsen (1989), who generalized the result. A proof of Theorem 9.1 using Theorem 8.1 is found in Fischer (1983a).

For most practical purposes in applied research, it suffices to consider special cases of Theorem 9.1. Suppose the design comprises a set of treatment groups, $G = \{G_1, \ldots, G_s\}$, such that the matrix $\boldsymbol{Q}_G = ((q_{gj}))$, $g = 1, \ldots, s$; $j = 1, \ldots, m$, $s \geq m$, has rank $r(\boldsymbol{Q}_G) = m$. The design may, of course, contain the 'null-treatment' of a control group as one of the s treatment combinations. If, within each of these treatment groups G_g, at least one (0,1) response sequence has been observed for some (real) item I_{i_g}, and (1,0) for some (real) item I_{h_g}, implying that the corresponding V-items all belong to M_1, the homogeneous set of equations as part of the alternative system (AS) (9.10),

$$\sum_j q_{gj} y_j = 0, \qquad \text{for } g = 1, \ldots, s \text{ and } j = 1, \ldots, m,$$

has only the null-solution $\boldsymbol{y} = \boldsymbol{0}$, and hence the LLRA yields a unique CML solution $\hat{\boldsymbol{\eta}}$. We formulate this conclusion in Corollary 9.1.

Corollary 9.1 *Let the design comprise a set of treatment groups, $G = \{G_1, \ldots, G_s\}$, such that the dosage matrix $\boldsymbol{Q}_G = ((q_{gj}))$, $g = 1, \ldots, s$; $j = 1, \ldots, m$, with $s \geq m$, has rank $r(\boldsymbol{Q}_G) = m$, where m is the number of effect parameters $\eta_1, \ldots, \eta_m$. The CML equations (9.7) of the LLRA have a unique finite solution $\hat{\boldsymbol{\eta}}$ if, within each treatment group G_g, at least one response sequence (0,1) has been observed for some (real) item I_{i_g}, and at least one response sequence (1,0) for some (real) item I_{h_g}.*

9.4 Some Properties of the LLRA

The uniqueness results of Section 9.3 will help us to draw conclusions about the scale properties of measurement of change within the LLRA framework: if the dosage matrix $\boldsymbol{Q}$ is considered given, the η_j are unique under realistic assumptions about the data. Suppose first, for simplicity, that the treatments are two courses and that each person participates in either of them, so that all $q_{vj} \in \{0, 1\}$. 'Treatment B_j' means 'participation in course B_j throughout the observation period (T_1, T_2)'. If the uniqueness conditions of Theorem 9.1 or Corollary 9.1 are satisfied, the CML equations of the LLRA will yield unique numbers $\hat{\eta}_1$ and $\hat{\eta}_2$. This should not mislead one to believe, however, that the η-scale is an absolute scale: recall that, in the RM, the θ-scale is an *interval scale* (cf. Chapter 2); since essentially the η_j are differences $\theta'_{vi} - \theta_{vi}$, see (9.1)–(9.2), it follows that the η_j lie on a *ratio scale.*

The conclusion about the ratio scale properties of the η_j is supported by two more arguments. First, let there exist one more treatment, B_3, with effect parameter η_3; then the latter can be compared to the joint effect of B_1 and B_2 if both are applied simultaneously or successively, i.e., compared to $\eta_1 + \eta_2$. Combining two treatments and comparing them to a third one corresponds to the *concatenation operation* in fundamental measurement, which is an essential basis of ratio scales (Krantz, Luce, Suppes, & Tversky, 1971, Vol. 1, p. 2). Second, the postulate of specific objectivity applied to the comparison of treatment effects, combined with the principle of generalizability of effects over items, can be employed to justify the choice of a logistic link function for models of change with additive treatment effect parameters. Within the scope of this chapter, we only can give a sketch of these arguments, referring the reader interested in further details to Fischer (1987a).

Suppose that the effect of a treatment B can be fully characterized by one scalar parameter α, and that the state of a person at different time points T_t is fully characterized by a latent trait parameter ξ_t, the person's position on the trait. If the change induced by B is 'lawful', there should exist a function F that transforms the person's original state ξ into a new state ξ',

$$\xi' = F(\xi, \alpha), \tag{9.11}$$

where F: $\mathbb{R}^2 \to \mathbb{R}$ is assumed to be continuous and strictly increasing in both arguments. F is called the 'effect function'.

Our basic methodological postulate – essentially that of specific objectivity (SO) applied to the measurement of treatment effects – is the following: if treatment B with effect α is applied to two persons, S_v and S_w with respective initial states ξ_v and ξ_w, then a scientifically meaningful comparison of S_v and S_w, carried out by means of a function U, should yield the same outcome at both time points,

$$U(F(\xi_v, \alpha), F(\xi_w, \alpha)) = U(\xi_v, \xi_w), \tag{9.12}$$

independent of the size of the treatment effect α. The function U: $\mathbb{R}^2 \to \mathbb{R}$, used as a tool to compare the persons, is called a 'comparator function'. U is assumed to be continuous, strictly increasing in ξ_v, and strictly decreasing in ξ_w. (We assume F and U to be defined on the whole real axis.)

It can be shown (Fischer, 1987a) that under these assumptions (9.12) is equivalent to the following assumption: for any two jointly or subsequently applied treatments B_1 and B_2 with parameters α_1 and α_2, given to a person with state parameter ξ, there exists a continuous function $H(\alpha_1, \alpha_2)$, independent of ξ and strictly monotone in each argument, such that

$$\alpha_{12} = H(\alpha_1, \alpha_2), \tag{9.13}$$

where α_{12} denotes the combined effect of B_1 and B_2. So, (9.13) is an alternative formulation of SO as applied to the measurement of treatment effects. Assumption (9.13) may be interpreted as a *concatenation operation* for treatment effects and as such has important consequences for the measurement properties of the effect parameters α.

The central result derivable from (9.12) or (9.13) is then the following:

Theorem 9.2 *There exist continuous, bijective transformations k: $\xi \to \theta$ and m: $\alpha \to \eta$, and continuous, bijective functions f, u, and h, all $\mathbb{R} \to \mathbb{R}$, such that, for all $\xi, \alpha \in \mathbb{R}$,*

$$F(\xi, \alpha) = f(\theta + \eta), \tag{9.14}$$

$$U(\xi_v, \xi_w) = u(\theta_v - \theta_w), \tag{9.15}$$

$$H(\alpha_1, \alpha_2) = h(\eta_1 + \eta_2). \tag{9.16}$$

Proof The proof is similar to that of Lemma 2.4; see Fischer (1987a).

The simple additive and subtractive relations (9.14)–(9.16) imply that $\tilde{F} = \theta + \eta$ is a reaction function equivalent to F, $\tilde{U} = \theta' - \theta = \eta$ is a comparator equivalent to U, and $\tilde{H} = \eta_1 + \eta_2 =: \eta_{12}$ characterizes the concatenation of two treatments.

This result is a correlative of Lemma 2.4 in Chapter 2. By essentially the same arguments as in Section 2.3, it can further be concluded that, within a probabilistic framework, the link function is the logistic function in (9.1) and that the measurement of treatment effects occurs on a *ratio scale*. (This comes as no surprise, because the θ-parameters are measurable on an interval scale.) Note, however, that these arguments are valid only if no quantitative dosages $\neq 0$ and $\neq 1$ are involved.

Matters become more complicated if the dosages q_{vj} are allowed to be expressed in arbitrary units, such as 'mg of drug per day' or 'number of psychotherapy sessions within (T_1, T_2)'. Then it is obvious that the quotients $\eta_j : \eta_l$ are no longer invariant under changes of the dosage units, unless the units for

B_j and B_l happen to be the same (e.g., two medications expressed in mg). The interpretation of the η_j as ratio-scaled quantities is then limited to statements like '20 mg of drug A per day throughout (T_1, T_2) is x times as effective as two psychotherapy sessions per week throughout (T_1, T_2)'. Moreover, the usage of weights q_{vj} expressed as multiples of arbitrary units implies that a proportionality of effects and dosages holds; this assumption needs to be tested, though, see Section 9.5.

Regarding the asymptotic properties of the CML estimator $\hat{\boldsymbol{\eta}}$, we state the following theorem:

Theorem 9.3 *Let all η_j be finite. The CML estimator $\hat{\boldsymbol{\eta}}$ of the LLRA (9.1), (9.2), and (9.4) is asymptotically multivariate normal around $\boldsymbol{\eta}$ with asymptotic covariance matrix $\boldsymbol{I}^{-1}$ if (a) there exists a set of treatment groups $G = \{G_1, \ldots, G_s\}$ with $r(\boldsymbol{Q}_G) = m$, where the group sizes $n_g \to \infty$ for $g = 1, \ldots, s$, and (b) within each treatment group, the sequence of person parameters θ_{vi} satisfies the condition*

$$\sum_{v=1}^{n} \sum_{i=1}^{k} \exp(-|\theta_{vi}|) = \infty. \tag{9.17}$$

Note that (9.17) is condition (8.27) applied to the θ_{vi}. We refrain from giving a proof of this theorem here because the conditional likelihood function and CML equations (9.7) can be considered those of a logit model, which has been treated extensively in the statistical literature (cf. Cox, 1970, p. 88; Haberman, 1978, Vol. 1, p. 340). Large-sample properties have been discussed both for grouped and ungrouped data. A proof under the weaker assumption that, as $n \to \infty$, the distribution of the dosage vectors converges to a distribution with non-singular covariance matrix, can be found in Haberman (1974, pp. 358–369).

One may wonder whether large-sample properties are relevant in the present context, because sample sizes in treatment groups are mostly small or moderate. However, the *effective* sample sizes in the conditional LLRA are the numbers of item-score combinations (x_{vi1}, x_{vi2}) with $x_{vi1} + x_{vi2} = 1$, for persons $S_v \in G_g$; let these numbers be denoted N_g in contrast to the treatment group sizes n_g. Even if the treatment groups are of moderate sizes, the N_g become large whenever the test is of sufficient length k; for instance, if $n_g = 30$ and $k = 30$, N_g will be of the order of $n_g k/2 = 450$ (assuming all response probabilities to be near .50), which is sufficiently large for applying the asymptotic results.

9.5 Inference in the LLRA

The main goal of applying the LLRA is to test hypotheses on treatment effects (or other effect parameters). A very general approach to constructing a test statistic is the conditional likelihood ratio (LR) method, which compares the conditional likelihood of the data under an H_0 with that under an H_1 believed to be true: let H_0 be defined by subjecting the m effect parameters η_j to certain restrictions, $\eta_j = \mu_j(\alpha_1, \ldots, \alpha_{m'})$, where the μ_j are functions of a smaller number m' of parameters α_j, defined such that the reduction to the α_j is unique. For simplicity, we shall restrict our considerations to the simple case that the η_j are linear functions of the α_l,

$$\boldsymbol{\eta} = \boldsymbol{M}\boldsymbol{\alpha}, \tag{9.18}$$

where $\boldsymbol{M}$ is an $m \times m'$ matrix of rank m', with $m > m'$. Such linear hypotheses suffice for most practical purposes. The rank condition assures that (9.18) is uniquely solvable for $\boldsymbol{\alpha}$ if $\boldsymbol{\eta}$ is known. The α_l can then be called the 'basic' parameters of the LLRA. The conditional LR statistic for testing H_0 against H_1 is

$$-2\ln\lambda = -2\ln\left\{\frac{L(\boldsymbol{X}_1, \boldsymbol{X}_2|\boldsymbol{X}_1 + \boldsymbol{X}_2; H_0)}{L(\boldsymbol{X}_1, \boldsymbol{X}_2|\boldsymbol{X}_1 + \boldsymbol{X}_2; H_1)}\right\} \overset{\text{as.}}{\sim} \chi^2, \tag{9.19}$$

with $df = m - m'$. For a proof of the asymptotic distribution of (9.19), see Section 9.7 (Appendix). Some typical hypotheses testable by means of this LR statistic are listed below.

(i) *'No change'*, H_0: $\eta_j = \tau = \rho_{jl} = 0$ for all treatments B_j, B_l. In most applications, this hypothesis is rather implausible (and thus not very interesting), because change, e.g., in developmental studies, is likely to occur due to natural development of the children, or in treatment studies some change will be induced by the treatments.

(ii) *'No trend'*, H_0: $\tau = 0$. In developmental studies, τ is usually the largest and statistically most significant effect parameter – as one would expect anyway. Ascertaining that $\tau \neq 0$ is therefore no great feat. In clinical studies, however, it may be anything but trivial to determine whether or not the patients' state of health (or impairment) tends to improve or deteriorate if no treatment is given.

(iii) *'No interaction effects'*, H_0: $\rho_{jl} = 0$ for some or all pairs (B_j, B_l). Interactions between treatments is what researchers will expect in many cases: a medicinal treatment may be the prerequisite for an effective psychotherapy of depressive patients, information an essential requirement for effectively changing an attitude, Course A a necessity for being able to successfully take Course B, etc. Such dependencies between treatments can be expressed in terms of interaction effects ρ_{jl}.

(iv) *'Ineffective treatments'*, H_0: $\eta_j = 0$ for some or all B_j. Typical examples are, e.g., studies on the effect of a psychotherapy, η_2, in addition to a conventional medicinal treatment, η_1, e.g., when $\eta_1 \neq 0$ is known.

(v) *'Equality of treatment effects'*, H_0: $\eta_j = \eta_l = \eta_k = \ldots$ for some (sub)set of treatments. Typical examples of this H_0 arise in educational studies when two or more teaching methods are compared.

(vi) *'Generalizability over items (or symptoms)'*, H_0: $\eta_j^{(a)} = \eta_j^{(b)} = \eta_j^{(c)} = \ldots$ for some or all treatments B_j, where $\hat{\eta}_j^{(a)}, \hat{\eta}_j^{(b)}, \hat{\eta}_j^{(c)}, \ldots$ are effect parameters estimated from subsets of items (or symptoms), $J_a, J_b, J_c, \ldots$ For instance, in the treatment of psychosis a typical question would be whether certain treatment(s) are equally effective with respect to affective symptoms, J_a, cognitive disorders, J_b, social relations, J_c, etc. The analysis of this kind of generalizability amounts to investigating *interactions* between treatments and item domains, including problems like learning transfer. If this test turns out to be significant, one may estimate and interpret the effect parameters independently for each item domain J_h, $h = 1, \ldots, d$, or alternatively model the differences between domains by means of domain-specific effect parameters α_h, subject to the normalization condition $\sum_h \alpha_h = 0$, so that δ_v in (9.3) becomes $\delta_{vh} = \sum_j q_{vj}\eta_j + \alpha_h + \tau + \sum_{j<l} q_{vj}q_{vl}\rho_{jl}$.

(vii) *'Generalizability over person groups'*, H_0: $\eta_j^{(s)} = \eta_j^{(t)} = \eta_j^{(u)} = \ldots$ for some or all treatments B_j, where $P_s, P_t, P_u, \ldots$ are subgroups or subpopulations of persons. This H_0 is analogous to (vi) and can be reformulated as a question about interactions between treatments and persons (or person groups). Subjects are often grouped by age, sex, or overall score in a test, and treatment effects are compared between groups.

(viii) *'Proportionality of dosages and effects'*. To test this H_0, an H_1 has to be formulated allowing for independent effects of the same treatment if applied with different dosages: let the dosages be restricted to a manageable (small) number of discrete levels, $q_{vj} \in \{q_j^{(1)}, \ldots, q_j^{(u_j)}\}$, define a design vector $\boldsymbol{b}_{vj}$ per person S_v × treatment B_j, with elements $b_{vj}^{(l)} = 1$ if S_v has received treatment B_j with dosage $q_j^{(l)}$, and $b_{vj}^{(l)} = 0$ otherwise; and introduce one effect parameter $\eta_j^{(l)}$ per treatment B_j × dosage $q_j^{(l)}$. Then, H_1 is formulated by means of the equation $\delta_v = \sum_j \sum_l b_{vj}^{(l)} \eta_j^{(l)}$, while H_0 is still given by (9.4). Since under H_0 the parameters $\eta_j^{(l)}$ satisfy the simple relations $\eta_j^{(l)} = q_j^{(l)}\eta_j$, which are of the form (9.18), H_0 can again be tested against H_1 by means of the conditional LR statistic (9.19).

In the applications known to the present author, this particular H_0 has been tested only once: Heckl (1976) reports a study where speech-handicapped children were treated (= trained in speech and verbal abilities) once per fortnight, or once a week, or twice a week. The proportionality hypothesis had to be rejected. Treating the patients once per week was considerably more effective than the fortnightly treatment, but only slightly inferior to two treatment sessions per week.

The attractiveness of the asymptotic conditional LR tests lies in their universality and their simplicity once the models under consideration have been estimated via CML; the log-likelihood function at $\hat{\boldsymbol{\eta}}$ is a byproduct of the estimation of the model, and no further computations are needed for the LR test. This is not to say, however, that there are no competitors: sometimes hypotheses are even more easily tested by means of Wald-type statistics outlined in Chapter 5, especially if they involve only one parameter, such as in hypothesis (ii) above; then it suffices to compute the estimate of the standard error of the respective parameter, based on the inverse of the information matrix $\hat{\boldsymbol{I}}$, that is, on matrix (9.9) evaluated at the CML estimate $\hat{\boldsymbol{\eta}}$,

$$\hat{\sigma}(\hat{\eta}_j) = \sqrt{i^{jj}}, \tag{9.20}$$

with i^{jj} for the j-th diagonal element of $\hat{\boldsymbol{I}}^{-1}$ (corresponding to parameter η_j). Thus, the asymptotically standard normal statistic for testing H_0: $\eta_j = a$ is

$$z = \frac{\hat{\eta}_j - a}{\hat{\sigma}(\hat{\eta}_j)}. \tag{9.21}$$

More general Wald tests are described in some detail in Chapter 5.

The researcher applying both the LR statistic and the Wald-type statistic to the same H_0 will sometimes be faced with the disconcerting fact that their results differ, especially in cases of moderate or small samples. This is due to both tests being valid asymptotically, but not holding exactly in limited samples. In large samples, however, accordance of both procedures is to be expected except for rare cases where one test statistic just barely reaches or exceeds the critical value, while the other barely fails to reach that value.

An overview of many typical LLRA applications is given in Fischer (1991a). They range from assessing effects of early education on intelligence (Rop, 1977; Zeman, 1976) over those of employee-training seminars on behavior and personality (e.g., Schmied, 1987; Iby, 1987; Koschier, 1993), of group therapy on self-perception (Sommer, 1979; Witek, 1979), of communication on attitude (Hammer, 1978; Kropiunigg, 1979a,b; Barisch, 1989), to various forms of psychotherapy intended as a support for medicinal treatments (Zimprich, 1980; Widowitz, 1987; Mutschlechner, 1987; Doff, 1992), or other more specific forms of psychological interventions (Heckl, 1976; Rella, 1976; Glatz, 1977) – to mention some of the more important areas of application.

A warning is in order, however, concerning the practical application of tests (i) through (vii) as they were carried out in many cases: firstly, the asymptotic distribution of (9.19) is derived under the presupposition that H_1 is true. Researchers employing log-linear models in similar situations choose the saturated hypothesis as a H_1, which is trivially true; nested null-hypotheses are tested against this H_1 in hierarchical order. In the present model, however, a truly saturated hypothesis would often have too many parameters and thus might seem too complicated. Practical applications have therefore mostly used 'quasi-saturated' H_1 hypotheses that seemed complex and plausible enough to the researcher, even if they did not fit perfectly. The effect of this compromise on the results is not known. Secondly, the hypothesis tests (i) through (vii), if based on the same set of data, are mutually dependent. Again, the consequences of this are difficult to appraise.

9.6 An LLRA for the Measurement of Environmental Effects

An old problem for which no convincing solution yet exists is the nature–nurture question: to what extent are traits, like intelligence or personality factors, the result of genetic vs. environmental influences? One traditional approach is to test monozygotic twins reared apart (MZTA) and to estimate the *heritability*, h^2; however, that concept is open to severe criticism (cf. Layzer, 1974; Feldman & Lewontin, 1975; Jaspars & de Leeuw, 1980; Fischer, 1987b, 1993a). We shall show that the LLRA offers an interesting alternative methodology which, however, rests on a fundamentally different formulation of the research question. What are the effects of certain types of environment, or of well-defined environmental conditions, on the development of intelligence or personality traits? How generalizable are such environmental effects?

We start out from the assumption that a number of different *types* of environment, $U_1, \ldots, U_m$, have been defined, based on observable socio-economic and socio-cultural variables. (The definition of such types requires meaningful hypotheses and the observation of relevant variables, but lies outside the scope of our present considerations. Whenever one refers to environmental effects, however, one must have specific kinds or types of environment in mind that may or may not differ with respect to their effects on the studied behaviors.) Alternatively, types of environment may be replaced by any combinations of environmental conditions, such as income, level of schooling, profession of the foster parents, conditions of housing, etc. To each type (or environmental condition) U_j, one effect parameter η_j will be attached characterizing the effect of U_j on those latent traits θ_i that are at the base of the behaviors of interest.

'Behaviors of interest' will be operationalized by means of a set of items I_i, $i = 1, \ldots, k$, assumed to be chosen by the researcher according to his/her experience and hypotheses; for instance, the items may be tasks considered appropriate for monitoring the cognitive development of pre-school children. These items may again be multidimensional (see Section 9.2 on the multidimensionality of the LLRA).

The unobservable genetic endowment of a child S_v will be characterized, as far as the given behavioral domain is concerned, by a parameter vector $\boldsymbol{\theta}_v = (\theta_{v1}, \ldots, \theta_{vk})$, so that – as in the LLRA for measuring treatment effects –

parameters θ_{vi} are assigned 1–1 to items I_i within persons. Since monozygotic twins may be assumed to be genetically identical, one parameter $\boldsymbol{\theta}_v$ now characterizes a *pair of twins* rather than a single individual. We therefore index the pairs of twins, henceforth denoted $\mathcal{P}_v$, with index 'v', and assign parameters $\boldsymbol{\theta}_v$ to them. The two twins of pair $\mathcal{P}_v$ will be denoted $\mathcal{T}_{v1}$ and $\mathcal{T}_{v2}$, respectively, whereby the order of $\mathcal{T}_{v1}$ and $\mathcal{T}_{v2}$ is arbitrary.

The model states that

$$P(+|\mathcal{P}_v, I_i, \mathcal{T}_{v1}) = \frac{\exp[\theta_{vi} + \sum_j q_{v1j}\eta_j]}{1 + \exp[\theta_{vi} + \sum_j q_{v1j}\eta_j]}$$

$$= \frac{\exp(\mu_{vi})}{1 + \exp(\mu_{vi})}, \tag{9.22}$$

$$P(+|\mathcal{P}_v, I_i, \mathcal{T}_{v2}) = \frac{\exp[\theta_{vi} + \sum_j q_{v2j}\eta_j]}{1 + \exp[\theta_{vi} + \sum_j q_{v2j}\eta_j]}$$

$$= \frac{\exp[\mu_{vi} + \sum_j (q_{v2j} - q_{v1j})\eta_j]}{1 + \exp[\mu_{vi} + \sum_j (q_{v2j} - q_{v1j})\eta_j]}, \tag{9.23}$$

where

q_{v1j} and q_{v2j} are observed weights specifying the types of environments or combinations of environmental conditions prevailing for $\mathcal{T}_1$ and $\mathcal{T}_2$, respectively; the row vector with elements $q_{v2j} - q_{v1j}$, $j = 1, \ldots, m$, will be denoted by $\boldsymbol{q}_{v2} - \boldsymbol{q}_{v1}$;

θ_{vi} and η_j are genetic and environmental effects, respectively, and

$\mu_{vi} = \theta_{vi} + \sum_j q_{v1j}\eta_j$.

It is completely arbitrary which one of the twins of pair $\mathcal{P}_v$ is denoted $\mathcal{T}_{v1}$, and which $\mathcal{T}_{v2}$: although interchanging $\mathcal{T}_{v1}$ and $\mathcal{T}_{v2}$ changes the definition of μ_{vi}, this is automatically compensated by a change of the sign of $\sum_j (q_{v2j} - q_{v1j})\eta_j$. Note that the μ_{vi} in (9.22) and (9.23) can formally be considered as 'person parameters' in an LLRA with 'dosages' $q_{v2j} - q_{v1j}$ and 'treatment' effects η_j; 'persons' now correspond to pairs of twins, $\mathcal{P}_v$, 'dosages' are the *differences* of weights of environmental conditions between the two twins within pair $\mathcal{P}_v$, and 'treatment' is an environment U_j (or an environmental condition).

The present definition of dosage of environment U_j has an interesting consequence: suppose a constant $c \neq 0$ is added to all η_j, and $c\sum_j q_{v1j}$ is subtracted from all θ_{vi}. Then the probabilities (9.22) and (9.23) remain constant iff

$$\sum_j (q_{v1j} - q_{v2j}) = 0, \quad \text{for all } \mathcal{P}_v, \tag{9.24}$$

that is, whenever the sums of environmental dosages are the same for both twins in all pairs $\mathcal{P}_v$. This will hold, e.g., in the simple case where each individual is exposed to only one environment, or if the dosages are defined in terms of years spent in different environments. It is concluded that, if (9.24) holds, the environmental effects η_j are defined at most up to an additive constant. Since the dosages can also be multiplied by an arbitrary constant (e.g., dosages defined in terms of months rather than years), we conclude that, if (9.24) holds, the environmental effects are measurable on an *interval scale*.[4] In this (typical) case, one environment U_r has to be defined as the *reference environment*, with $\eta_r = 0$ for normalization; the unit of measurement of the effect parameters is defined via the dosage unit.

Although at first sight this loss of scale level (as compared to typical LLRA applications where ratio scale measurement is achieved) seems deplorable, it can easily be seen that the change of the scale properties is a logical consequence of the underlying problem structure: if, within each pair $\mathcal{P}_v$, both twins are exposed to their environments by the same amount, there is no basis at all for statements like 'one environment is twice as effective as another', since no child can be observed without any environmental influence. *The parameters η_j are therefore meaningful only in relations to some standard or reference environment.* This has an important consequence especially for the interpretation of the θ_{vi}: for any two pairs of twins, $\mathcal{P}_v$ and $\mathcal{P}_w$, the differences $\theta_{vi} - \theta_{wi}$ need no longer be invariant under admissible normalizations of the η_j, because for $\eta_j \to \eta_j + c$ one has $\theta_{vi} - \theta_{wi} \to \theta_{vi} - \theta_{wi} - c(\sum_j q_{v1j} - \sum_j q_{w1j})$. Hence, even if the θ_{vi} could be estimated for each pair $\mathcal{P}_v$,[5] the differences $\theta_{vi} - \theta_{wi}$ would not yield a basis for comparing the genetic endowment of different pairs of twins, unless $\sum_j (q_{v1j} - q_{w1j}) = 0$ for all $\mathcal{P}_v$. This is logical, however, because we cannot hope to determine the difference between two genetic endowments (with regard to the ability measured for item I_i) if we can only measure *differences* between environments, but do not have access to absolute measures of environmental effects – given that the genetic information in the chromosomes of the children is unobservable. We may even view these consequences as a kind of intrinsic validation of our model construction: whenever a model has been designed in a meaningful manner, the formal properties of the parameters will be in accordance with their substantive interpretation.

If, however, certain environmental conditions, such as preschool education programs, can be *manipulated experimentally*, it is possible to observe children with and without their influence, resulting in a higher (namely: ratio) scale level for the respective effect parameters. These conclusions about the scale level of environmental effects will be made more rigorous by means of Theorem 9.4

[4]The same conclusion results from another argument: any $a \neq 1$, $a > 0$, could be chosen in (2.22), see Chapter 2, i.e., another member of the family of Rasch models could be taken as the basis of the analysis.

[5]This is not the case in the multidimensional LLRA with only one item per latent dimension D_i; for estimating the θ_{vi}, several items per D_i would be required.

below.

CML estimation of the environmental effect parameters η_j can proceed in a straightforward manner by the method outlined in Section 9.3. Theorem 9.1 and Corollary 9.1 yield uniqueness results for the CML estimator $\hat{\boldsymbol{\eta}}$. The definition of the sets M_1, M_2, M_3, however, requires some extra care because of the arbitrariness in denoting either twin of pair $\mathcal{P}_v$ as $\mathcal{T}_{v1}$ or $\mathcal{T}_{v2}$, respectively. The 'treatment combinations' in Theorem 9.1 now become *combinations of environmental conditions* occurring within pairs of twins, characterized by vectors $\boldsymbol{q}_{v2} - \boldsymbol{q}_{v1}$. To remove the consequences entailed by the above-mentioned arbitrariness, we proceed as follows: let '$\prec$' denote an order relation for vectors; whenever there exist two pairs of twins $\mathcal{P}_v$ and $\mathcal{P}_w$ such that $\boldsymbol{q}_{v2} - \boldsymbol{q}_{v1} = -(\boldsymbol{q}_{w2} - \boldsymbol{q}_{w1})$ and $\boldsymbol{q}_{v2} - \boldsymbol{q}_{v1} \prec \boldsymbol{q}_{w2} - \boldsymbol{q}_{w1}$, then interchange $\mathcal{T}_{w1}$ and $\mathcal{T}_{w2}$. This has the consequence that no two pairs $\mathcal{P}_v$ and $\mathcal{P}_w$ with $\boldsymbol{q}_{v2} - \boldsymbol{q}_{v1} = -(\boldsymbol{q}_{w2} - \boldsymbol{q}_{w1})$ will remain. Now we can define the sets M_1, M_2, M_3.

Let M_1 be the set of all combinations of environmental conditions, i.e., of vectors $\boldsymbol{q}_{u2} - \boldsymbol{q}_{u1}$, for which the following conditions (a) and (b) hold true: (a) there exists a pair of twins $\mathcal{P}_a$ with vector $\boldsymbol{q}_{a2} - \boldsymbol{q}_{a1} = \boldsymbol{q}_{u2} - \boldsymbol{q}_{u1}$ and an item I_{i_a} such that the respective responses of $(\mathcal{T}_{a1}, \mathcal{T}_{a2})$ to I_{i_a} are (1,0); (b) there exists a pair of twins $\mathcal{P}_b$ with vector $\boldsymbol{q}_{b2} - \boldsymbol{q}_{b1} = \boldsymbol{q}_{u2} - \boldsymbol{q}_{u1}$ and an item I_{i_b} such that the respective responses of $(\mathcal{T}_{b1}, \mathcal{T}_{b2})$ are (0,1).

Similarly, let M_2 be the set of all vectors $\boldsymbol{q}_{u2} - \boldsymbol{q}_{u1}$ for which the following conditions (a) and (b) hold true: (a) there exists a pair of twins $\mathcal{P}_c$ with vector $\boldsymbol{q}_{c2} - \boldsymbol{q}_{c1} = \boldsymbol{q}_{u2} - \boldsymbol{q}_{u1}$ and an item I_{i_c} such that the responses of $(\mathcal{T}_{c1}, \mathcal{T}_{c2})$ to I_{i_c} are (1,0); (b) no pair of twins $\mathcal{P}_d$ exists with vector $\boldsymbol{q}_{d2} - \boldsymbol{q}_{d1} = \boldsymbol{q}_{u2} - \boldsymbol{q}_{u1}$ such that, for some item I_{i_d}, the respective responses of $(\mathcal{T}_{d1}, \mathcal{T}_{d2})$ are (0,1).

Finally, let M_3 be the set of all vectors $\boldsymbol{q}_{u2} - \boldsymbol{q}_{u1}$ for which the following conditions (a) and (b) hold true: (a) there exists a pair of twins $\mathcal{P}_e$ with vector $\boldsymbol{q}_{e2} - \boldsymbol{q}_{e1} = \boldsymbol{q}_{u2} - \boldsymbol{q}_{u1}$ and an item I_{i_e} such that the responses of $(\mathcal{T}_{e1}, \mathcal{T}_{e2})$ to I_{i_e} are (0,1); (b) no pair of twins $\mathcal{P}_f$ exists with vector $\boldsymbol{q}_{f2} - \boldsymbol{q}_{f1} = \boldsymbol{q}_{u2} - \boldsymbol{q}_{u1}$ such that, for some item I_{i_f}, the respective responses of $(\mathcal{T}_{f1}, \mathcal{T}_{f2})$ are (1,0).

The allocation of vectors $\boldsymbol{q}_{u2} - \boldsymbol{q}_{u1}$ to sets M_1, M_2, M_3 is easily done by computer, in samples with only few environments, even by hand. Note that each vector $\boldsymbol{q}_{u2} - \boldsymbol{q}_{u1}$ is element of at most one of these sets. If a vector, say $\boldsymbol{q}_{w2} - \boldsymbol{q}_{w1}$, does not belong to any of the three sets, this means that, in all pairs of twins with that vector, both twins $\mathcal{T}_{w1}$ and $\mathcal{T}_{w2}$ have given the same responses and thus are uninformative with respect to CML estimation. The data of these pairs of twins can be discarded. Such cases will rarely occur, though, if the test is of appropriate difficulty and long enough.

Based on the sets M_1, M_2, M_3, the uniqueness result can now be stated.

Theorem 9.4 *The CML equations (9.7) for the LLRA as applied to model (9.22)–(9.23), with dosages $q_{vj} = q_{v2j} - q_{v1j}$, have a unique solution iff the* *('alternative') system of linear inequalities (9.25),*

$$\begin{aligned} \langle \boldsymbol{q}_{a2} - \boldsymbol{q}_{a1}, \boldsymbol{y} \rangle &= 0, \quad \textit{for all } (\boldsymbol{q}_{a2} - \boldsymbol{q}_{a1}) \in M_1, \\ \langle \boldsymbol{q}_{b2} - \boldsymbol{q}_{b1}, \boldsymbol{y} \rangle &\geq 0, \quad \textit{for all } (\boldsymbol{q}_{b2} - \boldsymbol{q}_{b1}) \in M_2, \\ \langle \boldsymbol{q}_{c2} - \boldsymbol{q}_{c1}, \boldsymbol{y} \rangle &\leq 0, \quad \textit{for all } (\boldsymbol{q}_{c2} - \boldsymbol{q}_{c1}) \in M_3, \end{aligned} \tag{9.25}$$

where $\langle \cdot, \cdot \rangle$ denotes the inner vector product, has only the null-solution $\boldsymbol{y} = \boldsymbol{0}$.

Proof The result follows directly from Theorem 9.1. Alternatively, it can be deduced from Theorem 8.2 for the LBTL. □

For most practical purposes, it will suffice to consider special cases of Theorem 9.4: let the matrices with elements q_{a1j} and q_{a2j} with $(\boldsymbol{q}_{a2} - \boldsymbol{q}_{a1}) \in M_1$ be denoted $\boldsymbol{Q}_1^*$ and $\boldsymbol{Q}_2^*$, respectively. For the first set of equations in (9.25), $(\boldsymbol{Q}_2^* - \boldsymbol{Q}_1^*)\boldsymbol{y} = \boldsymbol{0}$, it follows that $\boldsymbol{y} = \boldsymbol{0}$ if the rank $r(\boldsymbol{Q}_2^* - \boldsymbol{Q}_1^*)$ equals m, the number of environments or environmental conditions. So we have the following sufficient condition for uniqueness of the CML solution:

Corollary 9.2 *The CML equations for model (9.22)–(9.23) have a unique solution if $r(\boldsymbol{Q}_2^* - \boldsymbol{Q}_1^*) = m$.*

This sufficient result is simple and should be useful in most applications. Set M_1 can be expected to comprise most pairs of twins if the test length is appropriate, such that, in most $\mathcal{P}_v$, twin $\mathcal{T}_{v1}$ has scored better than $\mathcal{T}_{v2}$ on some item(s), and *vice versa* on some other item(s). Then, if the rank condition $r(\boldsymbol{Q}_2^* - \boldsymbol{Q}_1^*) = m$ is satisfied, that is, if there is a sufficient variation of environmental conditions within M_1, the normalized CML solution is unique.

If, in particular, (9.24) holds, the rank of the matrix of the entire system (9.25) is $r(\boldsymbol{Q}_2 - \boldsymbol{Q}_1) \leq m - 1$, and hence there always exists a non-null solution of (9.25), i.e., $\boldsymbol{y}' = (y, y, \ldots, y)$ with any $y \neq 0$. This again illustrates what we already know: if (9.24) holds, there can be no unique solution unless the environmental effects are normalized. In that case the question is, does there exist a unique normalized CML solution $\hat{\boldsymbol{\eta}}$ with, e.g., $\eta_1 = 0$? Considering the meaning of $\boldsymbol{y}$ in (9.25) as a transformation $\boldsymbol{\eta} \to \boldsymbol{\eta} + \boldsymbol{y}$ under which the conditional likelihood function is nondecreasing (Fischer, 1983a), it can be shown that under assumption (9.24) Theorem 9.4 yields the following result:

Theorem 9.5 *Let (9.24) hold, and let the environmental effects be normalized by setting $\eta_1 = 0$. Then the CML equations for the LLRA as applied to model (9.22)–(9.23), with dosages $q_{vj} = q_{v2j} - q_{v1j}$, have a unique normalized solution $\hat{\boldsymbol{\eta}}$ iff system (9.25) has no non-null solution $\boldsymbol{y}$ satisfying $y_1 = 0$.*

An even simpler case arises if each twin has grown up in just one environment. Then all q_{v1j} and q_{v2j} are either 1 or 0, $\sum_j q_{v1j} = \sum_j q_{v2j} = 1$ for all $\mathcal{P}_v$, and there is a 1–1 correspondence between the $\hat{\eta}_j$ and the environments U_j. For all pairs of environments (U_j, U_l), for which a pair of twins $\mathcal{P}_v$ exists with $q_{v2j} = q_{v1l} = 1$, (9.25) implies

$$\begin{aligned} y_j = y_l \quad &\text{if } (\boldsymbol{q}_{v2} - \boldsymbol{q}_{v1}) \in M_1, \\ y_j \geq y_l \quad &\text{if } (\boldsymbol{q}_{v2} - \boldsymbol{q}_{v1}) \in M_2, \\ y_j \leq y_l \quad &\text{if } (\boldsymbol{q}_{v2} - \boldsymbol{q}_{v1}) \in M_3. \end{aligned} \tag{9.26}$$

The $\geq$ and $\leq$ relations in (9.26) exactly correspond to the direction of dilines in a directed graph G defined as follows: assign vertices V_j 1–1 to environments U_j, insert a diline $\overrightarrow{V_jV_l}$ if there exists a pair of twins $\mathcal{P}_v$ such that $q_{v1j} = q_{v2l} = 1$ and where, for some item I_i, $\mathcal{T}_{v1}$ has $x_{vi1} = 1$, and $\mathcal{T}_{v2}$ has $x_{vi2} = 0$; analogously, insert a diline $\overrightarrow{V_lV_j}$ if $x_{vh1} = 0$ and $x_{vh2} = 1$ for some item I_h. Hence, (9.26) will have only the solution $\boldsymbol{y} = (y, y, \ldots, y)$, which for $y_1 = 0$ is the null-solution, iff graph G is strongly connected. This yields the following corollary:

Corollary 9.3 *Suppose that each twin has grown up in only one environment and that effect parameters η_j are assigned 1–1 to the environments. Then the CML equations of model (9.22)–(9.23) have a unique normalized solution $\hat{\boldsymbol{\eta}}$ with $\hat{\eta}_1 = 0$ iff graph $\boldsymbol{G}$ is strongly connected.*

The latter result can alternatively be deduced from Corollary 9.3 derived for the RM or from the analogous theorem for the BTL (Zermelo, 1929; Ford, 1957; Harary, Norman, & Cartwright, 1965); see also Fischer (1987b, 1993) where this and related results are given.

When deducing conclusions about the scale properties of the $\hat{\eta}_j$ from Theorems 9.4 and 9.5, and Corollaries 9.2 and 9.3, it has to be borne in mind that 'unique' means 'unique given the respective model'. From Chapter 2, however, we know that the choice of $a = 1$ in (2.22) is arbitrary and thus can be replaced by any other value $a > 0$. Parameter estimates $\hat{\eta}_j$ determined uniquely according to the above Theorems and Corollaries therefore lie on ratio scales (rather than absolute scales). This, moreover, implies that parameter estimates $\hat{\eta}_j$ that are unique only after a normalization (e.g., $\eta_1 = 0$) are measures on an interval scale. This tallies with our earlier conclusions.

9.7 Appendix: Derivation of the Asymptotic Distribution of the Conditional LR Statistic (9.19)

Given that under regularity conditions (Theorem 9.3) the estimator $\hat{\boldsymbol{\eta}}$ is asymptotically multivariate normal around $\boldsymbol{\eta}$ with nonsingular asymptotic covariance matrix $\boldsymbol{I}^{-1}$, we shall show that the LR statistic (9.19) is asymptotically χ^2-distributed with $df = m - m'$. For simplicity, we treat the case that $k = 1$ and that there are m person groups of equal sizes, $N_g = N/m$, with $\sum_g N_g = N$. (This actually entails no loss of generality because, if the N_g were different, some observations could be discarded in order to make all N_g equal.)

The asymptotic normality of $\hat{\boldsymbol{\eta}}$ implies that the asymptotic density of $\hat{\boldsymbol{\eta}}$ under H_1 is

$$L_1 = C \exp\left[-\frac{1}{2}(\hat{\boldsymbol{\eta}} - \boldsymbol{\eta})' \boldsymbol{I} (\hat{\boldsymbol{\eta}} - \boldsymbol{\eta})\right],$$

and therefore

$$\ln L_1 = -\frac{1}{2}\sqrt{N}\,(\hat{\boldsymbol{\eta}} - \boldsymbol{\eta})' \frac{\boldsymbol{I}}{N} \sqrt{N}(\hat{\boldsymbol{\eta}} - \boldsymbol{\eta}) + \ln C. \tag{9.27}$$

Now let some notation be introduced:

$$\boldsymbol{d} = \sqrt{N}(\hat{\boldsymbol{\eta}} - \boldsymbol{\eta}), \quad \boldsymbol{\mathcal{I}} = \frac{\boldsymbol{I}}{N}, \quad \boldsymbol{v}' = \frac{1}{\sqrt{N}}\left(\frac{\partial \ln L_1}{\partial \eta_1}, \ldots, \frac{\partial \ln L_1}{\partial \eta_m}\right).$$

Then, (9.27) becomes

$$\ln L_1 = -\frac{1}{2}\boldsymbol{d}'\boldsymbol{\mathcal{I}}\boldsymbol{d} + \ln C. \tag{9.28}$$

Differentiating (9.28) with respect to $\boldsymbol{\eta}$ (using the rule $\frac{\partial}{\partial \boldsymbol{x}}(\boldsymbol{x}'\boldsymbol{A}\boldsymbol{x}) = 2\boldsymbol{A}\boldsymbol{x}$, cf. Rao, 1965, p. 71) yields

$$\boldsymbol{v} = \boldsymbol{\mathcal{I}}\boldsymbol{d}, \quad \text{or} \quad \boldsymbol{d} = \boldsymbol{\mathcal{I}}^{-1}\boldsymbol{v}. \tag{9.29}$$

Similarly, the density of $\hat{\alpha}$ under H_0 is

$$\ln L_0 = -\frac{1}{2}\sqrt{N}(\hat{\boldsymbol{\alpha}} - \boldsymbol{\alpha})' \frac{\boldsymbol{J}}{N} \sqrt{N}(\hat{\boldsymbol{\alpha}} - \boldsymbol{\alpha}) + \ln C_0, \tag{9.30}$$

where $\boldsymbol{J}$ is the respective $m' \times m'$ information matrix. Let some more notation be introduced:

$$\boldsymbol{f} = \sqrt{N}(\hat{\boldsymbol{\alpha}} - \boldsymbol{\alpha}), \quad \boldsymbol{\mathcal{J}} = \frac{\boldsymbol{J}}{N}, \quad \boldsymbol{w}' = \frac{1}{\sqrt{N}}\left(\frac{\partial \ln L_0}{\partial \alpha_1}, \ldots, \frac{\partial \ln L_0}{\partial \alpha_{m'}}\right).$$

Then, (9.30) becomes

$$\ln L_0 = -\frac{1}{2}\boldsymbol{f}'\boldsymbol{\mathcal{J}}\boldsymbol{f} + \ln C_0. \tag{9.31}$$

Differentiating (9.31) with respect to $\boldsymbol{\alpha}$ gives

$$\boldsymbol{w} = \boldsymbol{\mathcal{J}}\boldsymbol{f}, \quad \text{or} \quad \boldsymbol{f} = \boldsymbol{\mathcal{J}}^{-1}\boldsymbol{w}. \tag{9.32}$$

Then, from (9.28),

$$2[\ln L_1(\hat{\boldsymbol{\eta}}) - \ln L_1(\boldsymbol{\eta})] = \boldsymbol{d}'\boldsymbol{\mathcal{I}}\boldsymbol{d}, \tag{9.33}$$

and from (9.31),

$$2[\ln L_0(\hat{\boldsymbol{\alpha}}) - \ln L_0(\boldsymbol{\alpha})] = \boldsymbol{f}'\boldsymbol{\mathcal{J}}\boldsymbol{f}. \tag{9.34}$$

Assume that H_0: $\boldsymbol{\eta} = \boldsymbol{M\alpha}$ is true, so that $\ln L_1(\boldsymbol{\eta}) = \ln L_0(\boldsymbol{\alpha})$; differentiating of $\ln L_0$ with respect to $\boldsymbol{\alpha}$ gives, by means of the chain rule $\partial \ln L_0/\partial\boldsymbol{\alpha} = (\partial \ln L_0/\partial\boldsymbol{\eta})(\partial\boldsymbol{\eta}/\partial\boldsymbol{\alpha})$,

$$\boldsymbol{w}' = \boldsymbol{v}'\boldsymbol{M}. \tag{9.35}$$

Finally, differentiating (9.35) once more with respect to $\boldsymbol{\alpha}$, again by means of the chain rule, yields

$$\boldsymbol{\mathcal{J}} = \boldsymbol{M}'\boldsymbol{\mathcal{I}}\boldsymbol{M}. \tag{9.36}$$

Inserting (9.29) in (9.33),

$$2[\ln L_1(\hat{\boldsymbol{\eta}}) - \ln L_1(\boldsymbol{\eta})] = \boldsymbol{v}'\boldsymbol{\mathcal{I}}^{-1}\boldsymbol{\mathcal{I}}\boldsymbol{\mathcal{I}}^{-1}\boldsymbol{v} = \boldsymbol{v}'\boldsymbol{\mathcal{I}}^{-1}\boldsymbol{v}, \tag{9.37}$$

and similarly inserting (9.32) in (9.34), and using (9.35), yields

$$2[\ln L_0(\hat{\boldsymbol{\alpha}}) - \ln L_0(\boldsymbol{\alpha})] = \boldsymbol{w}'\boldsymbol{\mathcal{J}}^{-1}\boldsymbol{\mathcal{J}}\boldsymbol{\mathcal{J}}^{-1}\boldsymbol{w} = \boldsymbol{v}'\boldsymbol{M}\boldsymbol{\mathcal{J}}^{-1}\boldsymbol{M}'\boldsymbol{v}. \tag{9.38}$$

Combining (9.37) and (9.38), and assuming that $\ln L_1(\boldsymbol{\eta}) = \ln L_0(\boldsymbol{\alpha})$, i.e., that H_0 is true, the likelihood-ratio statistic becomes

$$-2\ln\lambda = \boldsymbol{v}'\boldsymbol{\mathcal{I}}^{-1}\boldsymbol{v} - \boldsymbol{v}'\boldsymbol{M}\boldsymbol{\mathcal{J}}^{-1}\boldsymbol{M}'\boldsymbol{v} = \boldsymbol{v}'(\boldsymbol{\mathcal{I}}^{-1} - \boldsymbol{M}\boldsymbol{\mathcal{J}}^{-1}\boldsymbol{M}')\boldsymbol{v}. \tag{9.39}$$

To show that (9.39) has an asymptotic χ^2-distribution with $m - m'$ degrees of freedom if H_0 is true, a result by Ogasawara and Takahashi (1951; see Rao, 1973, p. 188) turns out to be useful:

Lemma 9.1 *Let $\boldsymbol{Y}$ be a vector-valued random variable with multinormal distribution $N(\boldsymbol{\mu}, \boldsymbol{\Sigma})$. A necessary and sufficient condition for the quadratic form $(\boldsymbol{Y} - \boldsymbol{\mu})'\boldsymbol{A}(\boldsymbol{Y} - \boldsymbol{\mu})$ to have a χ^2-distribution is $\boldsymbol{\Sigma A \Sigma A \Sigma} = \boldsymbol{\Sigma A \Sigma}$, in which case the degrees of freedom equals the rank of $\boldsymbol{A\Sigma}$.*

Applying this condition to (9.39), thereby taking into account that by (9.29) the covariance matrix $\boldsymbol{\Sigma}$ of $\boldsymbol{v}$ in (9.39) is $\boldsymbol{\mathcal{I}}\boldsymbol{\mathcal{I}}^{-1}\boldsymbol{\mathcal{I}} = \boldsymbol{\mathcal{I}}$, and using (9.36), it is seen that

$$\begin{aligned} \boldsymbol{A\Sigma A} &= (\boldsymbol{\mathcal{I}}^{-1} - \boldsymbol{M}\boldsymbol{\mathcal{J}}^{-1}\boldsymbol{M}')\boldsymbol{\mathcal{I}}(\boldsymbol{\mathcal{I}}^{-1} - \boldsymbol{M}\boldsymbol{\mathcal{J}}^{-1}\boldsymbol{M}') = \\ \boldsymbol{\mathcal{I}}^{-1} &- \boldsymbol{M}\boldsymbol{\mathcal{J}}^{-1}\boldsymbol{M}' - \boldsymbol{M}\boldsymbol{\mathcal{J}}^{-1}\boldsymbol{M}' + \boldsymbol{M}\boldsymbol{\mathcal{J}}^{-1}\boldsymbol{M}'\boldsymbol{\mathcal{I}}\boldsymbol{M}\boldsymbol{\mathcal{J}}^{-1}\boldsymbol{M}' = \\ \boldsymbol{\mathcal{I}}^{-1} &- 2\boldsymbol{M}\boldsymbol{\mathcal{J}}^{-1}\boldsymbol{M}' + \boldsymbol{M}\boldsymbol{\mathcal{J}}^{-1}\boldsymbol{\mathcal{J}}\boldsymbol{\mathcal{J}}^{-1}\boldsymbol{M}' = \boldsymbol{\mathcal{I}}^{-1} - \boldsymbol{M}\boldsymbol{\mathcal{J}}^{-1}\boldsymbol{M}' = \boldsymbol{A}. \end{aligned} \tag{9.40}$$

Hence, $-2\ln\lambda$ is asymptotically χ^2-distributed. Equation (9.40), $\boldsymbol{A\Sigma A} = \boldsymbol{A}$, moreover means that the matrix $\boldsymbol{A\Sigma A\Sigma} = \boldsymbol{A\Sigma}$, which implies that the matrix

$$\boldsymbol{A\Sigma} = (\boldsymbol{\mathcal{I}}^{-1} - \boldsymbol{M}\boldsymbol{\mathcal{J}}^{-1}\boldsymbol{M}')\boldsymbol{\mathcal{I}} = \boldsymbol{E}_m - \boldsymbol{M}\boldsymbol{\mathcal{J}}^{-1}\boldsymbol{M}'\boldsymbol{\mathcal{I}}$$

is idempotent, i.e., its rank equals its trace. ($\boldsymbol{E}_m$ denotes the identity matrix with m rows and columns.) Hence,

$$\begin{aligned} tr(\boldsymbol{E}_m - \boldsymbol{M}\boldsymbol{\mathcal{J}}^{-1}\boldsymbol{M}'\boldsymbol{\mathcal{I}}) &= m - tr(\boldsymbol{M}\boldsymbol{\mathcal{J}}^{-1}\boldsymbol{M}'\boldsymbol{\mathcal{I}}) = \\ = m - tr(\boldsymbol{\mathcal{J}}^{-1}\boldsymbol{M}'\boldsymbol{\mathcal{I}}\boldsymbol{M}) &= m - tr(\boldsymbol{\mathcal{J}}^{-1}\boldsymbol{\mathcal{J}}) = m - m'. \end{aligned}$$

□

10

Dynamic Generalizations of the Rasch Model

Norman D. Verhelst and Cees A.W. Glas[1]

ABSTRACT An overview is given of the different approaches to modeling dynamic phenomena within the framework of the Rasch model. Two main classes of dynamic phenomena are discerned: in the first class, change of the latent ability between test sessions is considered; in the second class, change within test sessions is studied. The models of the first class are mainly covered by the detailed study given in Chapter 9, although some other models have been discussed in the literature. The second class comprises two parts: models which violate the assumption of local stochastic independence, and models which do not. This latter class is shown to be a generalization of mathematical learning models which were popular in the fifties and sixties.

10.1 Introduction

A central assumption of most IRT models is local stochastic independence. The exact meaning of this axiom is that response variables are conditionally independent of each other, the condition being the value of the latent variable θ. The main application of this axiom is the fact that the probability of a response pattern, originating from a single person, is written as the product of the probabilities of the item responses. This means that no extra assumptions need to be made with regard to the probability of a response pattern beyond the model assumptions about item responses. However, using the independence axiom at the level of a person seems to imply that the value of θ associated with this person remains constant during test taking. Dynamic phenomena, such as learning or change of opinion, essentially imply change, and at first sight are not compatible with the static character of the latent trait. So the study of dynamic phenomena using IRT models is a challenge, and the purpose of the present chapter is to give an overview of the way this problem has been coped with.

There are two different categories of problems whose solution is also quite different. In the first category, the measurement of change between two (or more) test sessions is central. In studying these problems, it is assumed that during each test session the latent trait of the respondents are fixed. The second problem is concerned with changes in the latent trait during test taking. This problem represents a direct attack on the assumption of local independence. There are two basic approaches to this problem. In the first one, the Rasch model (RM) can

[1]National Institute for Educational Measurement, P.O.-Box 1034, 6801 MG Arnhem, The Netherlands; fax: (+31)85-521356

be conceived of as a special case of a log-linear model. By adding parameters to this log-linear model, changes in the latent trait during test taking can be modeled. The other approach comes within the framework of mathematical learning theory, a branch of formal modeling in psychology which was flourishing in the sixties (Sternberg, 1963).

The problem of change between test sessions will be treated in Section 10.2. Sections 10.3 and 10.4 deal with the problem of change within the test session. In Section 10.3, models will be considered where the axiom of local independence is no longer valid. In Section 10.4, models which do preserve the axiom of local independence and at the same time handle change of the ability during testtaking are discussed.

10.2 Change Between Test Sessions

Throughout this section, a very simple paradigm of measuring change will be studied. Consider the case where n persons answer to k binary items at time t_0. After a certain period, during which some experimental manipulation may take place, the same persons take the same test again. In general terms, the research hypothesis states that, as a consequence of the manipulation, the latent variable of the respondents will be different at both time points. To avoid complicated notation, the response of subject S_v on item I_i will be denoted X_{vi} at the pretest, and Y_{vi} at the posttest. The simplest and also most restrictive model states that

$$P(X_{vi}=1|\theta_v)=\frac{\exp(\theta_v-\beta_i)}{1+\exp(\theta_v-\beta_i)}, \tag{10.1}$$

$$P(Y_{vi}=1|\theta_v)=\frac{\exp(\theta_v-\delta-\beta_i)}{1+\exp(\theta_v-\delta-\beta_i)}. \tag{10.2}$$

The change, represented by the parameter $-\delta$ in (10.2), can be interpreted in two ways: it can be considered as a change in the latent ability of every respondent, but δ can also be interpreted as a change in the difficulty of every item, because an increase in ability is equivalent to a decrease in difficulty. Depending on the interpretation chosen, different models which require a different way of analysis arise. Models which locate the change in the item parameters will be considered first.

Conceiving the items presented on the two occasions as items of a single test of length $2k$, and imposing the linear restriction

$$\beta_{i+k}=\beta_i+\delta, \quad \text{for } i=1,\dots,k, \tag{10.3}$$

immediately shows that the problem of change measurement can be solved using an LLTM. For details of estimation and testing, the reader is referred to Chapter 8. Of course, one should be careful to control for the threats against

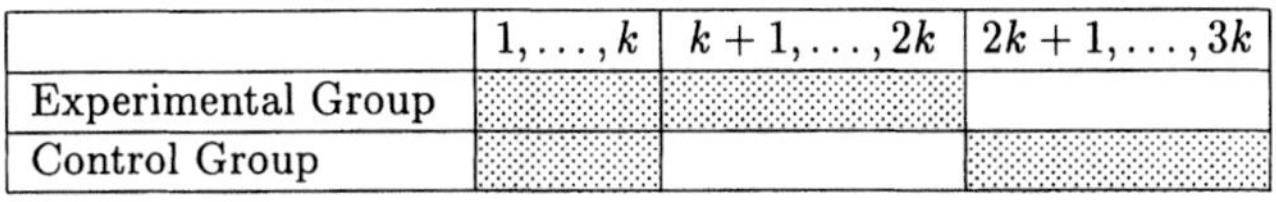
1, ..., k
k + 1, ..., 2k
2k + 1, ..., 3k
Experimental Group
Control Group

FIGURE 10.1. Incomplete design for controlling for memory effects (shaded = observed; blank = not observed)

the internal validity of the experiment. Finding a δ-estimate which differs significantly from zero does not automatically mean that the change is caused by the experimental treatment. Memory effects, for example, could be responsible for the change. In order to control for this source of invalidity, a control group might be tested at both time points. This results in an incomplete design as shown in Figure 10.1. The LLTM, given by (10.3), is expanded with

$$\beta_{i+2k} = \beta_i + \delta_2, \quad \text{for } i = 1, \ldots, k. \tag{10.4}$$

If the change is entirely due to memory effects or general history effects, then δ should be equal to δ_2. This equality can be tested statistically by imposing the extra restriction that $\delta = \delta_2$, and building a likelihood ratio test (see Chapter 9).

The generalizations of the above paradigm are straightforward. It is not necessary that the test forms used at pretest and posttest be identical (Fischer, 1989; see Chapter 9). If they are linked by at least one common item, the model is identified. Of course the restrictions (10.3) and (10.4) only apply to the common items, and the standard error of the change parameter δ will be larger with decreasing number of common items. Generalization to factorial designs is equally simple. Suppose a 2×2 factorial design with pretest and posttest. Associating one δ-parameter per cell, say δ_{ij}, for $i, j = 1, 2$, corresponds to a model with main effects and interactions, while a model with only main effects is given by the additional restrictions

$$\begin{aligned} \delta_{12} &= \delta_{11} + \delta_c, \\ \delta_{21} &= \delta_{11} + \delta_r, \\ \delta_{22} &= \delta_{11} + \delta_c + \delta_r, \end{aligned} \tag{10.5}$$

where δ_r and δ_c denote the effects of the (second) row and column, respectively.

Applying one of the LLTMs discussed above implies that not only the δ-parameters are to be estimated, but also the item difficulties β_i. Although this is not difficult, the LLTM implies that the k items in the test conform to the RM. If this is not the case, a specification error is made, which may make the estimates of the δ-parameters inconsistent to an unknown extent. One can, however, get rid of the assumption about the RM. By defining

$$\theta^*_{vi} = \theta_v - \beta_i,$$

(10.1) and (10.2) can be rewritten as

$$P(X_{vi}|\theta^*_{vi}) = \frac{\exp(\theta^*_{vi})}{1+\exp(\theta^*_{vi})}, \tag{10.6}$$

$$P(Y_{vi}|\theta^*_{vi}) = \frac{\exp(\theta^*_{vi}+\delta)}{1+\exp(\theta^*_{vi}+\delta)}. \tag{10.7}$$

From the assumption of local stochastic independence, it follows that a person-item combination can be conceived of as a 'virtual' person, such that (10.6) and (10.7) together define a simple RM with two items answered by nk persons. The assumption that the original k items measure the same latent trait is no longer necessary. This model is called the 'linear logistic model with relaxed assumptions' (LLRA, Fischer, 1977a, 1983a,b, 1989; see, Chapter 9; Fischer & Formann, 1982a). Although (10.6) and (10.7) correspond formally to a RM with two items, the generalization of this approach to more complicated designs is straightforward. The model given by (10.5), for example, is formally an LLTM for 5 virtual items in an incomplete design, where every virtual person has answered 2 items, the 'pretest item' and one of the four 'posttest items'.

A generalization of the LLRA in the context of measuring change has been proposed by Fischer (1989). In this model, called a hybrid model, it is possible to estimate changes even if pre- and posttest do not have common items and where the tests may be multidimensional. A condition, however, is that both tests contain at least one item measuring each dimension, and that all items measuring the same dimension form a Rasch scale when applied under the same experimental conditions.

The approach using the LLTM or the LLRA can be compared with a simple analysis of covariance. Both models imply that the change (or the impact of the experimental treatment) is independent of the initial level of ability and is constant for every person participating in the experiment. As a consequence, these models imply that the variance of the θ-values does not change through the treatment and that the correlation between pretest and posttest θ equals one. Andersen (1985) proposed an approach to cope with these objections. The θ-values associated with the persons in pre- and posttest are considered as a random sample from a bivariate normal distribution, with mean vector $\boldsymbol{\mu} = (\mu_1, \mu_2)$ and variance-covariance matrix $\boldsymbol{\Sigma}$. Andersen discusses only the case where the two tests comprise the same items and the item parameters are known or assumed to be known, e.g., from a previous calibration. So only five parameters have to be estimated: two means, two variances, and one covariance or correlation. Although Andersen reports difficulties in the estimation procedure, especially for the correlation, Glas (1989) succeeded in generalizing Andersen's approach in three respects: (i) the number of variates in the multivariate θ can be either three or two; (ii) the items used at different time points need not be identical, and (iii) item parameters and distribution parameters are estimated simultaneously using a multivariate version of the MML-procedure.

The estimate of the (average) amount of change in this approach is given by $\hat{\mu}_1 - \hat{\mu}_2$, and has a similar interpretation as the δ-parameter in (10.2), but at the same time differential growth is allowed for, since the correlation needs not to be unity, and the variances at pre- and post-test may differ. Although this approach seems more flexible (at least in the example with two time points), it has several drawbacks. In the first place, the estimation procedure needs numerical approximation of multiple integrals. Although generalizations of standard methods like Gauss-Hermite quadrature are easily implemented, there are no results on the numerical accuracy of these generalizations. Second, the generalization to factorial designs, like the one discussed above, seems impossible. In the 2×2 factorial design, five means have to be estimated, implying that a five-variate latent variable has to be used and that 10 correlation coefficients have to be estimated. But the observations are incomplete, because nobody is observed under two different experimental conditions, and therefore the corresponding correlation is inestimable, unless several equality constraints are imposed on the correlations. Third, the LLTM-approach allows for relatively simple ways of model testing. For example, one might suppose that the amount of change is dependent on the initial ability. Splitting the total sample in a low group and a high group on the basis of the pretest allows one to construct a likelihood ratio test to test the equality of the change parameter in both groups. A similar procedure is not possible in using the approach of Andersen and Glas, because the splitting represents a selection on the θ_1-variate and, as a consequence, the assumption of bivariate normality in the subgroups can no longer be valid.

A way out of these problems is to assume nothing about the bivariate distribution of the pre- and posttest abilities, but to estimate it entirely from the data. This approach is taken by Embretson (1991). For the simple pretest posttest setup, her model needs only a slight modification of (10.2):

$$P(Y_{vi} = 1|\theta_v) = \frac{\exp(\theta_v + \delta_v - \beta_i)}{1 + \exp(\theta_v + \delta_v - \beta_i)}, \tag{10.8}$$

meaning that at the posttest the ability of person S_v has changed by an amount δ_v, called modifiability. Notice that no structure whatsoever is imposed on the modifiabilities. It is easy to show that the number of items correct on the pretest (r_{1v}) and on the posttest (r_{2v}) are sufficient statistics for respectively θ_v and ($\theta_v + \delta_v$), so that CML-estimation of the item parameters is possible, given that there are some common items in pretest and posttest. Given these parameters, θ_v and δ_v are estimated by maximum likelihood. As a side result of this method, the (asymptotic) variance-covariance matrix of both estimators is estimated also. The advantage of this approach is that a detailed study can be made using the ability and modifiability estimates of the respondents. As an example, consider the correlation between the estimates of θ and δ. In a simulation study with zero correlation between the true values of θ and δ, Embretson found a correlation of $-.21$ in the estimates. But this correlation is biased through the estimation error in the estimates, and especially through their negative correlation. A rough correction, analogous to the correction for attenuation in classical test theory, can be constructed as follows. Let $\boldsymbol{\eta}_v = (\theta_v, \delta_v)$ and let $\hat{\boldsymbol{\eta}}_v$ be some estimator of $\boldsymbol{\eta}_v$. Using $\boldsymbol{\Sigma}(.)$ as a generic symbol for the variance-covariance matrix, the general expression for decomposing variances can be written as

$$\boldsymbol{\Sigma}(\hat{\boldsymbol{\eta}}) = E[\boldsymbol{\Sigma}(\hat{\boldsymbol{\eta}}|\boldsymbol{\eta})] + \boldsymbol{\Sigma}[E(\hat{\boldsymbol{\eta}}|\boldsymbol{\eta})]. \tag{10.9}$$

If the sample at hand is a random sample from the target population, the left-hand side of (10.9) is consistently estimated by the sample covariance matrix of the estimates. The first term in the right-hand side is estimated consistently from the average covariance matrix of the estimates. So the last term can be obtained from (10.9). If there exists an unbiased estimator $\hat{\boldsymbol{\eta}}$, the last term represents the covariance matrix of $\boldsymbol{\eta}$. In the RM, the ML estimate is biased (Lord, 1983a), and it is to be expected that this bias will not disappear in the present case. For the univariate case (the one-, two- and three-parameter logistic models), Warm (1989) has derived the so called weighted maximum likelihood estimator for θ and has shown that the resulting bias is of an order less than k^{-1}. Moreover, finite Warm estimates exist for zero and perfect scores (see also Chapter 4). The application of Warm's result to the present model is simple and straigtforward. Let $\xi_v = \theta_v + \delta_v$. The variance-covariance matrix $\boldsymbol{\Sigma}(\hat{\theta}_v, \hat{\xi}_v|\theta_v, \xi_v)$ is a diagonal matrix where the entries simply are the variances of the person parameters estimated from the pretest and posttest scores, respectively. Since $\boldsymbol{\eta}_v$ is a linear transformation of (θ_v, ξ_v), the variance-covariance matrix $\boldsymbol{\Sigma}(\hat{\boldsymbol{\eta}}_v|\boldsymbol{\eta}_v)$ is easily derived.

10.3 Changes During Testing: Interaction Models

In this section, models will be discussed where the axiom of local stochastic independence does not hold. Strictly speaking, these models are not item response models, but test response models, because the formal model is not specified at the item level, but directly at the test level. Two such models are known in the literature, viz., the class of log-linear RMs (Kelderman, 1984), and a class of models called 'conjunctive IRT kernels' (Jannarone, 1986). These two models, however, can be treated as special cases of a more general model, as will be shown in the sequel. A simple example may clarify the main idea. Assume $k = 2$, and a response pattern $\boldsymbol{x} = (x_1, x_2)$. The RM can be written as

$$P(\boldsymbol{x}|\theta) \propto \exp[r(\boldsymbol{x})\theta - \beta_1 x_1 - \beta_2 x_2], \tag{10.10}$$

where $r(\boldsymbol{x})$ is the simple raw score. The proportionality constant can easily be derived from the requirement that $\sum_{\boldsymbol{x}} P(\boldsymbol{x}|\theta) = 1$.

Notice that (10.10) or its straightforward generalization to the case of an arbitrary k, could have been taken as the definition of the RM. It is easy to deduce from (10.10) that the response variables X_1 and X_2 are stochastically independent, given θ. To construct a more general model than (10.10), we observe that the argument of the exponential function is linear in the score and the response variables. A possible generalization is to allow for products, and thus to model interactions (see also Chapter 6). For the case of two items, this results in

$$P(\boldsymbol{x}|\theta) \propto \exp[r^*(\boldsymbol{x})\theta - \beta_1 x_1 - \beta_2 x_2 - \beta_{12} x_1 x_2]. \tag{10.11}$$

The function r^* is called the score function. We leave it undefined for the moment, but only remark that it is a scalar function and does not depend on the β-parameters. The model defined by (10.11) comprises main effects of the items (β_1 and β_2) and a first order interaction effect (β_{12}). Since there are only two items, this model is saturated, but with more than two items, several first order interaction terms may be modeled, as well as a variety of higher order interactions. In this way, a family of interaction models is created, but the characteristics of it will depend on the specific score function chosen. In the family of models studied by Kelderman (1984), the score function is always the raw score, i.e., the number of correct items in the response pattern; in the family introduced by Jannarone (1986), the score function depends on the number and the order of the interactions which are part of the model. In the example given by (10.11), the score function is $r^*(\boldsymbol{x}) = x_1 + x_2 + x_1 x_2$. A general specification will be given later.

To formalize these ideas, let $J = \{1, \ldots, k\}$, let G be the power set of J, and R^* some subset of G. For example, let $k = 3$, and let $R^* = \{\{1\}, \{2\}, \{3\}, \{1,2\}, \{2,3\}\}$. The set R^* specifies the model within the family of models. Next define the corresponding set of ordered r-tuples $R = \{(m_1, \ldots, m_r)\}$, with $m_j < m_{j+1}$, for $j = 1, \ldots, r-1$, such that $(m_1, \ldots, m_r) \in R$ if and only if $\{m_1, \ldots, m_r\} \in R^*$. The set R only serves to properly index the elements of R^*. In the example, R is given by $\{(1), (2), (3), (1,2), (2,3)\}$, or in a shorthand notation omitting parentheses and commas between the elements of the tuples, $R = \{1, 2, 3, 12, 23\}$. The number of elements in R will be denoted by p. Furthermore, let $\boldsymbol{\beta}$ be a p-vector of parameters, one parameter per element of R. Let $T_R(\boldsymbol{X})$ be some scalar function of $\boldsymbol{X}$, the response pattern, which is independent of the parameters. The elements of R will be used as index of the elements of $\boldsymbol{\beta}$. In the example, there are five parameters, $\beta_1, \beta_2, \beta_3, \beta_{12}$, and β_{23}. The general logistic model, given R and T_R, is then defined by

$$P_{T_R,R}(\boldsymbol{x}|\theta) \propto \exp[t_R(\boldsymbol{x})\theta - \sum_R x_{m_1} \ldots x_{m_r} \beta_{m_1,\ldots,m_r}]. \tag{10.12}$$

From (10.12) it is clear that the general logistic model is an exponential family where $t_R(\boldsymbol{x})$ denotes the realization of $T_R(\boldsymbol{X})$, which is sufficient for θ, and in a sample of n persons, the statistic $\sum_v^n X_{vm_1} \ldots X_{vm_r}$ is sufficient for $\beta_{m_1,\ldots,m_r}$.

TABLE 10.1. Some special cases of the general logistic model

T	R	Denomination
$\sum_i^k x_i$	{1, ...,k }	Rasch Model
$\sum_i^k x_i$	Hierarchic Subsets of G	Loglinear Rasch Models (Kelderman, 1984)
$\sum_R X_{m_1,\ldots,m_r}$	Any Non-Empty Subset of G	Conjunctive IRT Kernels (Jannarone, 1986)
$\sum_i^k \alpha_i X_i$, ($\alpha_i \in \{1, 2, \ldots\}$)	$\{1, \ldots, k\}$	One Parameter Logistic Model (Chapter 14)

From the sufficiency of $T_R(\boldsymbol{x})$, it follows that CML-estimation is possible. The conditional probability of response pattern $\boldsymbol{x}$, given $t_R(\boldsymbol{x})$, is

$$P_{T_R,R}(\boldsymbol{x}|t_R(\boldsymbol{x}) = t) = \frac{\exp[-\sum_R x_{m_1} \ldots x_{m_r} \beta_{m_1,\ldots,m_r}]}{\tilde{\gamma}_t}, \tag{10.13}$$

where

$$\tilde{\gamma}_t := \tilde{\gamma}_t(\boldsymbol{\beta}) = \sum_{\boldsymbol{y}:t(\boldsymbol{y})=t} \exp\left[-\sum_R y_{m_1} \ldots y_{m_r} \beta_{m_1,\ldots,m_r}\right]. \tag{10.14}$$

Denoting by n_t the number of persons in the sample having score t, and defining $M(t, \boldsymbol{x})$ as the expected number of persons having response pattern $\boldsymbol{x}$, it readily follows from (10.13) that

$$\ln M(t, \boldsymbol{x}) = -\sum_R x_{m_1} \ldots x_{m_r} \beta_{m_1,\ldots,m_r} - \kappa_t, \tag{10.15}$$

where

$$\kappa_t = \ln \frac{n_t}{\tilde{\gamma}_t}.$$

This means that the model can be written as a log-linear model. However, the basic contingency table has dimension $k + 1$, one dimension for the score t, and one dimension for every item. Since the score is uniquely determined by the response pattern, many cells have zero counts per definition. These cells are called structural zeros, and the associated log-linear model is called a quasi log-linear model. Details on estimation procedures and testing can be found in Kelderman (1984).

Although the model defined by (10.12) is quite general, it is not necessarily identified for arbitrary T and R. In Table 10.1, some models which fit into this general framework and which have been studied to some extent, are listed. Although in Kelderman's approach it is also possible to include arbitrary subsets of G, Kelderman did only study hierarchic models. Hierarchic means that if $(m_1, \ldots, m_r) \in R$, then $(m_{i_1}, \ldots, m_{i_s}) \in R$ for all $\{m_{i_1}, \ldots, m_{i_s}\} \subset$

$\{m_1, \ldots, m_r\}$. This restriction is especially useful in building likelihood ratio tests. The advantage of Kelderman's approach to Jannarone's is the fact that in the estimation procedure for any model the structure of the basic contingency table remains the same, while in the approach of Jannarone, where T is directly dependent of R, the structure of the table depends on R.

To see how local independence is lost and how some parameters can be given a direct dynamic interpretation, a simple case of Kelderman's model will be investigated. Define $R = \{1, 2, \ldots, k, 12\}$; this means that the only deviation from the RM consists in a first order interaction between X_1 and X_2. Define $\varepsilon_{m_1,\ldots,m_r} = \exp(-\beta_{m_1,\ldots,m_r})$ and $\xi = \exp(\theta)$. Now partition the set all possible response patterns, $\{\boldsymbol{X}\}$, into four equivalence classes, where all patterns belonging to the same class have the same responses on items I_1 and I_2. It will be clear that, if the responses on items I_1 and I_2 are deleted from all response patterns, the four equivalence classes will be identical. So, since there are no interactions between either of the first two items and the remaining ones,

$$F_{ab} := P(X_1 = a, X_2 = b|\xi) \propto \varepsilon_1^a \varepsilon_2^b \varepsilon_{12}^{ab} \xi^{a+b}, \quad \text{for } a, b = 0, 1. \tag{10.16}$$

Using this definition, it follows immediately that

$$P(X_2 = 1|X_1 = 1, \xi) = \frac{F_{11}}{F_{11} + F_{10}} = \frac{\varepsilon_2 \varepsilon_{12} \xi}{1 + \varepsilon_2 \varepsilon_{12} \xi}, \tag{10.17}$$

$$P(X_2 = 1|X_1 = 0, \xi) = \frac{F_{01}}{F_{01} + F_{00}} = \frac{\varepsilon_2 \xi}{1 + \varepsilon_2 \xi}, \tag{10.18}$$

which implies that local independence is violated, unless $\varepsilon_{12} = 1$, i.e., $\beta_{12} = 0$. Taking the logit transformation of (10.17) and (10.18) and subtracting the second expression from the first one yields

$$\text{logit}[P(X_2 = 1|X_1 = 1, \xi)] - \text{logit}[P(X_2 = 1|X_1 = 0, \xi)] = -\beta_{12}. \tag{10.19}$$

If $\beta_{12} < 0$, this implies that item I_2 has become easier after a correct response on item I_1 than it is after an incorrect response, and thus this parameter can be interpreted as the difference of the amount of learning due to the different quality of the response on item I_1. Applying the same reasoning for the same definition of R in Jannarone's approach yields a difference between the two logit transformation equal to $\theta - \beta_{12}$, leading to the interpretation that in Jannarone's model the amount of learning depends on the initial value of the latent trait, a remarkable difference to Kelderman's model.

It may be interesting to point to an interpretation problem associated with (10.19). Using (10.16) and the same procedure as above, it follows for Kelderman's model that

$$\text{logit}[P(X_1 = 1|X_2 = 1, \xi)] - \text{logit}[P(X_1 = 1|X_2 = 0, \xi)] = -\beta_{12}, \tag{10.20}$$

which could be interpreted as a kind of 'backward' learning, which of course is not predicted by any learning theory. To understand this correctly, imagine that the population with latent value ξ is partitioned into two subpopulations, one subpopulation having a correct response to item I_2, and the other, a wrong response. If $\beta_{12} < 0$, a correct response to item I_1 makes item I_2 easier, so that people who could take advantage of this facilitating effect will be overrepresented in the former subpopulation, and this is precisely what formula (10.20) expresses. If the model is valid, and $\beta_{12} \neq 0$, then (10.20) says that item I_1 is biased with respect to (the response to) item I_2. Notice, however, that the models of Kelderman and Jannarone are no item response models: they do not predict what will happen if an item is eliminated from the test. It does not follow from their theory that, in the above example, the RM applies to the remaining $k-1$ items if item I_1 or item I_2 is eliminated.

10.4 Changes During Test Taking: Learning Models

In this section[2], two models will be discussed which are genuine learning models, and where local stochastic independence is nevertheless preserved. It will be shown how these two models are generalizations of existing mathematical learning models, a branch of mathematical psychology that flourished in the fifties and the sixties, see Sternberg (1963) for a review. The first model (Verhelst & Glas, 1993) elaborates on an idea of Fischer (1972, 1982, 1983a). In this approach the RM is combined with the missing data concept and with linear restrictions on the parameter space. Formally, it is an LLTM in an incomplete design. The basic idea conceives of an item as a collection of 'virtual items', one of which is supposed to be administered to each subject, depending, for example, on the response pattern on previous items. In the other model (Kempf, 1974, 1977a, 1977b), one or more parameters are associated with (subsets of) partial response patterns, up to, say, item I_{i-1}, and this parameter influences the response on item I_i.

Before discussing these models in more detail, some concepts of mathematical learning theory will be introduced. In mathematical learning theory, the control of change in behavior is generally attributed to two classes of events: one is the behavior of the responding subject itself; the other comprises all events that occur independently of the subject's behavior, but which are assumed to change that behavior. Models that only allow for the former class are called 'subject controlled'; if only external control is allowed, the model is 'experimenter controlled', and models where both kinds of control are allowed are labelled 'mixed models'. As an example of 'experimenter control', assume that during test taking the correct answer is given to the respondent after each response. If it is assumed

[2]This section is a shortened and slightly adapted version of Verhelst and Glas (1993). We wish to thank the Editor of Psychometrika for his kind permission to use this material.

that learning takes place under the influence of that feedback (generally referred to as reinforcement) independent of the correctness of the given response, the model is experimenter controlled; if it is assumed, however, that learning takes place only if the subject gave a wrong answer, the model is mixed. In the sequel it will be assumed that all controlling events can be binary coded, that the subject control can be modelled through the correctness of the responses on past items, and that experimenter control expresses itself at the level of the item. Let $\boldsymbol{X} = (X_1, \ldots, X_i, \ldots, X_k)$ be the vector of response variables, taking values 1 for a correct answer and zero otherwise, and let $\boldsymbol{Z} = (Z_1, \ldots, Z_i, \ldots, Z_k)$ be the binary vector representing the reinforcing event occurring after the response has been given. The variable $\boldsymbol{Z}$ is assumed to be independent of $\boldsymbol{X}$. The prototype of a situation where this independence is realized is the so-called prediction experiment where the subject has to predict a sequence of independent Bernoulli events, indexed by a constant parameter π (for example, the prediction of the outcome of coin tosses with a biased coin). The outcome itself is independent of the prediction, and it is observed frequently that in the long run the relative frequency of predicting each outcome matches approximately the objective probability of the outcomes (Estes, 1972). Therefore, it can be assumed that the outcome itself acts as a reinforcer and is the main determinant of the change in the behavior. Notice that in the example given, Z_i is independent of the vector of response variables $\boldsymbol{X}$. Therefore, any model that assumes control over the responses only through the outcomes is experimenter controlled.

Define the partial response vector $\boldsymbol{X}^i$, $(i > 1)$ as

$$\boldsymbol{X}^i = (X_1, \ldots, X_{i-1}), \tag{10.21}$$

and the partial reinforcement vector $\boldsymbol{Z}^i$, $(i > 1)$ as

$$\boldsymbol{Z}^i = (Z_1, \ldots, Z_{i-1}). \tag{10.22}$$

The model that will be considered, is in its most general form given by

$$P(X_i = 1|\theta, \boldsymbol{x}^i, \boldsymbol{z}^i) = \frac{\exp[\theta + \beta_i + f_i(\boldsymbol{x}^i) + g_i(\boldsymbol{z}^i)]}{1 + \exp[\theta + \beta_i + f_i(\boldsymbol{x}^i) + g_i(\boldsymbol{z}^i)]}, \tag{10.23}$$

where θ is the latent variable, β_i the 'easiness' parameter of item I_i, $\boldsymbol{x}^i$ and $\boldsymbol{z}^i$ represent realizations of $\boldsymbol{X}^i$ and $\boldsymbol{Z}^i$, respectively, and $f_i(.)$ and $g_i(.)$ are real-valued functions. Since the domain of these functions is discrete and finite, the possible function values can be conceived of as parameters. It is clear that the general model is not identified, because the number of parameters outgrows by far the number of possible response patterns. So some suitable restrictions will have to be imposed.

A general restriction, frequently applied in mathematical learning theory, is the requirement that the functions f_i and g_i are symmetric in their arguments, yielding models with commutative operators. Since the arguments are binary vectors, symmetry implies that the domain of the functions f_i and g_i can be restricted to the sum of the elements in the vectors $\boldsymbol{x}^i$ and $\boldsymbol{z}^i$, respectively. Defining the variables R_i and S_i as

$$R_i = \begin{cases} \sum_{j=1}^{i-1} X_j, & (i > 1), \\ 0, & (i = 1), \end{cases} \tag{10.24}$$

and

$$S_i = \begin{cases} \sum_{j=1}^{i-1} Z_j, & (i > 1), \\ 0, & (i = 1), \end{cases} \tag{10.25}$$

with realizations r_i and s_i, respectively, and assuming symmetry of the functions g_i and f_i, (10.23) reduces to

$$P(X_i = 1|\theta, r_i, s_i) = \frac{\exp[\theta + \beta_i + \delta_i(r_i) + \gamma_i(s_i)]}{1 + \exp[\theta + \beta_i + \delta_i(r_i) + \gamma_i(s_i)]} \tag{10.26}$$

where $\delta_i(0)$ and $\gamma_i(0)$ are defined to be zero for all i. If all δ and γ equal zero, no transfer takes place, and (10.26) reduces to the common RM (with β_i for the easiness of item I_i); if all δ are zero, and at least one of the γ is not, the resulting model is experimenter controlled; if all γ are zero, and at least one of the δ is not, the model is subject controlled, and in the other cases a mixed model results. Notice that (10.26) implies that no forgetting occurs: the influence of an equal number of correct responses or an equal number of positive reinforcements has the same influence on behavior, irrespective of their temporal distance to the actual response. This somewhat unrealistic assumption is the price to be paid for the elegant mathematical implications of commutative operators. In the sequel, however, a model will be discussed where this symmetry is at least partially abandoned.

Using the concept of incomplete data, it can be illustrated how the models discussed above fit into the ordinary RM. Only the subject controlled subcase of (10.26) will be considered in detail; that is, it is assumed that all γ are zero. Notice that it is assumed that the transfer taking place does not depend on the initial ability θ of the respondent, and any change in the ability (the increment $\delta_i(r_i)$) can be translated into a change of the difficulty of the items. Thus, the difficulty of an item is conceived of as being composed of an 'intrinsic' parameter $-\beta_i$ and a dynamic component $\delta_i(r_i)$. The latter is not constant, but depends on the rank number of the item in the sequence, as well as on the specific capability of the item to be susceptible to learning effects (hence the subscripted δ). These two effects can in principle be disentangled by experimental design, but to avoid overcomplicating the model, it will be assumed that the order of presentation is the same for all subjects.

TABLE 10.2. The transformation from real to virtual items for $k = 3$

Real Items			Virtual Items						Scores
1	2	3	(1,0)	(2,0)	(2,1)	(3,0)	(3,1)	(3,2)	
1	1	1	1	*	1	*	*	1	3
1	1	0	1	*	1	*	*	0	2
1	0	1	1	*	0	*	1	*	2
1	0	0	1	*	0	*	0	*	1
0	1	1	0	1	*	*	1	*	2
0	1	0	0	1	*	*	0	*	1
0	0	1	0	0	*	1	*	*	1
0	0	0	0	0	*	0	*	*	0

Let a 'real' item I_i be associated with a collection of 'virtual' items, denoted by the ordered pair (i, j), $j = 0, \ldots, i-1$. The virtual item (i, j) is assumed to be presented to all subjects who gave exactly j correct responses to the $i-1$ preceding 'real' items. Associated with response pattern $\boldsymbol{X}$ is a design vector $\boldsymbol{D}(\boldsymbol{X})$, with elements $D(\boldsymbol{X})_{ij}$, for $i = 1, \ldots, k; j = 0, \ldots, i-1$, defined by

$$D(\boldsymbol{X})_{ij} = \begin{cases} 1 & \text{if } R_i = j, \\ 0 & \text{otherwise.} \end{cases} \tag{10.27}$$

Response pattern $\boldsymbol{X}$ is transformed into a response pattern $\boldsymbol{Y}(\boldsymbol{X})$ with elements $Y(\boldsymbol{X})_{ij}$, for $i = 1, \ldots, k; j = 0, \ldots, i-1$, using

$$Y(\boldsymbol{X})_{ij} = \begin{cases} 1 & \text{if } D(\boldsymbol{X})_{ij} = 1 \text{ and } X_i = 1, \\ 0 & \text{if } D(\boldsymbol{X})_{ij} = 1 \text{ and } X_i = 0, \\ c & \text{if } D(\boldsymbol{X})_{ij} = 0, \end{cases} \tag{10.28}$$

where c is an arbitrary constant different from zero and one. In Table 10.2 the set of all possible response patterns $\boldsymbol{X}$ and the associated vectors $\boldsymbol{Y}(\boldsymbol{X})$ is displayed for $k = 3$, the constant c is replaced by an asterisk. It is easily verified that $\boldsymbol{Y}(\boldsymbol{X})$ and $\boldsymbol{D}(\boldsymbol{X})$ are unique transformations of $\boldsymbol{X}$.

The probability of an observed response pattern $\boldsymbol{x}$, that is, $\boldsymbol{y}(\boldsymbol{x})$ and $\boldsymbol{d}(\boldsymbol{x})$, can now be derived as follows. Let $\boldsymbol{\xi}$ be a $k(k+1)/2$ dimensional vector with elements ξ_{ij}, for $i = 1, \ldots, k; j = 0, \ldots, i-1$, where $\xi_{ij} = \beta_i + \delta_i(j)$. Then,

$$\begin{aligned} P(\boldsymbol{x}|\theta;\boldsymbol{\xi}) &= P(x_1|\theta;\boldsymbol{\xi}) \prod_{i>1} P(x_i|\boldsymbol{x}^i,\theta;\boldsymbol{\xi}) \\ &= P(x_1|\theta;\boldsymbol{\xi}) \prod_{i>1} \prod_{j=0}^{i-1} P(y(\boldsymbol{x})_{ij}|d(\boldsymbol{x})_{ij},\theta;\boldsymbol{\xi}) \end{aligned}$$

$$= \prod_{i=1}\prod_{j=0}^{i-1} \frac{\exp[y(\boldsymbol{x})_{ij} d(\boldsymbol{x})_{ij}(\theta + \xi_{ij})]}{[1 + \exp(\theta + \xi_{ij})]^{d(\boldsymbol{x})_{ij}}}$$

$$= \frac{\exp\left[\sum_{i=1}\sum_{j=0}^{i-1} y(\boldsymbol{x})_{ij} d(\boldsymbol{x})_{ij}(\theta + \xi_{ij})\right]}{\prod_{i=1}\prod_{j=0}^{i-1}[1 + \exp(\theta + \xi_{ij})]^{d(\boldsymbol{x})_{ij}}}. \tag{10.29}$$

The second line of the derivation is motivated by the fact that the response x_i is recoded to $y(\boldsymbol{x})_{ij}$ and the item administration variable $d(\boldsymbol{x})_{ij}$ is determined by the previous responses $\boldsymbol{x}^i$. The next line of the derivation follows from the assumption that, if $d(\boldsymbol{x})_{ij} = 1$, the RM holds, and

$$P(y(\boldsymbol{x})_{ij} = c | d(\boldsymbol{x})_{ij} = 0, \theta; \boldsymbol{\xi}) = 1.$$

Notice that (10.29) is equivalent to the original model. The advantage of writing the model in this way is that (10.29) is a simple generalization of the likelihood function for the RM to incomplete designs (see Fischer, 1981). Of course, a similar derivation may be made for experimenter controlled models, and a straightforward generalization yields a similar result for mixed models. Although the likelihood function for the experimenter controlled case is formally identical to (10.29), with the design vector being a function of $\boldsymbol{Z}$ instead of $\boldsymbol{X}$, the estimation problems are quite different, as will be discussed next.

To start with, a problem about identifiability should be clarified. Written in the original parametrization, numerator and denominator in the right hand member of (10.29) contain a factor $\theta + \beta_i + \delta_i(j)$ whose value is the same as $\theta^* + \beta_i^* + \delta_i^*(j)$, with $\beta_i^* = \beta_i + c + d_i$, $\delta_i^*(j) = \delta_i(j) - d_i$, and $\theta^* = \theta - c$ for arbitrary c and d_i, for $i = 1, \dots, k$, meaning that $k + 1$ restrictions should be imposed on the parameters to make the model identified. For example, β_1 and $\delta_i(0)$, for $i = 1, \dots, k$, can be set to zero, leaving

$$k + \frac{k(k+1)}{2} - (k+1) = \frac{(k-1)(k+2)}{2} \tag{10.30}$$

free item parameters.

Glas (1988b) has investigated the estimation problems in the RM for so-called multi-stage testing testing procedures, that is, designs where the sequence of tests administered is controlled by the sequence of test scores of the respondent. On the level of virtual items, the design used in the present model for analyzing data under a subject controlled model can be viewed as a limiting case of a multi-stage testing design where all tests consist of one item only, and the next test to be administered depends on the sum score R_i obtained on the preceding tests. The main result of Glas is the conclusion that, in the case of a multi-stage design, the CML estimation equations have no unique solution, while MML generally does yield consistent estimates.

With respect to CML estimation, the fact that the equations have no unique solution directly follows from the analogy between the multi-stage testing design considered by Glas and the present design. However, since the latter design is an extreme case of the former, the problem of obtaining CML estimates emerges more profound in the present case, as the following argument may reveal. The test administration design and the sum score are a sufficient statistic for θ. CML estimation amounts to maximizing the conditional likelihood function, where the condition is the sum score and the design jointly (see Glas, 1988b, for details). From inspection of Table 10.2 it is easily verified that, given the design (i.e., the location of the asterisks) and the sum score, the response vector $\boldsymbol{Y}(\boldsymbol{X})$ is completely determined. This means that the conditional sample space is a singleton, its likelihood is trivially identical to 1 and cannot be used to estimate the structural parameters, or any functions of structural parameters. The situation is totally different for experimenter controlled models, where the control over the design vector is completely independent of the responses. In that case, a sum score r obtained on a subtest of k virtual items can originate from $\binom{k}{r}$ different response patterns. So, conditioning on the design implies no restriction whatsoever on the sample space of $\boldsymbol{X}$, and CML can be applied straightforwardly. It should be stressed that the well-known theorems on the existence and uniqueness of CML estimates by Fischer (1981, 1983a) only apply to situations where, given the design, there is no restriction on the set of possible response patterns.

To show the applicability of the MML procedure, the missing data conception is not required. If θ is considered as a random variable with probability density function $g(\theta;\varphi)$ indexed by the parameter vector φ, the probability of observing a response pattern $\boldsymbol{x}$ is given by

$$P(\boldsymbol{x};\boldsymbol{\xi},\varphi) = \int_{-\infty}^{+\infty} P(\boldsymbol{x}|\theta;\boldsymbol{\xi})g(\theta;\varphi)d\theta, \tag{10.31}$$

where $P(\boldsymbol{x}|\theta;\boldsymbol{\xi})$ is defined by (10.29). This means that if, for all possible response patterns $\boldsymbol{x}$, $n_{\boldsymbol{x}}$ is the number of respondents producing $\boldsymbol{x}$, the counts $n_{\boldsymbol{x}}$ have a parametric multinomial distribution with index $n = \sum_{\boldsymbol{x}} n_{\boldsymbol{x}}$ and parameters $P(\boldsymbol{x};\boldsymbol{\xi},\varphi)$ for all binary k-vectors $\boldsymbol{x}$. The logarithm of the likelihood function is

$$\ln L(\boldsymbol{\xi},\varphi;\{\boldsymbol{x}\}) = \sum_{\boldsymbol{x}} n_{\boldsymbol{x}} \ln P(\boldsymbol{x}) = \sum_{v} \ln \int P(\boldsymbol{x}_v|\theta;\boldsymbol{\xi})g(\theta;\varphi)d\theta, \tag{10.32}$$

where $\{\boldsymbol{x}\}$ represents the data, and $\boldsymbol{x}_v$ is the response pattern of person S_v. Simultaneous maximization of (10.32) with respect to φ and $\boldsymbol{\xi}$ yields MML-estimates of the parameters.

It should be stressed that MML is not a mere technique of parameter estimation; it implies also an extension of the original measurement model as defined by (10.29). This extension is given by (10.31), which is an assumption about the distribution of the latent variable in the population, and at the same time about the sampling technique: using (10.31) implies a simple random sample from $g(\theta; \boldsymbol{\varphi})$. If in some application another sampling scheme is used, for example, clustered sampling where schools are randomly sampled, and then students are sampled within schools, (10.31) no longer holds if the intraclass correlation differs from zero, even if $g(\theta; \boldsymbol{\varphi})$ is the true probability density function for the population.

Although the choice of the distribution function of θ is in principle free, only a few distributions have been studied in the literature. The normal distribution is by far the most frequently used (Bock & Aitkin, 1981; Thissen, 1982; Mislevy, 1984), but see also Andersen and Madsen (1977) and Engelen (1989) for applications with other parametric distribution functions. Another approach is to specify as little as possible about the distribution of θ, rather to try to estimate the distribution from the data. Because no parametric distribution is assumed, this approach is called nonparametric MML or semi-parametric MML, the latter term referring to the fact that the model does have parameters, viz., the item parameters. It turns out, at least in the RM, that the distribution which maximizes the likelihood function is a discrete distribution with probability mass different from zero at a finite number, m, of support points (De Leeuw & Verhelst, 1986; Follmann, 1988; see also Chapter 13). Formally the support points and their associated mass are parameters to be estimated from the data, and jointly denote the parameter vector $\boldsymbol{\varphi}$ used in (10.32). More details on the estimation procedure with MML can be found in Glas and Verhelst (1989) and in Chapter 11. Verhelst and Glas (1993) discuss parametric as well as nonparametric MML-estimation for the present model.

Equation (10.31) defines restrictions on the parameters of the multinomial distribution. Without these restrictions there are $2^k - 1$ free parameters. It is clear that model (10.31) is not identified if the number of free parameters in $\boldsymbol{\xi}$ and $\boldsymbol{\varphi}$ jointly outgrows $2^k - 1$. The number of free parameters in the measurement model is given by (10.30). If a normal distribution is assumed for θ, two extra parameters are added. This means that a necessary condition for the existence of unique MML-estimates is that the inequality

$$\frac{(k-1)(k+2)}{2} + 2 \leq 2^k - 1 \tag{10.33}$$

must hold, and this is equivalent to $k \geq 3$. The problem of necessary and sufficient conditions for the existence and uniqueness of MML-estimates in this model is not solved, but the same is true for the RM.

Starting with the model defined by (10.29) and (10.32), it is possible to derive several interesting special cases by imposing linear restrictions on the parameters $\boldsymbol{\xi}$. Let $\boldsymbol{\eta}$ be an m-dimensional vector, $m < k(k+1)/2$, such that $\boldsymbol{\eta} = \boldsymbol{B}\boldsymbol{\xi}$, with $\boldsymbol{B}$ a (constant) matrix of rank m. Notice that the dimension of $\boldsymbol{\eta}$ is less than the number of virtual items, so that $\boldsymbol{\xi} = (\boldsymbol{B}'\boldsymbol{B})^{-1}\boldsymbol{B}'\boldsymbol{\eta}$, and the resulting model is an LLTM.

The following models are identified. (The normalization $\delta_0 = 0$ is assumed throughout.)

(i) The amount of learning depends on the number of successes on the preceding items, yielding the restriction

$$\xi_{ij} = \beta_i + \delta_j. \tag{10.34}$$

(ii) As a further restriction of the previous model, suppose that the amount of learning after each success is constant, yielding

$$\xi_{ij} = \beta_i + j\delta. \tag{10.35}$$

(iii) A two-operator model can be constructed by assuming that the amount of change in latent ability is not only a function of the number of previous successes, but also of previous failures. The most general version of a two-operator model with commutative operators is given by

$$\xi_{ij} = \beta_i + \delta_j + \varepsilon_{i-j-1}, \tag{10.36}$$

where $\varepsilon_0 = 0$. It can easily be checked that (10.36) is only a reparametrization of (10.29).

(iv) Analogously to (ii), it can be assumed that the amount of transfer is independent of the item, specializing (iii) further by imposing $\delta_j = j\delta$ and $\varepsilon_j = j\varepsilon$ for $j \neq 0$.

(v) It is possible to assume that the effect of an incorrect response is just the opposite of the effect of a correct response, by imposing the extra restriction $\delta = -\varepsilon$ on the model defined in (iv).

(vi) Finally, one could assume a limiting case where the amount of learning is the same irrespective of the correctness of the preceding responses. Formally, this can be modeled as a further restriction on (iv) by putting $\delta = \varepsilon$. In this case, however, the model is no longer identified if the rank order of presentation is the same for each respondent, because the parameter of each virtual item (i, j) is $\beta_i + (i-1)\delta$, and the value of δ can be freely chosen, because adding an arbitrary constant c to δ can be compensated for by subtracting $c(i-1)$, for $i > 1$, from β_i. Besides, in this case there is no more subject control. If, for example, the learning effect is caused by giving feedback after each response, or by the item text itself, or in generic terms by some reinforcer not under control of the respondent, the above model is also a limiting case of experimenter control, in the sense that there is no variability in the reinforcement schedule. So the solution to the identification problem is simple: introducing variability in the reinforcement schedule will solve the problem. For this restricted model, where the amount of learning increases linearly with the rank number of the item in the sequence, it suffices to administer the test in two different orders of presentation to two equivalent samples of subjects. Let, in the ordered pair (i, j) (i.e., the virtual item), i represent the identification label of the item and j the rank number in the presentation; then, $\xi_{ij} = \beta_i + (j-1)\delta$, and the model is identified if there is at least one i such that (i, j) and (i, j'), for $j \neq j'$, are virtual items.

It should be stressed that the cases (i) to (vi) are just relatively simple examples of learning models. They all imply that the susceptibility to learning is independent of the rank number of the item in the sequence. Model (ii), defined by (10.35), implies that, if $\delta > 0$, $\lim_{i\to\infty} P(X_i = 1|\theta) = 1$, for all (finite) values of β_i and θ. Such a prediction is not realistic in many situations where learning effects tend to level off due to habituation or satiation effects. A limit to the learning can be imposed by extending (10.35) to

$$\xi_{ij} = \beta_i + i^{-c} j\delta, \quad \text{for } c > 2. \tag{10.37}$$

10.4.1 The Relationship to Mathematical Learning Theory

Mathematical learning theory is an area that has gained much attention among mathematical psychologists in the early 1960's. Chapters 9, 10, and 17 of the Handbook of Mathematical Psychology (Luce, Bush, & Galanter, 1963) were entirely devoted to formal learning models and contain many interesting results. To get a good impression of the scope of these models, and of the problems that were recognized to be difficult, an example will be given. In a T-maze learning experiment, an animal is placed in a T-maze, where it has to choose between the left or right alley. Choosing one of these, it receives a food pellet, in the other, it receives nothing. In a simple learning model, it is assumed that (a) learning (i.e., a change in the tendency to choose the alley which yields the food reinforcer) occurs only on reinforced trials; (b) the 'inherent' difficulty of the situation is constant, and (c) there are no initial differences between the animals in the initial tendency to choose the reinforced alley. Of course this theory implies subject control. If the trials are identified with 'real items', then (b) and (c) imply that β_i is constant, say $\beta_i = \beta$, and that the initial ability θ is constant. The probability of a success on trial i, given r_i successes on previous trials, is given by

$$P(X_i = 1|\nu, R_i = j) = \frac{\nu\alpha^j}{1+\nu\alpha^j}, \tag{10.38}$$

where $\nu = \exp(\theta - \beta)$ and $\alpha = \exp(\delta)$, which is known as Luce's (1959) 'one-operator beta model'. Notice that (10.38) is a special instance of case (ii), defined by (10.35). If it is assumed that there is some learning following a non-reinforced trial, then

$$P(X_i = 1|\nu, R_i = j) = \frac{\nu\alpha_1^j\alpha_2^{i-j-1}}{1+\nu\alpha_1^j\alpha_2^{i-j-1}}, \tag{10.39}$$

with $\alpha_1 = \exp(\delta)$ and $\alpha_2 = \exp(\varepsilon)$, which is equivalent to Luce's 'two-operator beta-model'. So this model is just a special instance of case (iv) discussed above.

The assumptions of no variability in the difficulty parameters or in the initial θ are characteristic for the many learning models developed roughly between 1955 and 1970. The lack of variability in the difficulty parameters may be attributed mainly to the fact that most applications concerned experiments with constant conditions over trials, while the assumed constancy in initial ability was recognized as a problem: "... in most applications of learning models, it is assumed that the same values of the initial probability ... characterize all the subjects in an experimental group. ... It is convenience, not theory, that leads to the homogeneity assumption" (Sternberg, 1963, p. 99). The convenience has to be understood as the lack of tools at that time to incorporate individual differences as a genuine model component, and the rather crude estimation procedures used. Maximum likelihood methods were used, although rarely, only in conjunction with Luce's beta-model; but this model was by far less popular than the family of linear models introduced by Bush and Mosteller (1951), where the probability of a success can be expressed as a linear difference equation, while Luce's model can be written as a linear difference equation in the logit of success probability. Most estimation methods in the linear model were modified moment methods, frequently yielding problems, because it was clearly acknowledged that suppression of interindividual variability would underestimate the variance of almost every statistic: "... unless we are interested specifically in testing the homogeneity assumption, it is probably unwise to use an observed variance as a statistic for estimation, and this is seldom done" (Sternberg, 1963, p. 99).

The next example serves a triple purpose:

1. It is an example of a mixed model with non-commuting operators, that is, (10.23) applies, but not (10.26);

2. it illustrates the rich imagination of the older learning theorists and at the same time their struggle to handle rather complex mathematical equations; and

3. it yields a nice suggestion to construct a statistical test for the axiom of local stochastic independence in the RM.

The model is the 'logistic' variant of the 'one-trial perseveration model' of Sternberg (1959). The model was developed because the autocorrelation of lag 1 in the response pattern $\boldsymbol{X}$ in the T-maze experiment was larger than predicted by the theory of the one-operator linear model, suggesting that apparently there was a tendency to repeat the preceding response. Defining the choice of the non-rewarded response as a success, the model Sternberg proposed is

$$p_i = (1-b)a^{i-1}p_{i-1} + bX_{i-1}, \quad \text{for } i \geq 2,\ 0 < a, b < 1, \tag{10.40}$$

where $p_i = P(X_i = 1)$, a is a parameter expressing the learning rate, and b is a 'perseveration' parameter, expressing a tendency to repeat the previous response. The logistic analogue, mentioned but not analyzed in Sternberg (1963, p. 36), is given by

$$\text{logit}(p_i) = \theta + (i-1)\gamma + \delta X_{i-1}, \quad \text{for } i \geq 2, \tag{10.41}$$

where $\theta = \text{logit}(p_1)$ is treated as a constant. Equation (10.41) is readily recognized as a special case of (10.23), where $\beta_i = 0$, $g_i(\boldsymbol{Z}^i) = (i-1)\gamma$, and $f_i(\boldsymbol{X}^i) = \delta X_{i-1}$. Notice that $f_i(.)$ is not symmetrical, so (10.41) is not a special case of (10.26); moreover that (10.41) is more flexible than (10.40): by the restrictions put on the perseveration parameter b, tendencies to alternate the response require another model, while in (10.41) a positive δ expresses a perseveration tendency, and a negative δ expresses a tendency to alternate.

It is immediately clear that (10.41) violates the assumption of local stochastic independence. Now suppose attitude questionnaire data are to be analyzed by means of the RM, but there is some suspicion of response tendencies in the sense of, e.g., a tendency to alternate responses. Model (10.41) is readily adapted to this situation: set $\gamma = 0$, allow variation in the 'easiness' parameters β_i and in the latent variable θ. There are $2k-1$ virtual items: $(i,0)$, $(i,1)$ for $i > 1$, and $(1,1) \equiv (1,0)$, the second member of the ordered pairs being equal to the previous response. The assumption of local stochastic independence can be tested by means of a likelihood-ratio test using a restricted model where $\delta = 0$, but this is nothing else than the RM.

It is interesting to contrast the models discussed above with the dynamic item response model developed by Kempf (1974, 1977a, 1977b), where it is assumed that

$$P(X_i = 1|\boldsymbol{x}^i, \xi) = \frac{\xi + \tau(\boldsymbol{x}^i)}{\xi + \kappa_i}, \quad \text{for } \xi, \kappa_i > 0;\ 0 \leq \tau(\boldsymbol{x}^i) < \kappa_i, \tag{10.42}$$

where ξ is an ability parameter, κ_i is the difficulty parameter of item I_i, and $\tau(\boldsymbol{x}^i)$ represents the transfer as a result of the partial response vector $\boldsymbol{x}^i$. Defining $\varepsilon_i = \kappa_i^{-1}$ and dividing the numerator and denominator of (10.42) by κ_i yields

$$P(X_i = 1|\boldsymbol{x}^i, \xi) = \frac{[\xi + \tau(\boldsymbol{x}^i)]\varepsilon_i}{1 + \xi\varepsilon_i}, \tag{10.43}$$

which is equivalent to the RM if and only if the function τ is identically zero, i.e., if no transfer exists. The model defined in (10.43), which associates a new transfer parameter with each partial response vector, is hardly interesting, of course, but Kempf (1977b) demonstrates that the more restricted model defined by

$$P(X_i = 1|r_i, \xi) = \frac{\xi + \tau(r_i)}{\xi + \kappa_i}, \tag{10.44}$$

with r_i the partial sum score as defined in (10.24), has sufficient statistics for the person parameters, allowing for CML estimation of item and transfer parameters. Kempf (1977a, p. 306) notes that in the neighborhood of the maximum the likelihood surface may be very flat, causing instability and large standard errors of the parameter estimates. Apart from the numerical problems in the model, it again proves interesting to connect this model with mathematical learning theory. Let the assumption that the difficulty of the situation does not change be translated into $\kappa_i = \kappa$, and let ξ be constant over persons. Furthermore, let $\tau(0) = 0$. In order to satisfy the restriction that $0 \leq \tau(r) < \kappa$ for all r, it is assumed that $\tau(r) = a\tau(r-1) + (1-a)\kappa$, for $0 \leq a < 1$. Let p_i and p_{i-1} be the probability of a success on trial i and $i-1$, respectively; it then follows that

$$p_i = \begin{cases} ap_{i-1} + (1-a) & \text{if trial } i-1 \text{ is a success,} \\ p_{i-1} & \text{if trial } i-1 \text{ is a failure.} \end{cases} \tag{10.45}$$

This shows that Kempf's model is easily specialized such that the linear operator model by Bush and Mosteller (1951) results. Thus, the two basic approaches to mathematical learning theory, the linear operator model and the beta model, find natural generalizations in Kempf's model and the dynamic test model with incomplete design.

11

Linear and Repeated Measures Models for the Person Parameters

Herbert Hoijtink[1]

ABSTRACT Measurement instruments based on the Rasch model are often used to estimate the location θ of a person on a latent trait. This may be done once, or at several time points, or under several conditions. It is not unusual that researchers want to investigate hypotheses with respect to latent ability: Is the average arithmetic ability higher for men or for women? Can the attitude towards foreigners be predicted by age and socio-economic status? And, as a repeated measures example, does fear decrease after behavior therapy? The present chapter presents linear and repeated measures models for the person parameter, that can be used to investigate these kind of hypotheses.

11.1 Introduction

In principle, hypotheses with respect to the parameters of linear and repeated measures models for θ (henceforth called the person parameter) can be investigated using one of two methods. The first is to analyze estimates of θ using ANOVA, multiple regression, or models for repeated measures. An advantage of this method is its easy implementation employing standard software for multivariate statistics. A disadvantage is that the statistics of interest have to be corrected, since only estimates and not observed values of θ are available. The available correction formulas, however, are only asymptotically exact (Lord, 1983a; Hoijtink & Boomsma, 1991; see Chapter 4).

This chapter will discuss the second method of investigation of such hypotheses: directly modeling of θ as it appears in the marginal likelihood equation of the Rasch model (RM; see Chapter 3),

$$\ln L(G, \boldsymbol{\beta}|\boldsymbol{x}_1, ..., \boldsymbol{x}_n) = \sum_v \ln \int_\theta P(\boldsymbol{X}_v = \boldsymbol{x}_v|\theta, \boldsymbol{\beta})\, dG(\theta|\cdot), \tag{11.1}$$

where $G(\theta|\cdot)$ is the generic notation for the distribution function of θ (no specifications with respect to shape or parametrization are yet made). Note that here and in the sequel, unless indicated otherwise, all sums and products run from $v = 1, ..., n$, $i = 1, ..., k$, $q = 1, ..., Q$, or $t = 1, ..., T$. The probability of response vector $\boldsymbol{X}_v = \boldsymbol{x}_v$, conditional on the person location and the item difficulty parameters, is given by

[1]Rijksuniversiteit Groningen, Vakgroep Statistiek en Meettheorie, Grote Kruisstraat 2/I, 9712 TS Groningen, The Netherlands; e-mail: H.J.A.HOYTINK@PPSW.RUG.NL

$$P(\boldsymbol{X}_v = \boldsymbol{x}_v|\theta, \boldsymbol{\beta}) = \prod_i P(X_{vi} = x_{vi}|\theta) = \prod_i \frac{\exp[x_{vi}(\theta - \beta_i)]}{1 + \exp(\theta - \beta_i)}. \tag{11.2}$$

For the ANOVA/multiple regression kind of applications, the following linear model may be substituted for θ in (11.1):

$$\theta_v = \gamma_0 + \gamma_1 y_{1v} + \ldots + \gamma_P y_{Pv} + \epsilon_v = \psi_v + \epsilon_v, \tag{11.3}$$

where γ denotes a vector of regression coefficients, indexed $p = 1, ..., P$, relating the manifest predictors $\boldsymbol{y}$ to θ. The error of the prediction is denoted by ϵ. The predictors are either continuous variables, or dummy variables indicating group membership. Note that substitution of (11.3) in (11.1) implies replacement of $G(\theta|\cdot)$ by a related distribution function $H(\epsilon|\cdot)$. The simplest version of (11.3) is the model in which the best prediction of θ is given by the mean of its density function:

$$\theta_v = \gamma_0 + \epsilon_v. \tag{11.4}$$

If (11.4) is used, $G(\theta|\cdot)$ and $H(\epsilon|\cdot)$ only differ in their means, γ_0 and 0, respectively. Either the intercept γ_0 or the mean of $H(\epsilon|\cdot)$ is identifiable. This identification problem is solved constraining the mean of $H(\epsilon|\cdot)$ to zero.

Andersen (1980b), Mislevy (1985), and Verhelst and Eggen (1989) treat submodels of (11.3). The linear model (11.3) itself has been discussed by Zwinderman (1991b). The main distinction between their approaches concerns the cumulative distribution function of the error term $H(\epsilon|\cdot)$. Zwinderman uses $H(\epsilon|\sigma^2)$, a normal distribution with unknown variance, but alternatively one can use $H(\epsilon|\boldsymbol{\varepsilon}, \boldsymbol{\pi})$, a nonparametric density function with steps $\boldsymbol{\varepsilon}$ and weights $\boldsymbol{\pi}$. The rationale behind both methods will be discussed in the next section. Note that Zwinderman (1991c) also discusses models in which one or more of the predictors may be latent instead of manifest.

Repeated measures of θ can be modeled directly replacing the univariate cumulative density function $G(\theta|\cdot)$ by a multivariate function $G(\boldsymbol{\theta}|\cdot)$, where the number of dimensions equals the number of measures. Furthermore, the item responses will be labeled by the time point at which they are obtained. Models of this kind have previously been discussed by Andersen (1985), Glas (1989, 1992), and Mislevy (1985). They all use $G(\boldsymbol{\theta}|\boldsymbol{\mu}, \boldsymbol{\Sigma})$, a multivariate normal distribution with mean vector $\boldsymbol{\mu}$ and covariance matrix $\boldsymbol{\Sigma}$. In Chapter 8 and 9, the LLTM and LLRA models are presented. Both models differ from the models presented in this chapter, but can also be used for the analysis of repeated measures.

In Section 11.2, the distinction between measurement models and structural models (a generic name for linear and repeated measures models) on the one hand, and normal versus nonparametric error term density function on the other, will be discussed. Sections 11.3, 11.4, and 11.5 introduce linear models with a normal or nonparametric error term density function, and repeated measures models. In Section 11.6, EM and Newton-Raphson based estimation procedures will be briefly presented. Section 11.7 shows how likelihood ratio tests may be used to compare nested models. Section 11.8 gives a discussion of the contents of this chapter.

11.2 The Separation of Measurement Model from Structural Model

Item response models consist of two parts: the measurement model given by the item locations, and the structural model given by models for the person locations. Verhelst and Eggen (1989) note that in an ideal situation the calibration of the measurement model should be independent of the estimation of the parameters of the structural model.

Suppose we want to investigate whether the average arithmetic ability in the Netherlands differs between men and women. Then, either of the following two structural models might be appropriate: the density function of arithmetic ability is normal with the same mean for men and women; or, the density function is normal with a different mean for men and women. If the measurement model (the estimates of the item locations) is the same for both structural models, they can be compared using one of the methods described in Sections 11.3 and 11.4.

A measurement model can differ between structural models only as a result of specification errors made in one or more of the structural models. This has two consequences: since the measurement model differs between the structural models, they cannot be validly compared; and, since it is unknown which of the structural models is correct, it is impossible to decide which of the measurement models is correct.

For the RM, the measurement model can be separated from the structural model, using (3.13) in Chapter 3. The marginal likelihood equation (11.1) can be rewritten as

$$\begin{aligned}
&\ln L(G,\boldsymbol{\beta}|\boldsymbol{x}_1,...,\boldsymbol{x}_n)\\
&= \sum_v \ln \int_\theta P(\boldsymbol{X}_v=\boldsymbol{x}_v|r_v,\boldsymbol{\beta})\, P(R_v=r_v|\theta,\boldsymbol{\beta})\, dG(\theta|\cdot)\\
&= \sum_v \ln P(\boldsymbol{X}_v=\boldsymbol{x}_v|r_v,\boldsymbol{\beta}) \int_\theta P(R_v=r_v|\theta,\boldsymbol{\beta})\, dG(\theta|\cdot) \qquad (11.5)\\
&= \sum_v \ln P(\boldsymbol{X}_v=\boldsymbol{x}_v|r_v,\boldsymbol{\beta}) + \sum_v \ln \int_\theta P(R_v=r_v|\theta,\boldsymbol{\beta})dG(\theta|\cdot).
\end{aligned}$$

The first term of the last line of (11.5) renders CML estimation of the item parameters possible (Chapter 3). These estimates are independent of the structural model used since θ is conditioned out. Note that (11.5) is equivalent to the likelihood function of model (11.4) and can easily be generalized to apply to model (11.3), or to a repeated measures model (see below). Mislevy (1985), Verhelst and Eggen (1989), and Zwinderman (1991b) all suggest to estimate the item parameters by CML, and to consider them known when the parameters of the structural model have to be estimated.

Separation of measurement model and structural model cannot be achieved if MML is used to calibrate the measurement instrument (estimation is based on both terms in (11.5)). As a consequence, specification errors in either $G(\theta|\cdot)$ or $H(\epsilon|\cdot)$ (i.e., incorrect assumptions about either density) will lead to a bias of the item parameter estimates (see also Chapter 3). Both Glas (1989) and Zwinderman (1991a) illustrate that the MML-estimates of the item parameters will be biased if $G(\theta|\cdot)$ is incorrectly assumed to be normal (which would be the case, for example, if the density function of θ is normal for each gender, but one fails to recognize that it has a different mean for men and women).

Both De Leeuw and Verhelst (1986) and Pfanzagl (1993, 1994) show for model (11.4) that CML and MML estimates of the item parameters are asymptotically equivalent if $H(\epsilon|\cdot)$ is modeled using a nonparametric density function (De Leeuw and Verhelst, 1986; Follman, 1988; Lindsay, Clogg and Grego, 1991). Their result implies for model (11.4) that specification errors can be avoided using a nonparametric density function for $H(\epsilon|\cdot)$. It is not unlikely that these results can be extended to cover (11.3) as well. This has an important implication: structural models with nonparametric density functions for ϵ can validly be compared, treating CML estimates of item parameters as if they were the true parameters.

The previous line of argument does not imply that structural models with normal density functions have to be discarded. If the normality assumption is correct, item parameter estimates will not be affected by specification errors. It is then an advantage of the normal over the nonparametric model that the former is much easier to work with.

11.3 Linear Models for the Person Parameter: Normal Error-Term Density Functions

Model (11.3) may be substituted for θ in (11.2). This yields an equation that is basic whenever linear models for the person parameter are considered:

$$\begin{aligned} P(\boldsymbol{X}_v = \boldsymbol{x}_v|\psi_v, \epsilon_v, \boldsymbol{\beta}) &= P(\boldsymbol{X}_v = \boldsymbol{x}_v|r_v, \boldsymbol{\beta})\, P(R_v = r_v|\psi_v, \epsilon_v, \boldsymbol{\beta}) \\ &= \prod_i P(X_{vi} = x_{vi}|\psi_v, \epsilon_v, \beta_i), \end{aligned} \tag{11.6}$$

where, according to the RM,

$$P(X_{vi} = x_{vi}|\psi_v, \epsilon_v, \beta_v) = \frac{\exp(\psi_v + \epsilon_v - \beta_i)x_{vi}}{1 + \exp(\psi_v + \epsilon_v - \beta_i)}. \tag{11.7}$$

The marginal likelihood function of the linear model of interest, assuming that the cumulative density function $H(\epsilon|\sigma^2)$ is normal with mean μ equal to zero and variance σ^2, can be written as

$$\begin{aligned} &\ln L(\gamma, H|\boldsymbol{x}_1, ..., \boldsymbol{x}_n, \boldsymbol{y}_1, ..., \boldsymbol{y}_n, \boldsymbol{\beta}) = \ln L(\gamma, H|\cdot) \\ &= \ln \prod_v \int_\epsilon P(\boldsymbol{X}_v = \boldsymbol{x}_v|r_v, \boldsymbol{\beta})\, P(R_v = r_v|\psi_v, \epsilon, \boldsymbol{\beta})\, dH(\epsilon|\sigma^2) \\ &= \sum_v \ln P(\boldsymbol{X}_v = \boldsymbol{x}_v|r_v, \boldsymbol{\beta}) + \sum_v \ln \int_\epsilon P(R_v = r_v|\psi_v, \epsilon, \boldsymbol{\beta})\, dH(\epsilon|\sigma^2). \end{aligned} \tag{11.8}$$

Note that the first term of the last line of (11.8) does not depend on ψ or σ^2, the parameters of the structural model in which we are interested. As far as estimation of these structural parameters is concerned, the first term of the last line of (11.8) therefore may be treated as a constant (the item parameters are assumed to be known). Looking at the second term, it can then be seen that the distribution of the sufficient statistic for θ contains all the information required for estimating the parameters of the structural model.

In Section 11.6 the estimation of the structural parameters contained in (11.8) will shortly be discussed. Asymptotically correct standard errors of the estimates are obtained after inversion of the Fisher information matrix with respect to the parameters of interest. The interested reader is referred to Zwinderman (1991b). Note that Zwinderman (1991b) also provides the estimation equations and information matrix for the case where (11.8) is used to estimate the parameters of the measurement and structural model simultaneously.

11.4 Linear Models for the Person Parameter: Nonparametric Error-Term Density Functions

Note that (11.8) is easily generalized to an arbitrary $H(\epsilon|\cdot)$. Thus it can be inferred that the distribution of the sufficient statistic contains all the information necessary to estimate the parameters of the structural model, irrespective of the shape and parameterization of $H(\epsilon|\cdot)$. The paper by Lindsay (1983) implies that $H(\epsilon|\cdot)$ may be represented nonparametrically using a step function with steps $\boldsymbol{\varepsilon}$ and weights $\boldsymbol{\pi}$. It also implies that the nonparametric representation has at most $Q = n$ steps for model (11.3), and at most $Q = k + 1$ for model (11.4), if estimates of the steps and weights are obtained from (11.9) by means of maximum likelihood; this means that the addition of steps beyond a certain number will not lead to an increase in the likelihood. The maximum likelihood estimates may not always be unique, see Lindsay (1983) for a discussion of uniqueness.

The adjective 'nonparametric' means that no assumptions with respect to $H(\epsilon|\cdot)$ have to be made except for its existence. Note that the results of Lindsay (1983) apply to a very general class of models. De Leeuw and Verhelst (1986), Follman (1988), and Lindsay, Clogg and Grego (1991) discuss the nonparametric representation of $H(\epsilon|\cdot)$ for the case of the RM without substitution of a linear model for θ, that is, (11.4); see Chapter 3 for the details.

The likelihood function of the nonparametric model is

$$\ln L(\boldsymbol{\gamma}, \boldsymbol{\varepsilon}, \boldsymbol{\pi}|\boldsymbol{x}_1, ..., \boldsymbol{x}_n, \boldsymbol{y}_1, ..., \boldsymbol{y}_n, \boldsymbol{\beta}) = \ln L(\boldsymbol{\gamma}, \boldsymbol{\varepsilon}, \boldsymbol{\pi}|\cdot)$$
$$= \sum_v \ln P(\boldsymbol{X}_v = \boldsymbol{x}_v|r_v, \boldsymbol{\beta}) + \sum_v \ln \sum_q P(R_v = r_v|\psi_v, \varepsilon_q, \boldsymbol{\beta})\pi_q. \tag{11.9}$$

The summations in (11.9) through (11.11) run over $q = 1, ..., Q$. In Section 11.6, estimation of the parameters contained in (11.9) will be discussed briefly. Two restrictions on the parameter space are necessary. As usual the mean of the error term density function is restricted to zero:

$$\sum_q \varepsilon_q \pi_q = 0, \tag{11.10}$$

but also,

$$\sum_q \pi_q = 1. \tag{11.11}$$

Asymptotic standard errors of the estimates could be obtained from the inverse of the Fisher information matrix with respect to the parameters of interest. The necessary formulas, however, have not yet been derived.

For the structural model (11.4), the likelihood function (11.9) is formally equivalent to a latent class representation of the RM (see Chapter 13). The parameters $\theta_q = \gamma_0 + \varepsilon_q$, for $q = 1, ..., Q$, may be interpreted as the locations of the respective classes on the latent trait, the parameters $\boldsymbol{\pi}$ as the corresponding class weights. In Chapter 13, it is shown that the number of classes needed to saturate the likelihood function of the latent class RM is equal to $(k+1)/2$ if the number of items is odd, and $k/2 + 1$ if the number of items is even. In addition, if the number of items is even, one of the classes will be located at $-\infty$. This result corresponds with the results obtained by De Leeuw and Verhelst (1986), Follman (1988), and Lindsay, Clogg and Grego (1991).

Note that the latent class interpretation is also possible for the general linear model (11.3). There the nonparametric error term density function can be interpreted in terms of latent classes. Since $H(\epsilon|\boldsymbol{\varepsilon}, \boldsymbol{\pi})$ can also be interpreted as the nonparametric equivalent of an unknown but continuous density function $H(\epsilon|\cdot)$, it would be interesting to investigate how well the moments of $H(\epsilon|\boldsymbol{\varepsilon}, \boldsymbol{\pi})$ and $H(\epsilon|\cdot)$ correspond. Some simulations (n=200) done by the author indicate that for $k = 5$ the moments of the nonparametric estimate of $H(\epsilon|\cdot)$ are severely biased and have a large standard error. Using $k = 15$, both the bias and the standard error decreased substantially. The number of steps needed before the likelihood function stopped increasing was never more than $(k+2)/2$ for both models (11.3) and (11.4).

11.5 Repeated-Measures Models for the Person Parameters

Repeated measures models have been discussed by Andersen (1980b; 1985), Mislevy (1985), and Glas (1989; 1992). Note that the work of the last two authors is not specifically aimed at the analysis of repeated measures of the person parameter. They discuss more general models where the latent person parameter has a multivariate instead of a univariate density function.

The simplest appearance of repeated measures arises when the same set of items is presented to a sample of persons at different time points/under different conditions. More complex models have received attention in the literature: (partly) different sets of items at different time points/under different conditions (Glas, 1992); or, integration of a repeated measures model with a structural model like (11.3) (Mislevy, 1985). In this section however, attention will be restricted to the simplest model.

Note that (two) other approaches to the analysis of repeated measures exist that differ markedly from the approach taken by the authors mentioned above: the LLTM for the measurement of change, and the LLRA (see Chapter 9). The models to be discussed in this section assume that the responses at each time point can be explained by a person's ability at that time point (i.e., the RM). What is modeled is the shift in ability across time points. The LLTM for the measurement of change models the shift of the item parameters on one latent dimension across time points (in its simplest version). This shift can be interpreted as a time effect. The LLRA, on the other hand, is not a unidimensional IRT model: It assumes that each item response can be explained by an item-specific ability (i.e., one ability dimension for each item); what is modeled is the shift in this multidimensional ability space across time points.

The repeated measures model to be discussed here may be formalized as follows: Let a response vector $\boldsymbol{X}$ consist of $t = 1, ..., T$ vectors $\boldsymbol{X}^t$, containing the item-responses at time point t. Let $\boldsymbol{\theta}$ be a vector containing the person's parameter at time points $t = 1, ..., T$, which has a multivariate normal distribution $g(\boldsymbol{\theta}|\boldsymbol{\mu}, \boldsymbol{\Sigma})$ with mean vector $\boldsymbol{\mu} = [\mu^1, ..., \mu^T]$, and covariance matrix $\boldsymbol{\Sigma}$. Then, assuming the item parameters to be known and equal across time points, the probability of observing response vector $\boldsymbol{X}$ is given by

$$P(\boldsymbol{X} = \boldsymbol{x}) = \int_{\boldsymbol{\theta}} \prod_t P(\boldsymbol{X}^t = \boldsymbol{x}^t|\theta^t, \boldsymbol{\beta})\, dG(\boldsymbol{\theta}|\boldsymbol{\mu}, \boldsymbol{\Sigma}). \tag{11.12}$$

The likelihood function of the parameters of interest is then given by

$$\begin{aligned}
\ln L(\boldsymbol{\mu}, \boldsymbol{\Sigma}|\boldsymbol{x}_1, ..., \boldsymbol{x}_n, \boldsymbol{\beta}) &= \ln L(G|\cdot) \\
&= \ln \prod_v \int_{\boldsymbol{\theta}} \prod_t P(\boldsymbol{X}_v^t = \boldsymbol{x}_v^t|\theta^t, \boldsymbol{\beta})\, dG(\boldsymbol{\theta}|\boldsymbol{\mu}, \boldsymbol{\Sigma}) \\
&= \ln \prod_v \int_{\boldsymbol{\theta}} \prod_t P(\boldsymbol{X}_v^t = \boldsymbol{x}_v^t|r_v^t, \boldsymbol{\beta})\, P(R_v^t = r_v^t|\theta^t, \boldsymbol{\beta})\, dG(\boldsymbol{\theta}|\boldsymbol{\mu}, \boldsymbol{\Sigma}) \\
&= \sum_v \sum_t \ln P(\boldsymbol{X}_v^t = \boldsymbol{x}_v^t|r_v^t, \boldsymbol{\beta}) + \sum_v \ln \int_{\boldsymbol{\theta}} \prod_t P(R_v^t = r_v^t|\theta^t, \boldsymbol{\beta})\, dG(\boldsymbol{\theta}|\boldsymbol{\mu}, \boldsymbol{\Sigma}).
\end{aligned} \tag{11.13}$$

Note that the first term of the last line of (11.13) does not depend on either $\boldsymbol{\mu}$ or $\boldsymbol{\Sigma}$, the parameters of the structural model. As far as estimation of these parameters is concerned, this term may be treated as a constant (the item parameters are assumed to be known). The second term of the last line of (11.13) shows that the distribution of the sufficient statistic for θ at each time point t contains all the information that is necessary to estimate the parameters of the structural model.

Asymptotically correct standard errors of the estimates can be obtained from the inverse of the Fisher information matrix of the parameters of interest. The interested reader is referred to Mislevy (1985) or Glas (1989) for the details. Note that $G(\boldsymbol{\theta}|\boldsymbol{\mu}, \boldsymbol{\Sigma})$ can in principle be replaced by a nonparametric multivariate distribution. The author is not aware, however, of any work addressing this kind of model.

11.6 Estimation Procedures

For any of the models discussed above, direct estimation of the parameters of interest is possible using a (quasi) Newton-Raphson procedure that finds the parameter values maximizing the respective likelihood functions under the appropriate restrictions. Verhelst and Eggen (1989) explicitly discuss Newton-Raphson. A disadvantage of the Newton-Raphson procedure is that it may fail if the initial parameter estimates are inaccurate.

An alternative estimation procedure is given by the EM-algorithm (Dempster, Laird, and Rubin, 1977), an iterative two-step procedure which at least converges to a local maximum of the likelihood function. For the linear model with a normal error-term density function, and the repeated measures model, the EM-algorithm will always converge to a global maximum (see below).

In the first step (E-step), improved estimates of the marginal posterior density of each person parameter (the missing information as Dempster et al. call it) are computed, conditional upon the data and the current estimates of the structural model parameters. In the second step (M-step), the expected value of what Dempster et al. call the 'complete data likelihood', is maximized conditional on the current estimates of the marginal posterior density functions of each person parameter. The EM-algorithm iterates between both steps until the structural model parameters converge. The EM-algorithm is very popular; applications can be found in Glas (1989; 1992), Mislevy (1985), Verhelst and Eggen (1989), and Zwinderman (1991b).

The likelihood functions of the models discussed above are very similar. Rewriting (11.8), (11.9), and (11.13) in a compact form, we obtain for the normal model

$$\ln L(\gamma, H|\cdot) \propto \sum_v \ln \int_\epsilon P(r_v|\psi_v, \epsilon)\, dH(\epsilon; \sigma^2), \tag{11.14}$$

and for the nonparametric model,

$$\ln L(\gamma, H|\cdot) \propto \sum_v \ln \int_\epsilon P(r_v|\psi_v, \epsilon)\, dH(\epsilon; \boldsymbol{\varepsilon}, \boldsymbol{\pi}). \tag{11.15}$$

Note that restriction (11.10) may be effectuated constraining γ_0 to zero; restriction (11.11) can be effectuated using

$$\pi_Q = 1 - \sum_{q=1}^{Q-1} \pi_q. \tag{11.16}$$

For the repeated measures model we find

$$\begin{aligned} \ln L(\boldsymbol{\mu}, \boldsymbol{\Sigma}|\cdot) &\propto \sum_v \ln \int_{\boldsymbol{\theta}} \prod_t P(r_v^t|\theta^t)\, dG(\boldsymbol{\theta}|\boldsymbol{\mu}, \boldsymbol{\Sigma}), \\ &\propto \sum_v \ln \int_{\boldsymbol{\theta}} P(\boldsymbol{r}_v|\boldsymbol{\theta})\, dG(\boldsymbol{\theta}|\boldsymbol{\mu}, \boldsymbol{\Sigma}). \end{aligned} \tag{11.17}$$

All three likelihoods (11.14), (11.15), and (11.17) can be derived from the general form

$$\ln L(\gamma, \boldsymbol{\alpha}|\cdot) \propto \sum_v \ln \int_{\boldsymbol{\phi}} P(r_v|\gamma, \boldsymbol{\phi})\, dF(\boldsymbol{\phi}|\boldsymbol{\alpha}), \tag{11.18}$$

where, $\boldsymbol{\alpha}$ is a generic term for the parameters defining the cumulative distribution function F of $\boldsymbol{\phi}$, where $\boldsymbol{\phi}$ denotes the parameter to be integrated (either ϵ or $\boldsymbol{\theta}$).

The complete data likelihood of (11.18) is obtained assuming $\boldsymbol{\phi}$ to be observed instead of latent:

$$\ln L(\gamma, \boldsymbol{\alpha}|\cdot, \boldsymbol{\phi}) \propto \sum_v \ln P(r_v|\gamma, \boldsymbol{\phi})\, f(\boldsymbol{\phi}|\boldsymbol{\alpha}). \tag{11.19}$$

Dempster, Laird and Rubin (1977) and Little and Rubin (1987, p. 136) show that the EM-algorithm converges to the global maximum of (11.19) if the complete data likelihood $L(\boldsymbol{\gamma}, \boldsymbol{\alpha}|\cdot, \boldsymbol{\phi})$ (where the dot represents the data and the item parameters appropriate for the model at hand), not to be confused with the likelihood function (11.18), is a member of the exponential family. Note that this is the case for both the linear model with a normal error-term density function, and the repeated measures model.

In the M-step, the values of the parameters that maximize the expected value of the complete data likelihood have to be estimated. This is done conditional upon the current estimates (initial estimates, or estimates obtained from the previous iteration) of the marginal posterior density function of $\boldsymbol{\phi}$ for each person,

$$z(\boldsymbol{\phi}|r_v, \boldsymbol{\gamma}, \boldsymbol{\alpha}) = z_v(\boldsymbol{\phi}) = \frac{P(r_v|\boldsymbol{\gamma}, \boldsymbol{\phi}) f(\boldsymbol{\phi}|\boldsymbol{\alpha})}{\int_{\boldsymbol{\phi}} P(r_v|\boldsymbol{\gamma}, \boldsymbol{\phi})\, dF(\boldsymbol{\phi}|\boldsymbol{\alpha})}. \tag{11.20}$$

Let $\boldsymbol{z}$ denote $\{z_v(\boldsymbol{\phi})\}$ for $v = 1, ..., n$, then

$$E[\ln L(\boldsymbol{\gamma}, \boldsymbol{\alpha}|\cdot, \boldsymbol{\phi})\, |\boldsymbol{z}\,] = \sum_v \int_{\boldsymbol{\phi}} \ln P(r_v|\boldsymbol{\gamma}, \boldsymbol{\phi})\, f(\boldsymbol{\phi}|\boldsymbol{\alpha})\, dZ_v(\boldsymbol{\phi}). \tag{11.21}$$

In the E-step, the marginal posterior densities $\boldsymbol{z}$ are updated using the current estimates of the structural parameters in combination with (11.20). The EM-algorithm iterates between E-step and M-step to convergence of the structural parameter estimates.

The EM-algorithm converges slowly, but in the first few iterations the estimates improve rapidly. Consequently, the best optimization strategy may be to start with relatively crude initial estimates, do a few EM iterations (thus eliminating the problem of inadequate initial parameter estimates, since each EM cycle by definition increases the likelihood) and to finish with the (quasi) Newton-Raphson procedure. Several authors have advocated this strategy (see, for example, Follman, 1988, and Verhelst and Eggen, 1989).

11.7 Model Comparison

If the parameter space Ω_1 of $L_1(...)$ is a subspace of the parameter space Ω_2 of $L_2(...)$, then the statistic

$$LR = -2\,[\ln L_1(...) - \ln L_2(...)] \tag{11.22}$$

is asymptotically χ^2-distributed with degrees of freedom equal to the difference in dimensionality (number of parameters estimated) between Ω_1 and Ω_2 (Mood, Graybill, and Boes, 1974, pp. 440-442). This principle can be used to compare most of the models discussed so far: The likelihood $L_1(...)$ of any model obtained constraining one or more of the regression coefficients appearing in (11.3) to zero (i.e., no relation between the corresponding Y variable and θ), may be compared to the likelihood $L_2(...)$ of (11.3) without these constraints using (11.22). Note that models with a nonparametric error-term density function can only be compared using this method if the number of steps used for $H(\epsilon|\boldsymbol{\varepsilon}, \boldsymbol{\pi})$ is the same in each model.

The same principle can be used to compare different repeated measures models: the likelihood $L_1(...)$ obtained constraining some of the means in the vector $\boldsymbol{\mu}$ or (co)variances in $\boldsymbol{\Sigma}$ in (11.13) to equality, may be compared to the likelihood $L_2(...)$ of (11.13) without these constraints, using (11.22).

11.8 Discussion

The linear and repeated-measures models for the person parameter in the RM presented in this chapter have been developed mainly in the last decennium. (Note, however, that in Chapter 8 and 9 the LLTM and LLRA models are presented, providing an alternative to the repeated measures models addressed in this chapter.) The state of the art encompasses parameter estimation and model comparison for the counterparts (with latent dependent variables) of the classical analysis of (co)variance, repeated measures analysis, and multiple regression. It should be noted that all structural models discussed and most results obtained so far are also valid in the context of the OPLM model (Verhelst and Eggen, 1989; Chapter 14). Furthermore, generalizations such that models and results apply to the polytomous RM are also feasible (Glas, 1989; 1992).

Although the models have a sound theoretical foundation, it is not always easy to apply them to practical problems. For example, to avoid convergence to a local maximum, initial estimates should be relatively accurate if the linear model with a nonparametric error term density function is used; moreover, Glas (1992) notes that, due to practical problems with multivariate numerical intergration, it is difficult to analyze more than three repeated measures.

Interesting topics for future research will be multilevel models with the person parameter as the dependent variable, linear models where the person parameter functions as a predictor, residual analysis, and influence measures like a counterpart for Cook's distance.

A nice by-product of the methodology presented in this chapter is the possibility to use secondary characteristics of a person in addition to his/her response vector to obtain an estimate of latent ability. Bock and Aitkin (1981) present the Bayes expected a posteriori estimator (EAP) of θ, see also (4.20),

$$EAP = \frac{\int_\theta \theta\, P(\boldsymbol{X}_v|\theta)dG(\theta|\cdot)}{\int_\theta P(\boldsymbol{X}_v|\theta)\, dG(\theta|\cdot)}; \tag{11.23}$$

it is easily generalized to incorporate information from secondary person characteristics, see also (4.22),

$$EAP = \psi_v + \frac{\int_\epsilon \epsilon \, P(\boldsymbol{X}_v|\psi_v + \epsilon) \, dH(\epsilon|\cdot)}{\int_\epsilon P(\boldsymbol{X}_v|\psi_v + \epsilon) \, dH(\epsilon|\cdot)}. \tag{11.24}$$

See Klinkenberg (1992) for a paper where (11.24) is compared to (11.23) and to maximum likelihood estimators of θ.

12

The One Parameter Logistic Model

Norman D. Verhelst and Cees A. W. Glas[1]

ABSTRACT The One Parameter Logistic Model (OPLM) is an item response model that combines the tractable mathematical properties of the Rasch model with the flexibility of the two-parameter logistic model. In the OPLM, difficulty parameters are estimated and discrimination indices imputed as known constants. Imposing a very mild restriction on these indices makes it possible to apply CML to this model. Moreover, the model is an exponential family which makes it possible to construct easily a large class of statistical goodness-of-fit tests by means of the methods discussed in Chapter 5. In particular, since the discrimination indices are not estimated, but imputed by hypothesis, it is important to test the adequacy of their values. Two item-oriented tests which are especially powerful against misspecification of the discrimination indices are discussed in detail. An example is added.

12.1 Introduction

From a statistical point of view, the Rasch model (RM) is a null hypothesis. In the calibration phase, item parameters are estimated and one or more goodness-of-fit tests are performed. If these tests yield a significant result, the null hypothesis has to be rejected, meaning that the RM is not a valid model for the data at hand. In such a situation, the possible actions of the researcher are rather limited: one or more items may be eliminated in the hope that the RM is valid for the smaller set of items. Or, as an alternative explanation, the deviations might originate from a (small) subset of testees, which may be identified by the techniques described in Chapter 6. Eliminating these persons from the sample might then save the RM. However, eliminating persons from the sample may have far-reaching consequences with regard to the generalizability of the results. It is true, as was argued in Chapter 2, that the sample of persons does not need to be a random sample from the target population, but it is also true that the persons in the sample are not tested only for their own sake, but as representatives for some more or less accurately specified population. Elimination of persons from the sample may make the definition of the population for which the test is valid a rather cumbersome task. Similarly, elimination of items may have severe drawbacks. In general, inclusion of an item in a test is done for reasons which are well covered by the concept of content validity. Elimination on pure statistical grounds may afflict the content validity seriously. Moreover, construction of items in a professional setting, such as a testing bureau, may be

[1]National Institute for Educational Measurement, P.O.-Box 1034, 6801 MG Arnhem, The Netherlands; fax: (+31)85-521356

quite expensive and elimination of items for mere statistical reasons can conflict with other more content-oriented or even purely economic points of view.

A more drastic strategy, viz., turning to another model, also must be considered with care. The RM has appealing theoretical advantages which one might not be inclined to give up easily; for example, the existence of CML estimators for the item parameters and the sampling independence implied by it. These advantages have their price, though. The assumptions of the RM are rather strict and from a certain viewpoint not very realistic. If two items in a test consisting of Rasch homogeneous items have the same difficulty parameter, it can easily be shown that the item-test correlations of these two items should be equal in every population. Although item-test correlation is only a rough index of item discrimination power (Lord & Novick, 1968, p. 331), it has been used in classical test theory as a criterion for item analysis: eliminating the item with the lowest item-test correlation will in many cases increase the homogeneity of the test as measured for example by the KR-20 formula, see Gulliksen (1950, p. 379). (The technique proposed there is a bit more complicated, because it also takes the difference in item variances into account.) In the RM, a reasonable technique of item analysis is to eliminate items which yield significant U-tests or M-tests (see Chapter 5). The net result of this approach will generally be the elimination of the least and the best discriminating items. This shows very clearly the possible conflict arising from a rather restrictive model facing more wildly behaving data. The standard answer of many psychometricians is to choose a more flexible model that can handle unequal discriminating power of the items (Lord, 1980, p.190). A natural generalization of the RM, capable of handling unequal discriminations of the items, is the so-called two parameter logistic model (2PLM) or Birnbaum model (Birnbaum, Chapter 17 in Lord & Novick, 1968). The item response function of the 2PLM is

$$f_i(\theta) = \frac{\exp[\alpha_i(\theta - \beta_i)]}{1 + \exp[\alpha_i(\theta - \beta_i)]}, \quad \text{for } i = 1, \dots, k, \tag{12.1}$$

where $\alpha_i > 0$ is the discrimination parameter, and β_i the difficulty parameter of item I_i, for $i = 1, \dots, k$. Using the usual assumption of local stochastic independence, it is easily verified that the likelihood function, given a response pattern $\boldsymbol{x} = (x_1, ..., x_k)$, in the 2PLM is

$$L(\theta, \boldsymbol{\alpha}, \boldsymbol{\beta}; \boldsymbol{x}) = \prod_i f_i(\theta)^{x_i}[1 - f_i(\theta)]^{1-x_i} = \frac{\exp(\theta \sum_i \alpha_i x_i - \sum_i \alpha_i \beta_i x_i)}{\prod_i \{1 + \exp[\alpha_i(\theta - \beta_i)]\}}. \tag{12.2}$$

Notice that, if the α_i are known constants, the weighted score

$$r = \sum_i \alpha_i x_i \tag{12.3}$$

is a minimal sufficient statistic for θ. This means that all respondents having the same weighted score must have also the same estimate of θ. But since in the two parameter model the α_i are considered as unknown quantities to be estimated from the data, it follows that r is not a mere statistic, and hence it is impossible to use CML as an estimation method.

Also the application of other estimation methods presents a number of problems. For instance, simultaneous estimation of all parameters by maximization of (12.2) leads to an identification problem. To see this, notice first that two obvious identification constraints should be imposed on (12.1): the value of the item response function does not change if θ is transformed into $\theta^* = c(\theta + b)$, β_i into $\beta_i^* = c(\beta_i + d)$ and α_i into $\alpha_i^* = c^{-1}\alpha_i$, for arbitrary d and $c > 0$. To eliminate this indeterminacy, one could put $\beta_1 = 0$ and $\alpha_1 = 1$, for example, leaving $2(k-1)$ free item parameters. The likelihood equations for the θ-parameter are given by (see, e.g., Fischer, 1974, p. 276)

$$\sum_i \alpha_i x_{vi} = \sum_i \alpha_i f_i(\theta_v), \quad \text{for } v = 1, \ldots, n, \tag{12.4}$$

where the subscript v denotes the v-th observation in a sample of size n. Since $0 < f_i(\theta) < 1$ for all finite values of θ, it follows immediately that there is no finite solution of (12.4) if $\boldsymbol{x} = \mathbf{0}$ or $\boldsymbol{x} = \mathbf{1}$. This is, however, only a minor problem, which may be solved by excluding $\mathbf{0}$ and $\mathbf{1}$ from the sample space, leaving $2^k - 2$ response patterns in the restricted sample space. From (12.4) it follows immediately that $\hat{\theta}_v = \hat{\theta}_w$ whenever $\boldsymbol{x}_v = \boldsymbol{x}_w$. Since the parameter space of the discrimination parameters is $\mathbb{R}^{+(k-1)}$, there exist specific cases of (12.2) where all $2^k - 2$ response patterns yield a different weighted score (e.g., the case where $\alpha_i = 2^{i-1}$), and since the right hand member of (12.4) is strictly increasing in θ, there exist $2^k - 2$ different solutions of (12.4), or, in a finite sample, as many different estimates as there are different response patterns belonging to the restricted sample space.

The parametric multinomial model defined by

$$P(\boldsymbol{x}) = \frac{G(\boldsymbol{x})}{\sum\limits_{\boldsymbol{y} \in \mathcal{X}} G(\boldsymbol{y})}, \tag{12.5}$$

where $\mathcal{X}$ denotes the restricted sample space, and

$$G(\boldsymbol{x}) = \frac{\exp(\theta_{\boldsymbol{x}} \sum_i \alpha_i x_i - \sum_i \alpha_i \beta_i x_i)}{\prod_i \{1 + \exp[\alpha_i(\theta_{\boldsymbol{x}} - \beta_i)]\}}, \tag{12.6}$$

has the same likelihood equations and the same solutions (if they exist) as the 2PLM. But the parametric multinomial model only has $2^k - 3$ degrees of freedom in the restricted sample space, viz., the frequencies of the observed response patterns (under the restriction that their sum equals the sample size), but the full model contains $2^k - 2$ person parameters and $2(k-1)$ item parameters; so it is overparametrized and therefore not identified.

Another solution to the estimation problem in the 2PLM is to assume that the person parameters θ have a common distribution and to employ marginal maximum likelihood (MML, see Bock & Aitkin, 1984) estimation. However, also with MML the discrimination parameters α_i seem difficult to estimate (see, e.g., Mislevy & Bock, 1986, pp. 3–18), and therefore Mislevy (1986) introduced the assumption that the discrimination parameters have a common log-normal distribution. With these additional assumptions concerning the distribution of person and discrimination parameters, the model is no longer an exponential family, which further complicates the estimation and testing procedure. Therefore, an alternative will be introduced that combines the powerful mathematical properties of exponential family models, such as the RM, with the flexibility of the 2PLM.

The generalized one parameter model is formally identical to the 2PLM. The main difference is that the discrimination indices α_i are not considered as unknown parameters, but as fixed constants supplied by hypothesis. To emphasize this difference, the α_i will no longer be called parameters, but discrimination indices, and will be written as a_i. It may be useful to compare the parameter space of the One Parameter Logistic Model (OPLM) with that of the 2PLM. As mentioned above, one difficulty parameter may be fixed arbitrarily. Therefore, the item parameter space in the 2PLM is $\Omega = \mathbb{R}^{k-1} \times \mathbb{R}^{+k-1}$. On the contrary, the OPLM is a family of models where each member is characterized by a k-vector $\boldsymbol{a}$ of discrimination indices, and has $\Omega_{\boldsymbol{a}} = \mathbb{R}^{k-1}$ as its item parameter space. Obviously,

$$\Omega = \bigcup_{\boldsymbol{a}} \Omega_{\boldsymbol{a}}. \tag{12.7}$$

From (12.2) it follows that all a_i may be multiplied by an arbitrary positive constant. If, at the same time θ and all β_i are divided by the same constant, the resulting likelihood is not changed. Now assume that the 2PLM is valid with some vector of discrimination indices $\boldsymbol{a}^*$, the elements of which may be rational or irrational numbers. But every irrational number may be approximated to an arbitrary degree of precision by a rational number. Therefore, $\boldsymbol{a}^*$ may be replaced by a vector $\boldsymbol{a}$ which consists only of rational numbers, and therefore there exists a positive constant c such that each element of $c\boldsymbol{a}$ is a positive integer. Hence, (12.7) remains valid if the elements of $\boldsymbol{a}$ in the OPLM are restricted to the positive integers. In the sequel it will become clear that this makes the derivation of a CML estimation procedure possible.

It will be clear that, in order to judge the discriminating power of an item, one cannot use the magnitude of the discrimination index as such, but one has to take the ratio of the indices into account. If for a number of items all discrimination indices are relatively large, in the order of magnitude of one

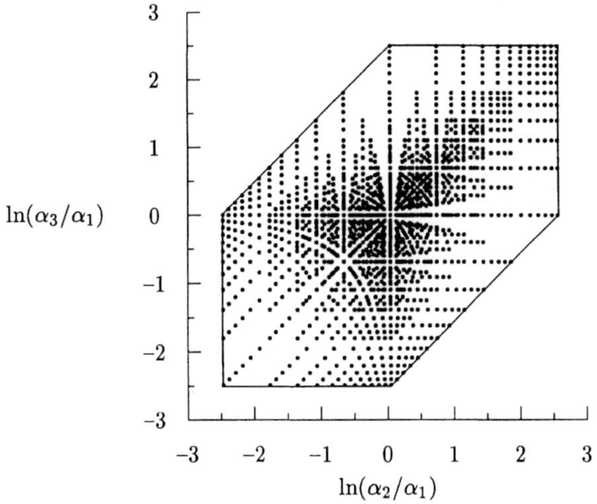

FIGURE 12.1. Parameter space for $k = 3$ and $\alpha_i \in \{1, \ldots, 12\}$; the convex hull represents the 2PLM with $\alpha_i \leq 12$.

hundred, say, all pairwise ratios will be close to one, and since analyses and statistical testing are done with finite samples, it is very likely that a model where all discrimination indices are equal will yield an equally good (or bad) fit. On the other hand, small discrimination indices, between one and five, say, yield ratios which may differ substantially from one. Therefore using different indices is important mainly if they are small in magnitude. This is convenient from a practical point of view: as will be shown below, combinatorial functions have to be evaluated for every possible weighted score $r_v = \sum_i a_i x_{vi}$, so the range of the scores is $\{0, 1, 2, .., \sum_i a_i\}$. In Figure 12.1 it is shown how the parameter space Ω is covered for $k = 3$ if the discrimination indices are restricted to $\{1, ..., 12\}$. The midpoint of the figure represents the RM, which is a special case of an OPLM, with all discriminations being equal.

12.2 Parameter Estimation

In this section, attention will mainly be paid to conditional maximum likelihood estimation of the difficulty parameters. It will be shown that this method is a quite straightforward generalization of the same method used in the RM, if a suitable parametrization is used. Define

$$\xi_v = \exp(\theta_v), \quad \text{for } v = 1, \ldots, n, \text{ and} \tag{12.8}$$

TABLE 12.1. The construction of $\tilde{\gamma}_3$ for $k = 4$ items with discrimination indices $a_1, \ldots, a_4$.

	a_1	a_2	a_3	a_4	Contribution
	1	1	1	2	to $\tilde{\gamma}_3$
Response Patterns	1	1	1	0	$\varepsilon_1\varepsilon_2\varepsilon_3$
	1	0	0	1	$\varepsilon_1\varepsilon_4$
	0	1	0	1	$\varepsilon_2\varepsilon_4$
	0	0	1	1	$\varepsilon_3\varepsilon_4$

$$\varepsilon_i = \exp(-a_i\beta_i), \quad \text{for } i = 1, \ldots, k. \tag{12.9}$$

Substituting (12.8) and (12.9) in (12.2), the likelihood function of a response pattern $\boldsymbol{x}_v = (x_{v1}, ..., x_{vk})$ becomes

$$L(\xi_v, \boldsymbol{\varepsilon}; \boldsymbol{x}_v) = \frac{\xi_v^{r_v} \prod_i \varepsilon_i^{x_{vi}}}{\prod_i (1 + \xi_v^{a_i} \varepsilon_i)} . \tag{12.10}$$

It is clear that the denominator and the first factor of the numerator of (12.10) will appear identically in the likelihood for all response patterns having the same weighted score r_v. Therefore, the likelihood function conditional on the score is

$$L_c(\boldsymbol{\varepsilon}; \boldsymbol{x}_v | r_v) = \frac{\prod_i \varepsilon_i^{x_{vi}}}{\tilde{\gamma}_{r_v}(\boldsymbol{\varepsilon})} , \tag{12.11}$$

the combinatorial function $\tilde{\gamma}_r(\boldsymbol{\varepsilon})$ being defined as

$$\tilde{\gamma}_r(\boldsymbol{\varepsilon}) = \begin{cases} \displaystyle\sum_{\boldsymbol{x}|r} \prod_i \varepsilon_i^{x_i}, & \text{if the sum contains at least one term,} \\ 0, & \text{otherwise,} \end{cases} \tag{12.12}$$

where the summation runs over all patterns $\boldsymbol{x}$ with score r. The subscript r is called the order of the $\tilde{\gamma}$-function.

Before deriving the likelihood equations, it is useful to take a closer look at the $\tilde{\gamma}$-functions and their partial derivatives. Consider the following example with four items with discrimination indices of respectively 1, 1, 1, and 2. To construct the $\tilde{\gamma}$-functions of order 3, we have to consider all response patterns yielding a score of 3. These response patterns are given in Table 12.1, together with their contribution to the $\tilde{\gamma}_3$-function. From the table it is clear that the $\tilde{\gamma}$-functions are not symmetric, because the arguments used are differentially weighted according to definition (12.9).

As an aside, it should be noted that, since the terms of the $\tilde{\gamma}$-function consist of products with an unequal number of factors, multiplying the ε-parameter with a positive constant c will have an unpredictable effect on the resulting $\tilde{\gamma}$-function. This means that the rule $\gamma_r(c\boldsymbol{\varepsilon}) = c^r\gamma_r(\boldsymbol{\varepsilon})$, which is valid for the symmetric functions γ, is not valid in this case.

Now consider the partial derivatives of the $\tilde{\gamma}$-function with respect to one of its arguments. As a first example, take the partial derivative with respect to ε_1. Of the four terms displayed in Table 12.1, only two contain this parameter, so that it is clear that the partial derivative is $\varepsilon_2\varepsilon_3 + \varepsilon_4$, which is $\tilde{\gamma}_2(\varepsilon_2, \varepsilon_3, \varepsilon_4)$. Next consider the partial derivative with respect to ε_4. From Table 12.1 it is easly derived that it is $\varepsilon_1 + \varepsilon_2 + \varepsilon_3$, and this is $\tilde{\gamma}_1(\varepsilon_1, \varepsilon_2, \varepsilon_3)$. These two examples show that the partial derivatives of a $\tilde{\gamma}$-function with respect to an argument ε_i again are $\tilde{\gamma}$-functions of the original arguments, but with ε_i excluded. The order of the resulting $\tilde{\gamma}$-functions equals the original order minus a_i. Applying the same notation as in preceding chapters by denoting the index of the excluded argument as superscript, the following rule should be obvious:

$$\frac{\partial\tilde{\gamma}_r(\boldsymbol{\varepsilon})}{\partial\varepsilon_i} = \tilde{\gamma}^{(i)}_{r-a_i}(\boldsymbol{\varepsilon}). \tag{12.13}$$

Applying this rule twice in succession, for $i \neq j$, gives

$$\frac{\partial^2\tilde{\gamma}_r(\boldsymbol{\varepsilon})}{\partial\varepsilon_i\,\partial\varepsilon_j} = \tilde{\gamma}^{(i,j)}_{r-a_i-a_j}(\boldsymbol{\varepsilon}). \tag{12.14}$$

Defining the random variable T_i, with realizations t_i, as

$$T_i = \sum_v X_{vi}, \quad \text{for } i = 1, \ldots, k, \tag{12.15}$$

and the vector $\boldsymbol{r} = (r_1, ..., r_n)$, the conditional likelihood function for the data matrix $\boldsymbol{X}$ can be written as

$$L_c(\boldsymbol{\varepsilon}; \boldsymbol{X}|\boldsymbol{r}) = \frac{\prod_i \varepsilon_i^{t_i}}{\prod_v \tilde{\gamma}_{r_v}(\boldsymbol{\varepsilon})}. \tag{12.16}$$

It proves useful to take partial derivatives with respect to the logarithm of the ε-parameters. Differentiating the logarithm of (12.16) in this way yields

$$\begin{aligned} \frac{\partial \ln L_c(\boldsymbol{\varepsilon}; \boldsymbol{X}|\boldsymbol{r})}{\partial \ln \varepsilon_i} &= \varepsilon_i \frac{\partial \ln L_c(\boldsymbol{\varepsilon}; \boldsymbol{X}|\boldsymbol{r})}{\partial\varepsilon_i} \\ &= t_i - \sum_v \frac{\varepsilon_i \tilde{\gamma}^{(i)}_{r_v-a_i}(\boldsymbol{\varepsilon})}{\tilde{\gamma}_{r_v}(\boldsymbol{\varepsilon})}, \end{aligned} \tag{12.17}$$

so that the likelihood equations result as

$$t_i - \sum_v \frac{\varepsilon_i \tilde{\gamma}^{(i)}_{r_v-a_i}(\boldsymbol{\varepsilon})}{\tilde{\gamma}_{r_v}(\boldsymbol{\varepsilon})} = 0, \quad \text{for } i = 1, \ldots, k. \tag{12.18}$$

This is a straightforward generalization of the CML estimation equations in the RM, see equations (3.12). The result can also be obtained in a quite different way. As (12.16) has the form of an exponential family likelihood, the likelihood equations are directly given by equating the sufficient statistics to their expected values, which are functions of the parameters (Andersen, 1980a, Section 3.2). Since we are dealing with conditional likelihoods, the likelihood equations will take the form

$$t_i = E(T_i|\boldsymbol{r}) = \sum_v \pi_{i|r_v}, \quad \text{for } i = 1, \ldots, k, \tag{12.19}$$

where $\pi_{i|r_v}$ denotes the probability of a correct response on item I_i given score r_v. Since the right-hand terms of (12.18) and (12.19) are equal for any score vector $\boldsymbol{r}$, the terms may be equated, which yields that, for any score r,

$$\pi_{i|r} = \frac{\varepsilon_i \tilde{\gamma}^{(i)}_{r-a_i}(\boldsymbol{\varepsilon})}{\tilde{\gamma}_r(\boldsymbol{\varepsilon})}, \quad \text{for } i = 1, \ldots, k. \tag{12.20}$$

A discussion of a general class of conditional likelihood functions, of which the conditional likelihood functions of the RM and the OPLM are special cases, can be found in Kelderman (1992). So, any method to solve the conditional likelihood equations in the RM can be applied here also. For details on these methods, see Chapter 3. The only extra complication is the evaluation of the $\tilde{\gamma}$-functions and their derivatives. For $i = 1, ..., k$, define the vector

$$\boldsymbol{\varepsilon}_{(i)} = (\varepsilon_1, ..., \varepsilon_i). \tag{12.21}$$

Introducing the formal definition $\boldsymbol{\varepsilon}_{(0)} \equiv 1$, the following double recurrence relation holds:

$$\tilde{\gamma}_r(\boldsymbol{\varepsilon}_{(i)}) = \varepsilon_i \tilde{\gamma}_{r-a_i}(\boldsymbol{\varepsilon}_{(i-1)}) + \tilde{\gamma}_r(\boldsymbol{\varepsilon}_{(i-1)}), \tag{12.22}$$

for $i = 1, ..., k$ and $r = 0, ..., \sum_{j=1}^{i} a_j$ (see also Kelderman, 1992). Note that, if some score r is impossible given the discrimination indices, the above recursion automatically yields $\tilde{\gamma}_r = 0$.

Again, this recursion is a direct generalization of the respective recursions for the RM, see (8.13). Although the use of (12.22) is computationally expensive, it gives numerically stable results. The loss of accuracy is at most $\log_{10}(2k)$ decimal digits, and only half of that number on the average (Verhelst, Glas, & Van der Sluis, 1984). Moreover, if the second-order derivatives of the $\tilde{\gamma}$-functions are needed, for example, in using the Newton-Raphson algorithm to find the solution of (12.18), clever bookkeeping may save a lot of time. As a simple example, consider the computation of $\tilde{\gamma}^{(k-1,k)}_r(\boldsymbol{\varepsilon})$. It is easily verified that this is equal to $\tilde{\gamma}_r(\boldsymbol{\varepsilon}_{(k-2)})$, and this quantity has been computed using (12.22). So, saving this result avoids a second identical computation. In order to generalize this idea, the parameters are partitioned into p equivalence classes, and the $\tilde{\gamma}$-functions per equivalence class are computed. If the g-th equivalence class of parameters is denoted by E_g, for $g = 1, .., p$, a slight generalization of (12.22) is

$$\tilde{\gamma}_r(E_g \cup E_h) = \sum_{j=0}^{r} \tilde{\gamma}_j(E_g)\tilde{\gamma}_{r-j}(E_h), \qquad (12.23)$$

for $g \neq h$. (See also the special case arising in the RM, recusion (8.12).) Applying this 'merging' formula repeatedly until the equivalence classes $E_1, ...E_{p-2}$ are merged, these results may be used to compute all $\tilde{\gamma}_r^{(i,j)}(\boldsymbol{\varepsilon})$ for all ε_i and ε_j belonging to E_{p-1} or E_p. This procedure can of course be applied to all pairs of equivalence classes, and there exist several shortcuts to avoid double computation of function values. The algorithm is described in detail in Verhelst and Veldhuijzen (1991), where it is also shown that the number of multiplications and additions is reduced with a factor 10 if $k = 100$ as compared to the straight application of (12.22), and the reduction increases with k.

The problem of the existence and uniqueness of the CML-estimates in the OPLM is difficult and not solved satisfactorily yet. Although a very general theorem by Jacobsen (1989) can be applied here, testing if the conditions put forward in the theorem are fulfilled in a given data set is a hard problem. In order to gain some insight into the problem, consider the case where an item is correctly answered by every respondent in the sample. It is well known from a theorem of Fischer (1981; see Corollory 8.3 in Chapter 8) that in this case the CML-estimates of the parameters do not exist for the RM. Such a case, however, is 'due to the data', meaning that by increasing the sample an incorrect response will eventually be given to this item. Another example is provided when the data are collected in an incomplete design that is not linked by common items. In such a case CML-estimates in the RM may exist, but they are not unique. But the researcher can repair this problem by collecting data in a linked design. In the OPLM, however, there exist models which never yield unique CML-estimates irrespective of the sample size or the data collection design. As a simple example, consider the case of a test with k items, and a model which assumes an even discrimination index for the first $k - 1$ items, and an odd value for a_k. It is clear that from any score r the response to the k-th item can be inferred with certainty: if the score is even, the response is wrong, and if the score is odd, the response is correct. Formally, this means that $\pi_{k|r}$ is either one or zero, and the likelihood equation for ε_k becomes

$$t_k = \sum_v \pi_{k|r_v} = \sum_v x_{vk} = t_k, \qquad (12.24)$$

which is trivially true for all values of $\boldsymbol{\varepsilon}$. This means that the above equation does not represent any restriction on the parameter space, such that there will exist infinitely many solutions. A further comment on this problem will be given in Section 12.5.

The estimation problem mentioned above occurs if CML is used, and may be avoided if another estimation method, such as marginal maximum likelihood (MML), is applied. However, it should be stressed that MML is not merely a 'method', but implies a quite different approach to measurement than implied by (12.2). In this expression, it is assumed that a distinct parameter θ_v is associated with every subject S_v in the sample. Such a model is sometimes called a functional model (De Leeuw & Verhelst, 1986). It is also possible to consider the sample of subjects as a sample from some population, and their associated θ-values as realizations of a random variable, which follows some distribution. So the measurement model is extended by some assumption about the distribution of the latent variable, and this extended model is denoted as a structural model. The most commonly used distribution is the normal distribution, and the purpose of the estimation procedure is to estimate the β-parameters and the two parameters of the distribution, the mean μ and the variance σ^2 jointly. The technical details of the procedure are identical in the OPLM and in the RM and can be found in Chapters 3 and 15.

12.3 Statistical Tests

The main difference between the OPLM and the 2PLM is the status of the discrimination indices: in the 2PLM, these indices are treated as unknown parameters, while in the OPLM they are treated as constants, fixed at the value they assume by hypothesis. Of course, statistical tests have to be built to test these hypotheses. This is the topic of the present section.

In a study which is in a sense a generalization of the OPLM to polytomous items, Andersen (1983, p. 127) points out that 'a correct set of category weights is critical for a correct analysis of the data'. It will be shown how the principles established in Chapter 5 can be applied to develop statistical tests that are especially sensible to misspecification of the discrimination indices. If the parameters are estimated using CML, the tests are generalized Pearson tests; in the case of MML-estimation the tests belong to the class of the Lagrange multiplier tests.

12.3.1 Theoretical and Computational Aspects of Generalized Pearson Tests

Recall from Chapter 5 that Pearson's original X^2 test statistic essentially compares observed and expected proportions in a multinomial model. The general approach of Chapter 5 was to provide a basis that allows to 'join classes'. The approach can be recapitulated as follows: let $\boldsymbol{\pi}$ be the M-vector containing the probabilities of all possible response patterns, and $\boldsymbol{p}$ the corresponding M-vector of observed proportions; moreover, let $\boldsymbol{U} = [\boldsymbol{T}_1|\boldsymbol{T}_2|\boldsymbol{Y}]$. The test will be based on the differences $\boldsymbol{U}'(\boldsymbol{p} - \hat{\boldsymbol{\pi}})$. A small example is constructed in Table 12.2. $\boldsymbol{T}_1$ contains as rows a subset of the response patterns; the reason why some

TABLE 12.2. The **U**-matrix for a four-item test

item	1	2	3	4	nontrivial scores						
a_i	1	2	2	3	2	3	4	5	6		
	$\boldsymbol{T}_1$				$\boldsymbol{T}_2$					$\boldsymbol{Y}$	
	0	1	0	0	1	0	0	0	0	2	0
	0	0	1	0	1	0	0	0	0	0	0
	0	0	0	1	0	1	0	0	0	0	0
	1	1	0	0	0	1	0	0	0	1	0
	1	0	1	0	0	1	0	0	0	0	0
	1	0	0	1	0	0	1	0	0	0	0
	0	1	1	0	0	0	1	0	0	0	0
	0	1	0	1	0	0	0	1	0	0	3
	0	0	1	1	0	0	0	1	0	0	0
	1	1	1	0	0	0	0	1	0	0	3
	1	1	0	1	0	0	0	0	1	0	3
	1	0	1	1	0	0	0	0	1	0	0

possible response patterns need not be considered, is discussed below. $\boldsymbol{T}_2$ contains as rows the indicator vectors of the weighted scores associated with the response patterns. It holds that $\boldsymbol{T}_1'(\boldsymbol{p} - \hat{\boldsymbol{\pi}}) = \boldsymbol{0}$ are the estimation equations for the item parameters, and $\boldsymbol{T}_2'(\boldsymbol{p} - \hat{\boldsymbol{\pi}}) = \boldsymbol{0}$ those for a set of dummy parameters for transforming the product-multinomial likelihood (12.16) into a multinomial likelihood (for details, see Chapter 5). Finally, $\boldsymbol{Y}$ is chosen such that $\boldsymbol{Y}'(\boldsymbol{p} - \hat{\boldsymbol{\pi}})$ produces the differences one is interested in. The choice of $\boldsymbol{Y}$ in Table 12.2 will be discussed later. Notice that $\boldsymbol{Y}'\boldsymbol{p}$ and $\boldsymbol{Y}'\boldsymbol{\pi}$ are linear combinations of the observed proportions and probabilities, respectively, for the response patterns listed as rows of $\boldsymbol{T}_1$.

Since all tests considered in this section are item oriented, i will be used as an index of the studied item. The test statistics to be developed in this section will essentially compare $\pi_{i|r}$ with their corresponding observed proportion $p_{i|r}$. But $\pi_{i|r}$ itself is a linear combination (the simple sum) of the probabilities of all response patterns yielding score r and a correct response on item I_i. Moreover, it may prove necessary to combine classes even further, since in the OPLM the number of different scores may be quite large.

Two comments are in order with respect to Table 12.2. (i) Although there are $2^4 = 16$ possible response patterns, four are omitted from the table, namely, those which yield scores of 0, 1, 15, or 16. Since each of these scores can be obtained only from a single response pattern, they are called trivial, because they do not contain any information on the item parameters. Occurrences of such patterns in the data set can be ignored without affecting the CML-estimates. Similarly, if r is a trivial score, $\pi_{i|r}$ and $p_{i|r}$ are both zero or both one, and therefore these patterns will not contribute to the test statistic. (ii) It is easily checked that the supermatrix $[\boldsymbol{T}_1|\boldsymbol{T}_2]$ satisfies the conditions (A) and (B) from Chapter 5: the columns of $\boldsymbol{T}_2$ add to the unit vector, and since the OPLM enhanced with a saturated multinomial model for the distribution of the nontrivial scores, is an exponential family, it follows directly from the theory in Chapter 5 that condition (A) is satisfied for $[\boldsymbol{T}_1|\boldsymbol{T}_2]$. So, from the theory of Chapter 5 it follows that

$$Q(\boldsymbol{U}) \;=\; n(\boldsymbol{p}-\hat{\boldsymbol{\pi}})'\boldsymbol{U}(\boldsymbol{U}'\hat{\boldsymbol{D}}_\pi\boldsymbol{U})^{-}\boldsymbol{U}'(\boldsymbol{p}-\hat{\boldsymbol{\pi}}), \tag{12.25}$$

with $\hat{\boldsymbol{D}}_\pi$ a diagonal matrix of the elements of $\hat{\boldsymbol{\pi}}$, is asymptotically χ^2-distributed with $df = \text{rk}(\boldsymbol{U}'\boldsymbol{D}_\pi\boldsymbol{U}) - q - 1$, where q is the number of free parameters estimated from the data, and 'rk' denotes the rank of a matrix. The degrees of freedom are easily checked: $\text{rk}[\boldsymbol{T}_1|\boldsymbol{T}_2]$ equals 1 plus the number of free parameters (three free item parameters and 4 dummy parameters), and the columns of $\boldsymbol{Y}$ are linearly independent of the columns of $[\boldsymbol{T}_1|\boldsymbol{T}_2]$. Hence, $df = 2$. So far, everything is just a recapitulation of the theory presented in Chapter 5.

Computing the quadratic form (12.25) may seem relatively simple for the example in Table 12.2, but in general it is not. The number M of rows of $\boldsymbol{U}$ grows exponentially with the number of items, so that, even with a moderate number of items, straightforward computation of (12.25) as it stands may become unfeasible. To simplify the computations, some restrictions will be imposed on the matrix $\boldsymbol{Y}$. These restrictions are the following.

(1) If the matrix $\boldsymbol{Y}$ has a non-zero entry in row g, then $\boldsymbol{T}_1(g,i) = 1$, that is, $\boldsymbol{Y}$ can only have non-zero entries for response patterns with a correct response on the studied item I_i. In the example of Table 12.2, $i = 2$.

(2) In each row of $\boldsymbol{Y}$, there is at most one non-zero entry. This ensures that $\boldsymbol{Y}'\boldsymbol{D}\boldsymbol{Y}$ is diagonal for every diagonal matrix $\boldsymbol{D}$.

(3) If $\boldsymbol{Y}(g,j) \neq 0$, then for all h with $r_h = r_g$ (i.e., the scores of the g-th and the h-th response pattern are equal) and $\boldsymbol{T}_1(h,i) = 1$, it holds that $\boldsymbol{Y}(h,j) = \boldsymbol{Y}(g,j)$. In Table 12.2, $\boldsymbol{Y}(8,2)$ is set to 3, and since $r_8 = r_{10}$ and $\boldsymbol{T}_1(10,i) = 1$, $\boldsymbol{Y}(10,2)$ must equal 3.

Conditions (1) and (3) ensure that the linear combinations $\boldsymbol{Y}'\boldsymbol{\pi}$ can be written as linear combinations of conditional probabilities $\pi_{i|r}$. Now let s denote the number of columns in $\boldsymbol{Y}$ and d the number of different nontrivial scores, i.e., the number of columns of $\boldsymbol{T}_2$. Further, let the nontrivial scores be ranked, and let $w(r)$ be the rank number of score r. Using the conditions (1) to (3) above, the information contained in $\boldsymbol{Y}$ can be represented in a $d \times s$ matrix $\boldsymbol{V}$, defined by $\boldsymbol{V}(w(r_l),q) \;=\; \boldsymbol{Y}(l,q)$, where the l-th response pattern has a '1' in the i-th place, i.e., $\boldsymbol{T}_1(l,i) = 1$, and r_l is the weighted score associated with that response pattern. The matrix $\boldsymbol{V}$ induced by the matrix $\boldsymbol{Y}$ of Table 12.2 is given in Table 12.3.

It is clear that the nonzero entries in the columns of $\boldsymbol{V}$ correspond to s nonintersecting subsets $G_q, q = 1,...,r$, of nontrivial scores. In the example, $G_1 =$

TABLE 12.3. V-matrix induced by Y of Table 12.2

Score r	$w(r)$	V	
2	1	2	0
3	2	1	0
4	3	0	0
5	4	0	3
6	5	0	3

$\{2,3\}$ and $G_2 = \{5,6\}$. In order to compute the quadratic form $Q(\boldsymbol{U})$, and not to have recourse to generalized inverses, an arbitrary column of $\boldsymbol{T}_1$ may be dropped. For notational convenience it will be assumed that the last column is dropped. The resulting matrix will be denoted also as $\boldsymbol{T}_1$. Partition $\boldsymbol{U}'\hat{\boldsymbol{D}}_\pi\boldsymbol{U}$ as

$$\boldsymbol{U}'\hat{\boldsymbol{D}}_\pi\boldsymbol{U} = \begin{pmatrix} \boldsymbol{T}'_1\hat{\boldsymbol{D}}_\pi\boldsymbol{T}_1 & \boldsymbol{T}'_1\hat{\boldsymbol{D}}_\pi\boldsymbol{T}_2 & \boldsymbol{T}'_1\hat{\boldsymbol{D}}_\pi\boldsymbol{Y} \\ \boldsymbol{T}'_2\hat{\boldsymbol{D}}_\pi\boldsymbol{T}_1 & \boldsymbol{T}'_2\hat{\boldsymbol{D}}_\pi\boldsymbol{T}_2 & \boldsymbol{T}'_2\hat{\boldsymbol{D}}_\pi\boldsymbol{Y} \\ \boldsymbol{Y}'\hat{\boldsymbol{D}}_\pi\boldsymbol{T}_1 & \boldsymbol{Y}'\hat{\boldsymbol{D}}_\pi\boldsymbol{T}_2 & \boldsymbol{Y}'\hat{\boldsymbol{D}}_\pi\boldsymbol{Y} \end{pmatrix}. \tag{12.26}$$

In the formulae to follow, the indices r and t denote scores, so the rank numbers w_r indicate the columns of $\boldsymbol{T}_2$; the indices g and h refer to the items and correspond to the $k-1$ columns of $\boldsymbol{T}_1$, and the indices p and q point to the s columns of $\boldsymbol{Y}$ or $\boldsymbol{V}$.

Let n be the number of persons with a nontrivial score, and let n_r denote the frequency of score r. Using the formal definition $\pi_{gg|r} = \pi_{g|r}$, the elements of the submatrices in (12.26) are given by

$$(\boldsymbol{T}'_1\hat{\boldsymbol{D}}_\pi\boldsymbol{T}_1)_{gh} = \frac{1}{n}\sum_r n_r\hat{\pi}_{gh|r}, \quad \text{for } g,h = 1,\ldots,k-1; \tag{12.27}$$

$$(\boldsymbol{T}'_2\hat{\boldsymbol{D}}_\pi\boldsymbol{T}_2)_{w_r,w_t} = \begin{cases} \dfrac{n_r}{n} & \text{if } r = t, \\ 0 & \text{if } r \neq t, \quad \text{for } w_r, w_t = 1,\ldots,d; \end{cases} \tag{12.28}$$

$$(\boldsymbol{T}'_1\hat{\boldsymbol{D}}_\pi\boldsymbol{T}_2)_{gw_r} = \frac{n_r}{n}\hat{\pi}_{g|r}, \quad \text{for } g = 1,\ldots,k-1;\, w_r = 1,\ldots,d; \tag{12.29}$$

$$(\boldsymbol{Y}'\hat{\boldsymbol{D}}_\pi\boldsymbol{Y})_{pq} = \begin{cases} \dfrac{1}{n}\displaystyle\sum_{r\in G_p} n_r\hat{\pi}_{i|r}v^2_{w_rp} & \text{if } p = q, \\ 0 & \text{if } p \neq q, \quad \text{for } p,q = 1,\ldots,s; \end{cases} \tag{12.30}$$

$$(\boldsymbol{Y}'\hat{\boldsymbol{D}}_\pi \boldsymbol{T}_1)_{pg} = \frac{1}{n}\sum_{r\in G_p} n_r \hat{\pi}_{ig|r} v_{w_r p}, \quad \text{for } g = 1,\ldots,k-1; \; p = 1,\ldots,s; \tag{12.31}$$

$$(\boldsymbol{Y}'\hat{\boldsymbol{D}}_\pi \boldsymbol{T}_2)_{pw_r} = \frac{n_r}{n}\hat{\pi}_{i|r} v_{w_r p}, \quad \text{for } p = 1,\ldots,s; w_r = 1,\ldots,d. \tag{12.32}$$

Since $(\boldsymbol{p} - \hat{\boldsymbol{\pi}})'[\boldsymbol{T}_1|\boldsymbol{T}_2] = \boldsymbol{0}'$, (12.25) can be written as

$$Q(\boldsymbol{U}) = n(\boldsymbol{p} - \hat{\boldsymbol{\pi}})'\boldsymbol{Y}[\boldsymbol{Y}'\hat{\boldsymbol{D}}_\pi \boldsymbol{Y} - \boldsymbol{Y}'\hat{\boldsymbol{D}}_\pi \boldsymbol{T}(\boldsymbol{T}'\hat{\boldsymbol{D}}_\pi \boldsymbol{T})^{-1}\boldsymbol{T}'\hat{\boldsymbol{D}}_\pi \boldsymbol{Y}]^{-1}\boldsymbol{Y}'(\boldsymbol{p} - \hat{\boldsymbol{\pi}}), \tag{12.33}$$

where $\boldsymbol{T} = [\boldsymbol{T}_1|\boldsymbol{T}_2]$. Using matrix partitioning and the shorthand notation $\boldsymbol{T}_{lm} = \boldsymbol{T}_l'\hat{\boldsymbol{D}}_\pi \boldsymbol{T}_m$, for $l, m = 1, 2$, it follows that

$$(\boldsymbol{T}'\boldsymbol{D}_\pi \boldsymbol{T})^{-1} = \begin{pmatrix} \boldsymbol{B}^{-1} & -\boldsymbol{B}^{-1}\boldsymbol{T}_{12}\boldsymbol{T}_{22}^{-1} \\ -\boldsymbol{T}_{22}^{-1}\boldsymbol{T}_{21}\boldsymbol{B}^{-1} & \boldsymbol{T}_{22}^{-1}\boldsymbol{T}_{21}\boldsymbol{B}^{-1}\boldsymbol{T}_{12}\boldsymbol{T}_{22}^{-1} + \boldsymbol{T}_{22}^{-1} \end{pmatrix} \tag{12.34}$$

with

$$\boldsymbol{B} = \boldsymbol{T}_{11} - \boldsymbol{T}_{12}\boldsymbol{T}_{22}^{-1}\boldsymbol{T}_{21}. \tag{12.35}$$

Using (12.27), (12.28), and (12.29), the elements of $\boldsymbol{B}$ can be written as

$$(\boldsymbol{B})_{gh} = \frac{1}{n}\sum_r n_r[\hat{\pi}_{gh|r} - \hat{\pi}_{g|r}\hat{\pi}_{h|r}] = \frac{1}{n}\mathcal{I}(\hat{\beta}_g, \hat{\beta}_h), \tag{12.36}$$

for $g, h = 1, ..., k-1$. This matrix (or its inverse) is commonly available as a result of the estimation procedure, because it is needed for the computation of the standard errors. By straightforward but tedious algebra it can be shown that (12.25) can be rewritten as

$$Q(\boldsymbol{U}) = n(\boldsymbol{p} - \hat{\boldsymbol{\pi}})'\boldsymbol{Y}[\boldsymbol{Y}'\hat{\boldsymbol{D}}_\pi \boldsymbol{Y} - \boldsymbol{F}'\boldsymbol{T}_{22}^{-1}\boldsymbol{F} - \boldsymbol{R}_i'\boldsymbol{B}^{-1}\boldsymbol{R}_i]^{-1}\boldsymbol{Y}'(\boldsymbol{p} - \hat{\boldsymbol{\pi}}), \tag{12.37}$$

where $\boldsymbol{R}_i$ is a $(k-1) \times s$ matrix with elements

$$(\boldsymbol{R}_i)_{gp} = \sum_{r\in G_p} \frac{n_r}{n} v_{w_r p}[\hat{\pi}_{ig|r} - \hat{\pi}_{i|r}\hat{\pi}_{g|r}], \tag{12.38}$$

for $g = 1,\ldots,k-1$ and $p = 1,\ldots,s$, and

$$\boldsymbol{F} = \boldsymbol{T}_2'\hat{\boldsymbol{D}}_\pi \boldsymbol{Y}. \tag{12.39}$$

Note that $\boldsymbol{F}'\boldsymbol{T}_{22}^{-1}\boldsymbol{F}$ is a diagonal $s \times s$ matrix, and the difference $\boldsymbol{Y}'\hat{\boldsymbol{D}}_\pi \boldsymbol{Y} - \boldsymbol{F}'\boldsymbol{T}_{22}^{-1}\boldsymbol{F}$ is given by

$$(\boldsymbol{Y}'\hat{\boldsymbol{D}}_\pi \boldsymbol{Y} - \boldsymbol{F}'\boldsymbol{T}_{22}^{-1}\boldsymbol{F})_{pp} = \sum_{r\in G_p} \frac{n_r}{n} v_{w_r p}^2[\hat{\pi}_{i|r}(1 - \hat{\pi}_{i|r})], \quad \text{for } p = 1,\ldots,s. \tag{12.40}$$

Define the quadratic form $Q^*(\boldsymbol{U})$ by dropping the last term between square brackets in (12.37), that is

$$\begin{aligned} Q^*(\boldsymbol{U}) &= n(\boldsymbol{p}-\hat{\boldsymbol{\pi}})'\boldsymbol{Y}[\boldsymbol{Y}'\hat{\boldsymbol{D}}_\pi\boldsymbol{Y} - \boldsymbol{F}'\boldsymbol{T}_{22}^{-1}\boldsymbol{F}]^{-1}\boldsymbol{Y}'(\boldsymbol{p}-\hat{\boldsymbol{\pi}}) \\ &= \sum_p^s \frac{[\sum_{r\in G_p} n_r v_{w_r p}(p_{i|r}-\hat{\pi}_{i|r})]^2}{\sum_{r\in G_p} n_r v_{w_r p}^2 \hat{\pi}_{i|r}(1-\hat{\pi}_{i|r})}, \end{aligned} \tag{12.41}$$

which is a slightly generalized form of the common Pearson X^2-statistic. However, in (12.41), the fact that the model parameters are estimated from the data is ignored, and therefore $Q^*(\boldsymbol{U})$ is probably not χ^2-distributed. The term $\boldsymbol{R}_i'\boldsymbol{B}^{-1}\boldsymbol{R}_i$ in (12.37) adds precisely the required correction to make $Q(\boldsymbol{U})$ asymptotically χ^2-distributed.

12.3.2 The S_i-Tests

Although restrictions are imposed to make the computations feasible, there remains a lot of freedom to define the $\boldsymbol{Y}$-matrix. In Table 12.3. the studied item is I_2. In the first column, deviations are taken for scores 2 and 3, and in the second column, for scores 5 and 6. It may easily be deduced from (12.25) that multiplication of a column of $\boldsymbol{Y}$ with a non-zero constant does not alter $Q(\boldsymbol{U})$. So the non-zero entries in column 2 may be replaced by 1. In the first column, however, the non-zero entries are not constant. The deviations originating from patterns with score 2 are weighted twice as much as those coming from patterns with score 3. This means that the test using $Q(\boldsymbol{U})$ will be more sensitive to model violations which cause deviations in patterns with score 2 than with score 3. Of course, one should have special reasons for using different weights. In this and the next subsection all non-zero entries of $\boldsymbol{Y}$ equal 1 in absolute value.

If a_i is misspecified – too high, say – the deviations $p_{i|r} - \hat{\pi}_{i|r}$ will tend to show a specific pattern as exemplified in Figure 12.2a.

As can be seen from (12.41), deviations belonging to the same group of scores G_p are added, and their sum is squared. So power will be lost if positive and negative deviations occur of about equal size and frequency in the same group. So if groups are formed consisting of neighboring scores, the tendency for negative and positive deviations to compensate will be small, and one may expect reasonable power of the test.

The exact distribution of $Q(\boldsymbol{U})$ is not known, however, and the χ^2-distribution is only a good approximation if the probabilities $\pi_{i|r}$ are not too extreme and the number of observations per group not too small. The computer program OPLM (Verhelst, Glas, & Verstralen, 1994) follows the rules listed below:

(1) The number of observations in each group is at least 30, i.e., $\sum_{r\in G_p} n_r \geq 30$, for $p = 1, ..., s$;

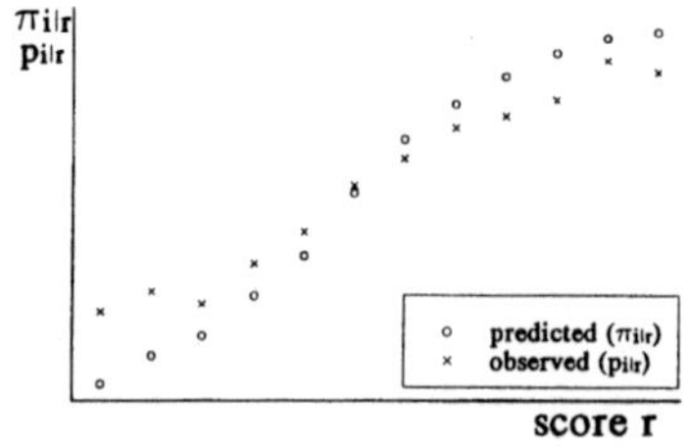

FIGURE 12.2a. Index a_i specified too high.

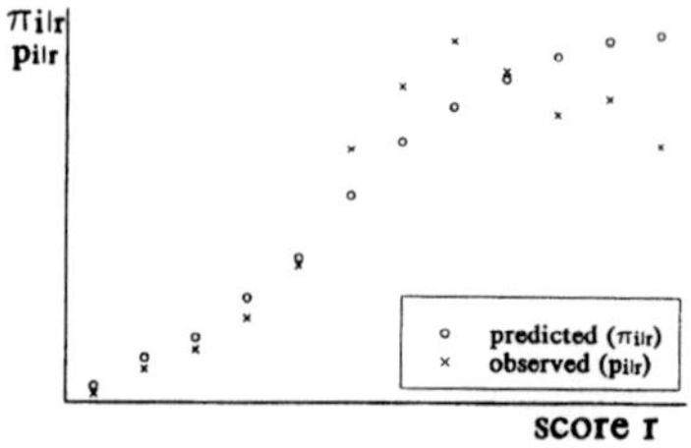

FIGURE 12.2b. Deviations not attributable to misspecification of a_i.

(2) The expected number of correct and incorrect responses in each group is at least 5, i.e., $\sum_{r\in G_p} n_r\hat{\pi}_{i|r} \geq 5$ and $\sum_{r\in G_p} n_r(1-\hat{\pi}_{i|r}) \geq 5$, for $p = 1, \ldots, s$. This means that the grouping will be item-dependent: easy items will tend to group more different scores at the high end than at the low end of the score range, and vice versa for the difficult items.

(3) To increase the power, groups of about equal size are formed, without violating the two preceding rules. Moreover, observations with the same score are always grouped together, because otherwise, the formulae (12.27) to (12.32) are no longer valid. For technical reasons, the number of groups is never larger than 8.

If every nontrivial score belongs to one of the groups, the columns of the matrix $\boldsymbol{Y}$ are linearly dependent on the columns of $\boldsymbol{T}$. This dependence disappears when an arbitrary column of $\boldsymbol{Y}$ is dropped. So, in general the number of degrees of freedom equals $s - 1$. The tests thus constructed are denoted as S_i tests. An example will discussed in Section 12.4.

12.3.3 The M_i-tests

Although the S_i-tests are powerful against misspecification of the discrimination indices, they easily detect also other systematic model violations. In Figure 12.2b an example is given of a systematic pattern which suggests a non-monotone item. Clearly, the S_i-test is sensitive to such deviations too. Moreover, if a_i is misspecified in the other direction (too low), the S_i-test is likely to give a significant result as well. So it may be useful to develop a test which is more specifically sensitive to misspecification of a_i and which moreover is directional, in that it gives a cue whether a_i is too high or too low. The rationale for the M_i-tests was proposed by Molenaar (1983) in the development of the U_i-tests.

Recall from Chapter 5 that the scores were partitioned in a low (L), medium (O), and high (H) group. The test statistic, apart from a normalizing denominator, is given by (5.24)

$$M_i^* = \sum_{r \in L} (p_{i|r} - \hat{\pi}_{i|r}) - \sum_{r \in H} (p_{i|r} - \hat{\pi}_{i|r}). \tag{12.42}$$

This test statistic can be written as $(\boldsymbol{p} - \hat{\boldsymbol{\pi}})'\boldsymbol{Y}$, where $\boldsymbol{Y}$ is a one-column matrix, whose nonzero entries equal 1 (in the L-group) or -1 (in the H- group). The test statistic $Q(\boldsymbol{U})$ based on this $\boldsymbol{Y}$ is asymptotically χ^2-distributed with $df = 1$, and hence its signed square root follows asymptotically the standard normal distribution. The sign of the square root corresponds to the sign of M_i^*, so that in cases where a_i was specified too high, the statistic will tend to be positive, and negative if a_i was too low. The only problem is to find a suitable definition of L, O, and H. A reasonable approach is to partition the n observations into K approximately equal subgroups, such that the score range for each of the subgroups is as small as possible. The group with the lowest scores constitutes L, the group with the highest scores H, and the remaining $K-2$ groups together O. The associated test statistic will be denoted $M2_i$ if $K = 2$, $M3_i$ if $K = 3$, etc. Another grouping that depends directly on the difficulty is constructed as follows. At the CML-estimates it holds that

$$\sum_r n_r p_{i|r} = \sum_r n_r \hat{\pi}_{i|r}. \tag{12.43}$$

Define $\overline{\pi}_i = n^{-1} \sum_r n_r \hat{\pi}_{i|r}$. A score r is allocated to L if $\hat{\pi}_{i|r} \leq \overline{\pi}_i$, and to H otherwise; O is empty. The test statistic thus constructed will be denoted M_i. In the program OPLM, M_i, $M2_i$, and $M3_i$ are implemented. The U_i-tests of Molenaar can of course be applied without any problem to the OPLM. It is instructive to compare U_i with the class of M-tests proposed in this section. In U_i the 25% lowest and the 25% highest scores are allocated to L and H, respectively. So, a relation can be sought between U_i and $M4_i$. In fact, with this definition of L and H, it can be shown (Verhelst & Eggen, 1989) that U_i can be written as the quadratic form $Q^*(\boldsymbol{U})$, defined in (12.41), where the weights $v_w(r)$ are given by

$$v_{w(r)} = \pm[n_r \hat{\pi}_{i|r}(1 - \hat{\pi}_{i|r})]^{-1/2} , \tag{12.44}$$

the sign being positive for L and negative for H. The weights used by Molenaar (1983) are inversely proportional to the estimated standard deviation of the deviations $n_r(p_{i|r} - \hat{\pi}_{i|r})$ and may look very appealing at first sight. However, one should be careful with these weights when applying U_i for very difficult or very easy items. With a difficult item, the conditional probability of a correct response given a low score, will tend to be very small, such that the weight associated with this score may be very large, and the use of the χ^2-distribution may become problematic. A detailed study on suitable weights in a variety of different conditions, however, is still lacking.

12.3.4 Item Oriented Lagrange Multiplier Tests

In principle, there is no objection to applying the methods of the preceding section if the parameters are estimated using MML. For the structural RM with a normal distribution of θ, Glas (1989) has shown that, with a slight modification of $[\boldsymbol{T}_1|\boldsymbol{T}_2]$, a matrix results which satisfies the conditions (A) and (B) of Chapter 5, and hence $Q([\boldsymbol{T}_1|\boldsymbol{T}_2|\boldsymbol{Y})]$ is asymptotically χ^2-distributed for arbitrary $\boldsymbol{Y}$. (The number of rows in both matrices equals the number of possible response patterns (namely, 2^k), and the number of columns in $\boldsymbol{T}_2$ equals d' (the number of different scores, including trivial ones). It is not difficult to show that this result also holds for the OPLM. However, since the enhanced model, constructed to apply MML (i.e., assuming a normal distribution of θ or some other distribution which does not predict the observed score distribution perfectly), is not an exponential family, it follows that $Q([\boldsymbol{T}_1|\boldsymbol{T}_2]) \neq 0$ in general, and therefore formula (12.33) cannot be used to compute the quadratic form. In fact, the matrix to be inverted in $Q(\boldsymbol{U})$ has dimension $k + d' + r - 2$, and no simplification has been found to simplify the computations. A further drawback of the method is that in general the construction of tests with $df = 1$ is impossible. The rank of $[\boldsymbol{T}_1|\boldsymbol{T}_2]$ is $k + d' - 1$, but the number of free parameters equals $k + 1$, so that in general $Q([\boldsymbol{T}_1|\boldsymbol{T}_2])$ is χ^2-distributed with $d' - 2$ degrees of freedom. Adding a one-column matrix $\boldsymbol{Y}$, as is done in the M-tests, automatically increases the degrees of freedom by 1, such that a one test with $df = 1$ is only possible in the trivial case with one item.

Fortunately, a test with $df = 1$ can be built using the Lagrange multiplier class of tests. (See Chapter 5 for a general introduction.) The rationale for the G_i-test is as follows: the OPLM is considered as a restricted model where only the item difficulties $\boldsymbol{\beta}$ and the distribution parameters $\boldsymbol{\lambda}$ are free to vary. Now consider a partition of the discrimination indices into two subsets, A_r and A_u. The discrimination indices belonging to A_r are treated as known constants, while the indices in A_u could in principle be unrestricted; so the parameters in the unrestricted case are $(\boldsymbol{\beta}, \boldsymbol{\lambda})$ and A_u. Now look at the special case where $A_u = \{a_i\}$. Using the notation of Chapter 5, $\boldsymbol{\eta}^* = (\boldsymbol{\beta}, \boldsymbol{\lambda}, a_i)$, and since the estimation is done under the restricted model, it holds that $\hat{\boldsymbol{\eta}}^* = (\hat{\boldsymbol{\beta}}, \hat{\boldsymbol{\lambda}}, a_i)$, i.e., a_i is fixed at its hypothetical value. It follows that the gradient of the log-likelihood function $\boldsymbol{b}(\hat{\boldsymbol{\eta}}^*)$ in general will not vanish. However, if the hypothetical value of a_i corresponds to its 'true' value, the quadratic form

$$G_i^2 = \boldsymbol{b}(\hat{\boldsymbol{\eta}}^*)'\mathcal{I}(\hat{\boldsymbol{\eta}}^*, \hat{\boldsymbol{\eta}}^*)^{-1}\boldsymbol{b}(\hat{\boldsymbol{\eta}}^*) \tag{12.45}$$

is asymptotically χ^2-distributed with $df = 1$. Since all the elements of $\boldsymbol{b}(\hat{\boldsymbol{\eta}}^*)$ but the last are zero, (12.45) can be rewritten as

$$G_i^2 = \frac{b(a_i)^2}{W_i}, \tag{12.46}$$

where

$$W_i = \mathcal{I}(a_i, a_i) - \mathcal{I}(a_i, \hat{\boldsymbol{\eta}}_1)\mathcal{I}(\hat{\boldsymbol{\eta}}_1, \hat{\boldsymbol{\eta}}_1)^{-1}\mathcal{I}(\hat{\boldsymbol{\eta}}_1, a_i), \tag{12.47}$$

with $\boldsymbol{\eta}_1 = (\boldsymbol{\beta}, \boldsymbol{\lambda})$; for the notation, see Section 5.5. The matrix $\mathcal{I}(\hat{\boldsymbol{\eta}}_1, \hat{\boldsymbol{\eta}}_1)^{-1}$ being available as a result of the estimation procedure, the computation of (12.46) is easy. Moreover, the algebraic sign of $b(a_i)$ gives an indication on how to change a_i to obtain a better fit. A positive sign points to a decrease, a negative value to an increase of a_i. So, the test statistic

$$G_i = \frac{|b(a_i)|}{\sqrt{W}_i}\text{sgn}(b(a_i)) \tag{12.48}$$

is asymptotically standard normally distributed and can be used to change the value of a_i in cases where it is significant. In order to get an impression of the amount of change needed for a_i, one might apply the univariate Newton-Raphson algorithm for a_i at $\hat{\boldsymbol{\eta}}^*$, yielding the simple result that the one-step change is given by

$$\frac{b(a_i)}{\mathcal{I}(a_i, a_i)}. \tag{12.49}$$

12.4 Examples

As a first example, an artificial data set of responses of 1000 respondents to 9 items will be analyzed. The θ-values of the respondents are drawn randomly from the standard normal distribution. The difficulty parameters are equally spaced between -2.0 and $+2.0$, and all discrimination indices but one are 1; the index a_5 is 2. The data set is analyzed using the RM, and some of the results are presented in Table 12.4. For each item, the test statistics S_i and $M2_i$ are computed, using (12.37) and (12.41), respectively, the latter being marked by an asterisk.

It is seen from these results that item I_5 is easily detected as a deviant item (S_5), and that a modification to a larger value of a_5 is suggested by $M2_5$. Besides, another interesting aspect about statistical testing becomes apparent from the table: the test statistics computed by means of formula (12.41) (i.e., S^* and $M2^*$) are systematically less extreme than the statistics computed using (12.37). Although this is not a mathematical necessity, it has been observed far more often than not in many analyses using both ways of testing. Moreover, and that is a reassuring finding, the difference between the result of (12.37) and (12.41) is usually small, although large differences have been observed incidentally. These findings, however, are not the result of a systematic study, and one should be

TABLE 12.4. Results of the analysis of the artificial data set using the RM

item	β_i	S_i	*df*	p	S_i^*	p^*	$M2_i$	$M2_i^*$
1	−2.14	.623	4	.960	.622	.961	−.44	−.32
2	−1.69	7.130	4	.129	6.957	.138	1.43	1.13
3	−1.09	7.779	4	.100	7.754	.101	.86	.75
4	.27	4.760	5	.446	4.689	.455	1.00	.98
5	−.01	37.018	5	.000	35.586	.000	−5.49	−5.39
6	.37	1.358	5	.929	1.348	.930	.88	.87
7	.79	2.567	5	.766	2.546	.770	.01	.01
8	1.55	6.103	4	.192	6.024	.197	2.08	1.84
9	1.94	3.427	4	.489	3.389	.495	.98	.81

careful not to take them for granted in the general situation. There is no formal proof on the asymptotic distribution of (12.41). Also, the reported p^*-values in Table 12.4 depend on an *ad hoc* decision: the degrees of freedom have been set equal to the *df* associated with (12.37).

The previous example is in several respects somewhat simplistic: the number of items is rather small, and the example was constructed in such a way that only one – precisely located and quite distinct – deviation from the RM was present. So, starting with the RM as a hypothesis makes sense. However, in practical applications things are usually more complicated. Staphorsius (1992) constructed a test for reading ability, comprising about 250 items. Responses were collected from over 10,000 respondents in a very complicated incomplete design. Applying the same strategy as in the above example proved to be a blind alley. Although the S_i and M_i-tests are locally sensitive to misspecification of the discrimination indices, they are not independent of each other. If the initial hypothesis (e.g., the RM) is wrong to a large extent, say for 25% of the items, many more of the statistical tests per item may yield a significant result, and the clear cue which resulted from Table 12.4 may altogether disappear. Therefore, the use of the S_i- and M_i-statistics as modification indices is sensible only if the initial hypothesis is not 'too wrong'.

From a purely statistical point of view, it does not matter where hypotheses come from. In practice, however, there are usually no theories wich predict to a sufficient degree of accuracy the discriminatory power of an item from its content or format; this means that the discrimination indices will have to be estimated from the observations. An estimation method which proved useful in practice (Verhelst, Verstralen, & Eggen, 1991) proceeds as follows. Using π_{ij} as a shorthand notation for $P(X_i = 1|\theta_j)$, the logit transformation of π_{ij} in the 2PLM yields

$$\text{logit}(\pi_{ij}) = a_i(\theta_j - \beta_i). \tag{12.50}$$

By introducing the assumption that the variance of θ in subgroups of respondents with the same number of items correct, is zero, the proportion p_{ij} of correct responses on item I_i in the subgroup having j correct responses is a consistent estimator of π_{ij}; therefore, the values of a_i, β_i, for $i = 1, ..., k$, and of θ_j, for $j = 1, .., k-1$, which minimize the quadratic loss function,

$$\sum_i^k \sum_{j=1}^{k-1} [\text{logit}(p_{ij}) - a_i(\theta_j - \beta_i)]^2, \tag{12.51}$$

are consistent estimators of the model parameters. But this extra assumption is not true in general, and therefore the estimates will not be generally consistent. The resulting estimates of the a-indices, however, are often good enough to serve as a first hypothesis for a series of the OPLM analyses. Details of the method can be found in the reference given.

A word of caution is in order about the level of significance α used in evaluating the S_i-statistics. If all statistics were independent, and the null hypothesis (i.e., the OPLM model) actually used were true, $100\alpha\%$ of the statistics should be expected to yield a significant result, but the probability that none is significant is $1 - (1 - \alpha)^k$. For $\alpha = .05$ and $k = 20$, this yields .64. So, if acceptance or rejection of the model as a whole is at stake, one should be careful not to base that decision on a single significant result at the nominal α-level, even if the tests are not independent. There are several methods to maintain an α-level test for the overall hypothesis, using multiple hypothesis testing. The best known is undoubtedly the Bonferroni method, but it yields in general a very conservative test. A method that is less conservative and can be applied for dependent tests is due to Hommel (1983). Let $p_{(i)}$ be the ordered significance level of the k S_i-statistics, i.e., $p_{(i)} \leq p_{(i+1)}$; reject the overall null hypothesis if and only if

$$p_{(i)} \leq \frac{i\alpha}{kC_k}, \quad \text{for } i = 1, \dots, k, \tag{12.52}$$

where

$$C_k = \sum_{i=1}^{k} \frac{1}{i} \,. \tag{12.53}$$

In the example of Table 12.4, $C_k = 2.829$. Thus, if α is chosen as .05, the smallest p-value (which is smaller than .0005) is to be compared to $.05/25.46 = 0.0020$, and so the hypothesis of a valid RM has to be rejected.

12.5 Discussion

Although the use of a model like the OPLM may seem old-fashioned to some psychometricians in view of the availability of the 2PLM, which seems to be much more flexible, one should be careful with the arguments put forward in the discussion. In Section 12.1 it was pointed out that joint ML-estimation of person and item parameters in the 2PLM is equivalent to assuming an over-parametrized multinomial model that is not identified. The consequence of this is that either unique estimates do not exist or, if they exist, that some restrictions hold between the estimates, entailing inconsistency of the estimators, because no restriction is imposed on the parameter space. For example, if $k = 2$, the restricted sample space is $\{(0,1),(1,0)\}$. If both n_{01} and n_{10} are different from zero, it can be shown by analytical solution of the likelihood equations, if α_1 is fixed at some positive constant, $c = \alpha_1$, that $c = \hat{\alpha}_2$, and $\hat{\theta}_{01} = \hat{\theta}_{10}$, and that the estimates of the difficulty parameters are the same as the JML-estimates in the RM. Notwithstanding this identification problem, the computer program LOGIST (Wingersky, Barton, & Lord, 1982), where this estimation procedure is implemented, is still used. Also in the MML-approach, there are problems with parameter estimation. The introduction of Bayesian estimators in the program BILOG, for example, does not reflect a genuine Bayesian philosophy, but is merely a technical method of solving a set of complicated equations, which otherwise – without the use of priors – often defy solution. (How else could the use of a non-Bayesian terminology like 'standard errors' and the use of statistical goodness-of-fit tests be explained?).

A still more important argument to use the OPLM instead of the 2PLM exists. A number of statistical item-oriented tests have been proposed in the literature for the 2PLM; for an overview, see Yen (1981). For all these tests, it has been claimed that the test statistic is asymptotically χ^2-distributed, but for none of them a proof was provided. Instead, strange arguments are used ("The χ^2-distribution is an approximation to the multinomial distribution"; Yen, 1981, p. 247) or safeguard formulations ("The test statistic is approximately χ^2-distributed") where it is not explained whether this approximation is due to the finiteness of the sample or whether there is also an 'asymptotic approximation'; for example, in the test proposed by Elliott, Murray, & Saunders (1977) the test statistic used is about half the S_i-statistic discussed in Section 12.3.2, no matter what the sample size is, while the (claimed) *df* is the same as for the S_i-test. Admittedly, developing well-founded statistical tests for the 2PLM is difficult, if at all possible, and seemed to be a good reason to look for a model with the same flexibility but within the realm of exponential families where the theory of statistical testing is well established and readily adapted to specific applications.

All this does not mean, of course, that all problems have been solved once and for all. A number of questions that need clarification are listed below.

(i) It is not known to what extent the χ^2-distribution is a useful approximation to the exact sampling distribution of $Q(\boldsymbol{U})$ as defined in (12.33). If a column of the $\boldsymbol{Y}$-matrix has only one non-zero entry, for example, the contribution of this column to the quadratic form is based on the observed frequency of only one response pattern. If the associated expected frequency is very small, use of the χ^2-distribution is possibly not justified.

(ii) Relatively little is known about the power of the S_i-tests. Experience with the model, however, has revealed that extreme items – very easy or very difficult ones – can be fitted with discrimination indices varying over a quite large range. The CML-estimate of the item parameter, however, varies systematically with the discrimination index: for easy items, the higher the discrimination index, the higher the estimate of the difficulty parameter. This needs not be a big problem as long as the item is administered to a population where the item remains easy. Use of the item in a less able population may reveal that the item does not fit any longer.

(iii) As mentioned in Section 12.2, one is not completely free in choosing the discrimination indices, because in some cases unique CML-estimates of the difficulty parameters do not exist whatever the sample size. Although there is a general theoretical result (the theorem of Jacobsen (1989), mentioned above), there is no algorithm to check the conditions in a specific application of the model. It may be reassuring, however, that the problem of non-unique solutions occurs – in our experience of many hundreds of analyses – mainly when the number of items is small.

(iv) From a statistical point of view there can be no objection to the treatment of the discrimination indices as known constants, i.e., as part of the statistical null hypothesis. In practice, however, it rarely happens that plausible hypotheses are available, such that reasonably well-fitting discrimination indices are usually found either by trial and error (based usually on the same data) or by some estimation method as the one described in Section 12.4. In either case, it is not true that the discrimination indices are 'known constants imputed by hypothesis', so that, strictly speaking, all statements about asymptotic distributions of the estimators or test statistics may not be true. Although it may be true that the effect of such a practice may be negligible if the number of sequential adaptations of the model is small in relation to the sample size, very little is known exactly in this respect.

13

Linear Logistic Latent Class Analysis and the Rasch Model

Anton K. Formann[1]

ABSTRACT As a special case of linear logistic latent class analysis, the paper introduces the latent class/Rasch model (LC/RM). While its item latent probabilities are restricted according to the RM, its unconstrained class sizes describe an unknown discrete ability distribution. It is shown that when the number of classes is (at least) half the number of items plus one, the LC/RM perfectly fits the raw score distribution in most cases and thus results in the same item parameter estimates as obtained under the conditional maximum likelihood method in the RM. Two reasons account for this equivalence. Firstly, the maximum likelihood (ML) method for the LC/RM can be seen to be a version of the Kiefer-Wolfowitz approach, so that the structural item parameters as well as the unknown ability distribution are estimated consistently. Secondly, through its moment structure the observed raw score distribution constrains the ability distribution. Irrespective of whether the ability distribution is assumed to be continuous or discrete, it has to be replaced by a step function with the number of steps (=classes) being half the number of items plus one. An empirical example (Stouffer-Toby data on role conflict) illustrates these findings. Finally, conclusions are drawn concerning the properties of the ML item parameter estimates in the LC/RM, and some relations to other methods of parameter estimation in the RM are mentioned.

13.1 Preliminary Remarks

In Rasch's (1960) model (RM) for analyzing the item-score matrix of subjects S_v, $v = 1, \ldots, N$, by items I_i, $i = 1, \ldots, k$, the probability of each entry is governed by two parameters, the first one, θ_v, characterizing subject S_v, and the other one, β_i, the item I_i. At first sight it would seem natural to try to estimate both sets of parameters simultaneously by maximizing the unconditional or joint likelihood of the data as a function of item and person parameters ('joint ML (JML) method'; see Chapter 3). However, if one desires to achieve a high precision of the item parameter estimates, one has to increase the number of persons. As a consequence, the number of parameters θ also increases whereby, for each θ, only a fixed number of observations, k, is available. This is, in fact, the classical situation of structural (=item) vs. incidental (=person) parameters where the ML estimates need not be consistent (Neyman & Scott, 1948).

Two rigorous solutions are available to deal with the problem of the inconsistency of the structural parameters in the presence of incidental parameters. The first one is the conditional ML (CML) method, already proposed by Rasch

[1]Department of Psychology, University of Vienna, Liebiggasse 5, A-1010 Vienna; fax: (0043)-1-4066422

(1960), which replaces the incidental person parameters by their sufficient statistics and conditions the θ out of the likelihood. The second solution was found in a completely different context even before the RM had been established: Kiefer and Wolfowitz (1956, p. 887) proved for their 'nonparametric' or 'semiparametric marginal ML method' "that, under usual regularity conditions, the ML estimator of a structural parameter is strongly consistent, when the (infinitely many) incidental parameters are independently distributed chance variables with a common unknown distribution function. The latter is also consistently estimated although it is not assumed to belong to a parametric class."

For a long time the consequences of this result concerning parameter estimation in the RM were not recognized, leading, among other approaches, to the attempt to estimate the item parameters under the assumption that the unknown ability distribution is a member of a given parametric family (referred to as the 'marginal ML method'). Meanwhile, the relevance of the Kiefer-Wolfowitz approach has been realized by many authors (de Leeuw & Verhelst, 1986; Follmann, 1988; Lindsay, Clogg, & Grego, 1991). Its essence is that these three seemingly different ML methods are strongly related to each other and, moreover, that the CML and the Kiefer-Wolfowitz estimators become identical under certain conditions. This will be shown in the following, whereby latent class analysis, as a nonparametric model, is chosen as the starting point of the considerations.

13.2 Unconstrained Latent Class Analysis for Dichotomous Data

Let the reaction x^*_{vi} of subject S_v to item I_i be coded by '1' in the case of a positive (right or affirmative) response, and by '0' in the case of a negative (wrong or disapproving) response. For k items, the N response vectors $\boldsymbol{x}^*_v = (x^*_{v1}, \dots, x^*_{vk})$ are multinomially distributed over the 2^k response patterns, so that the item-score matrix $\boldsymbol{X}^* = ((x^*_{vi}))$ can be replaced by the collection of response patterns $\boldsymbol{x}_s$ and their observed frequencies n_s, $s = 1, \dots, 2^k$.

Employing these data, we search for homogeneous groups of persons (denoted as latent classes) who have the same reaction tendencies with respect to the items. Let the number of classes be m and their incidence rates (class sizes) in the population π_t, $t = 1, \dots m$, with

$$\sum_t \pi_t = 1, \qquad 0 < \pi_t \leq 1. \tag{13.1}$$

Each homogeneous class C_t is characterized by two response probabilities per item I_i,

$$P(X^*_{vi} = 1|S_v \in C_t) =: p_{ti} \quad \text{and} \quad P(X^*_{vi} = 0|S_v \in C_t) = 1 - p_{ti}. \tag{13.2}$$

Since the responses are assumed to be stochastically independent within each class (Lazarsfeld, 1950; Lazarsfeld & Henry, 1968), the conditional probability of response pattern $\boldsymbol{x}_s = (x_{s1}, \ldots, x_{sk})$, given class C_t, equals the product

$$P(\boldsymbol{x}_s|C_t) = \prod_i p_{ti}^{x_{si}} (1 - p_{ti})^{1-x_{si}}. \tag{13.3}$$

The unconditional probability of response pattern $\boldsymbol{x}_s$ is obtained by weighting (13.3) with the latent class sizes and summing up all classes,

$$P(\boldsymbol{x}_s) = \sum_t \pi_t \prod_i p_{ti}^{x_{si}} (1 - p_{ti})^{1-x_{si}}. \tag{13.4}$$

To estimate the unknown parameters π_t and p_{ti} in unconstrained latent class analysis (LCA), the ML method is mostly used, starting from the log-likelihood

$$\ln L_{LCA} \propto \sum_s n_s \ln P(\boldsymbol{x}_s). \tag{13.5}$$

Having estimated the parameters, provided that all of them are identifiable (for the identifiability in latent class models, cf. McHugh, 1956, and Goodman, 1974; for related results, see, e.g., Teicher, 1963, and for an overview, Titterington, Smith, & Makov, 1985), the goodness of fit of unconstrained LCA is testable by means of standard statistical tests, such as Pearson and likelihood-ratio chi-squared statistics, against the unrestricted multinomial distribution. Thereby the number of degrees of freedom, *df*, is equal to the difference between the number of the response patterns minus 1, that is, $2^k - 1$, and the number of parameters to be estimated in the LCA, that is, $km + m - 1$,

$$df = 2^k - m(k + 1). \tag{13.6}$$

Applying an unconstrained LCA to scale subjects S_v means estimating the posterior class membership probabilities $P(C_t|\boldsymbol{x}_s)$ of the observed response pattern $\boldsymbol{x}_s$,

$$P(C_t|\boldsymbol{x}_s) = \pi_t \, \frac{P(\boldsymbol{x}_s|C_t)}{P(\boldsymbol{x}_s)}. \tag{13.7}$$

The optimal strategy of class assignment regarding classification errors is to allocate each response pattern $\boldsymbol{x}_s$ uniquely to the class C_{t_s} where the class membership probability is highest, that is,

$$P(C_{t_s}|\boldsymbol{x}_s) = \max_t \, P(C_t|\boldsymbol{x}_s). \tag{13.8}$$

13.3 Linear Logistic LCA and Rasch-Type Latent Class Models

Within the framework of LCA, the class sizes π_t and the latent response probabilities p_{ti} can be subjected to different kinds of constraints. One type is realized by means of linear logistic LCA (see Formann, 1982, 1985), which decomposes the log-odds of the latent response probabilities – only these are of interest here – into weighted sums of more elementary parameters λ_j, $j = 1, \dots, p$,

$$\ln[p_{ti}/(1-p_{ti})] = \sum_j q_{tij}\lambda_j. \tag{13.9}$$

The weights $\boldsymbol{Q} = (((q_{tij})))$, $t = 1, \dots, m$; $i = 1, \dots, k$; $j = 1, \dots, p$, are assumed to be given, so that the elementary parameters λ_j are conceived as fixed effects representing a hypothesis the researcher intends to test.

A simple special case of linear logistic LCA assumes 'located' items and classes. This means that both items and classes are assumed to be ordered, and the position of each item I_i on the underlying latent continuum is described by a single location parameter, β_i, and of each class C_t, by θ_t. Moreover, assuming that the β_i and θ_t are concatenated additively leads to the IC curves

$$p_{ti} = \exp(\theta_t + \beta_i)/[1 + \exp(\theta_t + \beta_i)], \tag{13.10}$$

yielding a model that may be called the 'latent class / Rasch Model' (LC/RM) because of its similarity to the RM. In both models, θ represents the ability (or attitude) of the respondents, in the RM for a single person, in the LC/RM for a homogeneous group of persons; and β describes the easiness (or attractivity) of item I_i. Again, as in the RM, the parameters of the LC/RM have to be normalized, for example, setting $\beta_k = 0$ or $\sum_i \beta_i = 0$.

In the LC/RM, the probability of response pattern $\boldsymbol{x}_s$ is

$$\begin{aligned} P(\boldsymbol{x}_s) &= \sum_t \pi_t \frac{\exp[\sum_i x_{si}(\theta_t + \beta_i)]}{\prod_i [1 + \exp(\theta_t + \beta_i)]} = \\ &= \exp(\sum_i x_{si}\beta_i) \sum_t \pi_t \frac{\exp(x_{s.}\theta_t)}{\prod_i [1 + \exp(\theta_t + \beta_i)]}, \end{aligned} \tag{13.11}$$

with $x_{s.} := \sum_i x_{si}$ for the raw score of response pattern $\boldsymbol{x}_s$; therefore the log-likelihood (13.5) becomes

$$\ln L_{LC/RM} \propto \sum_i x_{.i}\beta_i + \sum_s n_s \ln\left[\sum_t \pi_t \frac{\exp(x_{s.}\theta_t)}{\prod_i [1 + \exp(\theta_t + \beta_i)]}\right], \tag{13.12}$$

with $x_{.i} := \sum_s n_s x_{si}$ the item marginal of item I_i. Hence, the likelihood no longer depends on the item-score matrix $\boldsymbol{X}$, but is a function of the item marginals $(x_{.1}, \dots, x_{.k})$, the response patterns' raw scores and frequencies, as well as a function of the unknowns $\boldsymbol{\theta}, \boldsymbol{\beta}$, and $\boldsymbol{\pi}$. Since response patterns with the same raw score are equivalent with respect to the right-hand term ln[...] in (13.12), they can be collapsed, so that the log-likelihood becomes a function of the raw scores r, $r = 0, \ldots, k$, and their observed distribution $\boldsymbol{F} = (f_0, \ldots, f_k)$,

$$\ln L_{LC/RM} \propto \sum_i x_{.i}\beta_i + \sum_r f_r \ln \left[\sum_t \pi_t \frac{\exp(r\theta_t)}{\prod_i [1 + \exp(\theta_t + \beta_i)]} \right], \tag{13.13}$$

with $f_r = \sum_{\boldsymbol{x}_s | r} n_s$.

Next, let us consider the probability of raw score r, $P(r)$, which results from summing the response pattern probabilities (13.11) with raw score r:

$$\begin{aligned} P(r) &= \sum_{\boldsymbol{x}_s|r} P(\boldsymbol{x}_s) \\ &= \underbrace{\sum_{\boldsymbol{x}_s|r} \exp(\sum_i x_{si}\beta_i)}_{} \left[\sum_t \pi_t \frac{\exp(r\theta_t)}{\prod_i [1 + \exp(\theta_t + \beta_i)]} \right] \\ &= \gamma_r \left[\sum_t \pi_t \frac{\exp(r\theta_t)}{\prod_i [1 + \exp(\theta_t + \beta_i)]} \right], \end{aligned} \tag{13.14}$$

where $\gamma_r = \gamma_r(\epsilon_1, \ldots, \epsilon_k)$, with $\epsilon_i = \exp(\beta_i)$, is the elementary symmetric function of order r as in the CML method in the RM. Combining (13.11) and (13.14) gives a result well-known for the RM (see Chapter 3): for each response pattern $\boldsymbol{x}_s$ with score $x_{s.} = r$, the conditional probability $P(\boldsymbol{x}_s \mid r)$ does not depend on $\boldsymbol{\theta}$,

$$P(\boldsymbol{x}_s | x_{s.} = r) = \frac{P(\boldsymbol{x}_s)}{P(r)} = \gamma_r^{-1} \exp(\sum_i x_{si}\beta_i). \tag{13.15}$$

Note that this also applies to the corresponding class-specific probabilities, i.e., $P(\boldsymbol{x}_s | x_{s.} = r, C_t) = P(\boldsymbol{x}_s | C_t)/P(r | C_t) = \gamma_r^{-1} \exp(\sum_i x_{si}\beta_i)$.

According to (13.14),

$$\sum_t \pi_t \frac{\exp(r\theta_t)}{\prod_i [1 + \exp(\theta_t + \beta_i)]} = \frac{P(r)}{\gamma_r}, \tag{13.16}$$

which, inserted into the log–likelihood (13.13), gives

$$\begin{aligned} \ln L_{LC/RM} &\propto \sum_i x_{.i}\beta_i - \sum_r f_r \ln \gamma_r + \sum_r f_r \ln P(r) \\ &= \ln L_{RM/CML} \qquad\qquad + \ln L_F. \end{aligned} \tag{13.17}$$

So the log-likelihood of the LC/RM consists of two additive terms: the first one is identical to the conditional log-likelihood of the RM, $\ln L_{RM/CML}(\boldsymbol{X} | \boldsymbol{F}; \boldsymbol{\beta})$, which is the log-likelihood of the item marginals $(x_{.1}, \ldots, x_{.k})$ given the raw score distribution $\boldsymbol{F} = (f_0, \ldots, f_k)$; the second one is the log-likelihood of the raw score distribution $\boldsymbol{F}$, $\ln L_F(\boldsymbol{\pi}, \boldsymbol{\beta}, \boldsymbol{\theta})$.

Another consequence of the model structure is that the posterior class membership probabilities (13.7), $P(C_t|\boldsymbol{x}_s)$, $t = 1, \ldots, m$, of all response patterns $\boldsymbol{x}_s$ with the same raw score r, are identical,

$$P(C_t|\boldsymbol{x}_s, x_{s.}=r) = P(C_t|\boldsymbol{x}_{s^*}, x_{s^*.}=r) = P(C_t|r) =$$

$$= \frac{\pi_t \dfrac{\exp(r\theta_t)}{\prod\limits_i [1+\exp(\theta_t+\beta_i)]}}{\sum\limits_l \pi_l \dfrac{\exp(r\theta_l)}{\prod\limits_i [1+\exp(\theta_l+\beta_i)]}}. \tag{13.18}$$

Hence, in the LC/RM persons are not assigned to classes according to their response patterns, but according to their raw scores r, $r = 0, \ldots, k$. The raw score can therefore be considered a sufficient statistic for class membership. In addition to the class membership probabilities, the score parameter θ_r^* of each score can be estimated by means of the equation

$$\theta_r^* = \sum_t \theta_t P(C_t|r), \quad \text{for } r = 0, \ldots, k. \tag{13.19}$$

While θ_t represents the position of latent class C_t on the latent continuum (ability or attitude), the posterior mean θ_r^* gives the position of all persons with response patterns that have r positive answers. It is interesting to note that this contrasts with both the conditional and unconditional ML approaches in the RM, where person parameter estimates can be obtained only for $r = 1, \ldots, k-1$ (see, however, the remarks on 13.27).

The goodness of fit of the m-class LC/RM can be tested twice, firstly against the unrestricted multinomial distribution, and secondly, against the conventional m-class solution. Since the number of parameters under the m-class LC/RM is $2m + k - 2$ ($m-1$ class sizes π_t, m parameters θ_t, and $k-1$ parameters β_i), the df is

$$df = 2^k - 2m - k + 1 \tag{13.20}$$

for the Pearson and likelihood-ratio chi-squared statistics of the LC/RM against the unrestricted multinomial. For the likelihood-ratio test of the m-class LC/RM against the ordinary m-class model, should the latter hold,

$$df = (m-1)(k-1). \tag{13.21}$$

13.4 Rasch-Type Latent Class Models and the CML Parameter Estimates in the RM

Returning to the log-likelihood of the m-class LC/RM, it is evident that maximizing (13.17) maximizes the log-likelihood of the raw score distribution, $\ln L_F$, with respect to $\boldsymbol{\beta}$, $\boldsymbol{\pi}$, and $\boldsymbol{\theta}$. The number of parameters π and θ equals $2m-1$ and increases with the number of classes, whereas the number of item parameters remains $k-1$ independently of m; therefore, the larger the number of classes, the better the fit to the observed raw score distribution, however, only up to a certain saturation point. This optimal fit of $\ln L_F$ is attained if the observed score distribution, $\boldsymbol{F}$, and its expectation, $\boldsymbol{E} = (e_0, \ldots, e_r, \ldots, e_k)$, are identical, that is, if

$$f_r = e_r = NP(r), \quad \text{for } r = 0, \ldots, k, \tag{13.22}$$

or equivalently if

$$P(r) = f_r/N = \gamma_r \sum_t \pi_t \frac{\exp(r\theta_t)}{\prod_i [1+\exp(\theta_t+\beta_i)]}, \quad \text{for } r = 0, \ldots, k. \tag{13.23}$$

If this coincidence is possible for, say, m^* classes, then this m^*-class LC/RM will be equivalent to the conditional RM that conditions explicitly on the observed raw score distribution, and the item parameter estimates under the m^*-class LC/RM will be identical to the CML solution.

Proof To prove this equivalence, assume that (13.22) holds. From the log-likelihood (13.17) it follows that the ML equations of the LC/RM (m^*) take the form

$$\begin{aligned}
\frac{\partial \ln L_{LC/RM(m^*)}}{\partial \beta_i} &= \frac{\partial \ln L_{RM/CML}}{\partial \beta_i} + \frac{\partial \ln L_{F(m^*)}}{\partial \beta_i} = 0, \\
\frac{\partial \ln L_{LC/RM(m^*)}}{\partial \pi_t} &= \frac{\partial \ln L_{F(m^*)}}{\partial \pi_t} = 0, \\
\frac{\partial \ln L_{LC/RM(m^*)}}{\partial \theta_t} &= \frac{\partial \ln L_{F(m^*)}}{\partial \theta_t} = 0.
\end{aligned} \tag{13.24}$$

Since

$$\begin{aligned}
\frac{\partial \ln L_{F(m^*)}}{\partial \beta_i} &= \frac{\partial}{\partial \beta_i}\left[\sum_{r=0}^{k-1} f_r \ln P(r) + f_k \ln[1 - \sum_{r=0}^{k-1} P(r)]\right] = \\
&= \sum_{r=0}^{k-1} \frac{f_r}{P(r)} \frac{\partial P(r)}{\partial \beta_i} + \frac{f_k}{P(k)} \frac{\partial}{\partial \beta_i}[1 - \sum_{r=0}^{k-1} P(r)] \\
&= \sum_{r=0}^{k-1} \left[\frac{f_r}{P(r)} - \frac{f_k}{P(k)}\right] \frac{\partial P(r)}{\partial \beta_i},
\end{aligned} \tag{13.25}$$

the derivatives $\partial \ln L_F(m^*)/\partial \beta_i$ become zero when $f_r/P(r) = N$ for $r = 0, \ldots, k$; hence, under the LC/RM, the ML equations for $\boldsymbol{\beta}$ reduce to the CML equations under the RM,

$$\frac{\partial \ln L_{LC/RM(m^*)}}{\partial \beta_i} = \frac{\partial \ln L_{RM/CML}}{\partial \beta_i} = 0. \tag{13.26}$$

It is concluded that both $\ln L_{LC/RM(m^*)}$ and $\ln L_{RM/CML}$ result in the same solution for $\boldsymbol{\beta}$ if the observed raw score distribution equals the expected one. □

Due to the side condition $\sum_r P(r) = 1$, the expected raw score distribution $\boldsymbol{E} = (e_0, \ldots, e_k)$ consists of k independent elements. Given $\boldsymbol{\beta}$, they are functionally related to $2m-1$ parameters $\boldsymbol{\pi}$ and $\boldsymbol{\theta}$ of the m-class LC/RM. It is therefore to be expected (and will be shown in Section 13.6) that $2m-1$ must at least equal k to enable the LC/RM to fit the raw score distribution perfectly. From this it follows that a lower bound m^* for m is

$$m^* = \begin{cases} (k+1)/2 & \text{for } k \text{ odd,} \\ k/2+1 & \text{for } k \text{ even.} \end{cases} \tag{13.27}$$

If k is odd, all parameters may be identifiable, since then $2m^* - 1 = k$. If k is even, then $2m^* - 1 = k + 1$, that is, the number of π and θ exceeds the number of independent elements of the raw score distribution by 1, so that one parameter will be unidentifiable. By setting $\theta_1 = -\infty$, corresponding to $p_{1i} = 0$ for $i = 1, \ldots, k$, or $\theta_{m^*} = \infty$, corresponding to $p_{m^*i} = 1$ for $i = 1, \ldots, k$, this ambiguity can be removed; for possible alternative restrictions, see the numerical example in Section 13.5. However, then the score parameters (13.19) can no longer be computed. If m is chosen to be greater than the lower bound m^*, the number of unidentifiable parameters increases (each additional class adds two further parameters), and no improvement of fit is possible. So, for each m-class LC/RM with $m \geq m^*$ fitting the raw score distribution, the *df* for the goodness-of-fit tests against the unrestricted multinomial distribution of the response patterns is

$$df = 2^k - 2k. \tag{13.28}$$

13.5 A Numerical Example

The data of Stouffer and Toby (1951) on role conflict (tendency towards universalistic vs. particularistic values) will be used to demonstrate several aspects of the application of unrestricted LCAs and LC/RMs with increasing numbers of classes. The data are given in Table 13.1 and refer to a sample of $N = 216$ respondents. All of the following calculations are based on the FORTRAN program LCALIN for parameter estimation in linear logistic LCA (Formann, 1984, pp. 215–252).

As can be seen in Table 13.2, unlike the bad fit of the unconstrained one-class solution of LCA, which corresponds to the assumption of the independence of the four items, the unconstrained two-class solution fits well. This, as well as the observation that the classes and the items can be considered to be ordered, justify the application of the two-class LC/RM from the substantive point of view: all item latent probabilities p_{ti} of the first class are lower than those of the second class, and the order of the item latent probabilities of the first class

TABLE 13.1. Stouffer and Toby (1951) data on role conflict; response patterns and raw scores, their observed and expected frequencies for $m = 1, 2, 3$ classes assuming the LC/RM, as well as posterior class membership probabilities, $P(C_t|r)$, and score parameters, θ_r^*, for the two-class LC/RM.

Resp. Pattern	Frequencies Obs.	Expected $m=1$	$m=2$	$m=3$	Raw Score	Frequencies Obs.	Expected $m=1$	$m=2$	$m=3$	$m=2$ $P(C_1\|r)$	$P(C_2\|r)$	θ_r^*
0000	20	7.54	16.69	20.00	0	20	7.54	16.69	20.00	1.000	.000	−.58
0001	2	3.39	2.40	2.02								
0010	9	7.98	9.01	7.57	1	55	47.58	61.89	55.00	.998	.002	−.57
0100	6	7.54	8.34	7.01								
1000	38	28.67	42.14	38.40								
0011	2	3.59	1.34	1.34								
0101	1	3.39	1.24	1.24								
0110	4	7.98	4.67	4.64	2	63	86.83	58.99	63.00	.961	.039	−.46
1001	7	12.89	6.29	6.79								
1010	24	30.31	23.60	25.43								
1100	25	28.67	21.85	23.56								
0111	1	3.59	1.29	1.18								
1011	6	13.63	6.53	6.45	3	36	60.42	36.55	36.00	.498	.502	1.03
1101	6	12.89	6.04	5.98								
1110	23	30.31	22.69	22.39								
1111	42	13.63	41.86	42.00	4	42	13.63	41.86	42.00	.039	.961	2.50

coincides with that of the second class, except for the negligible violation of that order by the items I_2 and I_3; see Table 13.3a.

The fit of this restricted model is also excellent. So it is not surprising that the 8 item latent probabilities of the unconstrained two-class model are reproduced almost perfectly by 5 parameters ($\beta_1, \beta_2, \beta_3, \beta_4, \theta_1, \theta_2$, with $\beta_4 = -\beta_1 - \beta_2 - \beta_3$, so that $\sum_i \beta_i = 0$) of the LC/RM; compare both solutions in Table 13.3. From the parameter estimates for $\boldsymbol{\theta}$, $\hat{\theta}_1 = -.58$, $\hat{\theta}_2 = 2.62$, and the posterior class membership probabilities (13.18), $P(C_t|r)$, $t = 1, 2$; $r = 0, \ldots, k$, the score parameters θ_r^* were estimated using (13.19). They can be found together with the class membership probabilities in Table 13.1, and reveal that scores from 0 to 2 correspond approximately to the same position on the latent attitude. Contrary to this, the scores $r = 3$ and $r = 4$ are well-distinguished from each other and from the scores $r = 0, 1, 2$. From this it can be seen that the scores r are nonlinearly related to their score parameters θ_r^*, which is not the case in the RM: for $r = 1, 2, 3$, the person parameters are –1.35, 0.02, and 1.35 when estimated via ML based on the CML item parameter estimates.

TABLE 13.2. Likelihood-ratio statistic (L^2) for some models for the Stouffer and Toby (1951) data on role conflict.

Model	Number of Classes	Parameters	L^2	*df*
LCA(1) = LC/RM(1)	1	4	81.08	11
LCA(2)	2	9	2.72	6
LC/RM(2)	2	6	2.91	9
LC/RM(3)=RM/CML	3	7	1.09	8
LCA(2) vs. LC/RM(2)	–	–	0.19	3

Given the very good fit of the two-class models, the analysis would usually stop here. But in order to see the effect of increasing the number of classes, consider now the three-class LC/RM solution of the Stouffer and Toby data.

- Observe first that the fit becomes slightly better than it is under the two-class LC/RM, and that for $m = 3$ and $k = 4$ the model is overparameterized by one parameter, so that $df = 8$ (instead of $df = 7$; see Table 13.2). Whereas the item parameters $\boldsymbol{\beta}$ are uniquely determined, except for the side condition $\sum_i \beta_i = 0$, the parameters $\boldsymbol{\pi}$ and $\boldsymbol{\theta}$ need one restriction. A host of such restrictions is possible, among others the two mentioned in the preceding section ($p_{1i} = 0$ for $i = 1, \ldots, k$; or $p_{mi} = 1$ for $i = 1, \ldots, k$), and also $\sum_t \theta_t = 0$, or, equivalently, $\theta_m = -\sum_{t=1}^{m-1} \theta_t$. Fixing one class size π_t to a predetermined value π_t^* is also possible; this, however, only applies with some limitations since π_t^* cannot take all values within (0,1). Four equivalent solutions for the Stouffer and Toby data are shown in Table 13.3 for the three-class LC/RM with different types of restrictions. Each one of these solutions exhibits exactly the same goodness of fit, but has different $\boldsymbol{\pi}$ and $\boldsymbol{\theta}$ and, consequently, also different latent response probabilities. Because of this indeterminacy of $\boldsymbol{\pi}$ and $\boldsymbol{\theta}$, it does not make much sense to compute the score parameters θ_r^* when k is even. Therefore, they are given for the two-class LC/RM in Table 13.1.

- Second, for $m = 3$ the expected raw score distribution becomes identical to the observed one, but this is clearly not true with respect to the expected and the observed distributions of the response patterns. Otherwise, L^2 should be 0, but in fact it is 1.09; see Table 13.2, and for a direct comparison of the expected and the observed distributions of the response patterns, Table 13.1.

- Third, the estimates of the item parameters under the three-class LC/RM are numerically identical to those found in estimating the item parameters for the RM by CML; to make comparisons easy, the item parameter

TABLE 13.3. Stouffer and Toby (1951) data on role conflict; class sizes and item latent probabilities (a) for the two-class unconstrained LCA, (b) for the two-class LC/RM, and (c) for some equivalent three-class LC/RMs.

	Class	Class Size	Latent Probabilities for Items I_1	I_2	I_3	I_4	θ_t
a)	C_1	.720	.714	.329	354	.132	–
	C_2	.280	.993	.939	.926	.769	–
b)	C_1	.718	.716	.333	.350	.126	–.58
	C_2	.282	.984	.925	.930	.780	2.62
c)	C_1	.039	0*	0*	0*	0*	–
	C_2	.720	.768	.377	.395	.148	–.37
	C_3	.241	.990	.950	.953	.844	3.07
	C_1	.264	.539	.176	.187	.058	–1.41
	C_2	.571	.849	.506	.525	.228	.16
	C_3	.165	1*	1*	1*	1*	–
	C_1	.157	.444	.127	.136	.040	–1.79
	C_2	.643	.814	.443	.462	.187	–.09
	C_3	.200*	.995	.975	.977	.919	3.81
	C_1	.064	.206	.045	.049	.014	–2.92+
	C_2	.706	.780	.393	.411	.157	–.30+
	C_3	.230	.992	.956	.959	.863	3.22+
Sample			.792	.500	.514	.310	

Notes: * denotes fixed parameters, + side condition $\sum_t \theta_t = 0$.

estimates are given for the two- and three-class LC/RM (=RM/CML), as well as for the score groups. Andersen's (1973b) likelihood-ratio test ($L^2 = 0.63$, $df = 6$; cf. Table 13.4) supports the findings of the other statistical tests summarized in Table 13.2, namely that the Stouffer and Toby data conform to the RM to a very high degree. Note the further interesting by-product, that even the more parsimonious two-class LC/RM provides sufficient fit for the Stouffer and Toby data. Hence, the three-class LC/RM (=RM/CML) can be approximated by its two-class analogue without any remarkable loss of accuracy (for the same finding for the Bock & Lieberman, 1970, data, see Formann, 1984, pp. 115–118, and Follmann, 1988) and with the advantage of no indeterminacies concerning $\boldsymbol{\pi}$ and $\boldsymbol{\theta}$.

13.6 Statistical Background: The Moment Problem

The preceding numerical example has shown that in the case of one particular set of data it was actually possible to obtain the CML solution to the item parameters of the RM by applying the LC/RM with the number of classes

TABLE 13.4. Stouffer and Toby (1951) data on role conflict; item parameter estimates for the two- and three-class LC/RM as well as for the score-groups using the CML method (RM/CML).

	Easiness Parameter Estimates				$\ln L_{RM/CML}$
	I_1	I_2	I_3	I_4	
LC/RM ($T = 2$)	1.51	−0.11	−0.04	−1.36	–
LC/RM ($T = 3$) = RM/CML	1.57	−0.13	−0.06	−1.38	−169.916
RM/CML ($r = 1$)	1.56	−0.29	0.12	−1.39	−50.263
RM/CML ($r = 2$)	1.59	−0.13	−0.13	−1.33	−83.948
RM/CML ($r = 3$)	1.68	−0.11	−0.11	−1.46	-35.389

claimed in (13.27). However, this need not be true in general: degenerate LC/RM solutions may occur where (at least) one class size tends to zero, so that the *df* given in (13.28) is no longer correct; thereby, the LC/RM and the CML estimates may be identical (Verhelst, personal communication), or not identical (Follmann, 1988). Therefore, in the following the exact conditions will be studied that make the LC/RM and CML solutions identical.

The key question to be answered is: what conditions have to be satisfied for the observed raw score distribution to become identical with its expectation?

Define

$$w_r = \frac{f_r}{N\,\gamma_r\,P(r=0)} = \frac{f_r}{N\gamma_r \sum\limits_t \pi_t\{\prod\limits_i [1+\exp(\theta_t+\beta_i)]\}^{-1}}, \quad \text{for } r = 0,\ldots,k, \tag{13.29}$$

and

$$\lambda_t = \pi_t \frac{P(r=0|\theta_t)}{P(r=0)} = \frac{\pi_t\{\prod\limits_i [1+\exp(\theta_t+\beta_i)]\}^{-1}}{\sum\limits_l \pi_l\{\prod\limits_i [1+\exp(\theta_l+\beta_i)]\}^{-1}}, \quad \text{for } t = 1,\ldots,m. \tag{13.30}$$

Using the transformation

$$\xi_t = \exp(\theta_t), \quad \theta_t \in \mathbb{R},\ \xi_t \in [0,\infty), \tag{13.31}$$

(13.23) yields

$$w_r = \sum_t \lambda_t \xi_t^r, \quad \text{for } r = 1,\ldots,k, \tag{13.32}$$

where

$$w_0 = \sum_t \lambda_t = 1. \tag{13.33}$$

The system of equations (13.32) reiterates the old Stieltjes' (1894) moment problem. Let $w_0 = 1, w_1, w_2, \ldots$ be a sequence of real numbers. Is there a distribution function $H(y)$ defined on the interval $[0,\infty)$ such that the moments m_g,

$g = 0, 1, 2, \ldots$, of the random variable Y,

$$m_g = \int_0^\infty y^g dH(y), \quad \text{for } g = 0, 1, 2, \ldots, \tag{13.34}$$

equal $w_0, w_1, w_2, \ldots$?

Let Y be a random variable, whose moments m_g, for $g = 0, 1, 2, \ldots$, exist; let h and l be nonnegative integers, and $u_0, \ldots, u_h$ arbitrary real numbers. Then, the expectation

$$\mathcal{E}\left[\left(\sum_{g=0}^{h} u_g Y^{l+g}\right)^2\right] \tag{13.35}$$

is always nonnegative. This expectation equals the quadratic form

$$Q = \sum_{g,i=0}^{h} u_g u_i m_{g+i+2l}, \tag{13.36}$$

of the moments m_{g+i+2l}, for $g, i = 0, \ldots, h$, in $u_0, \ldots, u_h$. Q must be positive semi-definite because the expectation in (13.35) is always ≥ 0; this implies that all principal minors of the determinants D_{2l}, for $l \geq 0$,

$$D_{2l} = \begin{vmatrix} m_{2l} & m_{2l+1} & \cdots & m_{2l+h} \\ m_{2l+1} & m_{2l+2} & \cdots & m_{2l+h+1} \\ \vdots & \vdots & & \vdots \\ m_{2l+h} & m_{2l+h+1} & \cdots & m_{2l+2h} \end{vmatrix}, \tag{13.37}$$

are nonnegative. This results in a system of inequalities,

$$\begin{vmatrix} m_{2l} & \cdots & m_{2l+h} \\ \vdots & & \vdots \\ m_{2l+h} & \cdots & m_{2l+2h} \end{vmatrix} \geq 0,$$

$$\vdots \tag{13.38}$$

$$\begin{vmatrix} m_{2l} & m_{2l+1} \\ m_{2l+1} & m_{2l+2} \end{vmatrix} \geq 0,$$

$$|m_{2l}| \geq 0,$$

for all $l, h \geq 0$, which are satisfied by the moment sequence of the random variable Y. However, because of the redundancy inherent in this system of inequalities, it suffices to consider the so-called Hankel determinants (cf., e.g., Naas & Schmid, 1961, pp. 193-195) D_g and D'_g,

$$D_g = \begin{vmatrix} w_0 & w_1 & \cdots & w_g \\ w_1 & w_2 & \cdots & w_{g+1} \\ \vdots & \vdots & & \vdots \\ w_g & w_{g+1} & \cdots & w_{2g} \end{vmatrix}, \qquad g = 0, 1, 2, \ldots, \tag{13.39}$$

$$D'_g = \begin{vmatrix} w_1 & w_2 & \cdots & w_{g+1} \\ w_2 & w_3 & \cdots & w_{g+2} \\ \vdots & \vdots & & \vdots \\ w_{g+1} & w_{g+2} & \cdots & w_{2g+1} \end{vmatrix},$$

which necessarily have to be nonnegative, making the real numbers $w_0 = 1, w_1, w_2, \ldots$ a moment sequence. While this necessary condition is easy to prove, a necessary and sufficient condition for $w_0 = 1, w_1, w_2, \ldots$ to constitute a moment sequence is rather hard to derive. Therefore, only the main results are given in the following; the reader interested in further details is referred to Stieltjes (1894), Hamburger (1920), Shohat and Tamarkin (1943), Karlin and Studden (1966), and Krein and Nudel'man (1977).

For the moment problem to have a solution in $[0, \infty)$ with respect to the distribution function $H(y)$ in (13.34) (which comprises (13.32) as a special case), it is necessary and sufficient that the following conditions hold:

(a) $$D_g > 0, \quad \text{and} \quad D'_g > 0, \quad g = 0, 1, 2, \ldots, \tag{13.40}$$

if the distribution function H has infinitely many points of increase;

(b) $$\begin{aligned} &D_0 > 0, \ldots, D_h > 0, \quad D_{h+1} = D_{h+2} = \cdots = 0, \\ &D'_0 > 0, \ldots, D'_h > 0, \quad D'_{h+1} = D'_{h+2} = \cdots = 0, \end{aligned} \tag{13.41}$$

if H has $h + 1$ points of increase not including $y = 0$; and

(c) $$\begin{aligned} &D_0 > 0, \ldots, D_h > 0, \quad D_{h+1} = D_{h+2} = \cdots = 0, \\ &D'_0 > 0, \ldots, D'_{h-1} > 0, \quad D'_h = D'_{h+1} = \cdots = 0, \end{aligned} \tag{13.42}$$

if H has $h + 1$ points of increase, one of them being equal to the lower boundary $y = 0$.

In cases (b) and (c), the solution is unique; in case (a), there exists a manifold of solutions.

If only the first k real numbers $w_0 = 1, w_1, \ldots, w_k$ are given, as is the case in the LC/RM with k items, the determinants containing elements $w_{k+1}, w_{k+2}, \ldots$ are undefined. For the existing determinants D_g and D'_g defined in (13.39), the following is implied: D_g can be positive up to $h = k/2$, and D'_g can be positive up to $h' = (k-1)/2$ for k odd, and up to $h'' = (k-2)/2 = h - 1$ for k even. From this, together with (13.41) and (13.42), it follows that the (maximum) number of points of increase of the distribution function H (= the number of classes) has to be $m^* = h' + 1 = (k-1)/2 + 1 = (k+1)/2$ for k odd, and $m^* = h'' + 2 = (k-2)/2 + 2 = k/2 + 1$ for k even, respectively, where in the latter case the first point of increase is at the lower boundary $y = 0$. See also (13.27) and the empirical example in Section 13.5, where some further solutions equivalent to $y = 0$ are mentioned.

Thus, to check whether the expected raw score distribution, $\boldsymbol{E}$, of the m^*-class LC/RM will exactly fit the observed raw score distribution, $\boldsymbol{F}$, the following procedure can be employed. Compute the estimates of the determinants in (13.39), $\hat{D}_g$, for $g = 0, \ldots, h = k/2$, and $\hat{D}'_g$, for $g = 0, \ldots, h^*$, with $h^* = h'$ for k odd and $h^* = h''$ for k even, and

$$\hat{w}_r = \frac{f_r}{N\hat{\gamma}_r\hat{P}(r=0)} = \frac{f_r}{\hat{\gamma}_r f_0}, \qquad \text{for } r = 0, \ldots, k, \tag{13.43}$$

see (13.29), assuming that the $\gamma_r(\hat{\boldsymbol{\beta}})$ are available. If all determinants $\hat{D}_g$ and $\hat{D}'_g$ are positive, then the moment problem (13.34) is solvable for the discrete distribution function $(\lambda_1, \ldots, \lambda_{m^*})$, and $\boldsymbol{E}$ will be equal to $\boldsymbol{F}$(regular case). If not all determinants $\hat{D}_g$ and $\hat{D}'_g$ are positive, then – possibly due to sampling errors – $\boldsymbol{E}$ may be equal to $\boldsymbol{F}$ for some $m < m^*$ (irregular cases; see also the remarks at the beginning of Section 13.6). For a similar check and further diagnostics, see Lindsay, Clogg, and Grego (1991).

From a practical point of view, however, it seems more appealing to employ the LC/RM assuming $m = 1, \ldots, m^*$ classes; then it is directly seen whether the likelihood increases with an increasing number of classes, reaching its maximum at $m = m^*$ (regular case) or at some $m < m^*$ (irregular case), and whether the expected and the observed raw score distributions coincide.

13.7 Consequences and Relations to Other Methods of Parameter Estimation in the RM

From the identity of ML item parameter estimates of the m^*-class LC/RM and the CML estimates in the RM it immediately follows that all their properties and uniqueness conditions are the same: existence and uniqueness (Fischer, 1981), and consistency and asymptotic normality (Andersen, 1970; Pfanzagl, 1994). Moreover, it is to be expected that for large samples, i.e., for $N \to \infty$, the m^*-class LC/RM will almost surely fit the raw score distribution, because for $N \to \infty$ sampling errors become irrelevant so that the regular case becomes more and more probable. Finally, the ability distribution, expressed by the class sizes and estimated by ML, is estimated consistently; see Kiefer & Wolfowitz (1956) and the remarks in Section 13.1.

The results reported here are strongly related to the marginal ML (MML) method of parameter estimation in the RM. This method assumes that subjects are randomly sampled from a population with continuous ability distribution $H(\theta)$ – therefore, the underlying model is sometimes called the 'random-effects RM' in contrast to the original 'fixed-effects RM' –, so that the marginal probability of response pattern $\boldsymbol{x}$ is given by

$$P(\boldsymbol{x}) = \int_{-\infty}^{\infty} p(\boldsymbol{x}|\theta) dH(\theta). \tag{13.44}$$

The item parameters are then estimated by applying the ML method to this marginal distribution. The ability distribution $H(\theta)$, or at least the family to which it belongs, has to be assumed to be known (for example, normal; see Bock & Lieberman, 1970; Sanathanan & Blumenthal, 1978), or it has to be estimated as a discrete distribution with an arbitrary number of support points, along with the item parameters (cf. Bock & Aitkin, 1981, for the normal ogive model, as well as Thissen, 1982, and Mislevy, 1984, for the RM).

Following this latter approach, papers by de Leeuw and Verhelst (1986), Follmann (1988), and Lindsay, Clogg and Grego (1991) have shown that for the step function by which the unknown ability distribution is approximated not more points of increase are required than half the number of items plus 1. While de Leeuw and Verhelst (1986) arrived at this result assuming an unknown continuous ability distribution $H(\theta)$ with infinitely many points of increase, Lindsay, Clogg and Grego (1991) started out from the LC/RM of Formann (1984, 1989; see also Clogg, 1988), which can also be considered a random-effects model, however, with a discrete ability distribution.

When fitting the raw score distribution, the m^*-class LC/RM gives the optimal representation of the distribution of sufficient statistics for the person parameters. This follows from the decomposition of $\ln L_{LC/RM}$, see (13.17) and (13.24), which clearly also holds for the marginal likelihood derived from the marginal response pattern probabilities (13.44). The MML method must therefore lead to a suboptimal representation of the raw score distribution whenever the assumed ability distribution cannot generate the observed raw score distribution. This applies especially when the ability distribution is parameterized as a member of a certain family of distributions; see de Leeuw and Verhelst (1986). For a further interesting alternative to the marginal and the conditional ML methods, combining the CML approach with the estimation of the ability distribution, see Andersen and Madsen (1977).

Note the further connection with the 'extended (random-effects) RM' developed by Tjur (1982), which reformulates the ordinary RM as a specific loglinear model; see also Kelderman (1984). This model was criticized by Cressie and Holland (1983, p. 136) because it ignores the inequality constraints following from the moment structure of the raw score distribution. Consequently, "... it is possible to obtain a $p(\boldsymbol{x})$ satisfying ..." the loglinear model "... which does not even satisfy the condition of local non-negative dependence ...". This critique brought into discussion the moment structure later used by Kelderman (1984), de Leeuw & Verhelst (1986), Follmann (1988), and Lindsay, Clogg and Grego (1991), and eventually led to the present state of knowledge concerning the strong interrelations between the CML and MML methods in the RM and the latent class approach.

This knowledge, however, is primarily of theoretical importance. From a practical point of view, that is, from the point of view of parameter estimation, very efficient algorithms are available today for computing CML estimates in the RM; see Chapters 3 and 8. Parameter estimation by means of restricted latent class models cannot therefore be recommended as a serious rival any more, even if the EM algorithm (Dempster, Laird, & Rubin, 1977; Wu, 1983), which is widely used for that purpose (see, e.g., Formann, 1992), has proved to be a very useful tool in all cases of incomplete data. The EM algorithm seems to be comparatively stable with respect to convergence, but it sometimes is very slow in terms of the number of iterations needed.

14

Mixture Distribution Rasch Models

Jürgen Rost and Matthias von Davier[1]

ABSTRACT This chapter deals with the generalization of the Rasch model to a discrete mixture distribution model. Its basic assumption is that the Rasch model holds within subpopulations of individuals, but with different parameter values in each subgroup. These subpopulations are not defined by manifest indicators, rather they have to be identified by applying the model. Model equations are derived by conditioning out the class-specific ability parameters and introducing class-specific score probabilities as model parameters. The model can be used to test the fit of the ordinary Rasch model. By means of an example it is illustrated that this goodness-of-fit test can be more powerful for detecting model violations than the conditional likelihood ratio test by Andersen.

14.1 Introduction

The chapter deals with a generalization of the Rasch model (RM) to a discrete (finite) mixture distribution model. Stated briefly, the mixed Rasch model (MRM) assumes that the RM does not hold for the entire population, but does so within (latent) subpopulations of individuals which are not known beforehand. Section 14.1 describes the general structure of discrete and continuous mixture models. The RM itself being a special kind of a continuous mixture model, its generalization to the MRM assumes discrete mixtures of continuous mixture distributions. The central properties of the MRM, including parameter estimation, are outlined in Section 14.2. One of the most relevant applications of the MRM is the testing of fit of the ordinary RM. Some related results are presented in Section 14.3. The unrestricted MRM does not require any assumption about the ability distributions within the classes. This entails, however, a large number of model parameters. The latter can be reduced by imposing restrictions on the latent score distributions as proposed in Section 14.4. The mixture generalization has also been developed for ordinal RMs, like the partial credit or the rating scale model. These extensions of the MRM, however, are described in Chapter 20.

14.2 Continuous and Discrete Mixture Distribution Models

The basic assumption of mixture distribution models (MDMs) is that the distribution of a (possibly vector valued) observed random variable is not adequately

[1]Institut für die Pädagogik der Naturwissenschaften an der Universität Kiel, Olshausenstraße 62, D-24098 Kiel; e-mail: ROST@IPN.UNI-KIEL.DE

described by a single uniform probability function, but rather by a number of conditional probability functions. The variable on which the probabilities are conditioned is referred to as the mixing variable. The probability function of a discrete[2] variable X is formally defined as

$$P(X = x) = \int_{-\infty}^{\infty} P(X = x|\theta)\, dF(\theta), \tag{14.1}$$

where θ is a continuous mixing variable, and $F(\theta)$ its probability distribution. The unconditional probability $P(X = x)$ is obtained by integrating over the density of the mixing variable (Everitt and Hand, 1981). The mixing variable can also be discrete, which reduces the integral structure to a sum,

$$P(X = x) = \sum_{c=1}^{C} P(c)\, P(X = x|\theta = c). \tag{14.2}$$

In this case, the unconditional probability is a weighted sum, where the weights $P(c)$ are the probabilities that the mixing variable has value c, and the terms are summed over all values of c. Parameter θ in such a discrete MDM defines a (usually small) number of components of the mixture or classes of individuals. Therefore, the weights $P(c)$ define the relative size of the components of the mixture or class sizes. Item response models (Lord, 1980) are usually MD models, either continuous or discrete. The observed random variables described by IRT models are the response vectors of the individuals, $\boldsymbol{x} = (x_1, \ldots, x_k)$, in the present case with binary components $x_i \in \{0, 1\}$. The mixing variable θ is referred to as the 'latent variable'. The conditional probabilities are often defined as the logistic function, e.g., for the RM,

$$P(\boldsymbol{X} = \boldsymbol{x}|\theta) = \prod_{i=1}^{k} \frac{\exp[x_i(\theta + \beta_i)]}{1 + \exp(\theta + \beta_i)}, \tag{14.3}$$

where β_i is an easiness parameter of item I_i. Written in this way, the RM is a continuous MDM, because θ is a real valued latent variable. In the RM, however, it is not necessary to know or to specify the distribution of the mixing variable, $F(\theta)$, as is done in other IRT models.

Since the sum score of a response vector, $R = \sum_{i=1}^{k} X_i$, is a sufficient statistic for estimating θ, the classification of individuals according to their sum scores r is equivalent to the classification according to their estimated trait values $\hat{\theta}$ (but not their true trait values θ). Hence, the RM can also be written as a discrete MDM,

$$P(\boldsymbol{X} = \boldsymbol{x}) = P(r_{\boldsymbol{x}})\, P(\boldsymbol{X} = \boldsymbol{x}|r_{\boldsymbol{x}}), \tag{14.4}$$

[2]Densities of continuous variables and their mixtures are not treated here.

where

$$P\left(\boldsymbol{X}=\boldsymbol{x} \mid r_{\boldsymbol{x}}\right)=\frac{\exp \left(\sum_{i=1}^{k} x_{i} \beta_{i}\right)}{\gamma_{r}(\exp (\boldsymbol{\beta}))}, \tag{14.5}$$

and $\gamma_r(\exp(\boldsymbol{\beta}))$ are the elementary symmetric functions of order r of the item parameters $\exp(\beta_i)$. The summation over r can be dropped, because any vector $\boldsymbol{x}$ occurs under condition of a single score, namely, $r_{\boldsymbol{x}}$.

This formalization of the RM as a discrete MDM has the advantage that no restrictive assumptions about the distribution of the mixing variable have to be made. The probabilities $P(r)$ can be estimated directly by the relative frequencies of observed scores in the sample, $P(r) = n(r)/N$, with obvious notation. The conditional probabilities are independent of the θ.

Latent class models (Lazarsfeld, 1950; Lazarsfeld and Henry, 1968) can be viewed as the simplest discrete MDMs for item responses. They are based on two assumptions, that is, local independence of the item responses and constant response probabilities for all individuals within the same class,

$$P(\boldsymbol{X}=\boldsymbol{x})=\sum_{c=1}^{C} \pi_{c} \prod_{i=1}^{k} \pi_{i \mid c}^{x_{i}}\left(1-\pi_{i \mid c}\right)^{1-x_{i}}, \tag{14.6}$$

where one type of model parameters are *class size parameters* $\pi_c = P(c)$ as the mixing proportions of the MDM, and the other type are the conditional item probabilities $\pi_{i|c} = P(x_i = 1|c)$.

Many relations exist between the RM and latent class analysis (see Chapters 11 and 13 of this volume). The following section deals with the generalization of the RM to a 'really' discrete MDM by introducing both a continuous mixing variable and a discrete one.

14.3 The Dichotomous Mixed Rasch Model

Assuming that the *conditional* probabilities in a discrete mixture are defined by the RM, i.e., are themselves continuous mixtures, we have

$$P(\boldsymbol{X}=\boldsymbol{x} \mid c)=\int_{-\infty}^{\infty} \prod_{i=1}^{k} \frac{\exp \left[x_{i}\left(\theta_{c}+\beta_{i c}\right)\right]}{1+\exp \left(\theta_{c}+\beta_{i c}\right)} d F_{c}\left(\theta_{c}\right) \tag{14.7}$$

and

$$P(\boldsymbol{X}=\boldsymbol{x})=\sum_{c=1}^{C} \pi_{c} P(\boldsymbol{X}=\boldsymbol{x} \mid c). \tag{14.8}$$

Again, the continuous variables θ_c are conditioned out. This is done by introducing the elementary symmetric functions γ_{rc} of order r within class c, see (14.5), as well as the conditional score probabilities $\pi_{r|c} = P(r|c)$, so that

$$P(\boldsymbol{X} = \boldsymbol{x}|c) = \pi_{r|c} \frac{\exp\left(\sum_{i=1}^{k} x_i \beta_{ic}\right)}{\gamma_r\left(\exp\left(\boldsymbol{\beta}_c\right)\right)}. \tag{14.9}$$

This model is the MRM (Rost, 1990). It has three kinds of parameters, the class size parameters π_c, the latent score probabilities $\pi_{r|c}$, and the class-specific item parameters β_{ic}. The score probabilities are called 'latent' because they are not directly estimable by means of the observed frequencies; rather they are part of the latent class structure, that is, they depend on which components the observed score frequencies are mixed from:

$$P(r) = \sum_{c=1}^{C} \pi_c \pi_{r|c}. \tag{14.10}$$

Notice that the number of components or classes, C, is not a model parameter; it must either be assumed *a priori*, or be 'estimated' by comparing the fit of the model under different numbers of classes.

The normalization conditions of the parameters are as follows. The class sizes add up to 1 minus the probability of the two extreme scores,

$$\sum_{c=1}^{C} \pi_c = 1 - P(r=0) - P(r=k), \tag{14.11}$$

so that there are $C-1$ of them. This normalization condition will be explained later. The item parameters have to be normalized within each class,

$$\sum_{i=1}^{k} \beta_{ic} = 0,$$

so that there are $C(k-1)$ independent parameters. The latent score probabilities, however, are somewhat tricky to count. The probabilities of the two extreme scores $r=0$ and $r=k$ cannot be unmixed, i.e., estimated conditionally upon the latent classes. The reason is that the conditional probabilities of the corresponding response vectors, $\boldsymbol{x}_0 = (0, \ldots, 0)$ and $\boldsymbol{x}_k = (1, \ldots, 1)$, respectively, are identical for all classes:

$$\begin{aligned} P(\boldsymbol{X} = \boldsymbol{x}_0|c, r) &= 1 \quad \text{and} \\ P(\boldsymbol{X} = \boldsymbol{x}_k|c, r) &= 1 \quad \text{for all c.} \end{aligned} \tag{14.12}$$

Hence, it cannot be decided empirically which part of the observed frequencies $n(r=0)$ and $n(r=k)$ belongs to which class. Fortunately, this does not affect the estimation of the class-specific item parameters, as these two patterns do not contribute to the item parameter estimation anyway (as in the ordinary RM). However, the class size parameters have to be normalized accordingly.

In the normalization condition (14.11), the probabilities $P(r=0)$ and $P(r=k)$ have to be treated as model parameters, which can be estimated by $\hat{p}(r=0)=n(r=0)/N$ and $\hat{p}(r=k)=n(r=k)/N$.

Returning to the question of the number of independent score parameters, there are two parameters for the extreme scores that are independent of the number of classes. The remaining $k-1$ score probabilities in each class add up to one, i.e.,

$$\sum_{r=1}^{k-1} \pi_{r|c} = 1 \quad \text{for all } c \in \{1, \ldots, C\},$$

so that the total number of independent score parameters is $2 + C(k-2)$.

The parameter estimation can be done by means of an EM-algorithm (Dempster, Laird, & Rubin, 1977) with conditional maximum likelihood estimation of the item parameters within each M-step as follows: in the E-step, expected pattern frequencies are estimated for each latent class on the basis of preliminary estimates (or starting values) of the model parameters,

$$\hat{n}(\boldsymbol{x}, c) = n(\boldsymbol{x}) \frac{\pi_c P(\boldsymbol{x}|c)}{\sum_{c=1}^{C} \pi_c P(\boldsymbol{x}|c)}, \tag{14.13}$$

where $n(\boldsymbol{x})$ denotes the observed frequency of vector $\boldsymbol{X} = \boldsymbol{x}$, and $\hat{n}(\boldsymbol{x}, c)$ an estimate of the portion of that frequency for class c. The conditional pattern frequencies $P(\boldsymbol{x}|c)$ are defined by (14.9).

In the M-step, these proportions of observed pattern frequencies are the basis for calculating better estimates of the π_c, $\pi_{r|c}$, and β_{ic}. These parameters can be estimated separately for each latent class by maximizing the log-likelihood function in class c, see (14.9),

$$\ln L_c = \sum_{\boldsymbol{x}} \hat{n}(\boldsymbol{x}|c) \left[\ln \pi_{r|c} + \sum_{i=1}^{k} x_i \beta_{ic} - \ln \left[\gamma_r (\exp \beta_c) \right] \right]. \tag{14.14}$$

Setting the first partial derivatives with respect to β_{ic} to zero yields the estimation equations for the item parameters within class c,

$$\hat{\beta}_{ic} = \ln \frac{n_{ic}}{\sum_{r=0}^{k} m_{rc} \gamma_{r-1}^{(i)} / \gamma_r}, \tag{14.15}$$

where n_{ic} denotes preliminary estimates of the number of individuals with response 1 on item I_i in class c, m_{rc} estimates the number of individuals with score r in class c (both calculated by means of $\hat{n}(\boldsymbol{x}, c)$ obtained in the previous M-step), and $\gamma_{r-1}^{(i)}$ the elementary symmetric function of order $r-1$ of all item parameters $\exp(\beta_i)$ except that of item I_i (c.f. Rasch, 1960, and Chapter 3). The elementary symmetric functions at the right-hand side of equation (14.15) are calculated by means of preliminary item parameter estimates, and new (better) estimates are obtained from the left-hand side of equation (14.15). Only one iteration of this procedure suffices, because it is performed in each M-step of the EM-algorithm and, hence, convergence to the maximum-likelihood solution is achieved in the course of the EM-procedure.

The estimators of the score probabilities and class sizes are obtained explicitly as

$$\hat{\pi}_{r|c} = \frac{m_{rc}}{n_c} \tag{14.16}$$

and

$$\hat{\pi}_c = \frac{n_c}{N}, \tag{14.17}$$

where n_c denotes the number of individuals in class c, computed on the basis of $\hat{n}(\boldsymbol{x}, c)$.

The latent trait parameters θ_c in (14.7) do not appear in these equations, because they were eliminated by conditioning on their sufficient statistics r_c. They can, however, be estimated using the final estimates of the item and score parameters. This is done by maximizing the intra-class likelihood,

$$\begin{aligned}\ln L_c &= \sum_v \ln P(\boldsymbol{x}_v|c) \\ &= \sum_v r_v\theta_{vc} + \sum_i n_{ic}\beta_{ic} - \\ &\quad \sum_v \sum_i \ln(1 + \exp(\theta_{vc} + \beta_{ic})),\end{aligned} \tag{14.18}$$

with respect to the unknown trait parameters θ_{vc}, which only depend on the score r_v of individual S_v. The number of individuals in class c solving item I_i, n_{ic}, needs not to be known, as this term vanishes upon differentiation of (14.18) with respect to the trait parameters θ_{vc}. The estimation equations are

$$r_v = \sum_{i=1}^{k} \frac{\exp(\theta_{vc} + \beta_{ic})}{1 + \exp(\theta_{vc} + \beta_{ic})}, \tag{14.19}$$

which can be solved iteratively (cf. Fischer, 1974, and Chapter 3).

Hence, the class-dependent trait parameters of model (14.7) can be estimated in a second step of the estimation procedure, making use of the conditional item parameter estimates obtained in the first step, $\hat{\beta}_{ic}$.

To each individual as many trait (or ability) parameter estimates are assigned as there are latent classes. These can be interpreted as conditional trait estimates, i.e., individual S_v has the trait value θ_{vc} given that he/she belongs to class c. However, these conditional abilities of a single individual S_v do not differ much between classes since the estimates mainly depend on r_v which, of course, is the same in all classes. The estimates are slightly influenced by the range and variation of the item parameters β_{ic} within classes. But even this (relatively weak) effect depends only on the item parameter values β_{ic} occurring in (14.19) and does not depend on *which* items were solved by the individual.

On the other hand, class membership of an individual depends very little on his/her sum score, but strongly depends on which items have been solved. According to the Bayes theorem, the probability of being member of class c given pattern $\boldsymbol{x}$ is

$$P(c|\boldsymbol{x}) = \frac{\pi_c P(\boldsymbol{x}|c)}{P(\boldsymbol{x})} = \frac{\pi_c P(\boldsymbol{x}|c)}{\sum_h \pi_h P(\boldsymbol{x}|h)}, \tag{14.20}$$

where $P(\boldsymbol{x}|c)$, as defined by (14.9), is mainly a function of $\sum_{i=1}^{k} x_i\beta_{ic}$, the sum of parameters of all items solved by the individual.

Hence, an individual has the highest probability of membership in that class for which the sum of the parameters of the items solved by that individual, is highest.

As a result, the MRM enables one to separate the quantitative and the qualitative aspect of a response pattern: the continuous mixing variable mainly depends on *how many* items have been solved, the discrete mixing variable mainly depends on *which* items have been solved.

14.4 Testing the Fit of the RM by Means of the MRM

Many goodness-of-fit tests for the RM concentrate on its central property of specifically objective comparisons among items and among persons. Accordingly, item parameter estimates have to be constant – except for sampling error – across different subsamples of individuals. The goodness-of-fit test by Andersen (1973), for instance, splits the sample of individuals according to their sum scores r. A conditional likelihood ratio statistic is applied to evaluate the differences of item parameter estimates in these score groups (see Chapter 5). From a mixture models perspective, the Andersen statistic tests the null-hypothesis of a one-class RM against the alternative hypothesis of a manifest mixture RM, where the score is the mixing variable. In this special case, the marginal likelihood ratio (MLR) turns out to be the same as the conditional likelihood ratio (CLR) proposed by Andersen.

The marginal likelihood of the RM, L_m, splits into the conditional likelihood, L_c, multiplied by that of the score distribution, see (14.4),

$$\begin{aligned} L_m &= \prod_{\boldsymbol{x}} P(\boldsymbol{x}) \\ &= \prod_{r} P(r)^{n(r)} \prod_{\boldsymbol{x}|r} P(\boldsymbol{x}|r) \end{aligned}$$

$$= \left[\prod_r P(r)^{n(r)}\right] \mathrm{L}_c ; \tag{14.21}$$

hence, in the case of a manifest score mixture, MLRs can be reduced to conditional likelihood ratios (CLRs),

$$\mathrm{MLR} = \frac{L_m}{\prod_r L_m^{(r)}} = \frac{\left[\prod_r P(r)^{n(r)}\right] L_c}{\prod_r \left[P(r)^{n(r)} L_c^{(r)}\right]} = \mathrm{CLR}\,, \tag{14.22}$$

where $L_m^{(r)}$ is the marginal likelihood of all response patterns with score r, and $L_c^{(r)}$ is the corresponding conditional likelihood.

In the case of other manifest mixtures or even latent mixtures, the relationships between MLRs and CLRs are more complicated and cannot be treated here.

As far as the detection of lack of fit of the RM (in terms of different item parameter estimates in different subpopulations) is concerned, the score-splitting of the sample represents only one of an infinite number of manifest mixing variables. It has been shown by various authors (Gustafsson, 1980; Stelzl, 1979; Van den Wollenberg, 1982) that this model test may fail to detect given model violations.

Nevertheless, focussing on the central property of RMs (constant item parameters across subgroups of individuals) may be the best way of testing the model. The only problem is the identification of that partition of the sample which best represents a given heterogeneity of the individuals. This partition is found by applying the mixed Rasch model: the latent classes of individuals represent those latent subpopulations that maximize person homogeneity within classes, and heterogeneity between classes. As a result, classes of individuals are obtained where the item parameter estimates differ maximally between the classes.

This leads to the conjecture that the likelihood ratio test for the one-class solution against a two-class solution of the MRM is at least as powerful as a likelihood ratio test based on any other manifest two-group partition of the sample.

The following artificial example supports this conjecture. A Rasch-homogeneous data set with 5 items and 1000 persons was mixed with random response data of 200 persons, each with a $P = 0.5$ response probability for each item. Of course, even the second component of this mixture fits an RM where all item and trait parameters are equal. Table 14.4 gives some goodness-of-fit statistics of different models for these data. It can be seen from the likelihood ratio χ^2-statistic that the RM does not fit.

The same is true for the manifest mixture of the score groups (MMscore). The MLR between the RM and the MMscore model would be identical to the Andersen CLR statistic and, in this case, would not indicate a violation of the RM ($\chi^2 = 6.1$, df $= 4$). Contrasting with this, when the 'true partition' of the

TABLE 14.1. Goodness-of-fit statistics of some models for an artificial 2-component mixture data set. AIC is the Akaike Information Criterion $-2\ln L + 2\text{NP}$, where NP denotes the number of parameters.

	$\ln L$	NP	AIC	LR–2	df
RM	-3537.6	9	7093.2	47.7	22
MMscore	-3534.6	13	7095.2	41.6	18
MMtrue	-3529.6	17	7093.3	31.8	14
MRM_2	-3525.6	17	7085.3	23.8	14
MRM_3	-3520.8	25	7091.6	14.0	6
Msaturated	-3513.8	31	7089.6	–	–

sample is known and is used as a basis for the MLR (MMtrue), one obtains $\chi^2 = 15.9$ with df $= 8$, which is just significant at $p = 0.05$. 'True partition' means that the item parameters have been estimated for both manifest components for the mixture, i.e., for 1000 and 200 individuals, respectively. The marginal likelihood of such a mixture based on manifest variables is

$$L_m = \prod_{v=1}^{N} \left[\sum_{c=1}^{2} \pi_c P(\boldsymbol{x}_v|c) \right], \tag{14.23}$$

where the conditional pattern probabilities $P(\boldsymbol{x}_v|c)$ in (14.9) are restrained to equal the parameter estimates obtained from the manifest subsample of individuals.

MRM_2 denotes the 2-latent-classes MRM; it is seen that the likelihood is even higher than for the MMtrue model. Although the MRM_2 model does not fit the data well ($\chi^2 = 23.8$ with df $= 14$ is just significant at $p = 0.05$), the likelihood ratio between the RM and the MRM_2-model clearly indicates misfit of the RM ($\chi^2 = 23.9$, $df = 8$).

The main reason for the unsatisfactory fit of the two-class MRM is that too many parameters, in particular score parameters, have to be estimated under this model. This reduces the degrees of freedom and causes the χ^2-statistic to be significant. One way of making the model more parsimonious with respect to its score parameters is outlined in the next section.

To refute the conjecture that the LR test between the RM and the 2-class MRM indicates misfit of the RM due to capitalization on chance, a simulation study was carried out. The 1- and 2-class MRM LR-statistics were analyzed for datasets of 5 dichotomous items. 100 computations with 1000 individuals each were carried out with two different sets of item parameters and 5 different distributions of ability parameters. In two cases, the χ^2 was significant at the $p = 0.05$ level in favor for the 2-class MRM. The mean of the empirical χ^2-values was

8.77 and, hence, slightly higher than the expectation of the χ^2-distribution with 8 degrees of freedom. The empirical standard deviation of the likelihood ratio was 3.43, which is slightly smaller than $\sqrt{16} = 4.00$. Thus there is no indication that this model test refutes the null-hypothesis more often than expected.

We conclude that the MRM provides a simple and very powerful tool for detecting lack of fit of the RM. Instead of trying a series of critera for splitting the sample into subgroups of individuals, the MRM gives that latent partition where the item parameters differ the most. If this partition fails to detect misfit of the ordinary RM, there is good reason to cling to the assumption of person homogeneity.

14.5 Modeling the Latent Score Distributions

In the MRM, it is necessary to estimate the latent score distribution for each latent class. This means that the number of parameters becomes large, because for each additional latent class as many score parameters $\pi_{r|c}$ are to be estimated as there are new item parameters. Whilst the latter are important for defining the 'quality' of a new class, the former only are needed for fitting the frequency distribution of the scores which respresent the latent continuous distribution of the trait.

On the other hand, it does not make sense to restrict a latent score distribution to a particular shape like, e.g., the binomial distribution. The computer software for the MRM (MIRA, Rost & von Davier 1992; WINMIRA, von Davier 1994) provides an option for a 2-parameter approximation of the latent score distributions which has proved to be very efficient in modeling score distributions. It is defined by the logistic function

$$\pi_{r|c} = \frac{\exp\left(\frac{r}{k}\tau_c + \frac{4r(k-r)}{k^2}\delta_c\right)}{\sum_{s=0}^{k} \exp\left(\frac{s}{k}\tau_c + \frac{4s(k-s)}{k^2}\delta_c\right)}, \tag{14.24}$$

where τ represents the location of the distribution, and δ its dispersion. Accordingly, τ has a coefficient which is a linear function of r, and δ has a quadratic coefficient. Both coefficients are normalized, such that the parameter values of τ and δ describe the location and dispersion characteristics of the score distribution independently of k, i.e., of the number of items. This is illustrated in Figure 14.1 where the score distribution for each parameter set is shown for $k = 13$ and $k = 88$ items.

Figure 1a shows a well-centered ($\tau = 0$) score distribution with a small but positive δ, $\delta = 3$, indicating a relatively high dispersion. A negative τ shifts the distribution to the left, and a higher δ makes the distribution steeper (Figure 1b). In the case of an extreme dislocation, $\tau = 9$, the dispersion looks 'truncated'

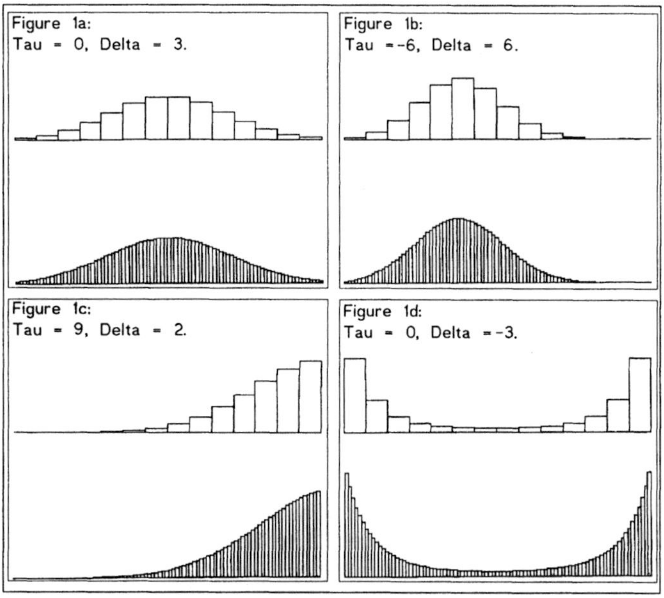

FIGURE 14.1. The score distributions for four different sets of parameters and 15 and 90 items each.

(Figure 1c), and a negative δ makes the distribution u-shaped (Figure 1d).

Hence, this kind of distribution is much more flexible for smoothing a discrete probability distribution of scores than, e.g., by assuming a normal density for the latent trait variable.

Table 14.2 shows the goodness-of-fit statistics for the RM and the 2- and 3-classes MRM, where the score distributions are restricted in this way within all classes. It turns out that the two-class solution of the MRM fits the data well, the χ^2 is non-significant.

14.6 Applications

Most applications of the MRM have been done using its polytomous generalization (see Chapter 20). A typical domain of application of the dichotomous MRM are spatial reasoning tasks, where it is assumed that two different solution strategies may lead to correct item responses. These solution strategies could be, e.g., analytic versus holistic, or pattern-matching versus mental rotation strategies.

The idea behind these applications is that two or more groups of individuals can be separated by their preferences for one of these strategies, or even by a

TABLE 14.2. Goodness-of-fit statistics for the 1-, 2- and 3-classes MRM for the simulated data (cf. Section 12.4). AIC is the Akaike Information Criterion $-2\ln L + 2\text{NP}$, where NP denotes the number of parameters.

	$\ln L$	NP	AIC	LR- 2	df
RM	-3537.8	6	7087.5	48.0	25
MRM2	-3525.7	13	7077.4	23.8	18
MRM3	-3521.6	20	7083.1	15.6	6

switching between the two strategies. *A priori* hypotheses have to be formulated about the relative difficulties of all items when solved using either of these strategies. The kind of strategy is then reflected by the profile of class-specific item parameters of the MRM.

Rost and von Davier (1993) found a 2-class solution for the cubes-tasks of the IST (Intelligenz-Struktur-Test; Amthauer, 1970). Köller, Rost, and Köller (1994) compare the two classes of the MRM with the results of an analysis by means of ordinary latent class analysis. Mislevy and Verhelst (1990) and Mislevy et al. (1991) apply a similar model to spatial reasoning tasks.

Part III

Polytomous Rasch Models and their Extensions

15

Polytomous Rasch Models and their Estimation

Erling B. Andersen[1]

ABSTRACT In this chapter, the polytomous Rasch model is introduced, based on the original formulation by Georg Rasch in the 1960 Berkeley Symposium on Mathematical Statistics and Probability. The various versions of the basic model, suggested in the literature, are briefly mentioned and compared. The main part of the chapter deals with estimation problems, theoretically as well as numerically. The connection to the RC-association models is discussed, and finally the theory is illustrated by means of data from a Danish psychiatric rating scale.

15.1 The Polytomous Rasch Model

When Georg Rasch in 1960 published his book (Rasch, 1960) introducing the Rasch model (RM), he was at the same time working on the extension to cases with polytomous items. The polytomous RM was introduced in the Berkeley symposium, held in 1960, in connection with an attempt to view psychological testing as a general measurement problem in psychology. The model presented by Rasch in Berkeley (Rasch, 1961) was as follows:

Let X_{vi} be the response of individual S_v on item I_i. The probability of response h, $h = 0, \ldots, m$, is

$$P(X_{vi} = h) = \frac{\exp[\phi_h\theta_v + \psi_h\alpha_i + \chi_h\theta_v\alpha_i + \omega_h]}{\sum_{l=0}^{m}\exp[\phi_l\theta_v + \psi_l\alpha_i + \chi_l\theta_v\alpha_i + \omega_l]}. \tag{15.1}$$

In (15.1) the vectors $(\phi_0, \ldots, \phi_m)$ and $(\psi_0, \ldots, \psi_m)$ are scoring functions for the categories, while θ_v is a uni-dimensional person parameter, and α_i a uni-dimensional item parameter. Rasch writes, however, that the model can easily be extended to multi-dimensional person parameters and item parameters, if the products $\phi_h\theta_v$ and $\psi_h\alpha_i$ are vector products.

In (15.1) different scorings are applied to the item parameters and the person parameters, but it is claimed in the last section of the Berkeley paper that, in order to satisfy basic measurement principles, one must have $\phi_h = \psi_h$. In most applications, a common scoring of the categories when estimating item parameters and when estimating person parameters is in fact almost always assumed.

[1]Institute of Statistics, University of Copenhagen, Studiestræde 6, DK-1455 Copenhagen K; e-mail: USIEBA@PC.IBT.DK

It is noteworthy that the model, as formulated by Rasch, contained a product term $\chi_h\theta_v\alpha_i$. The reason for introducing the product term was that, if it is assumed that there exist sufficient statistics for the α_i when the θ_v are known, and *vice versa,* then the bilinear form in the exponent in (15.1) follows. Rasch quickly decided, however, that only the case $\chi_h = 0$ is of interest and, as he put it, shows several features of considerable interest.

It is an interesting historical note that the model proposed by Rasch contains the Birnbaum 2-parameter logistic model as a special case if the observations are dichotomous, where without loss of generality we may put $\phi_0 = 1$, $\phi_1 = 0$, $\chi_0 = \chi$, $\chi_1 = 0$, and $\omega_0 = \omega_1 = 0$. Inserting this in (15.1), the term $\theta_v + \alpha_i + \chi\theta_v\alpha_i$ can be rewritten as

$$\left(\theta_v + \frac{\alpha_i}{1 + \chi\alpha_i}\right)(1 + \chi\alpha_i),$$

i.e., one obtains a Birnbaum model with item parameter $-\alpha_i/(1+\chi\alpha_i)$ and item discriminating power $(1+\chi\alpha_i)$. However, this model was never further discussed by Rasch.

In order to give a manageable exposition of the polytomous RM, we shall concentrate on the following two versions of (15.1):

$$P(X_{vi} = h) = \frac{\exp[\phi_h(\theta_v + \alpha_i) + \omega_h]}{\sum_{l=0}^{m} \exp[\phi_l(\theta_v + \alpha_i) + \omega_l]} \qquad (15.2)$$

and

$$P(X_{vi} = h) = \frac{\exp[\phi_h\theta_v + \beta_{ih}]}{\sum_{l=0}^{m} \exp[\phi_l\theta_v + \beta_{il}]}. \qquad (15.3)$$

The reason for introducing (15.3) is that, while preserving the unidimensionality of θ_v is important in order to make results interpretable, such a restriction is not necessary as regards the item parameters. Hence, we replace $\phi_h\alpha_i + \omega_h$ by the matrix β_{ih}, with k rows and $m+1$ columns. If one does not insist, however, on unidimensionality of the person parameter θ, a very general form of the polytomous Rasch model is

$$P(X_{vi} = h) = \frac{\exp[\theta_{vh} + \beta_{ih}]}{\sum_{l=0}^{m} \exp[\theta_{vl} + \beta_{il}]}. \qquad (15.4)$$

This model, of course, is obtained from (15.1), if χ_h and ω_h are 0, and ϕ_h, θ_v, ψ_h and α_i are all vectors, as suggested as a possibility by Rasch in 1960. The matrix formed by the ϕ_h as rows, and that formed by the ψ_h as rows, must then both be unit matrices.

Within the framework (15.3) a number of useful models have been suggested, which all satisfy the basic property of the RM, namely, that item parameters can be estimated using conditional maximum likelihood (CML) methods.

Andrich (1978a, 1982a) suggested the Rating Scale Model (RSM), which was based on a result in Andersen (1977), viz., that only the scoring

$$\phi_h = h, \quad h = 0, \ldots, m,$$

satisfied certain requirements. In Andrich's formulation of the RSM, it holds that

$$\phi_h \theta_v = h\theta_v \tag{15.5}$$

and

$$\beta_{ih} = h\alpha_i - \sum_{p=1}^{h} \tau_p. \tag{15.6}$$

Andrich gave an interpretation of these constraints based on a model for choosing response h by reaching latent threshold number h. The constraints (15.5) and (15.6) were based on the assumption that the individuals have equal discriminations between thresholds for different items. Andrich developed also other models, assuming unequal discriminations or equal distances between thresholds. Today it is usually model (15.3) combined with (15.5) and (15.6) which is denoted the RSM-model.

Fischer and Parzer (1991a) suggested to extend the RSM-model to a Linear Rating Scale Model (LRSM) by allowing the item parameters to be linear functions of a smaller set of 'basic' parameters. Fischer and Parzer worked with (15.2), but replaced the α_i by

$$\beta_i = \sum_{l=1}^{p} w_{il}\alpha_l + c, \tag{15.7}$$

where $p < m$, the w_{il} are known weights, and the α_l are a new set of basic parameters.

Masters (1982) suggested the Partial Credit Model (PCM) where different categories give the respondents different credits. For the PCM the probability of response h is (15.3) with

$$\beta_{ih} = \sum_{l=0}^{h} \alpha_{il} \tag{15.8}$$

and $\omega_h = 0$. Here α_{ih} is the credit awarded for response h.

Another linear model was suggested by Fischer (1973, 1983a) as a Linear Logistic Test Model (LLTM). Fischer treated the dichotomous case, but a corresponding Linear Partial Credit Model (LPCM) was formulated by Fischer and Ponocny (1994) for the polytomous case, combining (15.3) with

$$\beta_{ih} = w_{i0} + \sum_{l=1}^{p} w_{il}\alpha_{lh}. \tag{15.9}$$

This model is also discussed in Chapter 19. Here the weights w are known constants, and the α again are a smaller set of basic parameters.

In the RSM, the category weights $\phi_0, \ldots, \phi_m$ are equidistant and can, therefore, be chosen as $\phi_h = h$ without loss of generality. But there are other attractive models that can be derived from (15.2) or (15.3). In Andersen (1983) it was argued that one obtains a rather general model by assuming that the ϕ are of *arbitrary*, but known form. The basic reason for this view was that only for known ϕ it is possible to estimate the item parameters by a CML procedure. In fact, in both (15.2) and (15.3) the minimal sufficient statistic for θ_v is

$$t_v = \sum_{i=1}^{k} \sum_{h=0}^{m} \phi_h \mathcal{I}\{X_{vi} = h\},$$

where $\mathcal{I}\{\ldots\}$ is the indicator function, or equivalently,

$$t_v = \sum_{i=1}^{k} \phi_{h_{vi}}, \tag{15.10}$$

where h_{vi} is the observed response category for S_v on item I_i.

Hence, only if the category weights $\phi_0, \ldots, \phi_m$ are known, it is possible to make calculations based on the conditional likelihood given the scores $t_1, \ldots, t_N$. In order to stress the fact that the category weights are known, we shall use the notation $w_0, \ldots, w_m$ for the ϕ, thus avoiding greek letters. In conclusion we write the person parameter term in (15.2) or (15.3) as $w_h\theta_v$.

As regards the item parameters, there is no reason to impose restrictions. We may even assume that ϕ_h and α_i are vectors. Then (15.3) reduces to (15.2) for

$$\beta_{ih} = \sum_{t} \phi_{ht}\alpha_{it} + \omega_h;$$

in matrix form this is

$$\boldsymbol{\beta} = \boldsymbol{\alpha}\boldsymbol{\phi}' + \mathbf{1}\boldsymbol{\omega}', \tag{15.11}$$

where $\mathbf{1}$ is a k-vector of ones. If the α, ϕ, and ω after proper normalization can be identified from (15.3), the likelihood equations pertaining to the β, and the algorithm for solving them, can then be modified in standard ways to obtain estimates for the α, ϕ, and ω.

For these reasons, the version of the polytomous RM primarily treated in the following sections is (15.3) with $\phi_h = w_h$, i.e.,

$$P(X_{vi} = h) = \frac{\exp[w_h\theta_v + \beta_{ih}]}{\sum_{l=0}^{m} \exp[w_l\theta_v + \beta_{il}]}. \tag{15.12}$$

15.2 CML Estimation and the γ-Functions

As mentioned briefly, the separability of parameters, stressed by Rasch as the main justification for the model, is connected with the use of conditional likelihoods. The item parameters are thus estimated based on the conditional likelihood, given minimal sufficient statistics for the person parameters. The minimal sufficient statistic for θ_v in model (15.3) is (15.10), which for known category weights becomes

$$t_v = w_{h_{v1}} + w_{h_{v2}} + \ldots + w_{h_{vk}}, \tag{15.13}$$

where $(h_{v1}, \ldots, h_{vk})$ is the observed response pattern of S_v. The score on which we condition, is accordingly the sum of the category weights corresponding to the observed responses.

In order to derive the conditional likelihood, we first look at the likelihood function. Due to the exponential form, we choose to derive the logarithm of the likelihood function.

The denominator in (15.12) does not depend on the observed response h, which means that it is only a function of the vector $\boldsymbol{\beta}_i = (\beta_{i0}, \ldots, \beta_{im})$ and θ_v. Hence we may put

$$C(\theta_v, \boldsymbol{\beta}_i) = \sum_{l=0}^{m} \exp[w_l \theta_v + \beta_{il}]. \tag{15.14}$$

It is convenient to write the actual responses for individual S_v as the selection vector $(x_{vi0}, \ldots, x_{vim})$, where $x_{vih} = 1$ if the response is h, and 0 otherwise. By local independence we get

$$\begin{aligned}\ln L &= \sum_v \sum_i \sum_h (w_h \theta_v + \beta_{ih}) x_{vih} - \sum_v \sum_i \ln C(\theta_v, \boldsymbol{\beta}_i) \\ &= \sum_v \theta_v \sum_h w_h x_{v.h} + \sum_i \sum_h \beta_{ih} x_{.ih} - \sum_v \sum_i \ln C(\theta_v, \boldsymbol{\beta}_i),\end{aligned}$$

where

$$x_{v.h} = \sum_i x_{vih} \quad \text{and} \quad x_{.ih} = \sum_v x_{vih}.$$

Thus, according to standard results for the exponential family, the score

$$r_v = \sum_h w_h x_{v.h}$$

is minimal sufficient for θ_v. Similarly, the item totals $x_{.ih}$ are minimal sufficient for the item parameters β_{ih}. It is then well-known from exponential family theory (cf., e.g., Andersen, 1991, p. 43), that the ML estimates are obtained as the simultaneous solution of the equations (15.15) and (15.16),

$$r_v = E[R_v], \quad \text{for } v = 1, \ldots, n, \tag{15.15}$$

with

$$R_v = \sum_h w_h X_{v.h},$$

and

$$x_{.ih} = E[X_{.ih}], \quad \text{for } i = 1, \ldots, k \text{ and } h = 0, \ldots, m. \tag{15.16}$$

These estimates suffer from a serious drawback. Martin-Löf (1973), see also Andersen (1970), showed that the ML-estimates for the item parameters are inconsistent when n approaches infinity and k and m are fixed. At the same time the individual parameter estimates can only take a finite number of values and cannot accordingly approach the true values.

As an alternative, Rasch (1960) suggested to estimate the item parameters by a conditional maximum likelihood (CML) method, where the conditioning is with respect to the sufficient statistics for the individual parameters $\theta_1, \ldots, \theta_n$. Under suitable conditions on the variability of the θ in the population, these CML estimates are in fact consistent, as shown in Andersen (1970, 1973c).

The conditional distribution of the matrix $\{X_{.ih}\}$, $i = 1, \ldots, k$, $h = 0, \ldots, m$, given the minimal sufficient statistics $r_1, \ldots, r_n$ for $\theta_1, \ldots, \theta_n$, is again an exponential family. The item totals are sufficient for the β also in this conditional model. Hence, the CML estimates are – again by standard exponential family theory – obtained as solutions of the equations

$$x_{.ih} = E[X_{.ih}|r_1, \ldots, r_n], \quad \text{for } i = 1, \ldots, k \text{ and } h = 0, \ldots, m, \tag{15.17}$$

where the mean values $E[\ldots|r_1, \ldots, r_n]$ refer to the conditional distribution given the scores $r_1, \ldots, r_n$.

The derivation of these equations is straightforward. Since

$$X_{.ih} = \sum_{v=1}^{m} X_{vih},$$

we only need to derive the mean value of X_{vih}, given the individual score r_v. The joint probability of $\{X_{vih}\}$, for $i = 1, \ldots, k$ and $h = 0, \ldots, m$, is

$$\begin{aligned} f(x_{v10}, \ldots, x_{vkm}) &= P(X_{v10} = x_{v10}, \ldots, X_{vkm} = x_{vkm}) \\ &= \exp\Big(r_v\theta_v + \sum_i \sum_h \beta_{ih} x_{vih}\Big) \prod_i C^{-1}(\theta_v, \boldsymbol{\beta}_i). \end{aligned} \tag{15.18}$$

Hence, the marginal probability of the score $r_v = \sum_i \sum_h w_h x_{vih}$ is

$$f(r_v) = \exp(r_v\theta_v) \sum_{(r_v)} \exp\Big(\sum_i \sum_h \beta_{ih} x_{vih}\Big) \prod_i C^{-1}(\theta_v, \boldsymbol{\beta}_i), \tag{15.19}$$

where the summation (r_v) is over all feasible response patterns $\{x_{vih}\}$ for which the score is r_v. The sum in (15.19) is denoted

$$\gamma_{r_v}(\boldsymbol{\beta}) = \sum_{(r_v)} \exp\Big(\sum_i \sum_h \beta_{ih} x_{vih}\Big), \tag{15.20}$$

defining the so-called γ-functions. With the γ-notation, the conditional probability of the response $(x_{v10}, \ldots, x_{vkm})$, given r_v for individual S_v, is obtained from (15.18) and (15.19) as

$$f(x_{v10}, \ldots, x_{vkm} | r_v) = \frac{\exp(\sum_i \sum_h \beta_{ih} x_{vih})}{\gamma_{r_v}(\boldsymbol{\beta})}. \tag{15.21}$$

From this expression it follows immediately that

$$P(X_{vih} = 1 | r_v) = \exp(\beta_{ih}) \frac{\sum_{(r_v - w_h)} \exp(\sum_{j \neq i} \sum_h \beta_{jh} x_{vjh})}{\gamma_{r_v}(\boldsymbol{\beta})},$$

since we have to sum (15.21) over all response patterns $\{x_{vih}\}$ for which the score is r_v and $x_{vih} = 1$. Obviously, if $x_{vih} = 1$, the score is $r_v - w_h$ on the remaining $k - 1$ items. We have thus arrived at the important result

$$P(X_{vih} = 1 | r_v) = \exp(\beta_{ih}) \frac{\gamma^{(i)}_{r_v - w_h}(\boldsymbol{\beta})}{\gamma_{r_v}(\boldsymbol{\beta})}, \tag{15.22}$$

where $\gamma^{(i)}$ means that the γ-function is evaluated with item I_i omitted. It is now clear that

$$E[X_{.ih} | r_1, \ldots, r_n] = \sum_v E[X_{vih} = 1 | r_v]$$
$$= \exp(\beta_{ih}) \sum_r n_r \frac{\gamma^{(i)}_{r - w_h}(\boldsymbol{\beta})}{\gamma_r(\boldsymbol{\beta})},$$

where n_r is the number of individuals with $r_v = r$, and accordingly that the CML equations are

$$x_{.ih} = \exp(\beta_{ih}) \sum_r n_r \frac{\gamma^{(i)}_{r - w_h}(\boldsymbol{\beta})}{\gamma_r(\boldsymbol{\beta})}, \quad \text{for } i = 1, \ldots, k \text{ and } h = 0, \ldots, m. \tag{15.23}$$

It is important to note that (15.23) has been derived without assuming any structure in the matrix $\{\beta_{ih}\}$ of item parameters and only assuming that the category scores are known. This explains the fact that so many models for the polytomous case allow for CML estimation of item parameters.

In principle the CML equations can be derived for any set $w_0, \ldots, w_m$ of category weights. If, however, the category weights have irregular values, it may well happen that there is only one possible response pattern for each value of the score. If this is the case, the conditional probability (15.21) has the value 1, and no inference can be drawn from the conditional likelihood. Even if there are only a few score values for which (15.21) has nondegenerate values, the conditional likelihood function may contain too little information to be of practical use. Hence, the weights must allow a reasonably large number of response patterns for most score values. It is obvious that the equidistant scoring $w_h = h$, as assumed in the rating scale model (15.4)–(15.5), satisfies this requirement. But other scorings based on, or equivalent to, integer scores, may also be applicable. One such example with $m = 3$ would be

$$\phi_3 = 5, \quad \phi_2 = 3, \quad \phi_1 = 2, \quad \text{and} \quad \phi_0 = 0,$$

where the middle categories are one step closer to each other than the extreme categories are to the middle categories.

For the dichotomous RM, the conditions for existence of solutions to the likelihood equations are known in a form easy to verify from the item totals (Fischer, 1981). For the polytomous case the conditions are known in principle either in the original Barndorff-Nielsen form (Barndorff-Nielsen, 1978; cf. Andersen, 1991, p. 45), or in the form given by Jacobsen (1989). It is not known yet, however, whether these conditions can be brought into such a form that the existence of solutions can be verified from the item totals as easily in the polytomous case as in the dichotomous case.

In order to remove indeterminacies in model (15.12), we need to introduce $k+1$ constraints. A possible choice is to put

$$\beta_{im} = 0, \quad \text{for } i = 1, \ldots, k,$$

and

$$\sum_i \sum_h w_h \beta_{ih} = 0.$$

There are thus $k(m+1) - k - 1 = km - 1$ unconstrained parameters to be estimated from (15.23).

15.3 Technical and Numerical Matters Concerning the Solution of the CML Equations

Most computer algorithms so far have used a Newton-Raphson procedure, which is fast in the sense that it usually requires few iterations, and in the sense that the quantities involved are ratios of γ-functions, which are easily obtained from convenient recurrence relations, cf. (15.26) below. If we write (15.23) as

$$F_{ih}(\boldsymbol{\beta}) = \ln x_{.ih} - \beta_{ih} - G_{ih}(\boldsymbol{\beta}) = 0, \tag{15.24}$$

where $G_{ih}(\boldsymbol{\beta}) = \ln \sum_r n_r \gamma^{(i)}_{r-w_h}(\boldsymbol{\beta})/\gamma_r(\boldsymbol{\beta})$, the Newton-Raphson procedure is based on the algorithm

$$\boldsymbol{\beta}^{(N+1)} = \boldsymbol{\beta}^{(N)} + [F_{ih}(\boldsymbol{\beta}^{(N+1)}) - F_{ih}(\boldsymbol{\beta}^{(N)})]\boldsymbol{D}^{-1}(\boldsymbol{\beta}^{(N)}),$$

where $\boldsymbol{\beta}^{(N)}$ and $\boldsymbol{\beta}^{(N+1)}$ are vectors of estimates obtained in iterations N and $N+1$, respectively, and $\boldsymbol{D}$ is the matrix of derivatives

$$\{D_{ih.jk}(\boldsymbol{\beta}^{(N)})\} = \left\{ \frac{\partial F_{ih}(\boldsymbol{\beta}^{(N)})}{\partial \boldsymbol{\beta}_{jk}} \right\}. \tag{15.25}$$

From (15.20), isolating β_{ih}, there follows the recursion formula

$$\gamma_r(\boldsymbol{\beta}) = \sum_{h=1}^{m} \exp(\beta_{ih})\gamma^{(i)}_{r-w_h}(\boldsymbol{\beta}) + \gamma^{(i)}_r(\boldsymbol{\beta}), \tag{15.26}$$

which is a generalization of the well-known formula for the dichotomous case, where $m = 1$, $w_1 = 1$, and $w_0 = 0$. Hence, the derivatives of $\gamma_r(\boldsymbol{\beta})$ are given by

$$\frac{\partial \gamma_r(\boldsymbol{\beta})}{\partial \beta_{ih}} = \exp(\beta_{ih})\gamma^{(i)}_{r-w_h}(\boldsymbol{\beta}).$$

Since $G_{ih}(\boldsymbol{\beta})$ is a function of γ-ratios, and furthermore $F_{ih}(\boldsymbol{\beta})$ is a function of $G_{ih}(\boldsymbol{\beta})$, we conclude that the elements of $\boldsymbol{D}$ are also functions of γ-ratios.

It is tempting to compute the γ-ratios from (15.26) by dividing both left and right hand side by $\gamma_r(\boldsymbol{\beta})$ and solving for $\gamma^{(i)}_r(\boldsymbol{\beta})/\gamma_r(\boldsymbol{\beta})$. This, however, involves subtractions, and it was a common experience, when first tried by Georg Rasch and the author in the early sixties, that rounding errors quickly became a serious problem. In 1966 (Andersen, 1966; cf. also Andersen, 1972), the author solved this problem by using an algorithm where normed versions of the $\gamma^{(i)}_r(\boldsymbol{\beta})$ were solved by successive additions based on (15.26) as

$$\gamma^{(i)}_r(\boldsymbol{\beta}) = \sum_{h=0}^{m} \frac{\exp(\beta_{ih})\gamma^{(ij)}_{r-w_h}(\boldsymbol{\beta})}{\gamma^{(ij)}_r(\boldsymbol{\beta})}, \tag{15.27}$$

in which $\gamma^{(ij)}_r(\boldsymbol{\beta})$ is computed by omitting both item I_i and item I_j. Formula (15.27) can be used iteratively to compute the γ-functions, because, assuming $w_m > w_{m-1} > \ldots > w_1 > w_0 = 0$,

$$\gamma^{(1\ldots k)}_r(\boldsymbol{\beta}) = \begin{cases} 0 & \text{for } r > 0, \\ 1 & \text{for } r = 0. \end{cases}$$

As a side profit, this algorithm produces the γ-ratios

$$\frac{\gamma^{(ij)}_r(\boldsymbol{\beta})}{\gamma_r(\boldsymbol{\beta})},$$

which are key elements of the matrix $\boldsymbol{D}$. With a limited number of items and a corresponding limited number of categories, the full Newton-Raphson procedure is fast and efficient.

If the standard errors of the CML estimates are not required, a very direct method is to update the estimates as

$$\beta_{ih}^{(N+1)} = \ln x_{.ih} - \ln\left[\sum_r n_r \frac{\gamma_{r-w_h}^{(i)}(\boldsymbol{\beta}^{(N)})}{\gamma_r(\boldsymbol{\beta}^{(N)})}\right]. \tag{15.28}$$

This method is often fast and reliable. It still requires the computation of γ-ratios, though. The method is related to the iterative proportional fitting of marginals, which is widely used for log-linear models, since (15.28) can also be written as

$$\exp\{\beta_{ih}^{(N+1)}\} = \frac{x_{.ih}}{\sum_r \frac{n_r\gamma_{r-w_h}^{(i)}(\boldsymbol{\beta}^{(N)})}{\gamma_r(\boldsymbol{\beta}^{(N)})}}.$$

15.4 Marginal ML Estimation

The conceptual and philosophical justification of the conditional approach described in Section 15.2 was, according to Georg Rasch, the independence of the solution from any knowledge of the latent population density. Models where a latent population density was assumed to exist and where the item parameters were estimated simultaneously with the parameters of the assumed latent density, did not appear until in the late seventies.

If the item parameters are estimated jointly with the parameters of the latent population density, and the estimation method used is maximum likelihood, the procedure is known as marginal maximum likelihood (MML) estimation.

If the latent parameter θ varies in the population according to the latent density $\varphi(\theta)$, the probability of response h for individual S_v on item I_i will, according to (15.12) and with the notation (15.14), be

$$P(X_{vi} = h) = \int_{-\infty}^{+\infty} \exp[w_h\theta + \beta_{ih}]C^{-1}(\theta, \boldsymbol{\beta}_i)\varphi(\theta)d\theta. \tag{15.29}$$

If the category weights $w_0, \ldots, w_m$ are known, the likelihood will depend on the item parameters β_{ih} and whatever parameters the latent density contains. This model was introduced and discussed by Andersen and Madsen (1977), Sanathanan and Blumenthal (1978), Bartholomew (1980), and Bock and Aitkin (1981). Both the normal distribution and the logistic distribution have been suggested as parametric families to describe $\varphi(\theta)$. Irrespective of the choice of one or the other, it is necessary to apply numerical integration algorithms in order to solve the likelihood equations numerically. In practice this does not seem to be a problem as long as the chosen density does not possess heavy tails. Both the normal distribution and the logistic distribution thus work well with the RM.

An MML estimation, based on (15.29), is somewhat easier to handle than CML estimation, except for numerical problems, which we return to in the next section. The likelihood function is obtained from (15.29) by multiplication over v as

$$L = \prod_v P(X_{v1} = x_{v1}, \ldots, X_{vk} = x_{vk}) = \prod_v f(x_{v1}, \ldots, x_{vk}), \tag{15.30}$$

where $(x_{v1}, \ldots, x_{vk})$ is the response of individual S_v, and each of the x_{vi} can take the values $h = 0, \ldots, m$. Hence, the likelihood is equal to

$$L = \prod_{\boldsymbol{z}} [f(z_1, \ldots, z_k)]^{n(\boldsymbol{z})} = \prod_{\boldsymbol{z}} \pi_{\boldsymbol{z}}^{n(\boldsymbol{z})}, \tag{15.31}$$

where $\boldsymbol{z} = (z_1, \ldots, z_k)$ is a typical response pattern, and $\mathrm{n}(\boldsymbol{z})$ the number of individuals with response pattern $\boldsymbol{z}$. Hence, the product is to be taken over all k^{m+1} possible response patterns.

By local independence, i.e., independence of the responses given the level θ_v of the latent variable, we have for an arbitrary response pattern $(z_1, \ldots, z_k)$

$$f(z_1, \ldots, z_k) = \int_{-\infty}^{+\infty} \prod_{i=1}^{k} f(z_i|\theta)\varphi(\theta)d\theta.$$

Hence, inserting $h = z_1, z_2, \ldots, z_k$ in (15.29), we get, with $f(z_i|\theta) = P(X_{vi} = z_i)$,

$$\begin{aligned} \pi_{\boldsymbol{z}} &= f(z_1, \ldots, z_k) = \int_{-\infty}^{+\infty} \prod_{i=1}^{k} \left\{\exp[w_{z_i}\theta + \beta_{iz_i}]C^{-1}(\theta, \boldsymbol{\beta}_i)\right\} \varphi(\theta)d\theta \\ &= \int_{-\infty}^{+\infty} \exp[\theta \sum_i w_{z_i} + \sum_i \beta_{iz_i}] \prod_i C^{-1}(\theta, \boldsymbol{\beta}_i)\varphi(\theta)d\theta. \end{aligned} \tag{15.32}$$

In this expression, the sum $\sum_i w_{z_i}$ is the score corresponding to response pattern $(z_1, \ldots, z_k)$.

The MML estimation is now straightforward by ordinary theory for the parametric multinomial distribution when inserting (15.32) in (15.31). If

$$\alpha = (\beta_{10}, \ldots, \beta_{k,m-1}, \mu, \sigma^2)$$

is the combined vector of item parameters and the parameters of the normal latent density,

$$\varphi(\theta) = (\sqrt{2\pi}\sigma)^{-1} \exp\left[-\frac{1}{2}\left(\frac{\theta - \mu}{\sigma}\right)^2\right],$$

the likelihood equations for a normal latent density become

$$\sum_{\boldsymbol{z}} n(\boldsymbol{z}) \frac{\partial \ln \pi_{\boldsymbol{z}}}{\partial \alpha_j} = \sum_{\boldsymbol{z}} \frac{n(\boldsymbol{z})}{\pi_{\boldsymbol{z}}} \frac{\partial \pi_{\boldsymbol{z}}}{\partial \alpha_j} = 0, \quad \text{for } j = 1, \ldots, km+1. \tag{15.33}$$

It has been discussed whether an MML estimation is possible without any assumptions about the form of the latent density. One could term this a *distribution-free* estimation of the latent distribution. Since the score is the minimal sufficient statistic for the latent variable, it seems intuitively clear that this is only possible with an approximation to the latent density in the form of a discrete distribution with at most as many values with positive probability as there are obtainable values of the score. That this in fact is true, and what conditions it imposes on the class of latent distribution, has been studied in details by several authors for the dichotomous case. An excellent survey is found in De Leeuw and Verhelst (1986); see also Chapter 13. To give the flavor of the result, consider the log-likelihood for the MML estimation, derived from (15.31) and (15.32),

$$\ln L = \sum_{\boldsymbol{z}} n_{\boldsymbol{z}} \sum_i \beta_{iz_i} + \sum_{\boldsymbol{z}} n_{\boldsymbol{z}} \ln \left[\int_{-\infty}^{+\infty} \exp(r\theta) \prod_i C^{-1}(\theta, \beta_i) \varphi(\theta) d\theta \right]. \tag{15.34}$$

Equation (15.34) can be brought on the following form:

$$\ln L = \sum_i \sum_h \beta_{ih} x_{.ih} - \sum_r n_r \ln \gamma_r(\boldsymbol{\beta}) + \sum_r n_r \ln \pi_r,$$

where

$$\pi_r = P(\text{score} = r) = \gamma_r(\boldsymbol{\beta}) \int_{-\infty}^{+\infty} \exp(r\theta) \prod_i C^{-1}(\theta, \boldsymbol{\beta}_i) \varphi(\theta) d\theta.$$

A 'distribution free' estimation of $\varphi(\theta)$ now consists of estimates

$$\hat{\pi}_r = n_r / n,$$

which maximize the log-likelihood if there exists a density $\varphi(\theta)$ such that

$$\hat{\pi}_r = \gamma_r(\hat{\boldsymbol{\beta}}) \int_{-\infty}^{+\infty} \exp(r\theta) \prod_i C^{-1}(\theta, \hat{\boldsymbol{\beta}}_i) \varphi(\theta) d\theta,$$

for all values of r, where $\hat{\beta}$ are the CML estimates. For a more complete discussion of this problem, we refer to the paper by De Leeuw and Verhelst. Note that these arguments can also be used as a way to characterize the RM with a random latent variable. They noted that, for given values of π_r, $\phi(\theta)$ is only identifiable at $k/2+1$ points if k is even, and at $(k+1)/2$ points if k is odd.

Cressie and Holland (1983) noted (again for the dichotomous case) that the response probabilities $\pi_{\boldsymbol{z}}$ only have the representation (15.32) for a given latent distribution $F(\theta)$, where in their notation

$$dF(\theta) = \varphi(\theta)d\theta,$$

if

$$\pi_r = \int_{-\infty}^{+\infty} \exp(\theta_r)dG(\theta),$$

i.e., the r'th moment of a certain positive random variable with cumulativ distribution $G(\theta)$.

15.5 Technical and Numerical Matters Concerning the Solution of the MML Equations

Equations (15.33) can be solved by a Newton-Raphson method. Their solution involves, it seems, taking second derivatives of the response pattern probabilities $\pi_{\boldsymbol{z}}$. The derivatives of (15.33) are, however,

$$\sum_{\boldsymbol{z}} \frac{n(\boldsymbol{z})}{\pi_{\boldsymbol{z}}} \frac{\partial^2 \pi_{\boldsymbol{z}}}{\partial \alpha_j \partial \alpha_i} - \sum_{\boldsymbol{z}} \frac{n(\boldsymbol{z})}{\pi_{\boldsymbol{z}}^2} \frac{\partial \pi_{\boldsymbol{z}}}{\partial \alpha_j} \frac{\partial \pi_{\boldsymbol{z}}}{\partial \alpha_i},$$

and the mean value of this expression is equal to

$$-\sum_{\boldsymbol{z}} \frac{n}{\pi_{\boldsymbol{z}}} \frac{\partial \pi_{\boldsymbol{z}}}{\partial \alpha_j} \frac{\partial \pi_{\boldsymbol{z}}}{\partial \alpha_i},$$

because

$$E\Big[\sum_{\boldsymbol{z}} \frac{n}{\pi_{\boldsymbol{z}}} \frac{\partial^2 \pi_{\boldsymbol{z}}}{\partial \alpha_j \partial \alpha_i}\Big] = \sum_{\boldsymbol{z}} n \frac{\partial^2 \pi_{\boldsymbol{z}}}{\partial \alpha_j \partial \alpha_i} = 0,$$

due to

$$\sum_{\boldsymbol{z}} \pi_{\boldsymbol{z}} = 1.$$

So, if we use mean values of the second deratives in the Newton-Raphson procedure, or in fact the 'Fisher scoring method', we only need to find first order derivatives of the response probabilities.

The likelihood equations for an MML estimation can be brought on a form which is similar to the basic equations for the EM algorithm. In fact, if r is the score for a given individual, n_r the number of individuals with score r, and

$$C_r^{-1}(\boldsymbol{\beta}) = \int_{-\infty}^{+\infty} \exp(r\theta) \prod_i C^{-1}(\theta, \boldsymbol{\beta}_i)\varphi(\theta)d\theta,$$

then the application of (15.32) and (15.33) yields the equations

$$x_{.ih} - \sum_r \frac{n_r}{C_r(\boldsymbol{\beta})} \int_{-\infty}^{+\infty} \frac{\exp(w_h\theta + \beta_{ih})}{C(\theta, \boldsymbol{\beta}_i)} \exp(r\theta) \frac{\varphi(\theta)}{\prod_i C(\theta, \boldsymbol{\beta}_i)} d\theta = 0, \quad (15.35)$$

$$\sum_r \frac{n_r}{C_r(\boldsymbol{\beta})} \int_{-\infty}^{+\infty} \theta \exp(r\theta) \frac{\varphi(\theta)}{\prod_i C(\theta, \boldsymbol{\beta}_i)} d\theta = n\mu, \quad (15.36)$$

and

$$\sum_r \frac{n_r}{C_r(\boldsymbol{\beta})} \int_{-\infty}^{+\infty} \theta^2 \exp(r\theta) \frac{\varphi(\theta)}{\prod_i C(\theta, \boldsymbol{\beta}_i)} d\theta = n(\sigma^2 + \mu^2). \quad (15.37)$$

However, since

$$P(X_{vi} = h) = \exp(w_h\theta + \beta_{ih}) C^{-1}(\theta, \boldsymbol{\beta}_i),$$

it follows that

$$E[X_{.ih}|\theta] = n \exp(w_h\theta + \beta_{ih}) C^{-1}(\theta, \boldsymbol{\beta}_i).$$

Hence, since

$$f(\theta|r) = \exp(r\theta) C_r^{-1}(\boldsymbol{\beta}) \prod_i C^{-1}(\theta, \boldsymbol{\beta}_i)\varphi(\theta),$$

(15.35) can be brought on the form

$$x_{.ih} = n^{-1} \sum_r n_r E[E(X_{.ih}|\theta)|r]. \quad (15.38)$$

Similarly, (15.36) and (15.37) can be brought on the form

$$\sum_r n_r E[\theta|r] = n\mu \quad (15.39)$$

and

$$\sum_r n_r E[\theta^2|r] = n(\sigma^2 + \mu^2). \quad (15.40)$$

While (15.39) and (15.40) can be used directly to solve for μ and σ^2 when the left hand sides in (15.39) and (15.40) are computed, the β cannot be derived from the right hand side in (15.38). A partial EM algorithm is, therefore, possible, where the EM algorithm is applied to (15.40) and (15.39), but (15.38) has to be solved by the Newton-Raphson method for given values of μ and σ^2, or by a suitable marginal fitting algorithm. Equations (15.38), (15.39), and (15.40) show that the only numerical problem in programming an algorithm to obtain the estimates, apart from standard techniques for solving equations iteratively, is the numerical evaluation of a number of simple integrals, namely $E[\theta|r]$ and $E[\theta^2|r]$ for all score values r, and the integrals $E[E[X_{.ih}|\theta]|r]$ for all values of i, h, and the score r. On modern fast computers this is not really a problem, and the integrals are well-behaved due to the fast decreasing tails of the normal density.

15.6 The Connection to the RC-Association Model

The RC-association model was developed by Leo Goodman around 1980 (Goodman, 1979b, 1981). As referenced by Goodman, it was in fact already suggested by Georg Rasch (Rasch, 1966) as a latent structure model for polytomous items. Rasch derived the RC-association model in the following way. Consider the polytomous RM (15.12) for an individual with latent parameter θ, assume we are only considering the response to one item, and let the item parameters be $\beta_0, \dots, \beta_m$. Assume further that the latent parameter θ has population density $\varphi(\theta)$. The response X to the item in question then has probability

$$P(X = h|\theta) = \exp(w_h\theta + \beta_h)C^{-1}(\theta, \boldsymbol{\beta}),$$

where $C(\theta, \boldsymbol{\beta})$ is given by (15.14). The joint density of X and θ then becomes

$$P(X = h, \theta) = \exp(w_h\theta + \beta_h)C^{-1}(\theta, \boldsymbol{\beta})\varphi(\theta).$$

Assume now that the individuals are grouped into a number of θ-intervals: $T_1, \dots, T_K$. The joint probability of response h and θ being in interval T_k then is

$$P_{hk} = P(X = h, \theta \in T_k) = \int_{T_k} \frac{\exp(w_h\theta + \beta_h)}{C(\theta, \boldsymbol{\beta})}\varphi(\theta)d\theta. \tag{15.41}$$

The purpose of any grouping of θ-values of course is to collect individuals who in the given context are similar. The grouping thus should entail that the response probability $P(X = h|\theta)$ has a moderate variation within each θ-interval. This means that the approximation

$$P_{hk} \simeq \exp(w_h\theta_k + \beta_h)C^{-1}(\theta_k, \boldsymbol{\beta})\varphi(\theta_k),$$

where θ_k is an average value of θ in T_k, should be valid.

Introducing the parameters

$$\alpha_k = \ln \varphi(\theta_k) - \ln C(\theta_k, \boldsymbol{\beta}),$$

$$\mu_k = \theta_k,$$

and

$$\nu_h = w_h,$$

(15.41) can be written as

$$P_{hk} = \exp\{\mu_k \nu_h + \alpha_k + \beta_h\}, \tag{15.42}$$

where the assignment of parameters to greek letters corresponds to Goodman's (1981) paper, introducing the name RC-association model. It follows that Goodman's RC-association model coincides with a polytomous RM as a description of the interaction between a latent variable and a categorical background variable. Note here that the score values μ_k and ν_h, which in (15.42) describe the interaction, are an average value of the latent variable in θ-interval T_k and the score value w_h for category h, respectively. One point often stressed by Rasch was that, since μ_k is an average θ-value, one should expect the category scores for θ-intervals to remain constant if the categorical variable is derived as grouped or ungrouped scores from a latent structure model and then compared through a RC-association model to several different other categorical variables.

15.7 Estimation of the Person Parameter

It is often overlooked that the main reason for applying a latent structure model is not to estimate the item parameters, but to estimate the value of the latent variable for a given individual or for a group of individuals. For this problem there are two possibilities. If nothing is assumed regarding the latent distribution $\varphi(\theta)$, the estimate of θ is obtained from the logarithm of the probability of the response r given θ, i.e., from (15.19) with $r_v = r$ and $\theta_v = \theta$:

$$\ln f(r) = r\theta + \ln \gamma_r(\boldsymbol{\beta}) - \sum_i \ln[C(\theta, \boldsymbol{\beta}_i)].$$

The ML estimate of θ is, therefore, due to (15.13), given by

$$r - \sum_i \sum_h w_h \exp(w_h\theta + \hat{\beta}_{ih}) C^{-1}(\theta, \hat{\boldsymbol{\beta}}_i) = 0, \tag{15.43}$$

where the CML estimates for the β are inserted. This equation is easily solved, e.g., by means of the Newton-Raphson method. The solution of (15.43) is the ML-estimate of θ.

If the latent density $\varphi(\theta)$ has been estimated, an estimate of θ can be derived as the expected *a posteriori* value of θ, i.e.,

$$\hat{\theta} = E[\theta|\text{data}],$$

which we shall call the expected *a posteriori* (EAP) estimate. Since the score r is sufficient for θ, we only need to derive the conditional mean value of θ given the score r. It is implicit in (15.36) that

$$E[\theta|r] = C_r^{-1}(\boldsymbol{\beta}) \int_{-\infty}^{+\infty} \theta \exp(r\theta) \prod_i C^{-1}(\theta, \boldsymbol{\beta}_i)\varphi(\theta)d\theta. \tag{15.44}$$

TABLE 15.1. Item totals and score group totals

Items i	Response Categ. 0	1	2
1	409	235	154
2	507	149	142
3	600	126	72
4	630	101	67
5	412	213	173
Scores r	n_r		
0	24		
1	12		
2	27		
3	25		
4	34		
5	49		
6	73		
7	81		
8	108		
9	101		
10	264		
Total	798		

Hence, the EAP estimate of θ is obtained by computing the right-hand side in (15.44) for estimated values of β_i and the parameters of $\varphi(\theta)$, i.e. with the MML estimates inserted.

15.8 An Example

To illustrate the various estimation procedures, we present the results from a Danish rating scale application. In order to show as many details as possible, we have selected 5 of the items, each with 3 response categories. Table 15.1 gives the item totals and the number of individuals in each score group. The scores were obtained, according to the rating scale model, by choosing the equidistant weights $h = 2, 1$, and 0.

The item parameter estimates $\hat{\beta}_{ih}$ with standard errors in parentheses are given in Table 15.2. These values are the solutions of (15.17). Note that $\hat{\beta}_{i2} = 0$.

Consider next an MML estimation of the parameters with a normal latent density. The joint estimates of μ and σ^2 for a normal latent density and the item parameter estimates are shown in Table 15.3. Fortunately, there are only minor differences between the item parameter estimates in Table 15.3 and those in Table 15.2.

For moderate-size designs like the present, it is possible to estimate in the multinomial distribution (15.31). Table 15.4 thus shows, for each response pattern with an expected frequency larger than 3, how well the multinomial distribution over response patterns fits the model when the MML-estimates for β_i and

TABLE 15.2. CML item parameter estimates and standard errors

Items i	Response Categories h 0	1
1	−1.360	0.025
	(0.140)	(0.111)
2	−0.616	−0.204
	(0.133)	(0.124)
3	1.043	0.800
	(0.149)	(0.163)
4	1.340	0.745
	(0.155)	(0.174)
5	−1.530	−0.242
	(0.140)	(0.110)

TABLE 15.3. MML-estimates of all parameters jointly

Item i	Response Categories h 0	1
1	−1.369	0.001
	(0.142)	(0.112)
2	−0.621	−0.183
	(0.135)	(0.125)
3	1.064	0.807
	(0.147)	(0.156)
4	1.371	0.754
	(0.152)	(0.166)
5	−1.555	−0.270
	(0.142)	(0.111)
	$\hat{\mu} = 1.723$	$\hat{\sigma}^2 = 3.344$
	(0.098)	(0.334)

the parameters μ and σ^2 of $\varphi(\theta)$ are used in the response pattern probabilities (15.32). Note how a purely 'questionnaire-technical' effect, which is very often found, shows in this data set: the two extreme response pattens (00000) and (22222) have positive residuals, indicating an over-representation of respondents as compared with the expected numbers under the model. This is followed by negative residuals, indicating under-representation of respondents for all shown response patterns different from the two extreme patterns by 1, i.e., response patterns (00001), (00100), (01000), (10000), (12222), (22122), and (22212).

The *a priori* and *a posteriori* latent parameter estimates are compared in Table 15.5. The *a posteriori* estimates are obtained using a normal latent density with mean 1.723 and variance 3.344.

TABLE 15.4. Observed number, expected number, and standardized residual for each response pattern with expected value > 3.

Response Patterns	Observed Numbers	Expected Numbers	Standardized Residuals
00000	264	247.21	3.650
00001	30	50.82	−3.792
00002	7	9.31	−0.834
00010	3	7.58	−1.797
00011	2	3.84	−0.977
00100	10	10.88	−0.294
00101	7	5.50	0.671
01000	17	21.78	−1.193
01001	20	11.02	2.943
01002	4	3.73	0.148
02000	5	3.66	0.738
02001	3	3.42	−0.235
10000	41	55.36	−2.533
10001	29	28.00	0.222
10002	7	9.47	−0.870
10010	8	4.18	1.954
10011	11	3.90	3.737
10100	11	6.00	2.159
10101	7	5.60	0.623
10102	2	3.05	−0.624
11000	7	12.00	−1.579
11001	18	11.20	2.213
11002	7	6.12	0.378
11101	8	3.61	2.397
12000	3	3.72	−0.392
12001	5	5.61	−0.271
12002	9	4.62	2.139
12102	9	3.30	3.256
12222	0	3.25	−1.893
20000	6	7.74	−0.679
20001	5	7.22	−0.884
20002	5	3.94	0.559
21000	2	3.10	−0.645
21001	2	4.66	−1.291
21002	6	3.84	1.147
22001	2	3.52	−0.846
22002	9	4.26	2.410
22012	2	3.10	−0.648
22102	1	4.45	−1.718
22112	4	4.83	−0.407
22122	5	7.27	−0.942
22202	5	4.00	0.530
22212	4	6.89	−1.230
22222	24	18.76	1.686

TABLE 15.5. *A priori* and *a posteriori* estimates of the person parameter values for each score group

	Person Parameters	
Scores	ML	EAP
r = 0	–	-2.125
1	−1.964	−1.428
2	−1.224	−0.917
3	−0.729	−0.494
4	−0.313	−0.108
5	0.078	0.269
6	0.474	0.662
7	0.910	1.099
8	1.443	1.629
9	2.247	2.349
10	–	3.485

TABLE 15.6. Item parameter estimates and category score parameter estimates under (15.45) with standard errors in parentheses

Items i	CML Estimates	MML Estimates
1	−0.536	−0.549
	(0.052)	(0.053)
2	−0.076	−0.089
	(0.050)	(0.051)
3	0.485	0.504
	(0.055)	(0.058)
4	0.672	0.696
	(0.061)	(0.063)
5	−0.545	−0.562
	(0.052)	(0.055)
Categories h	CML Estimates	MML Estimates
0	2.000	2.000
	–	–
1	0.421	0.481
	(0.109)	(0.101)
2	0.000	0.000
	–	–

The differences between the *a priori* and *a posteriori* values reflect the knowledge contained in the estimated mean and the estimated variance of the latent normal density. The positive value of $\hat{\mu}$ shifts the θ-values upwards, while the relative large variance has the effect of the shifts being moderate in magnitude.

Assume now that the β have the form (15.10) with only one term in the summation, i.e.,

$$\beta_{ih} = \alpha_i \phi_h + \omega_h. \tag{15.45}$$

TABLE 15.7. Estimated values of the product $\hat{\alpha}_i\hat{\phi}_h$ based on the MML estimates in Table 15.6.

	Categories h	
Items i	0	1
1	−1.098	−0.264
2	−0.178	−0.043
3	1.008	0.242
4	1.392	0.335
5	−1.124	−0.270

It is easy to verify that our previous normalizations,

$$\beta_{im} = 0$$

and

$$\sum_i \sum_h w_h \beta_{ih} = 0,$$

together with (15.45) imply that $\omega_m = 0$, $\phi_m = 0$, and $\sum_i \alpha_i = 0$, and that a further multiplicative constraint is required. We choose

$$\phi_0 = m - 1.$$

For $m = 2$ this means $\phi_0 = 2$, ϕ_1 is unconstrained, and $\phi_2 = 0$.

Table 15.6 gives the CML and MML estimates of the item parameters under constraints (15.45).

When we compare the values in Table 15.3, including the standard errors, with the products $\hat{\alpha}_i\hat{\phi}_h$ shown in Table 15.7, it is quite obvious that the parametric form (15.10) does not fit the data, although the deviations are of moderate size. The values of the category scores suggest that an integer scoring, like

$$\phi_0 = 4, \quad \phi_1 = 1, \quad \phi_2 = 0,$$

would fit the data better than the equidistant scoring implicit in the rating scale model.

16

The Derivation of Polytomous Rasch Models

Gerhard H. Fischer[1]

ABSTRACT This chapter, analogous to Chapter 2, derives polytomous Rasch models from certain sets of assumptions. First, it is shown that the multidimensional polytomous Rasch model follows from the assumption that there exists a vector-valued minimal sufficient statistic $\boldsymbol{T}$ for the vector-valued person parameter $\boldsymbol{\theta}$, where $\boldsymbol{T}$ is independent of the item parameters $\boldsymbol{\beta}_i$; the sufficient statistic $\boldsymbol{T}$ is seen to be equivalent to the person marginal vector. Second, it is shown that, within a framework for measuring change between two time points, the partial credit model and the rating scale model follow from the assumption that conditional inference on (or specifically objective assessment of) change is feasible. The essential difference between the partial credit model and the rating scale model is that the former allows for a change of the response distribution in addition to a shift of the person parameter, whereas the rating scale model characterizes change exclusively in terms of the person parameter $\boldsymbol{\theta}$. A family of power series models is derived from the demand for conditional inference which accomodates the partial credit model, the rating scale model, the multiplicative Poisson model, and the dichotomous Rasch model.

16.1 Derivation of the Multidimensional Polytomous RM from Sufficiency

The multidimensional polytomous Rasch model (polytomous RM; Rasch, 1965, 1967a; see also Chapter 15) is a direct extension of the dichotomous RM to cases where all items have the same m response categories C_h, $h = 0, \ldots, m$. Let

$\boldsymbol{\beta}_i = (\beta_{i0}, \ldots, \beta_{im})$ be an item parameter, where each component β_{ih} expresses the attractiveness of the respective response category C_h of item I_i. The attractiveness of C_h is assumed to be the same for all subjects. Similarly, let

$\boldsymbol{\theta}_v = (\theta_{v0}, \ldots, \theta_{vm})$ be a parameter expressing S_v's tendency towards, or predilection for, responses in each of the categories C_h, $h = 0, \ldots, m$. These preferences are assumed to be independent of the particular item presented.

The multidimensional polytomous RM is then defined by

$$P(X_{vih} = 1|\boldsymbol{\theta}_v, \boldsymbol{\beta}_i) = \frac{\exp(\theta_{vh} + \beta_{ih})}{\sum_{l=0}^{m} \exp(\theta_{vl} + \beta_{il})}, \tag{16.1}$$

[1]Department of Psychology, University of Vienna, Liebiggasse 5, A–1010 Wien, Austria; e-mail: GH.FISCHER@UNIVIE.AC.AT

where X_{vih}, for $v = 1, \ldots, n$ and $i = 1, \ldots, k$, are locally independent random variables with realizations $x_{vih} = 1$ if subject S_v chooses category C_h of item I_i, and $x_{vih} = 0$ otherwise. The model assumes that, for each item presented, S_v chooses one and only one of the $m + 1$ categories, implying that $\sum_h X_{vih} = 1$ holds for all S_v and I_i. Clearly, model (16.1) excludes interactions between items and persons.

The model is overparameterized. The following normalizations, however, make it identifiable:

$$\theta_{v0} = \beta_{i0} = 0, \qquad \text{for all } v = 1, \ldots, n \text{ and } i = 1, \ldots, k, \tag{16.2}$$

$$\sum_i \beta_{ih} = 0, \qquad \text{for } h = 1, \ldots, m. \tag{16.3}$$

The indeterminacy of (16.1) and the normalization by (16.2) show that the measurement of a latent tendency θ_{vh} occurs only *relative* to the subject's response tendency θ_{v0} as an anchor or zero point; and the attractiveness of response category C_h of item I_i is similarly measured relative to the attractiveness of category C_0 of that item. Any comparison between two persons S_v and S_w is in fact based on the difference of the relative sizes of their latent trait parameters in relation to their response tendencies for category C_0, $(\theta_{vh} - \theta_{v0}) - (\theta_{wh} - \theta_{w0})$; and the same holds for the comparison of two items. This makes the psychological interpretation of the parameters rather difficult.

It is immediately seen that the dichotomous RM is a polytomous RM (16.1) with $m = 1$. From (16.1) and (16.2) one obtains

$$P(X_{vih} = 1 | \boldsymbol{\theta}_v, \boldsymbol{\beta}_i) = \frac{\exp(\theta_{v1} + \beta_{i1})}{1 + \exp(\theta_{v1} + \beta_{i1})},$$

which is the probability of a 'correct response' (category C_1) in the dichotomous RM, and

$$P(X_{vih} = 0 | \boldsymbol{\theta}_v, \boldsymbol{\beta}_i) = \frac{1}{1 + \exp(\theta_{v1} + \beta_{i1})},$$

the complementary probability of an 'incorrect response' (category C_0), where θ_{v1} and β_{i1} are scalar person and item parameters, respectively.

Following Andersen (1973a), we now derive the multidimensional polytomous RM (16.1) from the assumption that for $\boldsymbol{\theta}_v$ there exists a nontrivial minimal sufficient statistic $\boldsymbol{T}$ that is independent of the item parameters $\boldsymbol{\beta}_i$. Let the following notation be introduced:

$\boldsymbol{X}_{vi} = (X_{vi0}, \ldots, X_{vim})$ is the vector-valued response variable of S_v to item I_i, and

$\boldsymbol{X}_{v.} = \sum_i \boldsymbol{X}_{vi} = (X_{v.0}, \ldots, X_{v.m})$ the person marginal vector, where $X_{v.h} = \sum_i X_{vih}$ denotes the number of responses S_v has given in category C_h.

Andersen's theorem can now be formulated.

Theorem 16.1 *Let the response variables* $\boldsymbol{X}_{vi}$ *be locally independent. If, for person parameter* $\boldsymbol{\theta}_v$, *there exists a vector-valued minimal sufficient statistic* $\boldsymbol{T}(\boldsymbol{X}_{v1}, \ldots, \boldsymbol{X}_{vk})$ *that is independent of the item parameters* $\boldsymbol{\beta}_i$, *then, upon appropriate rescaling of the parameters, the model attains form (16.1).*

Proof If $\boldsymbol{T}(\boldsymbol{X}_{v1}, \ldots, \boldsymbol{X}_{vk})$ is a vector-valued minimal sufficient statistic that is independent of the item parameters $\boldsymbol{\beta}_i$, then $\boldsymbol{T}$ must be the same for all values of the item parameters, such as $\boldsymbol{\beta}_1 = \boldsymbol{\beta}_2 = \ldots = \boldsymbol{\beta}_k$. In the latter case, however, it follows from a result of Bahadur (1954) that $\boldsymbol{T}$ must be symmetric in its arguments $\boldsymbol{X}_{v1}, \ldots, \boldsymbol{X}_{vk}$, i.e., the items may be permuted without changing $\boldsymbol{T}$. Hence, $\boldsymbol{T}$ must be equivalent to the person marginal vector, $\boldsymbol{X}_{v.}$. Therefore, we may assume without loss of generality that $\boldsymbol{T}$ is this vector of person marginal sums.

Define

$$\phi(\boldsymbol{x}_{vi}|\boldsymbol{\theta}_v, \boldsymbol{\beta}_i) = \ln L(\boldsymbol{x}_{vi}|\boldsymbol{\theta}_v, \boldsymbol{\beta}_i) - \ln L(\boldsymbol{x}_{vi}|\boldsymbol{\theta}_0, \boldsymbol{\beta}_i), \tag{16.4}$$

where $\boldsymbol{\theta}_0$ is an arbitrary fixed parameter value. The sufficiency and symmetry of $\boldsymbol{T}$ imply that, if $\boldsymbol{x}'_{v1}, \ldots, \boldsymbol{x}'_{vk}$ is a permutation of $\boldsymbol{x}_{v1}, \ldots, \boldsymbol{x}_{vk}$,

$$\sum_{i=1}^{k} \phi(\boldsymbol{x}_{vi}|\boldsymbol{\theta}_v, \boldsymbol{\beta}_i) = \sum_{i=1}^{k} \phi(\boldsymbol{x}'_{vi}|\boldsymbol{\theta}_v, \boldsymbol{\beta}_i). \tag{16.5}$$

Let $\boldsymbol{x}_{vi} = \boldsymbol{x}'_{v1}$, $\boldsymbol{x}_{v1} = \boldsymbol{x}'_{vi} =: \boldsymbol{x}_0$, and $\boldsymbol{x}_{vl} = \boldsymbol{x}'_{vl}$ for $l \neq 1$ and $l \neq i$; moreover, set $\boldsymbol{\beta}_1 =: \boldsymbol{\beta}_0$ fixed.[2] Then (16.5) implies that

$$\phi(\boldsymbol{x}_{vi}|\boldsymbol{\theta}_v, \boldsymbol{\beta}_i) = \underbrace{\phi(\boldsymbol{x}_{vi}|\boldsymbol{\theta}_v, \boldsymbol{\beta}_0)}_{=:\, f} + \underbrace{\phi(\boldsymbol{x}_0|\boldsymbol{\theta}_v, \boldsymbol{\beta}_i) - \phi(\boldsymbol{x}_0|\boldsymbol{\theta}_v, \boldsymbol{\beta}_0)}_{=:\, -\psi},$$

or, with obvious notation,

$$\phi(\boldsymbol{x}_{vi}|\boldsymbol{\theta}_v, \boldsymbol{\beta}_i) = f(\boldsymbol{x}_{vi}|\boldsymbol{\theta}_v) - \psi(\boldsymbol{\theta}_v, \boldsymbol{\beta}_i). \tag{16.6}$$

This holds for all items I_i, $i \neq 1$. Denoting $\ln L(\boldsymbol{x}_{vi}|\boldsymbol{\theta}_0, \boldsymbol{\beta}_i)$ by $g(\boldsymbol{x}_{vi}|\boldsymbol{\beta}_i)$, (16.4) and (16.6) imply

$$\ln L(\boldsymbol{x}_{vi}|\boldsymbol{\theta}_v, \boldsymbol{\beta}_i) = f(\boldsymbol{x}_{vi}|\boldsymbol{\theta}_v) + g(\boldsymbol{x}_{vi}|\boldsymbol{\beta}_i) - \psi(\boldsymbol{\theta}_v, \boldsymbol{\beta}_i). \tag{16.7}$$

Since $\boldsymbol{x}_{vi}$ can assume m values, viz., $(1, 0, \ldots, 0), \ldots, (0, \ldots, 0, 1)$, the functions $f(\boldsymbol{x}_{vi}|\boldsymbol{\theta}_v)$ and $g(\boldsymbol{x}_{vi}|\boldsymbol{\beta}_i)$ may each also take on m values; let these be denoted by $\theta_{v0}, \ldots, \theta_{vm}$ and $\beta_{i0}, \ldots, \beta_{im}$, respectively. With this notation, the likelihood function of the model becomes

$$\ln L(\boldsymbol{x}_{vi}|\boldsymbol{\theta}_v, \boldsymbol{\beta}_i) = \sum_{h=0}^{m} \theta_{vh} x_{vih} + \sum_{h=0}^{m} \beta_{ih} x_{vih} - \psi(\boldsymbol{\theta}_v, \boldsymbol{\beta}_i); \tag{16.8}$$

[2]The notation '$a =: b$' means that b is defined by a.

more concretely,

$$P(X_{vih} = 1) = \frac{\exp(\theta_{vh} + \beta_{ih})}{\exp[\psi(\boldsymbol{\theta}_v, \boldsymbol{\beta}_i)]}. \tag{16.9}$$

Considering that $\sum_h P(X_{vih} = 1) = 1$, (16.9) becomes (16.1), which is the desired result. □

Applications of the multidimensional polytomous RM have been scarce, especially outside the circle of Rasch's students in Denmark. Three possible reasons for this come to mind. First, the multidimensional parameterization of the model makes the interpretation of the latent trait parameters difficult. Second, since a separate item parameter β_{ih} is attached to each category per item, at least moderate samples of data are required for reliable estimation. Third, rather long tests are needed even for an approximate estimation of the vector-valued person parameters, and even then some component θ_{vh} of vector $\boldsymbol{\theta}_v$ may not be estimable at all, namely, if S_v has never chosen response category C_h.

One of the major applications is due to Fischer and Spada (1973) who undertook to test the validity of Rorschach's (1921) psychological theory of inkblot perception in contrast to the psychometric conception of Holtzman, Thorpe, Swartz, and Heron (1961) that was developed for a similar inkblot test. Since Rorschach presumed that different personality traits account, e.g., for whole (W), detail (D), and small detail (d) responses (within the Mode of Apperception), the *multidimensional polytomous* RM was chosen by Fischer and Spada to measure these latent behavior tendencies and, at the same time, to describe characteristics of the inkblots. Since the classical scoring of the Rorschach test is based on the person marginal vector $\boldsymbol{X}_{v.}$[3], and since the multidimensional polytomous RM follows from the sufficiency of $\boldsymbol{X}_{v.}$, it was a central argument of Fischer and Spada that this model is *the* appropriate tool for formalizing and empirically testing Rorschach's implicit psychometric hypotheses.

Unlike Rorschach, Holtzman et al. view W, D, and d responses as manifestations of different degrees on *one* latent trait termed Location (and similarly for other scoring variables); they use the statistic $\sum_h \phi_h x_{v.h}$ with integer-valued ϕ_h as a score of S_v for Location. If Holtzman et al.'s conception is the more adequate one for modeling the perception of inkblots, a *unidimensional polytomous* RM with ordered categories (where θ_v is a scalar, see (15.2), Chapter 15) should suffice to describe the data. Fischer and Spada used the polytomous RM (15.2), not assuming the ϕ_h to be integers, however, as in the rating scale or partial credit model, but estimated them from the data. (On this estimation problem, see Andersen, 1966, or Fischer, 1974, pp. 470–472). Several inkblot variables like Location, Form Definiteness, Form Appropriateness, Movement, etc., were analyzed, using Holtzman Inkblot Test data, and the corresponding variables of the Rorschach test, using a set of Rorschach test data. Indications were found that Holtzman et al.'s psychometric conception is generally the more adequate. As a

[3]Considering only one response per person and plate.

novelty in the literature on projective inkblot tests, the authors also succeeded in demonstrating how a revision of the scoring rules is feasible based on the empirical application of model (15.2).

16.2 The Derivation of the Partial Credit and the Rating Scale Model from the Demand for Conditional Inference

In this section, two unidimensional polytomous RMs (Rasch, 1961, 1965), namely, the partial credit model (PCM) and the rating scale model (RSM), will be derived from the assumption that conditional inference on change is feasible. This is the same as postulating that the separation of person from item parameters be possible. Consider the typical situation where persons express their attitude towards an object by means of a rating scale item, or rate their subjective feeling of well-being vs. ailment, at time points T_1 and T_2, before and after a therapeutic intervention. As in the models of Chapter 9, responses at T_1 and T_2 formally correspond to responses given to two different 'virtual' items, I_1^* and I_2^*. (In what follows, we refer to the above framework for measuring change, but the reinterpretation of the observations as responses to two different items I_1 and I_2 presented at one time point will always be readily available.) Let the ratings be expressed on a discrete $m+1$ point rating scale, that is, in the form of two integers, where $X_{v1h_1} = 1$ (in the notation of the last section) is, for simplicity, written as $H_1 = h_1$, and $X_{v2h_2} = 1$ as $H_2 = h_2$, respectively.

We assume that the state of any fixed person at time point T_1 can be characterized by a unidimensional parameter θ governing a discrete distribution of variable H_1 with probability function $P(j|\theta) > 0$, $j = 0, 1, \ldots$, with $\theta \in \mathbb{R}$, subject to $\sum_j P(j|\theta) = 1$ (omitting index v for convenience). In order to make the results more general and applicable to a wider range of situations, we follow Fischer (1991b) in assuming that m may become infinitely large. The idea underlying this generalization is that researchers are often interested in frequencies of critical events, such as epileptic seizures, migraine attacks, accidents, outbursts of temper, or the like, where there is no strict upper bound on m. The case where H_1 is a rating on an $m+1$ point scale with fixed m, however, can easily be considered as a special case of the present more general definition of H as an unbounded integer random variable.

For time point T_2, we similarly assume that H_2 follows a discrete distribution with probability function $P^*(j|\theta + \eta) > 0$, $j = 0, 1, 2, \ldots$, with $\eta \in \mathbb{R}$ and $\sum_j P^*(j|\theta + \eta) = 1$.[4] The distribution parameter $\theta + \eta$ implies that we measure the treatment effect η as the amount by which parameter θ changes between T_1 and T_2.

---

[4]We use the asterisk to distinguish virtual items I_i^* from real items I_i. In the present case, there are two virtual items, $I_1^* \equiv I_1$ and I_2^*. It is therefore logical to denote the probability function of H_1 by P, and that of H_2, by P^*.

That change be measurable as the difference between corresponding θ-parameters is not trivial. But as has been argued in Chapters 2 and 9 (see also Fischer, 1987a), the additive concatenation of latent trait parameters and treatment effects (or measures of change) is admissible, for all values of θ and η, under the assumption that specifically objective results about change are feasible, which moreover implies that the θ-scale has interval scale properties.

Besides local independence, we make two kinds of technical assumptions about the $P(j|\theta)$ and $P^*(j|\theta+\eta)$: firstly, they are supposed to be continuous functions of the respective parameters; secondly, each variable H_1 and H_2 will be assumed to be stochastically increasing in its respective parameter in the sense that $P(j+1|\theta)/P(j|\theta)$ and $P^*(j+1|\theta+\eta)/P^*(j|\theta+\eta)$ are strictly monotone increasing in θ, for all $j = 1, 2, \ldots$, and

$$\lim_{\theta\to-\infty}[P(j|\theta)/P(0|\theta)] = \lim_{\theta\to-\infty}[P^*(j|\theta)/P^*(0|\theta)] = 0, \tag{16.10}$$

for all $j > 0$, which implies

$$\lim_{\theta\to-\infty} P(0|\theta) = \lim_{\theta\to-\infty} P^*(0|\theta) = 1. \tag{16.11}$$

This monotonicity assumption is motivated as follows. We want to interpret θ as a measure of the person's proneness to the critical events in question, so an increase of θ should imply an increase in the probability of observing $j+1$ rather than j events. Moreover, monotonicity assures that θ can be deduced uniquely if at least two probabilities, say $P(0|\theta)$ and $P(1|\theta)$, are known; this will allow us to apply the results even to cases where the observed distribution is truncated. The assumption (16.10), which implies (16.11), is an analogue to the 'no guessing' assumption for dichotomous ICCs.

Under the said assumptions (local stochastic independence, continuity, and strict monotonicity), the main result of this section can now be formulated as follows:

Theorem 16.2 *If the conditional probability $P(h_1, h_2|h_1 + h_2 = h)$, for all h_1 and h_2, is a function $V(h_1, h_2; \eta)$ independent of θ, then $P(h|\theta)$ and $P^*(h|\theta+\eta)$ must be probability functions of power series distributions*

$$P(h|\theta) = \frac{m_h \mu^h}{\sum_{j=0}^{\infty} m_j \mu^j} \quad \text{and} \quad P^*(h|\theta+\eta) = \frac{m_h^*(\mu\delta)^h}{\sum_{j=0}^{\infty} m_j^*(\mu\delta)^j}, \tag{16.12}$$

with $\mu = \exp(c\theta)$, $c > 0$, $\delta = \exp(\eta)$, and $m_j > 0$, $m_j^* > 0$ for all j.

Proof Consider the case $h_1 = 0$, $h_2 = 1$. According to assumptions, the conditional probability $P(h_1, h_2|h_1 + h_2)$ is

$$\frac{P(0|\theta)P^*(1|\theta+\eta)}{P(0|\theta)P^*(1|\theta+\eta) + P(1|\theta)P^*(0|\theta+\eta)} = V(0, 1; \eta) > 0. \tag{16.13}$$

Dividing both numerator and denominator of the left-hand side of (16.13) by $P(0|\theta)P^*(1|\theta+\eta)$, and rearranging terms, yields

$$\frac{P(1|\theta)P^*(0|\theta+\eta)}{P(0|\theta)P^*(1|\theta+\eta)} = \frac{1-V}{V}. \tag{16.14}$$

Introducing the notation

$$\begin{aligned} \ln[P(1|\theta)/P(0|\theta)] &=: X(\theta), \\ \ln[P^*(1|\theta+\eta)/P^*(0|\theta+\eta)] &=: Z(\theta+\eta), \\ \ln[(1-V)/V] &=: -Y(\eta), \end{aligned}$$

where $X(.)$, $Z(.)$, and $Y(.)$ are continuous and strictly monotone, it is seen that (16.14) implies the functional equation

$$Z(\theta+\eta) = X(\theta) + Y(\eta), \tag{16.15}$$

which holds for all θ and η. Setting $\theta = 0$, (16.15) becomes

$$Z(\eta) = Y(\eta) + X(0), \tag{16.16}$$

and setting $\eta = 0$ similarly yields

$$Z(\theta) = X(\theta) + Y(0). \tag{16.17}$$

Moreover, $\theta = \eta = 0$ gives $Z(0) = X(0) + Y(0)$. Inserting (16.16) and (16.17) in (16.15),

$$\begin{aligned} Z(\theta+\eta) &= Z(\theta) - Y(0) + Z(\eta) - X(0) \\ &= Z(\theta) + Z(\eta) - Z(0), \end{aligned}$$

from which it follows that

$$Z(\theta+\eta) - Z(0) = Z(\theta) - Z(0) + Z(\eta) - Z(0),$$

which can be written as

$$\tilde{Z}(\theta+\eta) = \tilde{Z}(\theta) + \tilde{Z}(\eta), \tag{16.18}$$

with $\tilde{Z}(t) := Z(t) - Z(0)$. This is the well-known functional equation of Cauchy (cf. Aczél, 1966), the general continuous monotone increasing solution of which is $\tilde{Z}(t) = ct$, with $c > 0$. The definition of $\tilde{Z}(t)$ implies $Z(t) = ct + Z(0)$, which inserted in (16.16) gives

$$Y(\eta) = c\eta + Z(0) - X(0) = c\eta + Y(0);$$

similarly, from (16.17),

$$X(\theta) = c\theta + Z(0) - Y(0) = c\theta + X(0).$$

Remembering the definition of $X(\theta)$ as $\ln[P(1|\theta)/P(0|\theta)]$, it is seen that

$$\frac{P(1|\theta)}{P(0|\theta)} = \exp(c\theta)\exp[X(0)],$$

or, with $\exp[X(0)] =: d_1$ and $\exp(c\theta) =: \mu > 0$,

$$\frac{P(1|\theta)}{P(0|\theta)} = d_1\mu. \tag{16.19}$$

Similarly, from the definition of $Z(\theta+\eta)$ as $\ln[P^*(1|\theta+\eta)/P^*(0|\theta+\eta)]$, taken at $\eta = 0$, it follows that

$$Z(\theta) = \ln\left(\frac{P^*(1|\theta)}{P^*(0|\theta)}\right) = c\theta + Z(0),$$

whence

$$\frac{P^*(1|\theta)}{P^*(0|\theta)} = \exp(c\theta)\exp[Z(0)] = d_1^*\mu, \tag{16.20}$$

with $d_1^* := \exp[Z(0)]$.

We shall show that more generally

$$\frac{P(j|\theta)}{P(j-1|\theta)} = d_j\mu \quad \text{and} \quad \frac{P^*(j|\theta+\eta)}{P^*(j-1|\theta+\eta)} = d_j^*\mu\delta, \tag{16.21}$$

for $j = 1, 2, \ldots$, with $\delta = \exp(c\eta)$. The proof of this conjecture will be given by complete induction. Starting from the assumption that (16.21) is true for $j = 1, 2, \ldots, k$, we shall show that (16.21) holds for $j = k+1$ also.

Let $h_1 + h_2 = k+1$. The conditional probability of h_1 and h_2, given $h_1 + h_2 = k+1$, is

$$\frac{P(h_1|\theta)P^*(h_2|\theta+\eta)}{P(0|\theta)P^*(k+1|\theta+\eta) + .. + P(h_1|\theta)P^*(h_2|\theta+\eta) + .. + P(k+1|\theta)P^*(0|\theta+\eta)}$$

$$= V(h_1, h_2; \eta). \tag{16.22}$$

Dividing the numerator and denominator of (16.22) by $P(h_1|\theta)P^*(h_2|\theta+\eta)$,

$$\left[\frac{P(0|\theta)P^*(k+1|\theta+\eta)}{P(h_1|\theta)P^*(h_2|\theta+\eta)} + \frac{P(1|\theta)P^*(k|\theta+\eta)}{P(h_1|\theta)P^*(h_2|\theta+\eta)} + \ldots + 1 + \right.$$

$$\left. \ldots + \frac{P(k|\theta)P^*(1|\theta+\eta)}{P(h_1|\theta)P^*(h_2|\theta+\eta)} + \frac{P(k+1|\theta)P^*(0|\theta+\eta)}{P(h_1|\theta)P^*(h_2|\theta+\eta)}\right]^{-1} = V(h_1, h_2; \eta).$$

Inserting (16.21) for $j = 1, \ldots, k$,

$$\frac{P^*(k+1|\theta+\eta)}{P^*(k|\theta+\eta)} \frac{d_k^* \dots d_{h_2+1}^* \mu^{k-h_2} \delta^{k-h_2}}{d_{h_1} \dots d_1 \mu^{h_1}} + \frac{d_k^* \dots d_{h_2+1}^* \mu^{k-h_2} \delta^{k-h_2}}{d_{h_1} \dots d_2 \mu^{h_1-1}} + \dots + 1 + \dots$$

$$\dots + \frac{d_k \dots d_{h_1+1} \mu^{k-h_1}}{d_{h_2}^* \dots d_2^* \mu^{h_2-1} \delta^{h_2-1}} + \frac{d_k \dots d_{h_1+1} \mu^{k-h_1}}{d_{h_2}^* \dots d_1^* \mu^{h_2} \delta^{h_2}} \frac{P(k+1|\theta)}{P(k|\theta)} = V(h_1, h_2; \eta)^{-1}. \tag{16.23}$$

Denoting, in an obvious way, the $k+1$ functions of constants $d_1, \dots, d_k$ and $d_1^*, \dots, d_k^*$ in (16.23) by $c_0, \dots c_{k+1}$, one obtains

$$c_{k+1} \frac{P^*(k+1|\theta+\eta)}{P^*(k|\theta+\eta)} \frac{1}{\mu\delta} \delta^{k+1-h_2} + c_k \delta^{k-h_2} + \dots + 1 + \dots$$

$$\dots + c_1 \delta^{-(h_2-1)} + c_0 \frac{P(k+1|\theta)}{P(k|\theta)} \frac{1}{\mu} \delta^{-h_2} = V(h_1, h_2; \eta)^{-1},$$

which is a power series in δ. Rearranging terms yields

$$\frac{P^*(k+1|\theta+\eta)}{P^*(k|\theta+\eta)} \frac{1}{\mu\delta} \delta^{k+1-h_2} + \frac{c_0}{c_{k+1}} \frac{P(k+1|\theta)}{P(k|\theta)} \frac{1}{\mu} \delta^{-h_2} = W(h_1, h_2; \eta), \tag{16.24}$$

with

$$\frac{c_0}{c_{k+1}} = \frac{\prod_{j=1}^k d_j}{\prod_{j=1}^k d_j^*},$$

where the right-hand side of (16.24) depends on h_1, h_2, and η, but is independent of θ. Inserting (16.21), for $j = 1, \dots, k$, in (16.24),

$$\frac{P^*(k+1|\theta+\eta)}{P^*(0|\theta+\eta)(\mu\delta)^k \prod_{j=1}^k d_j^*} \frac{1}{\mu\delta} \delta^{k+1-h_2} +$$

$$+ \frac{\prod_{j=1}^k d_j}{\prod_{j=1}^k d_j^*} \frac{P(k+1|\theta)}{P(0|\theta)\mu^k \prod_{j=1}^k d_j} \frac{1}{\mu} \delta^{-h_2} = W(h_1, h_2; \eta),$$

$$\frac{P^*(k+1|\theta+\eta)}{P^*(0|\theta+\eta)} + \frac{P(k+1|\theta)}{P(0|\theta)} = \mu^{k+1} \Big(\prod_{j=1}^k d_j^* \Big) \; W(h_1, h_2; \eta) \delta^{h_2}. \tag{16.25}$$

Now let $\eta \to -\infty$, i.e., $\delta \to 0$. Then (16.25) yields

$$\lim_{\eta \to -\infty} \frac{P^*(k+1|\theta+\eta)}{P^*(0|\theta+\eta)} + \frac{P(k+1|\theta)}{P(0|\theta)} = \mu^{k+1} \Big(\prod_{j=1}^k d_j^* \Big) \lim_{\eta \to -\infty} [W(h_1, h_2; \eta) \delta^{h_2}]. \tag{16.26}$$

Since the limit of the left-most term in (16.26) is 0 by assumption, the limit on the right-hand side of (16.26) exists and must be > 0. Denote the right-hand limit, for convenience, as

$$\lim_{\eta \to -\infty} [W(h_1, h_2; \eta)\delta^{h_2}] = d_{k+1} \frac{\prod_{j=1}^{k} d_j}{\prod_{j=1}^{k} d_j^*},$$

with $d_{k+1} > 0$, so that (16.26) yields

$$\frac{P(k+1|\theta)}{P(0|\theta)} = \mu^{k+1} \prod_{j=1}^{k+1} d_j. \tag{16.27}$$

This implies

$$\frac{P(k+1|\theta)}{P(k|\theta)} = d_{k+1}\mu. \tag{16.28}$$

The general structure of $P(h|\theta)$ is therefore

$$P(h|\theta) = P(0|\theta)\mu^h \prod_{j=0}^{h} d_j, \tag{16.29}$$

with $d_0 = 1$, for $h = 0, 1, 2, \ldots$. From the trivial condition $\sum_j P(j|\theta) = 1$ it follows that

$$P(0|\theta) = \left[\sum_{j=0}^{\infty} \mu^j \prod_{l=0}^{j} d_l\right]^{-1}.$$

Inserting this in (16.29) it is finally seen that

$$P(h|\theta) = \frac{\mu^h \prod_{j=0}^{h} d_j}{\sum_{j=0}^{\infty} \mu^j \prod_{l=0}^{j} d_l} = \frac{m_h \mu^h}{\sum_{j=0}^{\infty} m_j \mu^j}, \tag{16.30}$$

with appropriately defined constants $m_j := d_0 d_1 \ldots d_j$.

Similarly, inserting (16.28) in (16.24) shows that

$$\frac{P^*(k+1|\theta+\eta)}{P^*(k|\theta+\eta)} \frac{1}{\mu\delta} \delta^{k+1-h_2} + \frac{c_0}{c_{k+1}} d_{k+1}\delta^{-h_2} = W(h_1, h_2; \eta), \tag{16.31}$$

which means that the left hand term of (16.31) cannot depend on θ; this implies that

$$\frac{P^*(k+1|\theta+\eta)}{P^*(k|\theta+\eta)} \frac{1}{\mu\delta} =: d_{k+1}^*. \tag{16.32}$$

By the same arguments as before we finally obtain

$$P^*(h|\theta+\eta) = \frac{(\mu\delta)^h \prod_{j=0}^{h} d_j^*}{\sum_{j=0}^{\infty}(\mu\delta)^j \prod_{l=0}^{j} d_l^*} = \frac{m_h^*(\mu\delta)^h}{\sum_{j=0}^{\infty} m_j^*(\mu\delta)^j}, \tag{16.33}$$

with $d_0^* = 1$ and $m_j^* = d_0^* d_1^* \dots d_j^*$. □

In statistical literature, the distribution in (16.30) and (16.33) is called the 'power series distribution'. It is an exponential family that has been studied in some detail (cf. Noack, 1950; Patil, 1965); it is well-known to admit conditional inference procedures (see Johnson & Kotz, 1969, Vol. 1, p. 36). Many elementary distributions, like the binomial or the Poisson, are special cases of it. The Poisson distribution, for instance, is obtained by setting $m_h = d_1^h/h!$ in (16.30), so that $P(h|\mu) = \exp(-d_1\mu)(d_1\mu)^h/h!$.

Fischer (1991b) derived (16.30) within the framework of measuring change of event frequencies. He furthermore showed that, within the family of power series models, there is only one psychometric sub-model of change satisfying a certain 'ignorability principle': the multiplicative Poisson model (see Rasch, 1960, 1973; on its application to the measurement of change, see Fischer, 1977b). In this context, the meaning of ignorability is the following. Suppose, for instance, that the model is applied to measuring effects of clinical treatment(s) given to migraine patients (for an empirical example, see Fischer, 1991b); the researcher observes the patients' frequencies of attacks within a certain period prior to the treatments, and within another period after the treatments. This means that only a more or less arbitrary section of the process is observed. The ignorability principle posits that ignoring the unobserved events should not bias the result about change (about the treatment effects). The multiplicative Poisson model is the only power series model that satisfies this assumption. At the same time, the Poisson distribution is the only power series distribution where the parameter μ is proportional to the expected number of events within the observation period, $E(H)$, that is, where the observed frequencies are directly interpretable as estimates of the latent proneness parameters.

Returning to the problem of modeling rating scale data where the number of response categories is fixed to $m+1$, or to multiple choice items with $m+1$ ordered response categories, it suffices to apply the assumptions about the $P(j|\theta)$ and $P^*(j|\theta+\eta)$ to the probability functions for $j = 0, 1, \dots, m$, setting $P(j|\theta) = P^*(j|\theta+\eta) = 0$ for $j > m$, which is equivalent to setting $m_j = m_j^* = 0$ for $j > m$. Moreover, assume that the person parameter is constant (= no change), $\eta = 0$; then it is immediately seen that model (16.30) and (16.33) becomes the PCM for $k = 2$ items, I_1 and I_2, with parameters β_{1h} and β_{2h}, for $h = 0, \dots, m$,

$$P(h|\theta, I_1) = \frac{m_h\mu^h}{\sum_{j=0}^{m} m_j\mu^j} = \frac{\exp[h\theta + \beta_{1h}]}{\sum_{j=0}^{m}\exp[j\theta + \beta_{1j}]}, \tag{16.34}$$

$$P^*(h|\theta, I_2) = \frac{m_h^*\mu^h}{\sum_{j=0}^{m} m_j^*\mu^j} = \frac{\exp[h\theta + \beta_{2h}]}{\sum_{j=0}^{m}\exp[j\theta + \beta_{2j}]}, \tag{16.35}$$

where $\ln m_h =: \beta_{1h}$ and $\ln m_h^* =: \beta_{2h}$.

Thus, we have now derived the characteristic structure of the PCM, (16.34)–(16.35), from the postulate that a conditional inference procedure is feasible. Theorem 16.2 has been proved for $k = 2$ items only, but the meaning of the result is more general because, if conditional inference as defined above is feasible for *all* test lengths k, then it must be possible also for $k = 2$, and this implies the PCM. Technically, the derivation should be generalizable analogous to the proof of Theorem 2.1, but this is beyond the scope of this chapter.

The PCM was studied by Masters (1982) and Wright and Masters (1982); see also Andersen (1983) and Chapter 15. The normalization conditions in the PCM for a test of length k are $\beta_{i0} = 0$ for $i = 1, \ldots, k$, and $\sum_i \sum_h \beta_{ih} = 0$. For a dichotomous test ($m = 1$), (16.34) and (16.35) subject to this normalization become

$$P(1|\theta) = \frac{\exp(\theta - \beta_1)}{1 + \exp(\theta - \beta_1)}, \quad P(0|\theta) = \frac{1}{1 + \exp(\theta - \beta_1)},$$

$$P^*(1|\theta) = \frac{\exp(\theta - \beta_2)}{1 + \exp(\theta - \beta_2)}, \quad P^*(0|\theta) = \frac{1}{1 + \exp(\theta - \beta_2)},$$

with $\beta_1 := -\beta_{11}$ and $\beta_2 := -\beta_{21}$. This shows that the dichotomous RM is a special case of the PCM (as one would expect).

Returning to the problem of measuring change between time points T_1 and T_2, the PCM resulting from (16.30) and (16.33) is

$$P(h|\theta, T_1) = \frac{\exp[h\theta + \beta_{1h}]}{\sum_{j=0}^{m} \exp[j\theta + \beta_{1j}]},$$

$$P(h|\theta + \eta, T_2) = \frac{\exp[h(\theta + \eta) + \beta_{2h}]}{\sum_{j=0}^{m} \exp[j(\theta + \eta) + \beta_{2j}]}.$$

This model is somewhat difficult to interpret for the following reasons: change is expressed both as a shift $\theta \to \theta + \eta$ and as a change of the parameters $\beta_{1h} \to \beta_{2h}$, that is, as a change of the latent parameter θ and of the *form* of the distribution $P(j|\theta)$. A treatment may, of course, have a differential effect on the relative attractiveness of the response categories in different items, but in most applications the researcher will probably prefer to exclude such a possibility. It may therefore seem more appropriate to assume that change resides solely in the θ-parameter, since θ is the only parameter characterizing individual differences. Hence, we consider the following case: assume that the distributions of H_1 and H_2 are identical except for the latent parameters θ and $\theta + \eta$. Then, under the same technical assumptions as before, the following result immediately derives from Theorem 16.2.

Corollary 16.1 *Let the assumptions of Theorem 16.2 hold, but let* H_1 *and* H_2 *be governed by the same probability function* $P(h|\theta)$*, except for a possible change of* θ*. If the conditional probability* $P(h_1, h_2 | h_1 + h_2 = h)$*, for all pairs* h_1, h_2*, is a function* $V(h_1, h_2; \eta)$ *independent of* θ*, then* $P(h_1|\theta)$ *and* $P(h_2|\theta + \eta)$ *must be probability functions of power series distributions*

$$P(h|\theta) = \frac{m_h \mu^h}{\sum_{j=0}^{\infty} m_j \mu^j} \quad \text{and} \quad P(h|\theta + \eta) = \frac{m_h (\mu\delta)^h}{\sum_{j=0}^{\infty} m_j (\mu\delta)^j}, \tag{16.36}$$

with $\mu = \exp(c\theta)$, $c > 0$, $\delta = \exp(\eta)$, *and* $m_j > 0$ *for all* j.

Proof The result follows from Theorem 16.2. □

As before, the reinterpretation of the time points as two items, I_1 and I_2, given at one time point, is obvious. Analogous to (16.34)–(16.35), one obtains

$$P(h|\theta, I_1) = \frac{\exp[h(\theta + \beta_1) + \omega_h]}{\sum_{j=0}^{m} \exp[j(\theta + \beta_1) + \omega_j]}, \tag{16.37}$$

$$P(h|\theta, I_2) = \frac{\exp[h(\theta + \beta_2) + \omega_h]}{\sum_{j=0}^{m} \exp[j(\theta + \beta_2) + \omega_j]}, \tag{16.38}$$

with $\beta_1 = 0$ (normalization), $\eta =: \beta_2$, $\ln m_h =: \omega_h$. This is the RSM discussed by Andrich (1978a, 1978b), which can also be interpreted as a special case of the PCM (see also Chapter 15).

For test length k, the normalization conditions in the RSM are $\omega_0 = \omega_m = 0$ and $\sum_i \beta_i = 0$. Again, for $m = 1$, the dichotomous RM results as a special case of the RSM.

17

The Polytomous Rasch Model within the Class of Generalized Linear Symmetry Models

Henk Kelderman[1]

ABSTRACT Polytomous Rasch models are derived from a new requirement for objective measurement, the (quasi-)interchangeability of measurement instruments. This principle is introduced as a symmetry restriction on the statistical distribution of a set of measurements. The restriction is formulated as a generalized linear model. An interesting property of this requirement is that it does not assume conditionally independent item responses. This new requirement is related to the assumption of a Rasch model: the model is shown to be equivalent to the conditional or (extended) Rasch model; if the assumption of local independence is added, it is shown to be closely related to the original Rasch model.

17.1 Introduction

In scientific psychology, the construction of objective measurement instruments is of utmost importance. This was recognized by psychometricians such as Thurstone (1925) and Rasch (1960), who tried to formulate statistical models as standards against which scientific data have to be judged in order to be accepted as objective measurements. Andrich (1989a, 1989b), in a discussion of the history of psychometric thinking, contrasts this approach to that of Likert (1932) or Lord (1952, 1980), who were more inclined to perceive a psychometric model as a tool for pragmatically summarizing test data. In this chapter, we follow the former approach and discuss a new set of requirements for test data.

In Chapters 2 and 16, the Rasch model (RM) is derived from various sets of assumptions that embody certain measurement requirements (see Table 2.1). These requirements all invoke person parameters and latent traits in one way or another. Andersen (1973a, Theorem 1) proved that, if there exist sufficient statistics for the person parameters that do not depend on the item difficulty parameters, the model must have the logistic form of the RM. To derive this property, he used the result of Bahadur (1954) that, for the special case of equal item difficulty, the sufficient statistic must be symmetric in its arguments. For example, in the dichotomous RM, the minimal sufficient statistic for the person parameter is the number of correct responses. Obviously, this statistic is symmetric in its arguments. That is, in computing this statistic, the items are

[1]Faculty of Psychology and Pedagogics, Department of Work and Organizational Psychology, Vrije Universiteit, De Boelelaan 1081c, 1081 HV, Amsterdam; fax: (+31)20-4448702

interchangeable.

Ten Have and Becker (1991) noted that the interchangeability property is a useful measurement requirement in its own right and formulated a loglinear model (LLM) that imposes certain symmetry restrictions on the statistical distribution of a set of measurements. Their LLM's are related to loglinear symmetry models by Mandanski (1963), who coined the term 'interchangeability'.

In the present chapter, we discuss the epistemological and substantive justification of this interchangeability principle and derive a *generalized linear model* (GLM) for *measurement interchangeability* (MI). This model does not involve latent variables and does not make any assumptions about item characteristic curves, local stochastic independence, or sufficient person statistics. First, we consider the *complete measurement interchangeability* (CMI) model. This CMI model requires that measurement instruments (items) of the same concept are fully interchangeable, without changing any joint distributions involving them.

Next, to allow for different items to have different difficulties, the CMI model is relaxed. This GLM is called the *quasi measurement interchangeability* (QMI) model. It is then proved that this GLM must be a log-linear model (LLM), if the interchangeability property is to be left intact.

The QMI model is then shown to be equivalent to a sum score model (Agresti, 1993; Duncan, 1984; Duncan & Stenbeck, 1987; Kelderman, 1984, 1989, 1992). This sum score model is, in turn, shown to be equivalent to a family of 'conditional' RMs (Kelderman & Rijkes, 1994). In the dichotomous case, these models are known as the 'extended' RM discussed by Cressie and Holland (1983) and Tjur (1982). As the QMI model, none of these RMs contain latent variables.

Finally, it is shown that if we add the assumption of local (or conditional) independence to the QMI model, and assume that the model describes a single person, the latent variable re-appears. The resulting model is the one originally formulated by Rasch.

17.2 The Interchangeability of Measurement Instruments

In his discussion of assumptions and requirements in measurement in the social sciences, Andrich (1989b) defends the primacy of theory over observation and thereby quotes Kuhn (1961, p. 193).

> "In textbooks, the numbers that result from measurements usually appear as the archetypes of the 'irreducible and stubborn facts' to which the scientist must, by struggle, make his theories conform. But in scientific practice, as seen through the journal literature, the scientist often seems rather to be struggling with facts, trying to force them to conformity with a theory he does not doubt. Quantitative facts cease to seem simply the 'given'. They must be fought for and with, and in this fight the theory with which they are to be

> compared proves the most potent weapon. Often scientists cannot get numbers that compare well with theory until they know what numbers they should be making nature to yield."

Similar remarks about the role of measurement in research were made earlier by Campbell (1928).

In scientific research, one is interested in finding certain invariant relationships between theoretical concepts. Because, in the social sciences, the same concept can often be operationalized by several, say k, different measurement instruments, $I_1, \ldots, I_k$, it is important that these instruments yield similar measurement results. In this paper we use as a criterion for 'similar measurement results' that the measurement instruments can be replaced by each other without changing the relationships to other variables. We call this *measurement interchangeability*.

In psychology, research programs can be found that pay explicit attention to measurement interchangeability. In cognitive psychology, several measurement instruments may be constructed to measure the same type of intelligence (Gardner, 1983; Guilford, 1967). For example, Guilford initially proposed a taxonomy of 120 different intelligences. For each intelligence, several tests were constructed and, on the basis of the patterns of correlations within and between concepts, intelligences were refuted and replaced by the broader or narrower concepts depending on their relations to other measurements of intelligence (see Kelderman, Mellenbergh, & Elshout, 1981).

Similarly, each intelligence test itself consists of a set of distinct problems presented to the individual. The problems are usually constructed in such a way that each item invokes a similar mental activity from the person. In item analysis, the correlations among the items and between items and other variables are often studied to assess whether the items fit in the test.

In some practical testing situations, where scientific validity of the measurements is not of primary interest, interchangeability of item responses may still be a desirable property. For example, in educational testing, item bias studies are conducted to determine the fairness of test items with respect to certain variables, such as ethnic background and gender. It is then considered a desirable property that the items have the same relation to these variables.

Loosely speaking, two measurements will be defined to be interchangeable if they have the same relationship to other variables relevant for the current theoretical framework. We define $\boldsymbol{x} = (x_1, \ldots, x_k)$ to be the vector of variables for which measurement interchangeability is studied, and $\boldsymbol{z} = (z_1, z_2, \ldots)$ a vector of other theoretically relevant variables. The researcher may have a particular set of variables in mind. To test interchangeability, the researcher may want to choose a smaller set of z–variables that (s)he thinks are most important. This choice may be such that the probability of finding deviations from interchangeability is maximized. Variables such as gender or ethnic background are commonly used to test the RM, because it is probable that some items are not interchangeable with respect to these variables. Before we formalize the interchangeability requirement as a basis for models for measurement data, we make two other simplifying choices.

The first choice is the measurement level of the $x-$ and $z-$variables. Because continuous measurement is rarely possible in the social sciences, we will limit ourselves to variables that take only *discrete* non-negative values where, for simplicity, the measurements x_i, $i = 1, \ldots, k$, can take the same number, say $m + 1$, of discrete values $x_i = 0, 1, \ldots, m$.

The second choice is the type of relations between the variables. Since, in the social sciences, measurement is generally done with error, we assume *stochastic* rather than deterministic relations. The random variables are denoted by the capital letters $\boldsymbol{X} = (X_1, \ldots, X_k)$ and $\boldsymbol{Z} = (Z_1, Z_2, \ldots)$. In the remainder of this chapter the separate $z-$variables do not play a role in the discussion. Therefore, we replace the vector $\boldsymbol{Z}$ by the scalar Z for simplicity, remembering that it may denote a joint variable composed of several variables. We can now formulate a statistical model for measurement interchangeability .

17.3 A General Linear Model for Complete Interchangeability of Measurement Instruments

A general tool for the description of statistical relations between discrete variables is the generalized linear model (GLM; Nelder & Wedderburn, 1972; McCullagh & Nelder, 1983). Let $p_{\boldsymbol{x}z}$ be the joint probability of $\boldsymbol{x}$ and z, then the fully saturated GLM for $\boldsymbol{x}$ and z is

$$\begin{aligned}\text{link } p_{\boldsymbol{x}z} &= \lambda + \sum_{i=1}^{k} \lambda_{x_i}^{X_i} + \lambda_z^Z + \sum_{i=1}^{k} \lambda_{x_i z}^{X_i Z} + \\ &\quad + \sum_{i<j} \lambda_{x_i x_j}^{X_i X_j} + \cdots + \lambda_{x_1 x_2 \ldots x_k}^{X_1 X_2 \ldots X_k} \\ &\quad + \sum_{i<j} \lambda_{x_i x_j z}^{X_i X_j Z} + \cdots + \lambda_{x_1 x_2 \ldots x_k z}^{X_1 X_2 \ldots X_k Z},\end{aligned} \tag{17.1}$$

for all values of $\boldsymbol{x}$ and z. On the left-hand side of (17.1), 'link' is a monotone increasing, differentiable function. It is called the link function, since it links the probability on the left-hand side to the linear model on the right-hand side. This linear model is an ANOVA type model with main and interaction effects of the variables $x_1, \ldots, x_k, z$, where the λ-parameters describe the general mean (λ), main effects (λ_z^Z, $\lambda_{x_i}^{X_i}$, for $i = 1, \ldots k$), first order interaction effects ($\lambda_{x_i z}^{X_i Z}$, $\lambda_{x_i x_j}^{X_i X_j}$, for $i, j = 1, \ldots k$), and so on. Superscript capital letters denote the random variables involved in the effects, and subscript letters their realizations. We will suppress superscript notation in the rest of the chapter, unless it is really necessary to avoid ambiguity, e.g., when certain numeric values are assigned to the realizations.

To make the parameters of the GLM (17.1) estimable, the λ-parameters are normalized in such a way that they are zero if one or more of their subscripts are zero. For example, if $k = 2$, these normalizations are

$$\begin{gathered}
\lambda_0^{X_1} = \lambda_0^{X_2} = \lambda_0^{Z} = 0, \\
\lambda_{0x_2}^{X_1X_2} = \lambda_{x_10}^{X_1X_2} = \lambda_{00}^{X_1X_2} = \lambda_{0z}^{X_1Z} = \lambda_{x_10}^{X_1Z} = 0, \\
\lambda_{00}^{X_1Z} = \lambda_{0z}^{X_2Z} = \lambda_{x_20}^{X_2Z} = \lambda_{00}^{X_2Z} = 0, \\
\lambda_{0x_2z}^{X_1X_2Z} = \lambda_{x_10z}^{X_1X_2Z} = \lambda_{x_1x_20}^{X_1X_2Z} = \lambda_{00z}^{X_1X_2Z} = \\
\lambda_{0x_20}^{X_1X_2Z} = \lambda_{x_100}^{X_1X_2Z} = \lambda_{000}^{X_1X_2Z} = 0.
\end{gathered} \tag{17.2}$$

By this normalization, the parameters are expressed as a difference with respect to the null category. This is called the *simple contrast.*

In the saturated model, the parameters can be computed directly from link $p_{\boldsymbol{x}z} = \Lambda_{\boldsymbol{x}z}$, say. For $k = 2$, we may express the λ-parameters in terms of link $p_{x_1x_2z}$ as follows:

$$\begin{aligned}
\lambda &= \text{link } p_{000}, \\
\lambda_{x_1} &= \text{link } p_{x_100} - \text{link } p_{000}, \\
\lambda_{x_2} &= \text{link } p_{x_200} - \text{link } p_{000}, \\
\lambda_{z} &= \text{link } p_{00z} - \text{link } p_{000}, \\
\lambda_{x_1x_2} &= \text{link } p_{x_1x_20} - \text{link } p_{x_100} - \text{link } p_{0x_20} + \text{link } p_{000}, \\
\lambda_{x_1z} &= \text{link } p_{x_10z} - \text{link } p_{x_100} - \text{link } p_{00z} + \text{link } p_{000}, \\
\lambda_{x_2z} &= \text{link } p_{0x_2z} - \text{link } p_{0x_20} - \text{link } p_{00z} + \text{link } p_{000}, \\
\lambda_{x_1x_2z} &= \text{link } p_{x_1x_2z} - \text{link } p_{x_1x_20} - \text{link } p_{x_10z} - \text{link } p_{0x_2z} \\
&\quad + \text{link } p_{x_100} + \text{link } p_{0x_20} + \text{link } p_{00z} - \text{link } p_{000}.
\end{aligned}$$

It is easily verified that these parameters satisfy (17.2) and that their sum is equal to link $p_{x_1x_2z}$. Note further that the normalization (17.2) makes the parameters uniquely identifiable, but does not impose a constraint on the $p_{x_1x_2z}$.

By setting constraints on the λ-parameters, non-saturated GLM's can be formulated. McCullagh and Nelder (1983) discuss methods for the estimation and goodness-of-fit testing of GLM's. If the link function is the natural logarithm, we have the subclass of loglinear models (LLM's). For this case, Goodman (1978), Bishop, Fienberg, and Holland (1975), and others, have described methods of estimation and testing.

Bishop et al. (1975, pp. 300–301) describe a non-saturated LLM where the variables are interchangeable (Mandanski, 1963). Bishop et al. refer to them as *symmetry models.* For three variables, the model of *complete symmetry* specifies the probabilities of (x_1, x_2, x_3) as,

$$p_{x_1x_2x_3} = p_{x_1x_3x_2} = p_{x_2x_1x_3} = p_{x_2x_3x_1} = p_{x_3x_1x_2} = p_{x_3x_2x_1},$$

for all $x_1, x_2, x_3 = 0, \ldots, m$. Mandanski refers to this model as the *interchangeability* hypothesis. It can be represented as a loglinear model,

$$\ln p_{x_1x_2x_3} = \lambda + \lambda_{x_1}^{X_1} + \lambda_{x_2}^{X_2} + \lambda_{x_3}^{X_3} + \lambda_{x_1x_2}^{X_1X_2} + \lambda_{x_1x_3}^{X_1X_3} + \lambda_{x_2x_3}^{X_2X_3} + \lambda_{x_1x_2x_3}^{X_1X_2X_3}, \quad (17.3)$$

with constraints

$$\lambda_a^{X_1} = \lambda_a^{X_2} = \lambda_a^{X_3}, \quad (17.4)$$

$$\lambda_{ab}^{X_1X_2} = \lambda_{ba}^{X_1X_2} = \lambda_{ab}^{X_1X_3} = \lambda_{ba}^{X_1X_3} = \lambda_{ab}^{X_2X_3} = \lambda_{ba}^{X_2X_3}, \quad (17.5)$$

$$\lambda_{abc}^{X_1X_2X_3} = \lambda_{acb}^{X_1X_2X_3} = \lambda_{bac}^{X_1X_2X_3} = \lambda_{bca}^{X_1X_2X_3} = \lambda_{cab}^{X_1X_2X_3} = \lambda_{cba}^{X_1X_2X_3}, \quad (17.6)$$

for all $a, b, c = 0, \ldots, m$. Note that the constraints (17.4) through (17.6) specify both equalities between the different model terms (e.g., $\lambda_{ab}^{X_1X_2} = \lambda_{ab}^{X_2X_3}$) and the symmetry of each of the model terms (e.g., $\lambda_{ab}^{X_1X_2} = \lambda_{ba}^{X_1X_2}$). A slightly more convenient way to express the constraints (17.5) and (17.6) is to vary the order of the superscripts rather than the subscripts,

$$\lambda_{ab}^{X_1X_2} = \lambda_{ab}^{X_2X_1} = \lambda_{ab}^{X_1X_3} = \lambda_{ab}^{X_3X_1} = \lambda_{ab}^{X_2X_3} = \lambda_{ab}^{X_3X_2}, \quad (17.7)$$

$$\lambda_{abc}^{X_1X_2X_3} = \lambda_{abc}^{X_1X_3X_2} = \lambda_{abc}^{X_2X_1X_3} = \lambda_{abc}^{X_2X_3X_1} = \lambda_{abc}^{X_3X_1X_2} = \lambda_{abc}^{X_3X_2X_1}. \quad (17.8)$$

Ten Have and Becker (1991) have generalized this to a model with several sets of variables, where both the interaction effects between variables of two sets and the main and interaction effects within each set are equal and symmetrical. For our purpose it suffices to consider models with complete symmetry within one set of X–variables extended with one Z–variable:

$$p_{x_1x_2\ldots x_kz} = p_{x_{(1)}x_{(2)}\ldots x_{(k)}z}, \quad (17.9)$$

for all permutations $(x_{(1)}, x_{(2)}, \ldots, x_{(k)})$ of elements of the vector $(x_1, x_2, \ldots, x_k)$. We will refer to this model as the *complete measurement interchangeability* (CMI) model. In the following theorem we relate the GLM (17.1) to the CMI model (17.9).

Theorem 17.1 *A necessary and sufficient condition for complete measurement interchangeability (17.9) is that the main and interaction effects in the GLM (17.1) have the constraints*

$$\lambda_{x_i}^{X_i} = \lambda_{x_i}^{X_{(i)}}, \tag{17.10}$$

$$\lambda_{x_i z}^{X_i Z} = \lambda_{x_i z}^{X_{(i)} Z}, \tag{17.11}$$

$$\lambda_{x_i x_j}^{X_i X_j} = \lambda_{x_i x_j}^{X_{(i)} X_{(j)}}, \tag{17.12}$$

$$\lambda_{x_i x_j z}^{X_i X_j Z} = \lambda_{x_i x_j z}^{X_{(i)} X_{(j)} Z}, \tag{17.13}$$

$$\vdots$$

$$\lambda_{x_1 \ldots x_k z}^{X_1 \ldots X_k Z} = \lambda_{x_1 \ldots x_k z}^{X_{(1)} \ldots X_{(k)} Z}, \tag{17.14}$$

for all permutations $(X_{(1)}, X_{(2)}, \ldots, X_{(k)})$ *of* $(X_1, X_2, \ldots, X_k)$.

Proof The sufficiency of the constraints (17.10) through (17.14) on the parameters of (17.1) for CMI (17.9), is obvious. To prove the necessity note that, from the GLM (17.1) and the bijectivity of the link function, we have

$$\begin{aligned} &\sum_{i=1}^{k} \lambda_{x_i}^{X_i} + \sum_{i<j} \lambda_{x_i x_j}^{X_i X_j} + \ldots + \lambda_{x_1 x_2 \ldots x_k}^{X_1 X_2 \ldots X_k} + \\ &\sum_{i=1}^{k} \lambda_{x_i z}^{X_i Z} + \sum_{i<j} \lambda_{x_i x_j z}^{X_i X_j Z} + \ldots + \lambda_{x_1 x_2 \ldots x_k z}^{X_1 X_2 \ldots X_k Z} = \\ = &\sum_{i=1}^{k} \lambda_{x_i}^{X_{(i)}} + \sum_{i<j} \lambda_{x_i x_j}^{X_{(i)} X_{(j)}} + \ldots + \lambda_{x_1 x_2 \ldots x_k}^{X_{(1)} X_{(2)} \ldots X_{(k)}} + \\ &\sum_{i=1}^{k} \lambda_{x_i z}^{X_{(i)} Z} + \sum_{i<j} \lambda_{x_i x_j z}^{X_{(i)} X_{(j)} Z} + \ldots + \lambda_{x_1 x_2 \ldots x_k z}^{X_{(1)} X_{(2)} \ldots X_{(k)} Z}. \end{aligned} \tag{17.15}$$

with constraints (17.2). To prove that equations (17.10) through (17.14) follow from (17.15), let $\boldsymbol{x}^{(s)} = (x_1, \ldots, x_s, 0, \ldots, 0)$ be a vector of responses, where the first s responses are non-zero, and the last $k - s$ responses are zero. Furthermore, denote a permutation of elements of $\boldsymbol{x}^{(s)}$ by $\boldsymbol{x}_{(.)}^{(s)} = (x_{(1)}^{(s)}, x_{(2)}^{(s)}, \ldots, x_{(k)}^{(s)})$. First consider the case $s = 1$.

For $z = 0$ and $s = 1$, we obtain from (17.15) and constraints (17.2),

$$\lambda_{x_1}^{X_1} = \lambda_{x_1}^{X_{(1)}}, \tag{17.16}$$

so equality (17.10) holds. For $z \neq 0$ and $s = 1$, we have from (17.15) and (17.2),

$$\lambda_{x_1}^{X_1} + \lambda_{x_1 z}^{X_1 Z} = \lambda_{x_1}^{X_{(1)}} + \lambda_{x_1 z}^{X_{(1)} Z}, \tag{17.17}$$

so that (17.16) and (17.17) imply

$$\lambda_{x_1 z}^{X_1 Z} = \lambda_{x_1 z}^{X_{(1)} Z}.$$

Similarly, solving (17.15) with constraints (17.2), for $s = 1, 2$, yields the equations (17.10) through (17.12), and solving for $s = 1, \ldots, k$ yields (17.10) through (17.14). □

Note that Theorem 17.1 is valid for any bijective link function, so we may choose link = ln to formulate the CMI model.

If a set if items (or tests) satisfies the CMI property for a certain choice of Z, all items will have the same statistical relationship to Z and to each other. In addition to this, all items will have the same marginal distribution of their responses, due to the complete symmetry (17.9) of the joint distribution. The latter may be too strict for our purpose, because our requirement is that the X–variables be interchangeable only in their interactions. In the next section, we relax the CMI model in such a way that the marginal item-score distributions may be different.

17.4 Loglinear Models for Quasi-Interchangeability of Measurement Instruments

Bishop, Fienberg, and Holland (1975, p. 303) describe a loglinear symmetry model that "preserves the one-dimensional margins". They refer to it as the model of *quasi-symmetry.* This quasi-symmetry model can be obtained by dropping constraints (17.4) on the main effect parameters of the complete symmetry LLM model. Ten Have and Becker (1991) have generalized this to a model with several sets of variables, where the interaction effects between variables of two sets are equal and symmetrical and where the interaction effects within each set are equal and symmetrical. They call this 'class-quasi exchangeability'. To obtain a *quasi measurement interchangeability* (QMI) model, we may drop the restriction (17.10) that the main effect parameters in the GLM (17.1) be equal to each other, but retain the restrictions (17.11) through (17.14) on the interaction parameters.

This model modifies the quasi-symmetry model by Bishop et al. by adding the Z variable and replacing the log-link function by a general link function. However, unlike in the CMI model, the choice of the link function in this GLM will make a difference. For example, link = identity states a different hypotheses about the structure of the data than link = log. So, to complete the specification of the QMI model, we have to make a decision about the link function which is in the spirit of measurement interchangeability.

A reasonable assumption would be that the differing main effect parameters of the X-variables influence the marginal distributions of X_i, $i = 1, \ldots, k$, but not the conditional distribution of Z, given x. That is, the conditional distribution of Z, given x, only depends on the main effect parameter for Z and the interchangeable parameters involving X and Z. Thus, with respect to the conditional distribution, the measurements X are interchangeable.

In the next theorem we prove that under this assumption the link function must be logarithmic as in the quasi-symmetry model and in the class-quasi exchangeability model. For simplicity, first rewrite the GLM (17.1) as

$$\text{link } p_{\boldsymbol{x}z} = \Psi_{\boldsymbol{x}} + \Psi_{\boldsymbol{x}z}, \tag{17.18}$$

where

$$\Psi_{\boldsymbol{x}} = \lambda + \sum_{i=1}^{k} \lambda_{x_i} + \sum_{i<j} \lambda_{x_i x_j} + \ldots + \lambda_{x_1 x_2 \ldots x_k}$$

and

$$\Psi_{\boldsymbol{x}z} = \lambda_z^Z + \sum_{i=1}^{k} \lambda_{x_i z} + \sum_{i<j} \lambda_{x_i x_j z} + \ldots + \lambda_{x_1 x_2 \ldots x_k z}.$$

Theorem 17.2 *The following two propositions about the GLM (17.1) and (17.18) are equivalent:*

(a) link $=$ ln, *and*

(b) $p_{z|\boldsymbol{x}} = f(\Psi_{\boldsymbol{x}z})$.

Proof First we prove that (a) implies (b). From (17.18), with link $=$ log,

$$\ln p_{\boldsymbol{x}z} = \Psi_{\boldsymbol{x}} + \Psi_{\boldsymbol{x}z},$$

we have,

$$p_{z|\boldsymbol{x}} = \frac{p_{\boldsymbol{x}z}}{p_{\boldsymbol{x}}} = \frac{\exp(\Psi_{\boldsymbol{x}} + \Psi_{\boldsymbol{x}z})}{\sum_y \exp(\Psi_{\boldsymbol{x}} + \Psi_{\boldsymbol{x}y}^{XZ})} = \frac{\exp(\Psi_{\boldsymbol{x}z})}{\sum_y \exp(\Psi_{\boldsymbol{x}y}^{XZ})} = f(\Psi_{\boldsymbol{x}z}),$$

which is (b).

To prove that (b) implies (a), first note that the link (link: $\mathbb{R} \to]0,1[$) is bijective, so that it has an inverse g ($g :]0,1[\to \mathbb{R}$). Secondly, let $z' \neq z$, then from (b) and (17.18) it follows that

$$\frac{p_{z|\boldsymbol{x}}}{p_{z'|\boldsymbol{x}}} = \frac{p_{z\boldsymbol{x}}}{p_{z'\boldsymbol{x}}} = \frac{g(\Psi_{\boldsymbol{x}} + \Psi_{\boldsymbol{x}z})}{g(\Psi_{\boldsymbol{x}} + \Psi_{\boldsymbol{x}z'})} = \frac{f(\Psi_{\boldsymbol{x}z})}{f(\Psi_{\boldsymbol{x}z'})}. \tag{17.19}$$

For $\Psi_{\boldsymbol{x}} = 0$, we have,

$$\frac{g(\Psi_{\boldsymbol{x}z})}{g(\Psi_{\boldsymbol{x}z'})} = \frac{f(\Psi_{\boldsymbol{x}z})}{f(\Psi_{\boldsymbol{x}z'})}, \tag{17.20}$$

and from (17.19) and (17.20),

$$\frac{g(\Psi_{\boldsymbol{x}} + \Psi_{\boldsymbol{x}z})}{g(\Psi_{\boldsymbol{x}} + \Psi_{\boldsymbol{x}z'})} = \frac{g(\Psi_{\boldsymbol{x}z})}{g(\Psi_{\boldsymbol{x}z'})}.$$

Moreover, for $\Psi_{\boldsymbol{x}z'} = 0$, this becomes

$$\frac{g(\Psi_{\boldsymbol{x}} + \Psi_{\boldsymbol{x}z})}{g(\Psi_{\boldsymbol{x}})} = \frac{g(\Psi_{\boldsymbol{x}z})}{g(0)}.$$

With $\tilde{g}(.) = \frac{g(.)}{g(0)}$, this can be written as

$$\tilde{g}(\Psi_{\boldsymbol{x}} + \Psi_{\boldsymbol{x}z}) = \tilde{g}(\Psi_{\boldsymbol{x}})\tilde{g}(\Psi_{\boldsymbol{x}z}),$$

which is a well-known functional equation (Aczél, 1966, p. 38). Its general solution under continuity and monotonicity assumptions is $\tilde{g}(u) = \exp(au)$, with $a > 0$. From this and the definition of $\tilde{g}$ we obtain $g(u) = \exp[\ln g(0) + au]$, so that (17.18) becomes a loglinear model where the parameters are a linear transformation of the original parameters. To obey the constraints (17.2), we transform the general mean parameter to $\ln g(0) + a\lambda$ and multiply the remaining parameters by a. □

(Remark: the proof of this theorem is analogous to that of Pfanzagl (1994, 257-258) regarding the logistic form of the dichotomous RM.)

Because condition (b) of Theorem (17.2) is a desirable property, in what follows we assume that the QMI model has link = ln. In the next section we give a somewhat simpler formulation of the QMI model.

17.5 Sum Score Models and Quasi-Interchangeability of Measurement Instruments

An alternative representation of the QMI requirement is the sum score model (Kelderman, 1984, 1989, 1992). To formulate the sum score model, let B_{hix_i} be the h-th category coefficient of response x_i of item I_i. In the general multidimensional latent trait model for polytomous items (Kelderman, 1989, 1992), this coefficient can take any non-negative integer value but, for now, let $B_{hix_i} = 1$, for $h = 1, \ldots, m$, if $x_i = h$, and $B_{hix_i} = 0$ otherwise. Thus, B_{hix_i} indicates each of the non-zero categories. Furthermore, let $t_h \equiv B_{h1x_1} + \ldots + B_{hkx_k}$ be the frequency with which category h occurs in response vector $(x_1, \ldots, x_k)$, and let $\boldsymbol{t} \equiv (t_1, \ldots, t_h, \ldots, t_m)$ be the vector of these frequencies. Note that t_0 can be obtained from $\boldsymbol{t}$ by the formula $t_0 = k - \sum_{h=1}^{m} t_h$, so that no B_{hi0} need be defined for this case. The following theorem relates the GLM (17.1) to the sum score GLM.

Theorem 17.3 *The GLM (17.1) with constraints (17.11) through (17.14) is equivalent to the sum score GLM,*

$$\text{link } p_{\boldsymbol{x}z} = \lambda + \lambda_z + \sum_{i=1}^{k} \lambda_{x_i} + \lambda_{\boldsymbol{t}} + \lambda_{\boldsymbol{t}z}, \tag{17.21}$$

where $\boldsymbol{t}$ is defined as above, with constraints $\lambda_0^{T} = \lambda_{0z}^{TZ} = \lambda_{t0}^{TZ} = 0$.

Proof To prove that the GLM (17.1) with constraints (17.11) through (17.14) imposed on the interaction parameters implies the sum score GLM (17.21), rewrite the GLM (17.1) as

$$\text{link } p_{\boldsymbol{x}z} = \lambda + \lambda_z + \sum_{i=1}^{k} \lambda_{x_i} + \Gamma_{x_1,x_2,\ldots x_k z}, \tag{17.22}$$

where the last term replaces the corresponding sum of terms. Note that, under constraints (17.11) through (17.14), the Γ-term is invariant for an arbitrary permutation of X-variables, because it is a sum of interaction terms which are themselves invariant for arbitrary permutations of the X-variables. So,

$$\Gamma_{x_1 x_2 \ldots x_k z} = \Gamma_{x_{(1)} x_{(2)} \ldots x_{(k)} z}. \tag{17.23}$$

Moreover, because the addition operation is symmetrical in its arguments, we have the same sum score vector $\boldsymbol{t}$ for different permutations of $(x_1, x_2, \ldots, x_k)$. Let $\boldsymbol{x}'$ be the response vector $\boldsymbol{x}$ sorted in increasing order. Obviously, the sum score vector $\boldsymbol{t}$ uniquely determines the ordered response vector:

$$\mathbf{x}' = (\underbrace{0, \ldots, 0}_{k - \sum_{h=1}^{m} t_h \text{ times}}, \underbrace{1, \ldots, 1}_{t_1 \text{ times}}, \underbrace{2, \ldots, 2}_{t_2 \text{ times}}, \ldots, \underbrace{m, \ldots, m}_{t_m \text{ times}}).$$

Thus, we can write,

$$\Lambda_{\boldsymbol{t}z} \equiv \Gamma_{0,\ldots,0,1,\ldots,1,2,\ldots 2,\ldots,m,\ldots,m,z} = \Gamma_{x_{(1)},x_{(2)},\ldots x_{(k)},z}, \tag{17.24}$$

for all permutations of $\boldsymbol{x}$. Furthermore, reparameterize Λ as

$$\begin{aligned} \lambda_{\boldsymbol{t}} &= \Lambda_{\boldsymbol{t}0}^{TZ}, \\ \lambda_{\boldsymbol{t}z} &= \Lambda_{\boldsymbol{t}z} - \Lambda_{\boldsymbol{t}0}^{TZ}, \end{aligned} \tag{17.25}$$

so that the sum score GLM (17.21) follows from (17.22), (17.24), and (17.25). To prove that the sum score GLM (17.21) implies the GLM with constraints (17.11) through (17.14)), note that (17.25) and the equivalence in (17.24) imply (17.23). Moreover, as in Theorem 17.1, the symmetry of $\Gamma_{x_1,x_2,\ldots x_k z}$ in z and constraints (17.2) of the λ-parameters imply that the λ-parameters satisfy constraints (17.11) through (17.14) in the GLM (17.1). Thus, the sum score GLM (17.21) is equivalent to the GLM with constraints (17.11) through (17.14)) with an unspecified link function. □

(Remark: For the loglinear model this theorem was proved by Agresti, 1993.)

Choosing link=log in the sum score GLM (17.21), we obtain the sum score LLM, which is equal to the QMI model. In the next section it is shown that the sum score LLM, and therefore the QMI model, is equivalent to the conditional RM.

17.6 Quasi-Interchangeability and the Conditional RM

Let θ_{vh} be person S_v's value on latent trait $h = 1, \ldots, r$, and let β_{ix_i} be a parameter describing the difficulty of response x_i to item I_i with identifying constraint $\beta_{i0} = 0$. Furthermore, let B_{hix_i}, for $h = 1, \ldots, r$; $i = 1, \ldots, k$; $x_i = 0, \ldots, m$ be discrete nonnegative category weights describing the degrees of dependence of response x_i to item I_i on the latent trait h. The multidimensional polytomous RM (see Agresti, 1993; Kelderman & Rijkes, 1994;) may then be written as

$$p_{x_i|v} = c_{iv}^{-1} \exp\left(\sum_{h=1}^{r} B_{hix_i}\theta_{vh} - \beta_{ix_i}\right), \tag{17.26}$$

where c_{iv} is the proportionality constant chosen such that $\sum_{x_i} p_{x_i|v} = 1$. Instances of (17.26) have been described by Andersen (1973), Duncan (1984), Duncan & Stenbeck (1987), Kelderman (in press), Kelderman & Rijkes (1994), Wilson and Adams (1993), Stegelmann (1983), Wilson and Masters (1993). For example, the partial credit model (Masters, 1982) is obtained by setting $B_{hia} = a$, for $h = 1$; $i = 1, \ldots, k$; $a = 0, \ldots, m$. Andersen's (1973a; see also Chapter 15) model is obtained as above by setting $r = m$, $B_{hix_i} = 1$, for $h = 1, \ldots, m$, if $x_i = h$ and $B_{hix_i} = 0$. Note that in (17.26) no latent trait θ_{vh}, for $h = 0$, is defined pertaining to the response $x_i = 0$, whereas Andersen (Chapter 15), Fischer (Chapter 16), and Glas and Verhelst (Chapter 18) do define this latent trait, but normalize it to zero, $\theta_{v0} \equiv 0$.

In this chapter, we only need a submodel of (17.26) where the number of latent traits is equal to the number of non-zero responses, i.e., $r = m$. Assuming that the responses are independent within each individual S_v, the probability of the joint response $\boldsymbol{x}$ is

$$p_{\boldsymbol{x}|v} = \prod_{i=1}^{k} p_{x_i|v} = c_v^{-1} \exp\left(\sum_{h=1}^{m} t_h\theta_{vh}\right) \exp\left(-\sum_{i=1}^{k} \beta_{ix_i}\right), \tag{17.27}$$

where t_h is defined as before, and c_v is the proportionality constant. If $\sum_{\boldsymbol{x}|\boldsymbol{t}}$ denotes summation over all response vectors $\boldsymbol{x}$ with sum score vector $\boldsymbol{t}$, the probability of this sum score vector can be written as,

$$p_{\boldsymbol{t}|v} = c_v^{-1}\gamma_{\boldsymbol{t}}(\boldsymbol{\beta}) \exp\left(\sum_{h=1}^{m} t_h\theta_{vh}\right), \tag{17.28}$$

where $\gamma_{\boldsymbol{t}}(\boldsymbol{\beta}) = \sum_{\boldsymbol{x}|\boldsymbol{t}} \exp\left(-\sum_{i=1}^{k}\beta_{ix_i}\right)$ is a generalization of the well-known elementary symmetric function with $\boldsymbol{\beta} = (\beta_{11}, \beta_{12}, \ldots, \beta_{1m}, \ldots, \beta_{k1}, \beta_{k2}, \ldots, \beta_{km})$.

Dividing (17.27) by (17.28) gives

$$p_{\boldsymbol{x}|\boldsymbol{t}v} = \frac{p_{\boldsymbol{x}\boldsymbol{t}|v}}{p_{\boldsymbol{t}|v}} = \frac{p_{\boldsymbol{x}|v}}{p_{\boldsymbol{t}|v}} = \frac{\exp\,(-\sum_{i=1}^{k}\beta_{ix_i})}{\gamma_{\boldsymbol{t}}(\boldsymbol{\beta})}, \tag{17.29}$$

where the second equality follows from the complete dependence of $\boldsymbol{t}$ on $\boldsymbol{x}$.

Although this model is defined for each individual, it no longer contains the individual parameters θ_{vh}. This implies that the same model holds also for the probability of a response $\boldsymbol{x}$ of a randomly selected person with sum score $\boldsymbol{t}$,

$$p_{\boldsymbol{x}|\boldsymbol{t}} = \frac{\exp\,(-\sum_{i=1}^{k}\beta_{ix_i})}{\gamma_{\boldsymbol{t}}(\boldsymbol{\beta})}. \tag{17.30}$$

Note, however that (17.30) does not necessarily imply (17.29).

This conditional RM is related to the QMI model through the following theorem.

Theorem 17.4 *The conditional RM and the sum score LLM are equivalent.*

Proof We first prove that the sum score LLM implies the conditional RM. From the sum score GLM (17.21), with link = ln, we obtain the joint probability of a sum score vector $\boldsymbol{t}$ and z as

$$\begin{aligned} p_{\boldsymbol{t}z} = \sum_{\boldsymbol{x}|\boldsymbol{t}} p_{\boldsymbol{x}z} &= \exp\,(\lambda + \lambda_z + \lambda_{\boldsymbol{t}z}) \sum_{\boldsymbol{x}|\boldsymbol{t}} \exp\,(\sum_{i=1}^{k}\lambda_{x_i}) \\ &= \exp\,(\lambda + \lambda_z + \lambda_{\boldsymbol{t}z})\gamma_{\boldsymbol{t}}(-\boldsymbol{\lambda}^X), \end{aligned} \tag{17.31}$$

so that the conditional probability of a response vector $\boldsymbol{x}$, given $\boldsymbol{t}$ and z, becomes

$$p_{\boldsymbol{x}|\boldsymbol{t}z} = \frac{p_{\boldsymbol{x}z}}{p_{\boldsymbol{t}z}} = \frac{\exp\,(\sum_{i=1}^{k}\lambda_{x_i})}{\gamma_{\boldsymbol{t}}(-\boldsymbol{\lambda}^X)}, \tag{17.32}$$

with $\boldsymbol{\lambda}^{\boldsymbol{X}} = (\lambda_1^{X_1}, \lambda_2^{X_1}, \ldots, \lambda_r^{X_1}, \ldots, \lambda_1^{X_k}, \lambda_2^{X_k}, \ldots, \lambda_r^{X_k})$, where the second equality follows from substitution of (17.21) with link = ln and (17.31). The right-hand side of (17.32) is clearly equivalent to the conditional RM with $\lambda_{x_i} = -\beta_{ix_i}$. The only difference is that the left-hand side of (17.32) involves z, whereas (17.29) involves v. If we assume that the variable z describes individual S_v completely, (17.32) is equivalent to the individual model (17.29); if z is constant or absent, (17.32) is equivalent to (17.30).

To prove that the conditional RM implies the sum score model, we notice from (17.29) and elementary probability calculus that

$$p_{\boldsymbol{x}z} = p_{\boldsymbol{x}tz} = p_{\boldsymbol{x}|tz}p_{tz} = \exp\left(\ln p_{tz} - \ln \gamma_t(\boldsymbol{\beta}) - \sum_{i=1}^{k} \beta_{ix_i}\right). \tag{17.33}$$

Let $\Lambda_{tz} = \ln p_{tz} - \ln(\gamma_t(-\boldsymbol{\lambda}^X))$, and reparameterize this in the usual hierarchical way

$$\Lambda_{tz} = \lambda + \lambda_t + \lambda_z + \lambda_{tz},$$

where,

$$\begin{aligned}
\lambda &= \Lambda_{00}^{TZ} \\
\lambda_t^T &= \Lambda_{t0}^{TZ} - \Lambda_{00}^{TZ} \\
\lambda_z^Z &= \Lambda_{0z}^{TZ} - \Lambda_{00}^{TZ} \\
\lambda_{tz}^{TZ} &= \Lambda_{tz}^{TZ} - \Lambda_{t0}^{TZ} - \Lambda_{0z}^{TZ} + \Lambda_{00}^{TZ}.
\end{aligned}$$

Inserting this in (17.33), replacing $-\beta_{ix_i}$ by λ_{x_i}, yields the sum score LLM. □

From Theorems 17.2, 17.3, and 17.4 it follows that *the requirement of QMI is equivalent to the requirement that the data follow the conditional RM.*

For the case of the dichotomous RM ($m = 1$) with z constant, several authors have noticed the equivalence of the sum score LLM and the conditional RM.

Tjur (1982) and Kelderman (1984) assume the sum score $\boldsymbol{t}$ either as fixed, or as having an unrestricted multinomial distribution with parameters $\{p_t\}$. Tjur (1982) calls the latter an 'extended random model' and shows that it is equivalent to the sum score LLM. Obviously, the sum score distribution in the extended random model is less restrictive than the distribution in (17.28). In the terminology of loglinear modeling (Bishop et al., 1975, Sec. 5.4; Haberman; 1979, Sec. 7.3), the sum score LLM can be viewed as a quasi-loglinear model for the incomplete Item 1 × Item 2 × ... × Item k × sum score × background-variable contingency table. The table has structural zero cells if the sum score is not consistent with the item responses (Kelderman, 1984).

If the sum score is considered fixed, the sum score LLM describes product-multinomial sampling of $\boldsymbol{X}$ given $\boldsymbol{t}$, whereas if the sum score is assumed to have an unrestricted multinomial distribution, the sum-score LLM describes a multinomial distribution of the response vector $\boldsymbol{X}$. Bishop et al. (1975, Sec. 3.3) show that, for both models, the kernel of the likelihood is identical. As a result, both models have the same set of likelihood equations for estimating the parameters and the same statistics for testing the goodness-of-fit. Both models may, therefore, be used interchangeably.

Cressie and Holland (1982) derive the sum score LLM from the RM along different lines. They integrate (17.27) over the distribution of the person parameters, ignoring moment constraints implied by this distribution. They call it the 'extended RM'. The moment constraints imply complicated inequality constraints on the sum score parameters. These constraints can be checked (Cressie & Holland, 1982; Hout, Duncan, & Sobel, 1987; Kelderman, 1984; Lindsay, Clogg, & Grego, 1991), but estimation under them is difficult. To estimate and test quasi-loglinear models such as the sum score LLM, standard methods are available, see Bishop et al. (1975, Chapter 5) or Haberman (1979, Chapter 10) for a complete account. The maximum likelihood equations of loglinear models can be solved by iterative methods, such as iterative proportional fitting (IPF) or Newton-Raphson. Kelderman (1992) describes an algorithm especially constructed for the analysis of sum score LLM's, which is implemented in the LOGIMO (LOGlinear Irt MOdeling) program (Kelderman & Steen, 1988). LOGIMO is a Pascal program running on a VAX/VMS or a 386 PC/MS-DOS system. It can be obtained from iec ProGAMMA, Box 841, 9700 AV Groningen, The Netherlands. In practice, testing is often done with respect to some group membership z (Andersen, 1971b), such as gender or age, or both, but many other theoretically relevant variables may be used. Item bias with respect with these variables may also be studied (Kelderman, 1989).

17.7 Quasi-Interchangeability and the RM

The RM (17.26) assumes that each measurement is fully explained by unobserved latent variables characterizing the individual. The QMI model differs from the RM in that it does not involve latent variables, does not assume conditional independence, and does not necessarily describe the individual. The following theorem relates both measurement models to each other.

Theorem 17.5 *If z fully describes the individual S_v, and if the measurements $\boldsymbol{x}$ are conditionally independent given z, the QMI model is equivalent to the RM (17.26) where the responses X_i are independent given θ_{vh}.*

Proof If the responses are conditionally independent given z, all terms in the QMI model (17.12) through (17.14) that describe interactions between item responses must be equal to zero. The model then becomes,

$$p_{\boldsymbol{x}z} = \exp\,(\lambda + \lambda_z^Z + \sum_{i=1}^{k} \lambda_{x_i}^{X_i} + \sum_{i=1}^{k} \lambda_{x_i z}^{X_1 Z}),$$

where, because of the interchangeability constraint (17.11), $\lambda_{x_i z}^{X_1 Z} = \lambda_{x_i z}^{X_i Z}$ for all $i = 1, \ldots, k$; $x_i = 0, \ldots, m$.

For one item, say Item I_1, the model becomes

$$p_{x_1 z} = \exp\,(\lambda + \lambda_z^Z + \lambda_{x_1}^{X_1} + \lambda_{x_1 z}^{X_1 Z}) \sum_{x_2=0}^{m} \cdots \sum_{x_k=0}^{m} \exp\,(\sum_{j=2}^{k} \lambda_{x_j}^{X_j} + \sum_{j=2}^{k} \lambda_{x_j z}^{X_1 Z}).$$

The conditional probability of x_1 given z then is

$$p_{x_1|z} = \frac{p_{x_1 z}}{\sum_{y=0}^{m} p_{yz}^{X_1 Z}} = \frac{\exp(\lambda_{x_1}^{X_1} + \lambda_{x_1 z}^{X_1 Z})}{\sum_{y=0}^{m} \exp(\lambda_{y}^{X_1} + \lambda_{yz}^{X_1 Z})}. \tag{17.34}$$

If z completely characterizes individual S_v, (17.34) is equivalent to the RM (17.26), with $\theta_{hv} = \lambda_{x_1 z}$, $z = v$, $m = r$, for $h = 1, \ldots, r$, $x_1 = 0, \ldots, m$, $h = x_1$, and $\beta_{1x_1} = -\lambda_{x_1}$.

Conversely, because (17.26) is equivalent to (17.34) under the assumption of conditional (local) independence, we obtain from (17.26) the joint probability of $\boldsymbol{x}$ and z as

$$\begin{aligned} p_{\boldsymbol{x}z} &= p_z \prod_{i=1}^{k} p_{x_i|z} \\ &= p_z \prod_{i=1}^{k} \frac{\exp(\lambda_{x_i}^{X_i} + \lambda_{x_i z}^{X_1 Z})}{\sum_{y=0}^{m} \exp(\lambda_{y}^{X_i} + \lambda_{yz}^{X_1 Z})} \\ &= \exp(\lambda + \lambda_z^Z + \sum_{i=1}^{k} \lambda_{x_i}^{X_i} + \sum_{i=1}^{k} \lambda_{x_i z}^{X_1 Z}), \end{aligned}$$

where

$$\lambda = \ln p_1^Z - \ln \prod_{i=1}^{k} \sum_{x_i=0}^{m} \exp(\lambda_{x_i}^{X_i} + \lambda_{x_i 1}^{X_1 Z})$$

and

$$\lambda_z^Z = \ln p_z - \ln \prod_{i=1}^{k} \sum_{x_i=0}^{m} \exp(\lambda_{x_i}^{X_i} + \lambda_{x_i z}^{X_1 Z}) - \lambda.$$

This completes the proof. □

Note that, in terms of the loglinear model parameters, the latent trait parameters of the RM can be interpreted as the interactions of the item categories with the individual. Interchangeability implies that these interactions are the same for all items.

17.8 Conclusion

It is shown that the conditional RM can be derived from the requirement that items are interchangeable in their relations to other variables and to each other when modeled within a loglinear model. For the dichotomous case, the model is equivalent to Cressie and Holland's (1983, p. 136) extended RM, of which they correctly remark that "The resulting set of probabilities ... are not, in general, true manifest probabilities for any latent trait model. However, this model can still be of practical use ...". We have shown that the model is still interesting in its own right when seen in the light of interchangeability. Also, it becomes quite clear what the difference is between the conditional (extended) RM and the original RM. The latter requires the possibility of statistical independence of the items given some (unobserved) variable, the former does not. This assumption is often disputed in item response theory (see, e.g., Jannarone, 1986; Holland, 1981). The present chapter shows that this assumption is not necessary in the QMI model or, equivalently, in the conditional RM.

18

Tests of Fit for Polytomous Rasch Models

Cees A. W. Glas and Norman D. Verhelst[1]

ABSTRACT In this chapter, a number of the tests of model fit for the Rasch model for dichotomous items presented in Chapter 5 are generalized to a class of IRT models for polytomous items. Again, the problem of evaluating model fit is solved in the framework of the general multinomial model, and it is shown that the four types of tests considered in Chapter 5 – generalized Pearson tests, likelihood ratio tests, Wald tests, and Lagrange multiplier tests – can all be adapted to the framework of polytomous items. Apart from providing global measures of overall model fit, test statistics must also provide information with respect to specific model violations. The last section of this chapter gives an example of a testing procedure focussing on the model violation of item bias or differential item functioning.

18.1 Introduction

In Chapter 5, an overview of the various approaches to testing the fit of the Rasch model for dichotomous items was presented. A taxonomy of tests was introduced based on three aspects: the specific assumptions and properties of the model tested, the type of statistic on which the test is based, and the mathematical sophistication of the procedure, particularly, the extent to which the (asymptotic) distribution of the statistic is known. This taxonomy can also be applied to the case of polytomous items. However, for several reasons the actual setup of this chapter significantly differs from that of Chapter 5. Firstly, the Rasch model for polytomous items has quite a few generalizations. Therefore, first a general expression encompassing most of these different models will be given. Another problem is the relative lack of testing procedures. For this reason, this chapter will contain more new than old material. Finally, the third aspect of the taxonomy, that is, the extent to which the (asymptotic) distribution of the statistic is known, has little relevance in the present case, because most existing procedures have firm statistical roots. This may be due to the circumstance that the greater complexity of the models for polytomous items thwarts simple approximations.

It will be shown that most of the mathematically well-founded statistics of Chapter 5 can be generalized to the case of polytomous items. The focus will mainly be on generalized Pearson statistics. The definitions of likelihood ratio statistics, Wald statistics, and Lagrange multiplier statistics given in Chapter 5

[1]National Institute for Educational Measurement, P.O.-Box 1034, 6801 MG Arnhem, The Netherlands; fax: (+31)85-521356

will hardly need adjustment. In the last section of this chapter, an example of a testing procedure focussing on item bias or DIF will be given. The statistics will apply to the framework of the OPLM for polytomous items, but the ideas on which the testing procedure is based apply to all models considered in this chapter.

18.2 A General Formulation of a Rasch Model for Polytomous Items

To code the response of a person to an item, let item I_i have $m_i + 1$ response categories C_h, indexed $h = 0, 1, ..., m_i$. The response to the item will be represented by an $(m_i + 1)$-dimensional vector $\boldsymbol{x}'_i = (x_{i0}, ..., x_{ih}, ..., x_{im_i})$, where x_{ih} is defined by

$$x_{ih} = \begin{cases} 1 & \text{if the response is in category } C_h,\ h = 0, .., m_i, \\ 0 & \text{otherwise.} \end{cases} \tag{18.1}$$

Consider a model where the probability of a response in category C_h, $h = 0, ..., m_i$, as a function of a vector of ability parameters $\boldsymbol{\theta}' = (\theta_1, ..., \theta_q, ..., \theta_Q)$ and a vector of the parameters of item I_i, $\boldsymbol{\beta}'_i = (\beta_{i1}, ..., \beta_{iu}, ..., \beta_{iU})$, is given by

$$\begin{aligned} P(X_{ih} = 1 | \boldsymbol{\theta}, \boldsymbol{\beta}_i) &= \frac{\exp(\sum_{q=1}^{Q} r_{ihq}\theta_q - \sum_{u=1}^{U} s_{ihu}\beta_{iu})}{\sum_{l=0}^{m_i} \exp(\sum_{q=1}^{Q} r_{ilq}\theta_q - \sum_{u=1}^{U} s_{ilu}\beta_{iu})} \\ &= \frac{\exp(\boldsymbol{r}'_{ih}\boldsymbol{\theta} - \boldsymbol{s}'_{ih}\boldsymbol{\beta}_i)}{\sum_{l=0}^{m_i} \exp(\boldsymbol{r}'_{il}\boldsymbol{\theta} - \boldsymbol{s}'_{il}\boldsymbol{\beta}_i)}, \end{aligned} \tag{18.2}$$

where $\boldsymbol{r}'_{ih} = (r_{ih1}, ..., r_{ihq}, ..., r_{ihQ})$ and $\boldsymbol{s}'_{ih} = (s_{ih1}, ..., s_{ihu}, ..., s_{ihU})$ are fixed 'score functions', which are part of the sufficient statistics. Notice that $\boldsymbol{r}_{ih}$ specifies the relation between the Q ability dimensions and response category C_h. In the same manner, $\boldsymbol{s}_{ih}$ defines the relation between the U parameters of item I_i and category C_h.

Introducing the matrices of score functions $\boldsymbol{R}_i = [\boldsymbol{r}_{i0}, ..., \boldsymbol{r}_{ih}, ..., \boldsymbol{r}_{im_i}]$ and $\boldsymbol{S}_i = [\boldsymbol{s}_{i0}, ..., \boldsymbol{s}_{ih}, ..., \boldsymbol{s}_{im_i}]$ and using (18.2), the probability of response $\boldsymbol{x}_i$ can be written as

$$P(\boldsymbol{x}_i | \boldsymbol{\theta}, \boldsymbol{\beta}_i) \propto \exp[\boldsymbol{x}'_i(\boldsymbol{R}'_i\boldsymbol{\theta} - \boldsymbol{S}'_i\boldsymbol{\beta}_i)]. \tag{18.3}$$

Let a test consist of k items. Using (18.3) and local stochastic independence, the probability of response pattern $\boldsymbol{x}$, $\boldsymbol{x}' = (\boldsymbol{x}'_1, ..., \boldsymbol{x}'_i, ..., \boldsymbol{x}'_k)$, given the ability and item parameters, can be written as

$$P(\boldsymbol{x} | \boldsymbol{\theta}, \boldsymbol{\beta}) \propto \exp[\boldsymbol{x}'(\boldsymbol{R}'\boldsymbol{\theta} - \boldsymbol{S}'\boldsymbol{\beta})], \tag{18.4}$$

where $\boldsymbol{R}$ is a matrix of score functions defined by $\boldsymbol{R} = [\boldsymbol{R}_1, ..., \boldsymbol{R}_i, ..., \boldsymbol{R}_k]$, and $\boldsymbol{S}$ a matrix of score functions $\boldsymbol{S} = [\boldsymbol{S}_1, ..., \boldsymbol{S}_i, ..., \boldsymbol{S}_k]$. A model of the universality of

(18.2), however, will generally not be identified. Therefore, in the specializations of (18.2) several restrictions need to be imposed; we shall return to this topic below.

One of the most desirable properties of the Rasch model (RM) for dichotomous items is that item parameters can be estimated using a conditional maximum likelihood (CML) procedure. This also holds for the class of models defined by (18.2). Loosely speaking, the property of parameter separation and the possibility of obtaining CML estimates of one set of parameters conditionally on minimal sufficient statistics for the other set, provides the justification for speaking of RMs for polytomous items. The feasibility of a CML procedure is shown as follows. Let $\{\boldsymbol{x}\}$ be the set of all possible response patterns. For all possible outcomes $\boldsymbol{x}$, a sufficient statistic $\boldsymbol{r}$ is defined by $\boldsymbol{r} = \boldsymbol{R}\boldsymbol{x}$, and for every possible $\boldsymbol{r}$, a set $\{\boldsymbol{x}|\boldsymbol{r} = \boldsymbol{R}\boldsymbol{x}\}$ is defined. Notice that $\bigcup_{\boldsymbol{r}}\{\boldsymbol{x}|\boldsymbol{r} = \boldsymbol{R}\boldsymbol{x}\} = \{\boldsymbol{x}\}$. In the sequel, it will become clear that the number of elements in the set of all possible values of $\boldsymbol{r}$, denoted by $\{\boldsymbol{r}\}$, must be substantially lower than the number of elements in $\{\boldsymbol{x}\}$. Therefore, the elements of $\boldsymbol{R}$ are usually integer-valued, although, strictly speaking, this is not necessary. The conditional probability of response pattern $\boldsymbol{x}$, given the associated value of $\boldsymbol{r}$, is given by

$$\begin{aligned} P(\boldsymbol{x}|\boldsymbol{r}, \boldsymbol{\beta}) &= \frac{\exp[\boldsymbol{x}'(\boldsymbol{R}'\boldsymbol{\theta} - \boldsymbol{S}'\boldsymbol{\beta})]}{\sum_{\{\boldsymbol{y}|\boldsymbol{R}\boldsymbol{y}=\boldsymbol{r}\}} \exp[\boldsymbol{y}'(\boldsymbol{R}'\boldsymbol{\theta} - \boldsymbol{S}'\boldsymbol{\beta})]} \\ &= \frac{\exp(-\boldsymbol{x}'\boldsymbol{S}'\boldsymbol{\beta})}{\sum_{\{\boldsymbol{y}|\boldsymbol{R}\boldsymbol{y}=\boldsymbol{r}\}} \exp(-\boldsymbol{y}'\boldsymbol{S}'\boldsymbol{\beta})} \\ &= \frac{\exp(-\boldsymbol{x}'\boldsymbol{S}'\boldsymbol{\beta})}{\gamma_{\boldsymbol{r}}}, \end{aligned} \tag{18.5}$$

where $\gamma_{\boldsymbol{r}}$ is a combinatorial function defined by

$$\gamma_{\boldsymbol{r}} = \sum_{\{\boldsymbol{y}|\boldsymbol{R}\boldsymbol{y}=\boldsymbol{r}\}} \exp(-\boldsymbol{y}'\boldsymbol{S}'\boldsymbol{\beta}). \tag{18.6}$$

The data can be viewed as counts $n_{\boldsymbol{x}}$, for all $\boldsymbol{x} \in \{\boldsymbol{x}\}$. The distribution of these counts is (product) multinomial, with probabilities defined by (18.5). Notice that these probabilities are functions of the item parameters only. Maximizing the likelihood function associated with this model produces the desired CML estimates.

In Table 18.1 a number of well known models for polytomous items are listed that are special cases of the general model (18.2). First, there is the unidimensional RM, which has been extensively studied by Rasch (1960), Andersen (1972, 1973c, 1977) and Fischer (1974). It is derived from the general model by setting Q and U equal to one, defining $r_{ih1} = h$ and $s_{ih1} = 1$, so that the parameters θ_q and β_{iu} of the general model are equated with the parameters θ and β_{ih} of the unidimensional RM, respectively. To identify the model, it is necessary to

TABLE 18.1. Overview of RMs for polytomous items

Model	$P(X_{ih} = x_{ih}\|\boldsymbol{\theta}, \boldsymbol{\beta}_i) \propto$
General	$\exp[x_{ih}(\sum_{q=1}^{Q} r_{ihq}\theta_q - \sum_{u=1}^{U} s_{ihu}\beta_{iu})]$
Unidimensional Rasch	$\exp[x_{ih}(h\theta - \beta_{ih})]$
Partial credit	$\exp[x_{ih}(h\theta - \sum_{u=0}^{h}\beta_{iu})]$
Rating scale	$\exp[x_{ih}(h\theta - \sum_{u=0}^{h}\beta_i + \alpha_u)]$
Binomial trials	$\exp[x_{ih}(h\theta - \sum_{u=0}^{h}\beta_i + \ln(u/(m_i - u + 1)))]$
OPLM	$\exp[x_{ih}(hr_i\theta - \beta_{ih})]$
Extended partial credit	$\exp[x_{ih}(r_{ih}\theta - \beta_{ih})]$
Multidimensional Rasch	$\exp[x_{ih}(\theta_h - \beta_{ih})]$
Extended multidimensional	$\exp[x_{ih}(r_{ih}\theta_h - \beta_{ih})]$
MPLT	$\exp[x_{ih}(\sum_{q=1}^{Q} r_{ihq}(\theta_q - \beta_{iq}))]$

set r_{i0q} and s_{i0u} equal to zero, that is, there are no item parameters for the zero category, and the latent ability continuum is unidimensional. The partial credit model by Masters (1982) is a reparameterization of the unidimensional RM and, consequently, it also fits the general model. The advantage of Masters' formulation is that the item parameters can be viewed as category bounds: β_{ih} is the point on the latent continuum where the probabilities of scoring in category j and $j-1$ are equal (see Masters, 1982).

The rating scale model (Rasch, 1961; Andrich, 1978a; Fischer & Parzer, 1991) is derived from the partial credit model by imposing linear restrictions, as is the binomial trials model (Andrich, 1978b). However, it can also be shown directly that the rating scale model fits the general model, without a reference to linear restrictions. Consider an item with three categories, that is, $m_i = 2$; moreover, set $Q = 1$ and $U = 3$. Let the item parameter vector of the general model, $(\beta_{i1}, \beta_{i2}, \beta_{i3})$, be specialized as the vector $(\beta_i, \alpha_1, \alpha_2)$ of item parameters of the rating scale model. The fit to the general model is shown by choosing the score functions $\mathbf{s}'_{i1} = (1,1,0)$ and $\mathbf{s}'_{i2} = (2,1,1)$. The relation between the person parameters of the general and the rating scale model is analogous to the case of the partial credit model. The binomial trials model fits the general model by imposing the extra restriction $\alpha_h = \ln[h/(m_i - h + 1)]$.

The OPLM for polytomous items (Verhelst, Glas, & Verstralen, 1994) has the same motivation as the OPLM for dichotomous items presented in Chapter 12: indices are introduced to account for differences in discrimination between the items. In Section 12.1 it was shown that, without (much) loss of generality, these discrimination indices may be taken to be integer valued. The extended partial credit model, which can be seen as a generalization of the OPLM, was developed by Wilson and Masters (1993) from an entirely different point of view. This model is motivated by the problem that item parameters in the partial credit model cannot be estimated if certain response categories are unobserved. The idea is as follows. If, for example, an item has 5 response categories $\{C_0, C_1, C_2, C_3, C_4\}$, and the third category is not responded to, the item format is transformed to 4 categories with weights {0,1,3,4}. If the first category is unobserved, the category weights will be {1,2,3,4}. In this way, the relative contribution of the various items to the sufficient statistic for ability, that is, the sum score, is not altered by the presence of unobserved categories.

The last three models of Table 18.1 are distinguished from the above models by the fact that the former have unidimensional person parameters, while the latter have multidimensional person parameters. In the multidimensional Rasch model (Rasch, 1961; Andersen 1972, 1973c, 1977; Fischer, 1974; see Chapters 15 and 16), it is assumed that, for all items, response categories with the same index are associated with the same response tendency, that is, with the same latent variable θ_h. Furthermore, it is assumed that every item relates to the same set of response tendencies and, therefore, $m_i = m$, for $i = 1, ..., k$. The model is derived from the general model by introducing $Q = U = m$ and $r_{ihq} = s_{ihu} = 1$, for $q = 1, ..., Q$. This model can be generalized to the last but one model of Table 18.1, where the discrimination of the items with respect to the response tendencies are mediated by a weight r_{ih}. The last model of Table 18.1 is the multidimensional polytomous latent trait model (MPLT) by Kelderman and Rijkes (1994). Here, the one-one relation between the latent dimensions and response categories is discarded, and it is assumed that the probability of a response in some category is a function of weighted differences between person and item parameters, which are pairwise associated with the latent dimensions, that is, every latent dimension has an associated item and a person parameter.

In the next sections, a number of testing procedures for the general class of IRT models defined by (18.2) will be discussed.

18.3 Generalized Pearson Tests

In Chapter 5, a general framework was presented for defining Pearson-type test statistics and deriving their asymptotic distribution. The class of so-called generalized Pearson tests is based on evaluating a linear function $\boldsymbol{d} = N^{1/2}\boldsymbol{U}'(\boldsymbol{p} - \boldsymbol{\pi}(\hat{\boldsymbol{\phi}}))$, where $\hat{\boldsymbol{\phi}}$ is a vector of model parameters evaluated using BAN estimates (e.g., ML estimates), $\boldsymbol{\pi}(\hat{\boldsymbol{\phi}})$ is the vector of the probabilities of the response patterns evaluated at $\hat{\boldsymbol{\phi}}$, and $\boldsymbol{p}$ is the associated vector of observed proportions. Generalized Pearson tests are based on the generalized Pearson statistic

$$Q = \boldsymbol{d}'\boldsymbol{W}^{-}\boldsymbol{d}, \tag{18.7}$$

with $\boldsymbol{W} = \boldsymbol{U}'\hat{\boldsymbol{D}}_\pi\boldsymbol{U}$, and $\hat{\boldsymbol{D}}_\pi$ a diagonal matrix of the elements of $\boldsymbol{\pi}(\hat{\boldsymbol{\phi}})$. The matrix $\boldsymbol{U}$ can be chosen in such a way that Q has power against specific model violations, which may also show in the so-called vector of deviates $\boldsymbol{d}$. Glas and Verhelst (1989) have derived two conditions for Q to be asymptotically χ^2-distributed. These two conditions, already given in Chapter 5, Section 5.2.1, are:

(A) the columns of $\boldsymbol{A}$, $\boldsymbol{A} = \boldsymbol{D}_\pi^{-1/2}[\partial\boldsymbol{\pi}/\partial\boldsymbol{\phi}']$, belong to $\mathcal{M}(\boldsymbol{D}_\pi^{1/2}\boldsymbol{U})$, where $\mathcal{M}(\boldsymbol{P})$ denotes the linear manifold of columns of $\boldsymbol{P}$;

(B) there exists a vector of constants $\boldsymbol{c}$ such that $\boldsymbol{U}\boldsymbol{c} = \boldsymbol{1}$, where $\boldsymbol{1}$ is a vector with all elements equal to 1.

If $\boldsymbol{\phi}$ is a full-rank parameterization of the model, the asymptotic χ^2-distribution of Q has rank$(\boldsymbol{U}'\boldsymbol{D}_\pi\boldsymbol{U})$ – order$(\boldsymbol{\phi})$ -1 degrees of freedom. If $\boldsymbol{\phi}$ is not a full-rank parametrization, order$(\boldsymbol{\phi})$ must be replaced by the number of free parameters in the model.

In the next three sections, some generalized Pearson tests for RMs for polytomous items will be discussed.

18.3.1 The Adaptation of the R_{1c}-Test to Polytomous Items

In Section 5.2.2, a number of tests were discussed, developed to have power against violations of the axiom of monotone increasing and parallel curves of item response functions. In the next three sections, some of these tests will be generalized to the general model (18.2) with a number of its special cases listed in Table 18.1.

Model (18.5) does not yet fit the framework of the multinomial model: the probabilities of the response patterns resulting in the same value $\boldsymbol{r}$ sum to one, that is, $\sum_{\{\boldsymbol{x}|\boldsymbol{R}\boldsymbol{x}=\boldsymbol{r}\}} P(\boldsymbol{x}|\boldsymbol{r},\boldsymbol{\beta}) = 1$, and as a consequence, the distribution function of the counts of the response patterns has a product-multinomial form. As in the case of dichotomous items, this problem is solved using a procedure by Birch (1963, see also Haberman, 1974). Let $\{\boldsymbol{r}\}$ be the set of all possible values of $\boldsymbol{r}$. For all $\boldsymbol{r} \in \{\boldsymbol{r}\}$, let $N_{\boldsymbol{r}}$ be the number of persons in the sample obtaining $\boldsymbol{r}$. Assume that $N_{\boldsymbol{r}}$, $\boldsymbol{r} \in \{\boldsymbol{r}\}$, has a multinomial distribution, indexed by the sample size N and the parameters $\omega_{\boldsymbol{r}}$ for all $\boldsymbol{r} \in \{\boldsymbol{r}\}$. Notice that the ML estimate of $\omega_{\boldsymbol{r}}$ is given by $\hat{\omega}_{\boldsymbol{r}} = n_{\boldsymbol{r}}/N$. Using (18.5), the probability of response pattern $\boldsymbol{x}$ is now be given by

$$P(\boldsymbol{x}|\boldsymbol{\omega},\boldsymbol{\beta}) = \frac{\omega_{\boldsymbol{r}} \exp(-\boldsymbol{x}'\boldsymbol{S}\boldsymbol{\beta})}{\gamma_{\boldsymbol{r}}}, \tag{18.8}$$

where $\boldsymbol{\omega}$ is the vector with elements $\omega_{\boldsymbol{r}}$ for all $\boldsymbol{r} \in \{\boldsymbol{r}\}$. It is easily verified that $\sum_{\{\boldsymbol{x}\}} P(\boldsymbol{x}|\boldsymbol{\omega},\boldsymbol{\beta}) = 1$, so that the model now fits the general multinomial model.

As with R_{1c} and S_i for dichotomous items, also in this version of the tests the sample of persons is partitioned into G subgroups. The definition of this partition is important for the power of the tests. For instance, if the practitioner suspects that subgroups, gender groups, or ethnic groups, say, can be distinguished, where the responses cannot be described by the same model, the partitioning can be based on this background variable. Furthermore, the partitioning of the sample of respondents can also be based on dividing the latent space into homogeneous ability regions. Consider the model (18.4). It is easily verified that this model is an exponential family. It is well-known (see, for instance, Andersen, 1980a, or Barndorff-Nielsen, 1978) that in this case ML estimation boils down to equating the realizations of the minimal sufficient statistics to their expected values. For the present development, the interest is only in the ability parameters, so the item parameters will be treated as fixed constants. It follows that the ML estimate of ability is the solution $\hat{\theta}_{\boldsymbol{r}}$ of the equation $\boldsymbol{r} = E(\boldsymbol{r}|\boldsymbol{\beta}, \boldsymbol{\theta})$. Via these estimates, the latent space is associated with observed variables, and a partition of the latent space is translated into a partition of the sample of respondents on the basis of observed variables. If a model has a unidimensional ability parameter, the method reduces to partitioning the ability continuum in G continuous disjoint regions and dividing the sample of respondents on the basis of the associated values of the sufficient statistic. In a multidimensional framework, of course, applying this principle becomes more complicated. This is beyond the scope of the present chapter; some considerations can be found in Glas (1989, Section 7.3.3).

Once the subgroups are defined, tests can be based on the difference between the counts of the numbers of persons belonging to score region g and responding in category C_h of item I_i, M_{gih}, and their CML expected values, $E(M_{gih}|\hat{\boldsymbol{\omega}}, \hat{\boldsymbol{\beta}})$, that is, the expected value given the frequency distribution of the respondents' values of the minimal sufficient statistic for ability and the CML estimates of the item parameters. These differences, which will be denoted

$$d^*_{gih} = m_{gih} - E(M_{gih}|\hat{\omega}, \hat{\beta}), \tag{18.9}$$

can be combined into two tests, a global one and an item-oriented one. The global statistic is defined by

$$R_{1c} = \sum_g \mathrm{d}'_g \boldsymbol{W}^-_g \boldsymbol{d}_g, \tag{18.10}$$

where $\boldsymbol{d}_g$ is a vector of the elements d_{gi}, $d_{gi} = d^*_{gi}/\sqrt{N}$, for $i = 1, ..., k$ and $h = 0, ..., m_i$, and $\boldsymbol{W}_g$ is the matrix of weights $\boldsymbol{U}'\hat{\boldsymbol{D}}_\pi \boldsymbol{U}$. Suppose $\boldsymbol{\phi}$ is the vector of free parameters in $\boldsymbol{\omega}$ and $\boldsymbol{\beta}$. Then R_{1c} has an asymptotic χ^2-distribution with $\sum_g \mathrm{rank}(\boldsymbol{W}_g) - \mathrm{order}(\boldsymbol{\phi}) - 1$ degrees of freedom.

A sketch of the proof of the statement concerning the asymptotic distribution is as follows. First, it will be convenient to write (18.8) as

$$P(\boldsymbol{x}|\boldsymbol{\omega},\boldsymbol{\delta}) = \frac{\omega_r \exp(-\boldsymbol{x}'\boldsymbol{\delta})}{\gamma_r}, \tag{18.11}$$

where

$$\boldsymbol{\delta} = \boldsymbol{S}'\boldsymbol{\beta}. \tag{18.12}$$

In what follows it will be immaterial whether $\sum_{i=1}^{k} m_i$ is smaller than, equal to, or greater then kU. In the first case, the model defined by (18.11) and (18.12) is derived from the model (18.8) by imposing linear restrictions on parameters of (18.8), in the last case it is the other way round. If $\sum_{i=1}^{k} m_i = kU$, (18.11) is a reparametrization of (18.8). Condition (A) of Chapter 5 entails the requirement that the columns of matrix $\boldsymbol{A}$, $\boldsymbol{A} = \boldsymbol{D}_\pi^{-1/2}[\partial\boldsymbol{\pi}/\partial\boldsymbol{\phi}']$, belong to the linear manifold $\mathcal{M}(\boldsymbol{D}_\pi^{1/2}\boldsymbol{U})$. But $[\partial\boldsymbol{\pi}/\partial\boldsymbol{\phi}'] = [\partial\boldsymbol{\pi}/\partial\boldsymbol{\delta}'][\partial\boldsymbol{\delta}/\partial\boldsymbol{\beta}'][\partial\boldsymbol{\beta}/\partial\boldsymbol{\phi}']$. Furthermore, (18.11) defines an exponential family and, therefore, using Lemma 1 of Section 5.2, it follows that $[\partial\boldsymbol{\pi}/\partial\boldsymbol{\delta}'] = \boldsymbol{D}_\pi\boldsymbol{T} - \boldsymbol{\pi}\boldsymbol{\pi}'\boldsymbol{T}$, where $\boldsymbol{T}$ is the matrix of sufficient statistics, which will be studied in some more detail in the sequel. The matrix $[\partial\boldsymbol{\beta}/\partial\boldsymbol{\phi}']$ is associated with the (linear) restrictions that need to be imposed on $\boldsymbol{\beta}$ to make the model identified, so $[\partial\boldsymbol{\beta}/\partial\boldsymbol{\phi}']$ will be a matrix of constants. In order to make Lemma 5.3 of Chapter 5 applicable to showing that condition (A) is fulfilled, the columns of $\boldsymbol{A}$, $\boldsymbol{A} = [\boldsymbol{D}_\pi^{1/2}\boldsymbol{T} - \boldsymbol{\pi}^{1/2}\boldsymbol{\pi}'\boldsymbol{T}][\partial\boldsymbol{\delta}/\partial\boldsymbol{\phi}']$ have to belong to $\mathcal{M}(\boldsymbol{D}_\pi^{1/2}\boldsymbol{U})$; this is true if the columns of $\boldsymbol{T}$ belong to $\mathcal{M}(\boldsymbol{U})$.

To study $\boldsymbol{T}$ in some more detail, consider Table 18.2. The example is a test of three items with two, three, and three response categories, respectively. For model (18.11), the response patterns are the sufficient statistics, and, therefore, they are entered in the first eight columns of the matrix $\boldsymbol{T}$ of Table 18.2. Moreover, in the example it is assumed that the unweighted sum score is the sufficient statistic for ability, but the generalization to a more complex sufficient statistic is straightforward. The sufficient statistics associated with the dummy score parameters ω_r, $r = 0, ..., 5$, are entered in the last six columns of $\boldsymbol{T}$.

Next it will be shown that the differences (18.9) can indeed be produced as a linear function $N\boldsymbol{U}'(\boldsymbol{p} - \hat{\boldsymbol{\pi}})$. Continuing the example of Table 18.2, consider the matrix $\boldsymbol{U}$ of Table 18.3. The R_{1c}-statistic defined by this matrix will be based on a partition of the score range in three regions, consisting of scores 0, 1, and 2, of score 3, and of scores 4 and 5, respectively. In the last column of Table 18.2, the elements of $\boldsymbol{\pi}$ are listed. If the inner product of the column of $\boldsymbol{U}$ associated with group one, item one, and category zero, i.e., the first column of $\boldsymbol{U}$, with $\boldsymbol{\pi}$ is evaluated, it is easily verified that this inner product constitutes a sum over the probabilities of the six response patterns with a sum score in the first region where the response to the first item is in the zero category. Using this principle, it can be verified that all differences (18.9) can be produced by multiplying a column of matrix $\boldsymbol{U}$ with $N(\boldsymbol{p} - \hat{\boldsymbol{\pi}})$. Furthermore, for the six columns of $\boldsymbol{U}$ associated with the dummy score parameters ω_r, taking the inner product with $N(\boldsymbol{p} - \hat{\boldsymbol{\pi}})$, gives zero. This is due to the fact that the counts of the numbers of

TABLE 18.2. An example of the matrix T

	T				
Item	1	2	3		
Cat.	01	012	012		
Score				012345	probability
0	10	100	100	100000	$\pi(10, 100, 100)$
1	10	100	010	010000	$\pi(10, 100, 010)$
1	10	010	100	010000	$\pi(10, 010, 100)$
1	01	100	100	010000	$\pi(01, 100, 100)$
2	01	010	100	001000	$\pi(01, 010, 100)$
2	10	100	001	001000	$\pi(10, 100, 001)$
2	10	010	010	001000	$\pi(10, 010, 010)$
2	10	001	100	001000	$\pi(10, 001, 100)$
2	01	100	010	001000	$\pi(01, 100, 010)$
3	01	010	010	000100	$\pi(01, 010, 010)$
3	10	010	001	000100	$\pi(10, 010, 001)$
3	10	001	010	000100	$\pi(10, 001, 010)$
3	01	100	001	000100	$\pi(01, 100, 001)$
3	01	001	100	000100	$\pi(01, 001, 100)$
4	01	010	001	000010	$\pi(01, 010, 001)$
4	01	001	010	000010	$\pi(01, 001, 010)$
4	10	001	001	000010	$\pi(10, 001, 001)$
5	01	001	001	000001	$\pi(01, 001, 001)$

persons obtaining r are modeled by a saturated multinomial distribution, and that the inner products of the column of $\boldsymbol{U}$ related to score r with $N\boldsymbol{p}$ and $N\hat{\boldsymbol{\pi}}$ both produce a count of the number of persons obtaining r.

Another important feature of the $\boldsymbol{U}$ matrix of Table 18.3 is that it imposes a block-diagonal structure on $\boldsymbol{U}'\boldsymbol{D}_\pi\boldsymbol{U}$, hence (18.10) can be written as a sum of quadratic forms. Finally, the proof that R_{1c} has an asymptotic χ^2-distribution can be completed by observing another feature of the exemplary matrix $\boldsymbol{U}$ of Table 18.3. It is easily verified that condition (B) given in Chapter 5, implying the existence of a vector of constants $\mathbf{c}$ such that $\boldsymbol{U}\mathbf{c} = \mathbf{1}$, is fulfilled, since the six columns of $\boldsymbol{U}$ associated with the dummy score parameters sum to a vector with all elements equal to unity. This generally holds for all matrices $\boldsymbol{U}$ constructed in this way.

18.3.2 The Adaptation of the S_i-Test to Polytomous Items

Apart from the global model test of the previous section, the practitioner should also have available an item oriented test for evaluating the contribution of specific items to a possible lack of model fit. To this end, in Section 5.2.3 the so-called S_i- and M-tests were introduced. These tests are readily generalized to the case

TABLE 18.3. An example of the matrix **U**; the cells left blank are equal to zero.

Item	1	2	3		1	2	3		1	2	3	
Cat.	01	012	012		01	012	012		01	012	012	
Score				012				3				45
0	10	100	100	100								
1	10	100	010	010								
1	10	010	100	010								
1	01	100	100	010								
2	01	010	100	001								
2	10	100	001	001								
2	10	010	010	001								
2	10	001	100	001								
2	01	100	010	001								
3					01	010	010	1				
3					10	010	001	1				
3					10	001	010	1				
3					01	100	001	1				
3					01	001	100	1				
4									01	010	001	10
4									01	001	010	10
4									10	001	001	10
5									01	001	001	01

of polytomous items; the former will be treated in this section, the latter is the topic of the next section.

The S_i-test is based on the same kind of partitioning of the sample of respondents into G homogeneous subgroups and on the same differences (18.9) as the R_{1c}-test. Two versions of the test will be discussed. For the first one, let the vector $\boldsymbol{d}_i$ have elements d_{gi}, $g = 1, ..., G$. Then S_i is defined as

$$S_i = \boldsymbol{d}_i' \boldsymbol{W}_i^- \boldsymbol{d}_i. \tag{18.13}$$

This version of the statistic is based on the matrix $\boldsymbol{U}$ of Table 18.4. As in the case of the S_i-test for dichotomous items, the validity of Condition (A) is assured by explicitly entering the sufficient statistics for the parameters in the matrix $\boldsymbol{U}$. This is done by constructing the matrix as $\boldsymbol{U} = [\boldsymbol{T}_1|\boldsymbol{T}_2|\boldsymbol{Y}]$, where $\boldsymbol{T}_1$ contains the sufficient statistics for $\boldsymbol{\delta}$, $\boldsymbol{T}_2$ those for $\boldsymbol{\omega}$, and $\boldsymbol{Y}$ the relevant contrasts. It can be verified that computing $N\boldsymbol{Y}'(\boldsymbol{p} - \hat{\boldsymbol{\pi}})$ does indeed produce the differences (18.9) for item I_i. Moreover, in Section 5.2.1 it was shown that if ML-estimates are used, $N[\boldsymbol{T}_1|\boldsymbol{T}_2]'(\boldsymbol{p}-\hat{\boldsymbol{\pi}}) = \boldsymbol{0}$; so, $[\boldsymbol{T}_1|\boldsymbol{T}_2]$ as a contrast does produce differences equal to zero, and it only influences the outcome of the statistic via the matrix of weights. In fact, it can be shown that the matrix of weights is equivalent to the

TABLE 18.4. An example of the matrix U; the entries left blank are equal to zero.

	T_1			T_2	Y		
Item	1	2	3				
Cat.	01	012	012				
Score				012345			
0	10	100	100	100000	100		
1	10	100	010	010000	100		
1	10	010	100	010000	010		
1	01	100	100	010000	100		
2	01	010	100	001000	010		
2	10	100	001	001000	100		
2	10	010	010	001000	010		
2	10	001	100	001000	001		
2	01	100	010	001000	100		
3	01	010	010	000100		010	
3	10	010	001	000100		010	
3	10	001	010	000100		001	
3	01	100	001	000100		100	
3	01	001	100	000100		001	
4	01	010	001	000010			010
4	01	001	010	000010			001
4	10	001	001	000010			001
5	01	001	001	000001			001

covariance matrix of the differences, and $[\boldsymbol{T}_1|\boldsymbol{T}_2]$ must be introduced to account for a reduction in the variance of the statistic due to the restrictions imposed by parameter estimation (see, for instance, Glas, 1989).

The cautious reader will notice that the format of $\boldsymbol{U}$ differs from the matrix $\boldsymbol{U}$ for the S_i-test given in Section 12.3.1 in Table 12.2. Firstly, for reasons of symmetry, also the parameters of the zero category are included. Secondly, some columns of the matrix $\boldsymbol{Y}$ lie in $\mathcal{M}(\boldsymbol{T})$. This should be taken into account when determining the degrees of freedom. Notice that the first and fourth column of $\boldsymbol{Y}$ add to the third column of $\boldsymbol{T}_1$, so these two columns associated with the zero category of item I_2 can be removed from $\boldsymbol{Y}$ without altering the statistic. Moreover, the second and third column of $\boldsymbol{Y}$ can be removed because the columns of $\boldsymbol{T}_2$ associated with item I_2 can be constructed from columns of $\boldsymbol{Y}$. It is a simple task of linear algebra to show that the matrix $\boldsymbol{U}$ resulting from these five deletions is of full column rank. So the statistic is based on a $\boldsymbol{Y}$ with four columns which are independent of $\boldsymbol{T}$ and, consequently, when applied in the context of the partial credit model and the OPLM, it has an asymptotic χ^2-distribution with $df = 4$.

TABLE 18.5. An example of matrix **U** for item I_2, where the low region consists of Categories C_0 and C_1 and the high region of Category C_2; the entries left blank are equal to zero.

	T_1			T_2	Y		
Item	1	2	3				
Cat.	01	012	012				
Score				012345			
0	10	100	100	100000	10		
1	10	100	010	010000	10		
1	10	010	100	010000	10		
1	01	100	100	010000	10		
2	01	010	100	001000	10		
2	10	100	001	001000	10		
2	10	010	010	001000	10		
2	10	001	100	001000	01		
2	01	100	010	001000	10		
3	01	010	010	000100		10	
3	10	010	001	000100		10	
3	10	001	010	000100		01	
3	01	100	001	000100		10	
3	01	001	100	000100		01	
4	01	010	001	000010			10
4	01	001	010	000010			01
4	10	001	001	000010			01
5	01	001	001	000001			01

The second version of the test is mainly applicable to models where the indices of the response categories are positively related to ability level, in the sense that a response in a higher category reflects a higher ability level, such as is the case in the partial credit model, the rating scale model, and the OPLM. For such models, the item categories can be dichotomized in a low and a high range.

Consider the example of Table 18.5. In this example, the set of categories $\{C_0, C_1, C_2\}$ of item I_2 is dichotomized in a low region $\{C_0, C_1\}$ and a high region $\{C_2\}$. The first, third, and fifth column of $\boldsymbol{Y}$ are associated with the low score region, the other columns of $\boldsymbol{Y}$ are associated with the high score region. Hence, $\boldsymbol{Y}$ has an entry equal to one in the first column if the response pattern has the response to item I_2 in Category C_0 or C_1, and an entry equal to one in the second column if the response to item I_2 is in Category C_2. So, the test is based on the differences between the observed and expected numbers of persons scoring low (or high) on item I_2 while obtaining a low (or high) test score.

TABLE 18.6. An example of the matrix **U** for the M statistic for item I_2

	$\boldsymbol{T}_1$			$\boldsymbol{T}_2$	$\boldsymbol{Y}$
Item	1	2	3		
Cat.	01	012	012		
Score				012345	
0	10	100	100	100000	0
1	10	100	010	010000	0
1	10	010	100	010000	0
1	01	100	100	010000	0
2	01	010	100	001000	0
2	10	100	001	001000	0
2	10	010	010	001000	0
2	10	001	100	001000	1
2	01	100	010	001000	0
3	01	010	010	000100	0
3	10	010	001	000100	0
3	10	001	010	000100	0
3	01	100	001	000100	0
3	01	001	100	000100	0
4	01	010	001	000010	0
4	01	001	010	000010	-1
4	10	001	001	000010	-1
5	01	001	001	000001	-1

18.3.3 The M-Tests for the OPLM for Polytomous Items

Also the M-test presented in Section 5.2.3 is readily generalized to polytomous models. Like the second version of the S_i-test, this test is mainly applicable to models where the indices of the response categories are positively related to ability level (for instance, the partial credit model, the rating scale model, and the OPLM). The item categories again are dichotomized in a low and a high range. Furthermore, the range of possible values of the sufficient statistic for ability, r, must be divided in a low, middle, and high range.

The quintessence of the procedure is to construct a statistic with $df = 1$. Therefore, the difference of the observed and expected numbers of responses in the high categories of an item in the group of respondents in the high range of r is subtracted from the difference of the observed and expected numbers of responses in the high categories of an item in the group of respondents in the low range of r. An example of the matrix $\boldsymbol{U}$ for this generalized Pearson test is shown in Table 18.6. Notice that the $\boldsymbol{Y}$ matrix has only one column, hence the one degree of freedom of the statistic.

18.3.4 The Generalization of the R_0- and R_{1m}-Tests

The three statistics discussed above apply to the context of CML estimation. In Section 5.2.2, two generalized Pearson statistics for the MML framework were introduced: the R_0-statistic for testing the assumption concerning the ability distribution, and the R_{1m}-statistic for testing the assumption of parallel item response curves. Both statistics are readily generalized to the framework of polytomous items. Consider the general model given by (18.4). This model is extended with the assumption that the Q-dimensional ability parameter $\boldsymbol{\theta}$ has a Q-variate distribution with density $g(\boldsymbol{\theta}|\boldsymbol{\lambda})$ parameterized by $\boldsymbol{\lambda}$. Accordingly, the probability of response pattern $\boldsymbol{x}$, given the item parameters $\boldsymbol{\beta}$ and the parameters $\boldsymbol{\lambda}$ of the ability distribution, can be written as

$$\begin{aligned} P(\boldsymbol{x}|\boldsymbol{\beta},\boldsymbol{\lambda}) &= \int .. \int \frac{\exp[\boldsymbol{x}'(\boldsymbol{R}'\boldsymbol{\theta} - \boldsymbol{S}'\boldsymbol{\beta})]}{z(\boldsymbol{\theta},\boldsymbol{\beta})} g(\boldsymbol{\theta}|\boldsymbol{\lambda}) d\boldsymbol{\theta} \\ &= \exp(-\boldsymbol{x}'\boldsymbol{S}'\boldsymbol{\beta}) \int .. \int \frac{\exp(\boldsymbol{x}'\boldsymbol{R}'\boldsymbol{\theta})}{z(\boldsymbol{\theta},\boldsymbol{\beta})} g(\boldsymbol{\theta}|\boldsymbol{\lambda}) d\boldsymbol{\theta}, \end{aligned} \tag{18.14}$$

where $z(\boldsymbol{\theta},\boldsymbol{\beta}) = \sum_{\{\boldsymbol{x}\}} \exp[\boldsymbol{x}'(\boldsymbol{R}'\boldsymbol{\theta} - \boldsymbol{S}'\boldsymbol{\beta})]$.

In addition to the parameter transformation $\boldsymbol{\delta} = \boldsymbol{S}'\boldsymbol{\beta}$, it proves useful to introduce the parameters

$$\tau_{\boldsymbol{r}} = \int .. \int \frac{\exp(\boldsymbol{r}'\boldsymbol{\theta})}{z(\boldsymbol{\theta},\boldsymbol{\beta})} g(\boldsymbol{\theta}|\boldsymbol{\lambda}) d\boldsymbol{\theta}, \tag{18.15}$$

for all $\boldsymbol{r} \in \{\boldsymbol{r}\}$. Let $\boldsymbol{\tau}$ be a vector with elements $\tau_{\boldsymbol{r}}$ for all $\boldsymbol{r} \in \{\boldsymbol{r}\}$. Then (18.15) can be written as the exponential family

$$P(\boldsymbol{x}|\boldsymbol{\delta},\boldsymbol{\tau}) = \exp(-\boldsymbol{x}'\boldsymbol{\delta})\tau_{\boldsymbol{r}} = \exp(-\boldsymbol{x}'\boldsymbol{\delta} + \boldsymbol{k}(\boldsymbol{x})' \ln \boldsymbol{\tau}), \tag{18.16}$$

where the vector $\boldsymbol{k}(\boldsymbol{x})$ has all elements equal to zero, except for the element where $\boldsymbol{R}\boldsymbol{x} = \boldsymbol{r}$, which is equal to one. Moreover, $\ln \boldsymbol{\tau}$ is understood to be a vector with elements $\ln \tau_{\boldsymbol{r}}$. Let $\boldsymbol{\xi}' = (\boldsymbol{\delta}', \boldsymbol{\tau}')$. Using Lemma 1 of Section 5.2 it follows that $[\partial\boldsymbol{\pi}/\partial\boldsymbol{\xi}'] = \boldsymbol{D}_\pi \boldsymbol{T} - \boldsymbol{\pi}\boldsymbol{\pi}'\boldsymbol{T}$, where $\boldsymbol{T}$ is the matrix of sufficient statistics. If $\boldsymbol{T}$ is partitioned as $[\boldsymbol{T}_1|\boldsymbol{T}_2]$, where $\boldsymbol{T}_1$ is associated with $\boldsymbol{\delta}$, and $\boldsymbol{T}_2$ with $\boldsymbol{\tau}$, it can be verified that $\boldsymbol{T}$ is equivalent to the matrix of sufficient statistics for the CML framework, that is, $\boldsymbol{T}_1$ has the response patterns as rows, and $\boldsymbol{T}_2$ has rows $\boldsymbol{k}(\boldsymbol{x})$, for all $\boldsymbol{x}$ in $\{\boldsymbol{x}\}$. An example was given in Table 18.2.

The R_0-statistic is defined as the generalized Pearson statistic based on the matrix of contrasts $\boldsymbol{U} = [\boldsymbol{T}_1|\boldsymbol{T}_2]$. To assure an asymptotic χ^2-distribution, the conditions (A) and (B) of Chapter 5 must be fulfilled. If $\boldsymbol{\phi}' = (\boldsymbol{\beta}', \boldsymbol{\lambda}')$, condition (A) requires that the columns of matrix $\boldsymbol{A}$, $\boldsymbol{A} = \boldsymbol{D}_\pi^{-1/2}[\partial\boldsymbol{\pi}/\partial\boldsymbol{\phi}']$, belong to the linear manifold $\mathcal{M}(\boldsymbol{D}_\pi^{1/2}\boldsymbol{U})$. But $[\partial\boldsymbol{\pi}/\partial\boldsymbol{\phi}'] = [\partial\boldsymbol{\pi}/\partial\boldsymbol{\xi}'][\partial\boldsymbol{\xi}/\partial\boldsymbol{\phi}'] = [\boldsymbol{D}_\pi \boldsymbol{T} - \boldsymbol{\pi}\boldsymbol{\pi}'\boldsymbol{T}][\partial\boldsymbol{\xi}/\partial\boldsymbol{\phi}']$, and $\boldsymbol{D}_\pi^{-1/2}[\partial\boldsymbol{\pi}/\partial\boldsymbol{\phi}'] = [\boldsymbol{D}_\pi^{1/2}\boldsymbol{T} - \boldsymbol{\pi}^{1/2}\boldsymbol{\pi}'\boldsymbol{T}][\partial\boldsymbol{\xi}/\partial\boldsymbol{\phi}']$ belongs to the manifold $\mathcal{M}(D_\pi^{1/2}T)$, as was shown in the CML framework. Therefore, the generalized Pearson statistic based on T has an asymptotic χ^2-distribution with degrees of freedom equal to rank$[T'D_\pi T]$ minus the number of free parameters in ϕ minus one.

The R_{1m}-statistic is based on the same matrix of contrasts U as the R_{1c}-statistic. Therefore, the proof that R_{1m} has an asymptotic χ^2-distribution follows the same line as that given for R_{1c}. However, there is an important difference between the two statistics. Consider the exemplary matrix U of Table 18.3. In the CML framework, taking the inner product of a column of U associated with a dummy score parameter ω_r with $N(p - \hat{\pi})$ results in a zero outcome. This is due to the fact that the counts of the number of persons obtaining r is modeled by a saturated multinomial distribution. In the framework of MML estimation, the inner product of a column of U associated with a parameter τ_r with $N(p - \hat{\pi})$ does not produce a zero deviate. As a result, the value of the statistic both depends on deviates constructed to evaluate the differences between predicted and observed item characteristic curves, and differences that reflect the degree to which the frequency distribution of the sufficient statistics for ability is properly modeled. This shows that, when the model is tested within the MML framework, the assumption concerning the measurement model and the assumption concerning the distribution of the latent ability parameters are hard to separate.

18.3.5 The Generalization to Incomplete Designs

In the above sections, it was assumed that all persons in the sample responded to the same set of items. It will now be sketched how the tests defined above can be adapted to cases of incomplete designs. To describe these test administration designs, let the items be indexed $i = 1, ..., k$, and the persons $v = 1, ..., N$. The test administration design will consist of so-called booklets indexed $b = 1, ..., B$, defined by the item indices $\{i|b\}$. A booklet therefore is a collection of items. Furthermore, $\{v|b\}$ will be the indices of the persons taking booklet b. It is assumed that for every two booklets b and b', $\{v|b\} \cap \{v|b'\} = \emptyset$. Let $\beta^{(b)}$ be the parameters of the items in booklet b.

For the MML framework, $\lambda^{(b)}$ will be the parameters of the ability distribution of the persons taking booklet b. In what follows, it will be assumed that the design is linked. A design is linked if all booklets b and b' in the design are linked, that is, if there exists a sequence of booklets $b, b^{(1)}, b^{(2)}, ..., b'$, such that any two adjacent booklets in the sequence have common items, or, in the case of MML estimation, are assumed to be administered to samples from the same ability distribution. In the former case, the item parameter vectors of the adjacent booklets $b^{(n)}$ and $b^{(m)}$, $\beta^{b(n)}$ and $\beta^{b(m)}$, have parameters in common, in the latter case, for the adjacent booklets $\beta^{b(n)}$ and $\beta^{b(m)}$, $\lambda^{b(n)} = \lambda^{b(m)}$.

To use the theory for the construction of generalized Pearson statistics, the models for incomplete designs have to be brought in the framework of the multinomial model. For the conditional model (18.5) this is done as follows. Let $\boldsymbol{x}^{(b)}$ be a response pattern, and let $\{\boldsymbol{x}^{(b)}\}$ be the set of all possible response patterns of booklet b. As in Section 18.2, scoring functions $\boldsymbol{R}$ and $\boldsymbol{S}$ are defined as $\boldsymbol{R} = [\boldsymbol{R}_i, \boldsymbol{R}_{i'}, \boldsymbol{R}_{i''}, ...]$ and $\boldsymbol{S} = [\boldsymbol{S}_i, \boldsymbol{S}_{i'}, \boldsymbol{S}_{i''}, ...]$, for all i, i', i'' in $\{i|b\}$, where the scoring functions $\boldsymbol{R}_i, \boldsymbol{R}_{i'}, \boldsymbol{R}_{i''}, ..., \boldsymbol{S}_i, \boldsymbol{S}_{i'}, \boldsymbol{S}_{i''}, ...$, etc., are as defined in Section 18.2. Using these definitions, the probability of response pattern $\boldsymbol{x}^{(b)}$ given item and ability parameters is

$$P(\boldsymbol{x}^{(b)}|\boldsymbol{\theta}, \boldsymbol{\beta}^{(b)}) \propto \exp[\boldsymbol{x}^{(b)\prime}(\boldsymbol{R}^{(b)\prime}\boldsymbol{\theta} - \boldsymbol{S}^{(b)\prime}\boldsymbol{\beta}^{(b)})]. \tag{18.17}$$

This is a straightforward generalization of (18.4). Let $\boldsymbol{r}^{(b)}, \boldsymbol{r}^{(b)} = \boldsymbol{R}^{(b)}\boldsymbol{x}^{(b)}$ be a score pattern on booklet b, and let $\{\boldsymbol{r}^{(b)}\}$ be the set of all possible score patterns $\boldsymbol{r}^{(b)}$. For all $\boldsymbol{r}^{(b)} \in \{\boldsymbol{r}^{(b)}\}$, let $N_{\boldsymbol{r}(b)}$ be the number of persons in the sample obtaining score pattern $\boldsymbol{r}^{(b)}$. It is assumed that $N_{\boldsymbol{r}(b)}$, for $\boldsymbol{r}^{(b)} \in \{\boldsymbol{r}^{(b)}\}$, has a multinomial distribution defined by the sample size N and the probabilities $\omega_{\boldsymbol{r}(b)}$. The ML estimate of $\omega_{\boldsymbol{r}(b)}$ is given by $\hat{\omega}_{\boldsymbol{r}(b)} = n_{\boldsymbol{r}(b)}/N$. Using (18.17), the probability of response pattern $\boldsymbol{x}^{(b)}$ is given by

$$P(\boldsymbol{x}^{(b)}|\boldsymbol{\omega}^{(b)}, \boldsymbol{\beta}^{(b)}) = \frac{\omega_{\boldsymbol{r}(b)} \exp(-\boldsymbol{x}^{(b)\prime}\boldsymbol{S}^{(b)\prime}\boldsymbol{\beta}^{(b)})}{\gamma_{\boldsymbol{r}(b)}}, \tag{18.18}$$

where $\boldsymbol{\omega}^{(b)}$ is a vector with elements $\omega_{\boldsymbol{r}(b)}$ for all $\boldsymbol{r}^{(b)} \in \{\boldsymbol{r}^{(b)}\}$, and $\gamma_{\boldsymbol{r}(b)}$ a combinatorial function defined by $\gamma_{\boldsymbol{r}(b)} = \sum_{\{\boldsymbol{x}^{(b)}|\boldsymbol{R}^{(b)}\boldsymbol{x}^{(b)}=\boldsymbol{r}^{(b)}\}} \exp(-\boldsymbol{x}^{(b)\prime}\boldsymbol{S}^{(b)\prime}\boldsymbol{\beta}^{(b)})$. It is easily verified that the probabilities (18.18) sum to one.

Tailoring the version of the model for the MML framework (18.14) to the multinomial model requires introducing an ancillary saturated multinomial model, this time for the numbers of persons administered booklet b, N_b, $b = 1, ..., B$. This multinomial model will have parameters N and κ_b, $b = 1, ..., B$. Then (18.14) generalizes to

$$P(\boldsymbol{x}^{(b)}|\boldsymbol{\beta}^{(b)}, \boldsymbol{\lambda}^{(b)}) = \kappa_b \int .. \int \frac{\exp[\boldsymbol{x}^{(b)\prime}(\boldsymbol{R}^{(b)\prime}\boldsymbol{\theta} - \boldsymbol{S}^{(b)\prime}\boldsymbol{\beta}^{(b)})]}{z_b(\boldsymbol{\theta}, \boldsymbol{\beta}^{(b)})} g_b(\boldsymbol{\theta}|\boldsymbol{\lambda}^{(b)}) d\boldsymbol{\theta}, \tag{18.19}$$

where $z_b(\boldsymbol{\theta}, \boldsymbol{\beta}^{(b)}) = \sum_{\{\boldsymbol{x}^{(b)}\}} \exp[\boldsymbol{x}^{(b)\prime}(\boldsymbol{R}^{(b)\prime}\boldsymbol{\theta} - \boldsymbol{S}^{(b)\prime}\boldsymbol{\beta}^{(b)})]$, and $g_b(\boldsymbol{\theta}|\boldsymbol{\lambda}^{(b)})$ represents the density of the ability distribution for those persons who responded to booklet b.

With the models brought in the framework of the multinomial model, the generalized Pearson tests can be adapted to incomplete designs. The generalization of R_0-, R_{1m}-, R_{1c}-, and S_i-statistics is accomplished by constructing a matrix of contrasts $\boldsymbol{U}$, which has the form

$$
\boldsymbol{U} = \begin{pmatrix} \boldsymbol{U}^{(1)} & & & & \\ & \ddots & & & \\ & & \boldsymbol{U}^{(b)} & & \\ & & & \ddots & \\ & & & & \boldsymbol{U}^{(B)} \end{pmatrix}, \tag{18.20}
$$

where $\boldsymbol{U}^{(b)}$ is the usual matrix for testing a certain contrast in the case of a complete design. As a consequence of the structure of the super-matrix (18.20), the complete statistic will be a sum of the statistics for the separate booklets in the design. Given the structure of the matrix $\boldsymbol{U}$, the proofs that these versions of R_0, R_{1m}, R_{1c}, and S_i have an asymptotic χ^2-distribution can readily be derived from the analogous proofs for a complete design. A sketch of the proof is as follows. Using the parametrizations defined in Sections 18.1 and 18.3.4, the partial derivatives of the probabilities (18.18) or (18.19) with respect to the dummy parameters have the form $\boldsymbol{D}_\pi \boldsymbol{T} - \boldsymbol{\pi}\boldsymbol{\pi}'\boldsymbol{T}$, with

$$
\boldsymbol{T} = \begin{pmatrix} \boldsymbol{T}^{(1)} \\ \vdots \\ \boldsymbol{T}^{(b)} \\ \vdots \\ \boldsymbol{T}^{(B)} \end{pmatrix}, \tag{18.21}
$$

where $\boldsymbol{T}^{(b)}$ is the matrix of the sufficient statistics for the response patterns of booklet b. In Section 18.3 it was shown that, in the case where a model is an exponential family, checking condition (A) of Chapter 5 boils down to verifying that the columns of $\boldsymbol{T}$ belong to the linear manifold of the columns of $\boldsymbol{U}$. Comparing (18.21) with (18.20) this can be verified for every booklet b separately; how this verification works for a single booklet can be derived from the instances for complete designs given above.

18.4 Likelihood Ratio and Wald Tests

In Sections 5.3 and 5.4, several likelihood ratio (LR) and Wald tests for the RM for dichotomous items were proposed. As will be shown below, these tests easily generalize to the class of models for polytomous items (18.2). The application of LR and Wald tests to dichotomous items generally boils down to postulating that a RM holds for a number of subgroups of the sample of respondents and to testing the hypothesis that these RMs are the same; that is, to testing the restriction that the parameters of the models for the subgroups are equal. As with dichotomous items, an LR test can be constructed within the CML framework, that is, a test based on the same rationale as R_{1c} and asymptotically equivalent to R_{1c}. As in Section 18.3.1, where R_{1c} is defined for polytomous items, the latent space is partitioned into G homogeneous regions, and respondents are assigned to G subgroups depending on the region to which the value of $\boldsymbol{\theta}_r$ belongs, where

$\boldsymbol{r}$ is the minimal sufficient statistic for the latent variable $\boldsymbol{\theta}$. Within the CML framework, the estimates of the item parameters in the subgroups, $\boldsymbol{\beta}_g$, $g = 1,..,G$, should be equal within chance limits. This gives rise to the LR statistic

$$LR = 2\Big(\sum_{g=1}^{G} \ln L_C(\boldsymbol{\beta}_g; \boldsymbol{X}_g) - \ln L_C(\boldsymbol{\beta}; \boldsymbol{X})\Big), \tag{18.22}$$

where $L_C(\boldsymbol{\beta}; \boldsymbol{X})$ is the conditional likelihood of the item parameters evaluated using the CML estimates based on all data $\boldsymbol{X}$, and $L_C(\boldsymbol{\beta}_g; \boldsymbol{X}_g)$ the likelihood function evaluated using the CML estimates of the item parameters based on the data $\boldsymbol{X}_g$ of subgroup g. This statistic has an asymptotic χ^2-distribution with *df* equal to the number of parameters estimated in the subgroups minus the number of parameters estimated in the total data set. Notice that it is irrelevant whether the data are collected in a complete or incomplete design: the partition of the latent space defines the equivalence classes of the sufficient statistic. In the case of an incomplete design, $L_c(\boldsymbol{\beta}_g; \boldsymbol{X}_g)$ will generally be a product over all booklets. One may, of course, also keep the booklets separated, for instance, because one suspects that different models apply to the responses of the subgroups of persons taking the different booklets. In that case one may opt for the LR statistic defined by

$$LR = 2(\sum_{b=1}^{B} \sum_{g=1}^{G_b} \ln L_c(\boldsymbol{\beta}_g^{(b)}; \boldsymbol{X}_g^{(b)}) - \ln L_c(\boldsymbol{\beta}; \boldsymbol{X})), \tag{18.23}$$

where $L_c(\boldsymbol{\beta}_g^{(b)}; \boldsymbol{X}_g^{(b)})$ is the likelihood function evaluated using the CML estimates of the item parameters based on the data $\boldsymbol{X}_g^{(b)}$ of subgroup g within booklet b. Since booklets need not necessarily consist of different items, definition (18.23) also includes the situation where all persons respond to the same items, and one wants to test the hypothesis that the model applies in two distinct sub-populations.

The above hypotheses can also be evaluated using asymptotically equivalent Wald statistics. To give the general expression for the Wald test, let $\boldsymbol{\phi}$ be a vector of parameters. Testing the hypothesis $\boldsymbol{h}(\boldsymbol{\phi}) = \boldsymbol{0}$ can be done using the statistic

$$\boldsymbol{W} = \boldsymbol{h}'(\hat{\boldsymbol{\phi}})[\boldsymbol{T}'(\hat{\boldsymbol{\phi}})\boldsymbol{\Sigma}\boldsymbol{T}(\hat{\boldsymbol{\phi}})]^{-1}\boldsymbol{h}(\hat{\boldsymbol{\phi}}), \tag{18.24}$$

where $\boldsymbol{\phi}$ is a vector of parameters, $\boldsymbol{T}(\boldsymbol{\phi}) = \partial \boldsymbol{h}(\boldsymbol{\phi})/\partial \boldsymbol{\phi}'$, and $\boldsymbol{\Sigma}$ the asymptotic covariance matrix of the parameters. All expressions involving parameters in (18.24) are evaluated using BAN estimates. An alternative to (18.22) is given by the Wald statistic based on $\boldsymbol{\beta}_g - \boldsymbol{\beta}_{g+1} = \boldsymbol{0}$, for $g = 1, ..., G-1$. Hence, in this case $\boldsymbol{h}(\boldsymbol{\phi})$ becomes $\boldsymbol{h}(\boldsymbol{\beta})' = (\boldsymbol{\beta}_1' - \boldsymbol{\beta}_2', \boldsymbol{\beta}_2' - \boldsymbol{\beta}_3', ..., \boldsymbol{\beta}_{g-1}' - \boldsymbol{\beta}_g')$. An alternative to (18.23) is the Wald statistic based on $\boldsymbol{\beta}_g^{(b)} - \boldsymbol{\beta}_{g+1}^{(b)} = \boldsymbol{0}$, for $g = 1, ..., G-1$ and $b = 1, ..., B$.

The asymptotic equivalence of R_{1c}-, LR, and Wald statistics does not imply that there are no considerations with respect to which one should be preferred. For, if the model does not hold, information about model violations differs from test to test. For instance, the Wald statistic is directly based on differences between parameter estimates weighted by their asymptotic variances, while the contribution of misfitting items to the value of the LR statistic is much harder to evaluate. The R_{1c} statistic is based on the weighted difference between observed and expected frequencies, and these differences may be informative with respect to the severity of the model violations. For instance, if the sample size is large, the power of the test will also be large, which may result in a significant outcome although the percentage of 'explained' responses may be quite high.

The possibilities of applying LR and Wald tests in the framework of MML estimation are limited compared to the possibilities of the CML framework. For instance, a statistic as (18.22) cannot properly be applied, because the counts $n_{\boldsymbol{r}}$, for all $\boldsymbol{r}$ in $\{\boldsymbol{r}\}$, are a sufficient statistic for $\boldsymbol{\theta}$, and splitting up the sample according to values of $\boldsymbol{r}$ will invalidate the estimates of the population parameters $\boldsymbol{\lambda}$ in the subgroups. This, of course, also applies to an MML analogue of (18.23).

In an incomplete design, the global validity of the model for the subgroups may be tested using

$$LR = 2\Big(\sum_{b=1}^{B} \ln L_m(\boldsymbol{\beta}^{(b)}, \boldsymbol{\lambda}^{(b)}; \boldsymbol{X}^{(b)}) - \ln L_m(\boldsymbol{\beta}, \boldsymbol{\lambda}; \boldsymbol{X})\Big), \tag{18.25}$$

with $\ln L_m(\boldsymbol{\beta}^{(b)}, \boldsymbol{\lambda}^{(b)}; \boldsymbol{X}^{(b)})$ the likelihood function evaluated using the MML estimates of the item and population parameters obtained from the data of booklet b, and $\ln L_m(\boldsymbol{\beta}, \boldsymbol{\lambda}; \boldsymbol{X})$ the likelihood function under the restriction that item and population parameters are the same in all booklets. On may also consider special cases where only the equality of item parameters or the equality of population parameters are tested. The Wald alternative of (18.25) is based on the hypotheses $\boldsymbol{\beta}^{(b)} - \boldsymbol{\beta}^{(b+1)} = \boldsymbol{0}$ and $\boldsymbol{\lambda}^{(b)} - \boldsymbol{\lambda}^{(b+1)} = \boldsymbol{0}$, for $b = 1, ..., B-1$.

18.5 An Example Concerning the Detection of Item Bias

A dichotomous item is defined to be biased if, conditional on ability level, the probability of a correct response differs between groups (Mellenbergh, 1982, 1983). Items may differ in difficulty, and groups in their capability to produce a correct response to the item, but that does not yet constitute item bias. An item is only considered biased if its difficulty differs between subjects with identical ability level belonging to different populations. Since item bias is an attribute related to items and populations, the terminology 'differential item functioning' (DIF) is often preferred. In the literature on DIF it is practice to distinguish between a reference population and a focal population. For instance, if bias

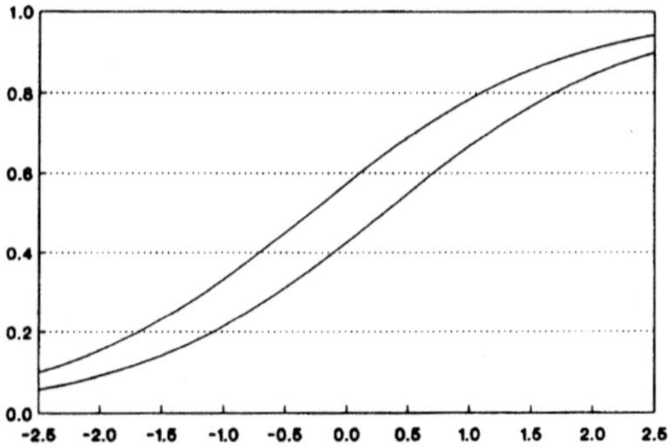

FIGURE 18.1. An example of a uniformly biased dichotomous item

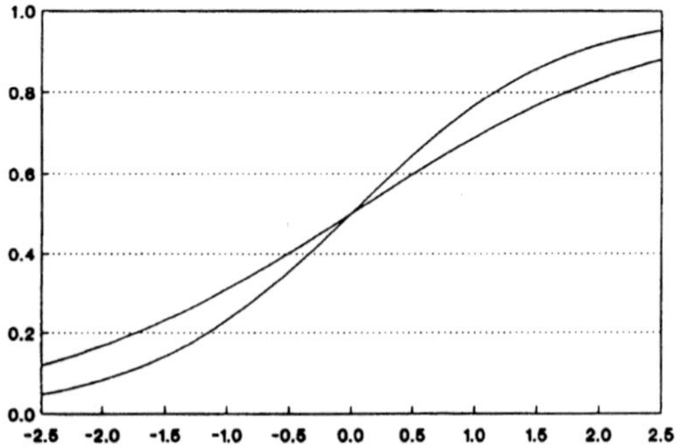

FIGURE 18.2. An example of a non-uniformly biased dichotomous item

induced by cultural differences is studied, usually the autochthonous respondents are the reference population, while the allochtonous respondents constitute the focal group.

The generalization to polytomous items is straightforward. A polytomous item can be considered biased if the set of probabilities of scoring in the various categories of the item, conditional on ability, differs between groups. For dichotomous items, a distinction is made between uniform bias and non-uniform bias (Mellenbergh, 1982, 1983). An item is uniformly biased if the probability of a correct response in the focal population is systematically higher or lower than the probability of a correct response in the reference population for all ability levels. An example of a uniformly biased item is shown in Figure 18.1. An item is non-uniformly biased if the order of the magnitudes of the probabilities of a correct response in the reference and focal population differs across ability levels. An example is given in Figure 18.2.

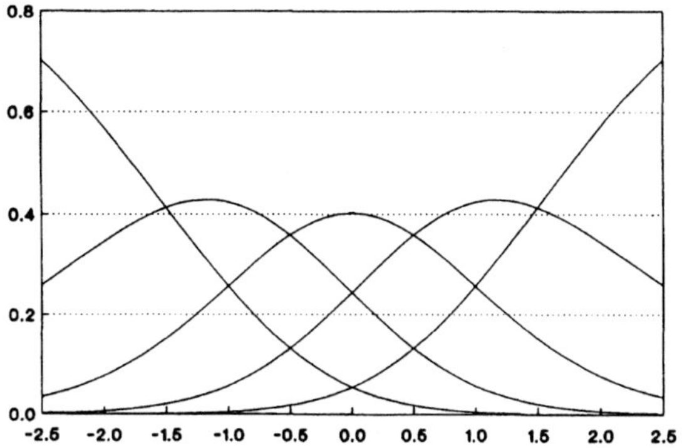

FIGURE 18.3. Item characteristic curves in the partial credit model

The bias in the example is still rather systematic, one group performs better at low ability levels, the other at high ability levels. The systematic patterns of Figure 18.1 and 18.2 can be effectively modeled by varying the location and slope parameters of the item response function. In practice, the pattern of bias can, of course, be much more irregular, and explicitly modeling the responses of both groups may not always be feasible.

Generalization of the concepts of uniform and non-uniform bias to cases of dichotomous items is complicated by the presence of more than one characteristic curve per item. Consider the example of Figure 18.3. The monotone decreasing curve relates to the probability of scoring in the zero category, the monotone increasing curve to the probability of scoring in the highest category, and the remaining single-peaked curves to the probabilities of scoring in other categories. The item characteristic curves conform to the partial credit model (see Table 18.1), but since only one item is considered, the curves also comply with the OPLM and the rating scale model. In the partial credit model, the parameters β_{ih}, $j = 1, ..., m_i$, are the boundary values where the probabilities of scoring in category C_{j-1} and category C_j are equal, that is, the parameters give the position on the x-axis where the curves of categories C_{j-1} and C_j cross.

Intuitively, the distinction between uniform and non-uniform bias in the case of dichotomous items is related to whether the item characteristic curves for the two populations cross. In the case of polytomous items, the number of characteristic curves and their dependence prevents this simple definition. For unidimensional polytomous models, such as the partial credit model, the rating scale model, the OPLM, or the extended partial credit model, one may call an item uniformly biased if the expected score on the item, given $\boldsymbol{\theta}$, is systematically higher or lower for the focal population. More important than the matter of uniform or non-uniform bias is the question whether the response behavior of both populations can be properly described by an IRT model. However, as with dichotomous items, practice shows that varying the location and discrimination parameters often suffices to model the response behavior of the different populations. An example will be given below.

Several techniques for detecting DIF have been proposed, most of them based on evaluating differences in response probabilities between groups conditional on some measure of ability. The most generally used technique is based on the Mantel-Haenszel statistic (Holland & Thayer, 1988), others are based on log-linear models (Kok, Mellenbergh, & Van der Flier, 1985), on IRT models (Hambleton & Rogers, 1989), or on log-linear IRT models (Kelderman, 1989). In the Mantel-Haenszel, log-linear, and log-linear IRT approaches, the difficulty level of the item is evaluated conditionally on unweighted sum scores. However, adopting the assumption that the unweighted sum score is a sufficient statistic for ability (together with some technical assumptions, which will seldomly be inappropriate) necessarily leads to adoption of the RM (Fischer, 1974, 1993b). However, with the exception of the log-linear IRT approach, the validity of the RM is not explicitly tested. Therefore, the approach suggested here consists of two steps:

(1) searching for an IRT model for fitting the data of the sample from the reference population, and, as far as possible, the sample from the focal population;

(2) evaluating the differences in response probabilities between the samples from the reference and focal population in homogeneous ability groups.

If biased items are detected, one may, in some instances, want to go further and estimate the impact of DIF. With respect to the impact, two additional steps might be taken:

(3) attempting to model the responses of the focal population to biased items;

(4) estimating the overall test results for the focal population if no DIF were present.

The example to be presented concerns part of the examination of the business curriculum for the Dutch higher secondary education, the 'HAVO level'. The example was part of a larger study of gender-based item bias in examinations in secondary education. Since the objective here is to illustrate the statistical procedures rather than to give an account of the findings with respect to gender-based item bias, no actual examples of biased items will be shown. For a detailed report of the findings the interested reader is referred to Bügel and Glas (1992). The analyses were carried out using a sample of 1000 boys and 1000 girls from the complete examination population. For convenience of presentation, the example will be limited to 10 items.

The first step of the procedure consisted of fitting the OPLM. For dichotomous data, the CML estimation procedure and the iterative procedure of postulating

TABLE 18.7. Summary of fit statistics for the reference group

Item	A		C	df	P	M	$M2$	$M3$
I_1	2	[:1]	11.724	7	.110	−0.294	−0.648	−0.039
		[:2]	6.685	7	0.462	−0.460	0.098	−0.584
I_2	3	[:1]	5.918	6	0.432	−1.390	0.716	0.587
		[:2]	6.346	7	0.500	−0.195	0.554	0.029
		[:3]	4.025	5	0.546	0.003	0.512	0.878
I_3	4	[:1]	9.685	5	0.085	1.543	2.476	3.615
		[:2]	1.624	6	0.951	0.893	0.750	0.167
I_4	2	[:1]	4.054	7	0.774	0.578	0.423	0.163
		[:2]	10.543	7	0.160	0.238	−0.309	−1.202
		[:3]	3.582	5	0.611	0.472	0.010	−0.634
I_5	2	[:1]	9.124	6	0.167	1.408	1.601	1.888
		[:2]	2.208	7	0.947	0.284	0.837	−0.631
		[:3]	5.140	7	0.643	−1.064	0.494	−0.928
I_6	3	[:1]	6.090	7	0.529	0.743	0.761	0.006
		[:2]	4.065	7	0.772	0.315	0.836	0.414
I_7	3	[:1]	5.873	7	0.555	−0.063	−0.961	0.286
		[:2]	15.456	6	0.017	0.528	−0.645	1.892
I_8	3	[:1]	6.971	5	0.223	−0.687	−0.361	−1.348
		[:2]	15.915	6	0.014	−1.473	−.427	−2.709
		[:3]	6.283	6	0.392	0.010	−0.002	−0.141
I_9	4	[:1]	6.359	6	0.384	0.120	−0.930	−0.779
		[:2]	1.958	6	0.923	−1.202	−0.913	−0.386
I_{10}	4	[:1]	2.321	4	.677	−0.187	−1.186	−0.158
		[:2]	2.575	5	0.765	−1.126	−0.794	−1.339
		[:3]	5.503	5	0.358	−0.653	−1.213	0.532

$R_{1c} = 75.18; df = 72; p = .38$

discrimination indices and testing these, are described in Chapter 12. For the polytomous version of the model, given in Table 18.1, the procedure is essentially the same, and the generalizations of the test statistics used to this end were given in the above sections. In order not to let DIF interfere with searching for a proper model, only the data of the reference group were used at first. The analyses were performed using the computer program OPLM (Verhelst, Glas, & Verstralen, 1994). In Table 18.7 a summary of the fit statistics for the final model are presented. The column labeled 'A' shows the discrimination indices. For computing the S_i- and M-statistics, the range of scores that can be obtained on a polytomous item is dichotomized into a range of low scores and a range of high scores. The division of the score range of an item into the scores $0, ..., j$ and $j+1, ..., m_i$ can be written in the shorthand notation $[:j+1]$, for $j = 0, .., m_i - 1$. OPLM computes the S_i- and M-statistics for all dichotomizations $[:1]$,...,$[:m_i]$.

TABLE 18.8. Hypothesis testing

Analysis	Model	R_{1c}	*df*	*P*
1	reference group	75.18	72	.375
2	focal group	127.28	72	.000
3	groups combined	356.75	168	.000
4	focal group, 9 adjusted index	59.98	72	.843
5	combined groups, 3 splitted	258.61	166	.000
6	combined groups, 9 splitted	279.55	166	.000
7	combined groups, 3 and 9 splitted	154.30	164	.697

The column labeled 'S' displays the values of the S_i-statistics, the next two columns give the *df* and the probability P of the outcome, respectively. In the last three columns, the values of the three versions of the M-statistic are given; these statistics have an asymptotic standard normal distribution. At the bottom of the table, the value of the R_{1c}-statistic is given; it can be seen that model fit is acceptable.

For the next two analyses, the discrimination indices found for the reference group remain unaltered. In the first analysis, CML parameter estimation and testing model fit were performed using only the sample of the focal population; in the next analysis, CML parameter estimation and testing model fit was carried out using the data of both groups simultaneously. The results of the computation of R_{1c} are shown in Table 18.8, in the rows denoted 'Analysis 2' and 'Analysis 3'. It can be seen that in both cases the model had to be rejected. The result of Analysis 2 shows that the discrimination indices of the reference population do not fit the focal population, even if the parameter estimates are obtained in this latter group. The results of Analysis 3 also indicate that the same model cannot properly describe the data in both groups simultaneously. To investigate whether this last result is due to DIF, Table 18.9 gives a summary of fit statistics for the combined focal and reference groups. The table has the same format as Table 18.7. It can be seen that the Items I_3 and I_9 are important sources of misfit. To further explore the hypothesis of DIF, the differences between observed and expected frequencies associated with the R_{1c}-statistic were examined. To compute this statistic, both the respondents of the reference and the focal group were divided into four subgroups on the basis of their weighted sum scores. The subgroups were formed in such a way that they had approximately the same number of respondents. The chosen score ranges and the resulting numbers of respondents in the subgroups are displayed in the first lines of Table 18.10. Furthermore, for all items and all categories the scaled deviates in the subgroups, that is, the differences (18.9) divided by their standard deviations, are shown. The columns labeled 'SS' give the sum of the squares of the scaled deviates for all item $\times$ category combinations. Notice that especially the magnitude of the

TABLE 18.9. Summary of fit statistics; focal and reference groups combined

Item	A		C	df	P	M	M2	M3
1	2	[:1]	28.189	14	0.013	−0.864	−0.791	−1.121
		[:2]	12.748	14	0.546	0.067	0.517	−1.236
2	3	[:1]	7.399	11	0.766	−0.070	0.183	1.079
		[:2]	13.011	14	0.526	−0.838	−0.625	−1.210
		[:3]	4.268	10	0.934	0.795	0.755	0.024
3	4	[:1]	107.862	12	0.000	2.315	.658	2.771
		[:2]	37.500	12	0.000	−1.438	0.548	−1.787
4	2	[:1]	8.121	14	0.883	−0.721	−0.351	−0.338
		[:2]	15.971	14	0.315	−0.131	−0.475	−0.610
		[:3]	15.665	10	0.110	−0.137	−1.084	−1.317
5	2	[:1]	11.393	12	0.496	1.428	−0.339	0.395
		[:2]	15.399	14	0.351	−1.318	−1.453	−1.701
		[:3]	10.520	14	0.723	−1.997	−1.455	−1.384
6	3	[:1]	10.486	14	0.726	0.358	1.505	0.543
		[:2]	11.375	14	0.656	0.442	1.264	0.518
7	3	[:1]	18.279	14	0.194	−0.438	−1.395	−0.066
		[:2]	18.005	12	0.116	1.376	−1.221	1.179
8	3	[:1]	9.410	10	0.494	−1.049	−0.234	−0.955
		[:2]	19.127	13	0.119	−1.341	−0.615	−1.566
		[:3]	8.080	12	0.779	−0.322	0.173	−0.609
9	4	[:1]	113.760	12	0.000	4.025	4.297	4.614
		[:2]	35.874	12	0.000	2.657	2.655	3.173
10	4	[:1]	14.893	9	0.094	−1.120	−1.083	−1.070
		[:2]	16.264	10	0.092	−1.642	−1.612	−2.343
		[:3]	24.262	11	0.012	−2.164	−2.123	−0.712

$R_{1c} = 356.75; df = 168; p = .00$

sums of squares of item I_3 is large compared to the rest of the values of the sums of squares. Moreover, for this item the scaled deviates are generally positive for the reference group and negative for the focal group. This indicates that this item favors the reference group, since this group produces more responses in categories C_h with $h > 0$ than expected from a model calibrated on both groups simultaneously. In the same manner, the item is unfavorable for the focal population, since this group produces less responses as expected in categories C_h with $h > 0$, and consequently more responses in category C_0. For item I_9 the pattern is far less clear.

From Analysis 2 in Table 18.8 it was anticipated that the discrimination index for item I_9 could be different for the two groups. Analysis 4 of Table 18.8 showed that this was indeed the case: adjusting the discrimination index from 4 to 2 for the focal group resulted in a good model fit.

TABLE 18.10. Scaled deviates evaluated using CML based on both groups simultaneously

		Reference Population					Focal Population				
		1	2	3	4		1	2	3	4	
	r	1-20	21-37	38-52	53-73		1-20	21-36	37-52	53-73	
	n	228	237	253	263	SS	240	246	239	252	SS
Item	h										
1	1	−0.1	−1.5	0.6	−0.6	3.2	1.3	−0.9	0.3	1.1	4.0
	2	−1.5	0.4	−0.9	−0.1	3.5	0.5	0.2	0.7	0.3	1.0
2	1	−0.9	−0.1	0.8	0.4	1.7	1.9	−0.2	−0.2	−2.0	7.8
	2	−0.4	−0.5	−1.2	−0.0	2.1	−0.2	−0.0	0.8	1.3	2.5
	3	−1.4	0.9	0.4	−0.8	3.8	−0.1	0.3	−0.0	0.2	0.1
3	1	5.1	2.8	0.0	−1.0	35.7	−2.9	−3.4	0.5	−1.0	21.6
	2	0.9	2.4	1.7	0.4	10.4	−1.8	−2.0	−3.2	1.0	18.9
4	1	1.9	−0.0	0.5	0.0	4.1	−0.4	−0.6	−1.7	0.6	4.0
	2	−2.0	−0.8	0.1	−0.3	4.8	−0.2	2.2	1.8	−1.3	10.3
	3	−0.8	−0.2	−1.0	−0.0	1.8	−0.8	−0.2	−0.8	2.0	5.6
5	1	1.8	0.4	1.0	−1.1	6.0	−0.8	−0.1	−0.8	−1.4	3.4
	2	0.0	−0.1	−0.8	−0.0	0.7	−0.6	0.8	−0.1	1.1	2.4
	3	−0.8	−1.1	−0.2	0.5	2.4	0.7	−0.8	1.4	0.5	3.4
6	1	−0.3	0.6	0.4	0.7	1.2	−0.4	0.2	−0.2	−1.3	2.1
	2	−0.4	−0.4	−1.3	−0.3	2.3	0.4	1.7	−0.1	0.7	3.7
7	1	−0.7	0.0	−1.0	1.1	2.8	−0.9	0.6	1.0	−0.5	2.6
	2	1.5	−2.8	0.6	−1.1	11.8	−0.2	0.8	0.8	0.6	1.9
8	1	1.8	−0.5	0.5	−1.1	5.2	0.2	−0.9	−1.0	0.6	2.4
	2	−2.3	1.0	0.4	−0.0	6.6	−0.1	0.2	0.6	−0.7	1.0
	3	−0.7	0.5	−1.3	0.7	3.2	1.2	−0.4	0.0	0.2	1.7
9	1	1.1	0.4	0.6	−1.0	3.0	1.4	−0.2	−1.9	0.1	6.0
	2	0.7	0.4	1.0	1.7	4.8	3.3	0.7	−1.0	−4.2	30.6
10	1	0.4	1.1	0.6	−0.6	2.4	−0.2	−0.4	−1.6	0.0	3.0
	2	−1.3	−0.0	0.2	0.8	2.6	−1.5	1.9	−0.4	−1.5	8.6
	3	−1.0	−2.0	−0.5	−0.4	5.8	0.6	−0.0	1.9	1.5	6.6
	SS	63.0	36.8	18.3	15.0	133.1	39.2	34.4	36.8	45.8	156.3
	R_{1c}	58.5	45.2	20.4	19.9	144.1	39.2	45.3	43.8	84.2	212.6

In the last three analyses of Table 18.8 it was attempted to construct a model fitting the data of both groups simultaneously. In Analysis 5, the parameters of item I_3 were allowed to differ between the reference and the focal group. The discrimination index was held constant. It can be seen that this did not result in an acceptable model fit. In Analysis 6, the same procedure was carried out for item I_9, with the distinction that the discrimination was set equal to 4 in the reference group and equal to 2 in the focal group. Again the results were unsatisfactory. Finally, in Analysis 7, both the item parameters and the discrimination indices of the items I_3 and I_9 were allowed to differ between groups, and it can be seen from the last line of Table 18.8 that this model proved to be acceptable.

The last step consisted of estimating the impact of the bias on the distribution of both the weighted and unweighted sum scores of the respondents. As a first step, the validity of the model of Analysis 7 extended with normal ability distributions for the reference and focal group was evaluated. Item and population parameters were estimated using MML. Computing the R_0 statistic resulted in $\chi^2 = 121.79$ ($df = 138$, $P = .83$), the outcome of R_{1m} was $\chi^2 = 267.83$ ($df = 303$, $P = .92$), so this enhanced model was not rejected. Next, for the focal population the expected frequency distribution $E(N_r|\boldsymbol{\beta}, \boldsymbol{\lambda})$ was computed in two ways: first with the parameters of the items I_3 and I_9 equal to the values found for the focal population, and then with these item parameters equal to the values found for the reference population. Of course, in both instances the estimates of the population parameters of the focal population were used. In this way the results of the focal population on an unbiased test were estimated, that is, the results if the item parameters had been the same for the reference and the focal population. The computations were carried out both for weighted and unweighted scores. In both instances, the mean of the expected frequency distribution of the focal population was lower for the biased test. So the bias did indeed have a negative result on the mean performance of the focal group.

18.6 Conclusion

In this chapter, several of the tests of model fit for the RM for dichotomous items presented in Chapter 5 were generalized to the class of IRT models for polytomous items defined by (18.2). It was shown that for this class of models a CML estimation procedure is feasible, and it was argued that this provides a justification for speaking of RMs for polytomous items. Another justification might be that some well-known models that go under the name 'RMs', such as the partial credit model, and the rating scale model, are special cases of (18.2). The analysis of and solutions to the problem of evaluating model fit proposed in Chapter 5 easily generalized to the case of polytomous items. Again the problem was solved in the framework of the general multinomial model and all four types of tests considered in Chapter 5, that is, generalized Pearson tests, likelihood ratio tests, Wald tests, and Lagrange multiplier tests, could be adapted to the framework of polytomous items. However, since the model (18.2) is quite general, only the general formats of the test statistics were given, while explicit examples referred to a special case of the model, the OPLM. Explicit expressions for all possible statistics for all special cases of the model given in Table 18.1 are beyond the scope of the present chapter, and much research in this area still needs to be done. The most important aspect of this research will be to construct model tests that have power against specific model violations. Furthermore, the information that comes with evaluating the statistics must be useful for diagnostic purposes. A good example is the R_{1c}-statistic: this tests has power against improper modeling of the item characteristic curves, and the deviates that are computed with the statistic explicitly show which discrimination indices need to be adjusted.

In the last section of this chapter, an example of a testing procedure focussing on the model violation of item bias or DIF was given. Only some of the stages of the process were demonstrated, but the glimpses shown may suffice to demonstrate that item analysis and model fitting is an iterative process aimed at disentangling the various sources of misfit. The example is only one of the many that could be given to show that the RM and its derivatives prove to be a powerful tool for solving measurement problems and answering research questions in the social sciences.

19

Extended Rating Scale and Partial Credit Models for Assessing Change

Gerhard H. Fischer[1] and Ivo Ponocny[2]

ABSTRACT The rating scale model (Rasch, 1961, 1965, 1967; Andrich, 1978a, 1978b) and the partial credit model (Masters, 1982; Andersen, 1983) are extended by incorporating linear structures in the item parameters or item × category parameters, respectively, yielding a 'linear rating scale model' (LRSM) and a 'linear partial credit model' (LPCM). For both models, CML estimation and CLR test methods for assessing the fit and for testing composite hypotheses on change are discussed, and typical repeated measurement designs are considered.

19.1 Introduction

This chapter deals with generalizations of the LLTM and the LLRA (cf. Chapters 8 and 9) to items with polytomous response formats. The most obvious use of such models lies in the domain of repeated measurement designs for assessing change and for testing hypotheses on the effects of treatments. Older related models for polytomous item responses are due to Fischer (1972a, 1977a, 1983b), but applications have been scarce (e.g., Hammer, 1978; Kropiunigg, 1979). More recently, Fischer and Parzer (1991) have formulated a 'linear rating scale model' (LRSM) which can be derived from the rating scale model (RSM) of Rasch (1961, 1965, 1967) and Andrich (1978a, 1978b) by embedding a linear structure in the item parameters β_i, analogous to the decomposition of item parameters in the LLTM. Similarly, the β_{ih} of the partial credit model (PCM; Masters, 1982; Andersen, 1983; see Sections 15.1 and 16.2) can be decomposed linearly, yielding a linear partial credit model (LPCM). The latter was first studied by Glas and Verhelst (1989) within the MML framework, and then by Fischer and Ponocny (1994) under the CML perspective.

The motivation for these extensions of well-known polytomous Rasch models (RMs) is traceable to the lack of appropriate polytomous models for the estimation of treatment effects and for hypothesis testing about the effect parameters: there is a demand for models for repeated responses to attitude or self-rating items, or ratings of clinical symptoms, in the presence of individual differences

[1]Department of Psychology, University of Vienna, Liebiggasse 5, A-1010 Vienna, Austria; e-mail: GH.FISCHER@UNIVIE.AC.AT

This research was supported in part by the Fonds zur Förderung der wissenschaftlichen Forschung under Grant No. P10118-HIS.

[2]Department of Psychology, University of Vienna, Liebiggasse 5, A-1010 Vienna, Austria; e-mail: IVO.PONOCNY@UNIVIE.AC.AT

in terms of latent person parameters. As will be seen, the said extensions of the polytomous unidimensional RM with integer scoring are well-suited for this type of psychometric research.

Rasch's (1961, 1965, 1967) polytomous unidimensional model was

$$P(X_{vih} = 1|S_v, I_i, C_0, \dots, C_m) = \frac{\exp[\phi_h(\theta_v + \beta_i) + \omega_h]}{\sum_{l=0}^{m} \exp[\phi_l(\theta_v + \beta_i) + \omega_l]}, \tag{19.1}$$

where

X_{vih}, for $v = 1, \dots, n$; $i = 1, \dots, k$; $h = 0, \dots, m$, are ('locally') independent random variables with realizations $x_{vih} = 1$ if person S_v chooses response category C_h of item I_i, and $x_{vih} = 0$ otherwise;

$\phi = (\phi_0, \dots, \phi_m)$ is the so-called 'scoring function' which attaches one weight ϕ_h to each of the categories C_h, $h = 0, \dots, m$, with $\phi_0 = 0$;

β_i is a scalar item 'easiness' or 'attractiveness' parameter (depending on the item domain);

θ_v is the position of person S_v on the respective unidimensional trait;

ω_h is the attractiveness of response category C_h, with $\omega_0 = 0$.

All models considered in this chapter posit that the latent trait be unidimensional (i.e., assume scalar person parameters θ_v). The item parameters are either scalar, as in (19.1), or vector valued, as in the polytomous multidimensional RM or in the PCM (see Chapters 15 and 16).

Rasch's (personal communication) initial intention regarding the scoring function had been to estimate the ϕ_h empirically, but he was aware that this entails serious difficulties. Andersen (1966; 1980a, p. 276) observed that the multiplicative concatenation of ϕ_h with θ_v and β_i creates both theoretical and practical problems for the estimation. (They are similar to those arising in the dichotomous 2PL or Birnbaum model.) If the ϕ_h are allowed to assume any real values on $[0, 1]$, for instance, the conditional likelihood of a response pattern, given the respective raw score r_v,

$$r_v = \sum_i \sum_h x_{vih}\phi_h, \tag{19.2}$$

may become degenerate (i.e., attain value 1 irrespective of the other parameters), so that the conditional likelihood no longer contains information about the item parameters. This problem is avoided if the ϕ_h are restricted to integer values, e.g., $\phi_h = h$, for $h = 0, \dots, m$. The polytomous unidimensional RM with these integer values ϕ_h is identical to the RSM proposed by Andrich (1978a, 1978b).

The latter author, however, reparameterizes the attractiveness parameters ω_h in the form

$$\omega_h = \sum_{l=1}^{h} \tau_l, \quad \text{for } h = 1, \ldots, m, \tag{19.3}$$

where the τ_l, for $l = 1, \ldots, m$, are uniquely determined as $\tau_l = \omega_l - \omega_{l-1}$ and are interpreted as parameters of thresholds between adjacent categories. This interpretation is motivated by the following observation (Masters, 1982, p. 152; Wright & Masters, 1982, p. 49): the conditional probability that person S_v chooses category C_h on item I_i, given that S_v chooses only from $\{C_{h-1}, C_h\}$, is

$$\begin{aligned} P(C_h|C_{h-1} \vee C_h; S_v, I_i) &= \frac{\exp[h(\theta_v + \beta_i) + \omega_h]}{\exp[(h-1)(\theta_v + \beta_i) + \omega_{h-1}] + \exp[h(\theta_v + \beta_i) + \omega_h]} \\ &= \frac{\exp(\theta_v + \beta_i + \omega_h - \omega_{h-1})}{1 + \exp(\theta_v + \beta_i + \omega_h - \omega_{h-1})} \\ &= \frac{\exp(\theta_v + \beta_i + \tau_h)}{1 + \exp(\theta_v + \beta_i + \tau_h)}, \end{aligned} \tag{19.4}$$

which constitutes a dichotomous RM with difficulty parameters $\beta_i + \tau_h$, for $h = 1, \ldots, m$. Within item I_i, the relative difficulty of passing from one category to the next depends only on the respective τ_h, which leads to their interpretation as threshold parameters.

A problem with the RSM is that each response category C_h must have the same relative attractiveness ω_h (or the same threshold parameter τ_h) in all items. Suppose we have, in a clinical self-rating study with four-point rating items, one item addressing the feeling of fatigue, and another referring to anxiety. Category C_1 ('slightly'), for instance, may in fact have the same meaning when applied to the feeling of fatigue and to that of anxiety, but as regards the highest category, C_3, it is possibly easier to admit to feeling 'very tired' than to being 'extremely afraid'.

Generalizations that take care of this problem and, at the same time, even make allowance for different numbers of response categories per item, were proposed by Masters (1982) and Andersen (1983). These authors introduced parameters β_{ih}, $h = 0, \ldots, m_i$, instead of $h\beta_i + \omega_h$, where $m_i + 1$ is the number of response categories of item I_i. Andersen's model is the somewhat more general one (see Chapter 15), but we limit the present considerations to Masters' PCM, defined by

$$P(X_{vih} = 1|S_v, I_i) = \frac{\exp(h\theta_v + \beta_{ih})}{\sum_{l=0}^{m_i} \exp(l\theta_v + \beta_{il})}, \tag{19.5}$$

for $i = 1, \ldots, k$ and $h = 0, \ldots, m_i$. All the other notation is as before. The parameters θ_v still represent the persons' positions on a unidimensional latent trait, but the items are characterized by parameter vectors (β_{ih}), for $i = 1, \ldots, k$

and $h = 0, \ldots, m_i$, so that the model is capable of accomodating item $\times$ category interactions and different numbers of categories between items.

Again, as in (19.3), the β_{ih} can be written as sums of threshold parameters,

$$\beta_{ih} = \sum_{l=1}^{h} \tau_{il}, \quad \text{for } h = 1, \ldots, m_i, \tag{19.6}$$

where the thresholds τ_{ih} are the points on the latent continuum where the 'category characteristic curves' $P(X_{i,h-1} = 1|\theta)$ and $P(X_{ih} = 1|\theta)$ intersect (Wright & Masters, 1982, p. 44).

To prevent an overparameterization of the PCM, a normalization of the parameters, e.g., as follows, will be assumed henceforth:

$$\beta_{i0} = 0, \quad \text{for } i = 1, \ldots, k, \text{ and} \tag{19.7}$$

$$\beta_{11} = 0 \quad \text{or} \quad \sum_{i=1}^{k} \sum_{h=1}^{m_i} \beta_{ih} = 0. \tag{19.8}$$

The raw scores, as in the RSM, are

$$r_v = \sum_{i=1}^{k} \sum_{h=1}^{m_i} h x_{vih}, \tag{19.9}$$

serving as minimal sufficient statistics for θ_v. Similarly, the item marginal sums

$$x_{.ih} = \sum_{v} x_{vih}, \quad \text{for } i = 1, \ldots, k \text{ and } h = 0, \ldots, m_i, \tag{19.10}$$

are the minimal sufficient statistics for the β_{ih}. Obviously, the RSM is a special PCM, with $\beta_{ih} = \beta_i + \omega_h$.

Within the CML approach, the number of parameters to be estimated for the PCM is $\sum_{i=1}^{k} m_i$ (the number of item $\times$ category parameters β_{ih}) minus 1 (due to the normalization (19.8)), that is, $\sum_{i=1}^{k} m_i - 1$. In the special case where all items have the same number of response categories, $m_i = m$, the result is $km - 1$. Therefore, the number of parameter estimates in the PCM increases considerably with m. In the RSM, on the other hand, this number is much smaller: $k - 1$ (item parameters) plus $m - 1$ (category parameters), that is, $k + m - 2$. While a PCM application with, for instance, $k = 25$ items and $m + 1 = 5$ categories requires the estimation of 99 parameters, the RSM needs only 27 parameter estimates. A consequence of this is that in the PCM the usual Newton-Raphson procedure for solving CML equations, requiring the computation and inversion of the matrix of second-order partial derivatives (the so-called 'Hesse matrix') in each iteration cycle, is no longer economical and had better be replaced by a more efficient procedure (described in Section 19.3).

19.2 LPCM and LRSM

Analogous to the definition of the LLTM as an extension of the RM, the item parameters of the PCM will now be subjected to the linear constraints (19.11),

$$\beta_{ih} = \sum_{j=1}^{p} w_{ihj}\alpha_j + hc, \tag{19.11}$$

for $i = 1, \ldots, k$ and $h = 1, \ldots, m_i$, where

α_j, for $j = 1, \ldots, p$, are the so-called 'basic parameters' of the model;

w_{ihj}, for $j = 1, \ldots, p$, are the weights of the basic parameters α_j for item parameter β_{ih}, with $w_{i0j} = 0$ for all i, j because of (19.7); these weights must be given *a priori* and are part of the model structure;

c is the normalization constant.

The multiplication of c by h in (19.11) is necessary in order to compensate any shift of the person parameter scale in (19.5), $\theta \to \theta + \delta$, by the change $c \to c - \delta$ of the normalization constant. The model defined in (19.5) together with (19.11) is denoted the 'partial credit model with linear constraints' or, more concisely, the 'linear partial credit model' (LPCM).

Although many different uses of the LPCM can be thought of, our main thrust is in describing change under the impact of treatments. Clearly, 'change' always resides in the persons; for formal convenience, however, we shall equivalently model change in terms of the item parameters. To that end, we again introduce the notion of 'virtual items' (V-items), in contrast to 'real items': suppose a person S_v takes a test at time point T_1, then undergoes some treatment(s), and finally takes the same test again at time point T_2. At T_1, the probability of S_v choosing category C_h of item I_i is, according to model (19.5), completely determined by θ_v and by the β_{ih}, $h = 0, \ldots, m_i$. At T_2, however, the treatment(s) may have changed (a) S_v's trait parameter θ_v and (b) the attractiveness of some (or all) response categories in some (or all) items. In order to account for both kinds of changes, we formally consider item I_i at T_2 as a new item, denoted I_i', with new item parameters β_{ih}'. So, one 'real item' I_i generates two 'virtual items' (V-items), I_i and I_i'. This conceptualization is immediately extended to any number of time points or testing conditions, $T_1, T_2, T_3, \ldots$, with items $I_i, I_i', I_i'', \ldots$, etc.

As a sufficiently general terminology and notation for a measurement of change framework, we henceforth distinguish the k (real) items I_i from u V-items, denoted by I_l^*, $l = 1, \ldots, u$, with V-item parameters β_{lh}^*. We shall apply the LPCM to these virtual items rather than to the real items. The V-items are then decomposed in the same manner as in (19.11),

$$\beta_{lh}^* = \sum_{j=1}^{p} w_{lhj}\alpha_j + hc, \tag{19.12}$$

for $l = 1, \ldots, u$ and $h = 1, \ldots, m_l$, with appropriately redefined weights w_{lhj}.

Supposing, for the above example with two time points and one treatment group, that a test of k items is used, we have $2k$ V-items, the first k of which are identical to the k real items,

$$\beta^*_{lh} = \beta_{lh} + hc,$$

for $l = 1, \ldots, k$ and $h = 1, \ldots, m_l$; for the other k V-items suppose, e.g., that the original item parameters have been changed by a weighted sum of the treatment effects η_j,

$$\beta^*_{lh} = \beta_{l-k,h} + h \sum_j q_{gj}\eta_j + hc,$$

for $l = k+1, \ldots, 2k$ and $h = 1, \ldots, m_l$, where q_{gj} is the dosage of treatment B_j given to group G_g of persons under consideration. (For $h = 0$, $\beta_{ih} = 0$ and $\beta^*_{lh} = 0$ holds according to the normalization conditions.)

If there is more than one treatment group, as will be the case in most well-designed studies on treatment effects, different person groups respond to different subsets of V-items. This means that in such studies it is natural to have structurally incomplete designs. The extension to more than two time points T_t with dosages q_{gjt} is obvious but cannot be treated within the scope of this chapter.

Having introduced the LPCM, we shall now show that the PCM, the LRSM, the RSM, the LLTM, and the RM, are special cases of the LPCM. (We assume $m_i = m$ here for simplicity.)

1. The PCM is obtained from (19.11) by setting $w_{ihj} = 1$ for $i = 1, \ldots, k$ and $h = 1, \ldots, m$, with $j = (i-1)m + h$; otherwise, $w_{ihj} = 0$. This makes the basic parameters α_j identical to the β_{ih} (except for the $\beta_{i0} = 0$).

2. The LRSM for a k-items test presented at one time point is obtained from (19.11) by restricting the β_{ih} to $\beta_{ih} = h\beta_i + \omega_h$, with $m_i = m$, where the β_i are subject to further linear constraints,

$$\beta_i = \sum_{j=1}^{p} w_{ij}\alpha_j + c, \quad \text{for } i = 2, \ldots, k, \tag{19.13}$$

and $\beta_1 = 0$ for normalization; (19.11) then becomes

$$\beta_{ih} = h \sum_j w_{ij}\alpha_j + \omega_h + hc. \tag{19.14}$$

For the measurement of change between T_1 and T_2, where the items are replaced by V-items with parameters β^*_{lh}, we similarly obtain from (19.12) that

$$\beta^*_{lh} = h\beta_l + \omega_h + hc,$$

for $l = 1, \ldots, k$ and $h = 1, \ldots, m$, and

$$\beta^*_{lh} = h\beta_{l-k} + h\sum_j q_{vj}\eta_j + \omega_h + hc,$$

for $l = k+1, \ldots, 2k$ and $h = 1, \ldots, m$, where q_{vj} is the dosage of treatment B_j given to person S_v up to time point T_2 at which he/she reacts to V-item I^*_l. (Note that each V-item I^*_l corresponds uniquely to one time point T_t.)

3. The RSM of Andrich (1978a, 1978b) for data obtained at a single time point results as a special case of (19.14), setting $w_{i,i-1} = 1$, and $w_{ij} = 0$ otherwise, for $i = 2, \ldots, k$ and $j = 1, \ldots, k-1$, combined with $w_{1j} = 0$, for $j = 1, \ldots, k-1$, thus setting $\beta_1 = 0$ for normalization.

4. Finally, both the dichotomous LLTM (cf. Chapter 8) and the dichotomous RM result as special cases of the LRSM and the RSM, respectively, for $m = 1$.

Note that (19.12) is very general: if change is understood merely as a shift of the person on the unidimensional latent trait θ, the weights of the treatment effects – as basic parameters in (19.12) – have to be restricted to $w_{lhj} = hq_{vjt}$ for $h = 0, \ldots, m_l$, for all items I^*_l given to persons S_v at time point T_t. This excludes interactions of treatments and items. Such interactions can be accomodated immediately, however, by means of an appropriate, more general choice of the weights. How this is done will be exemplified in Section 19.4 below.

19.3 CML estimation

In this chapter, the treatment of estimation and hypothesis testing in the LPCM is based exclusively on the CML approach. Since the attractive properties of CML have been pointed out in Chapters 2, 3, 8, and 15, we refrain from discussing them again here. Our main motivation in choosing CML is that rigorous asymptotic test procedures for composite hypotheses are available, based on the conditional likelihood ratio statistic, which do not require any computations beyond the CML estimation.

For the sake of simplicity, let us presume that there are treatment groups G_g of sample sizes n_g, so that all $S_v \in G_g$ undergo the same treatments with the same dosages. (The case of individual treatment combinations or individual dosages can be covered by setting $n_g = 1$.) Then, the conditional likelihood of the data has the following form, analogous to (3.9), (8.19), and (15.21),

$$L = \prod_g L_g = \prod_g \frac{\prod_{l=1}^{u}\prod_{h=1}^{m_l} {\epsilon_{lh}}^{s_{glh}}}{\prod_r {\gamma_{gr}}^{n_{gr}}}, \tag{19.15}$$

where

$\epsilon_{lh} = \exp(\beta^*_{lh})$, $l = 1, \dots, u$; $h = 0, \dots, m_l$;[3] this implies $\epsilon_{l0} = 1$ due to normalization (19.7);

s_{glh} denotes the number of persons in G_g who have chosen category C_h of V-item I_l^*, $s_{glh} = \sum_{S_v \in G_g} x_{vlh}$; if item I_l^* was not presented to group G_g, $s_{glh} = 0$ for $h = 0, \dots, m_l$;

n_{gr} is the number of persons in G_g who have raw score r, where r is defined as in (19.9);

γ_{gr} is the combinatorial function defined in (15.20), see Chapter 15, namely,

$$\gamma_{gr} = \sum_{\boldsymbol{y}_g | r} \prod_{l=1}^{u} \prod_{h=1}^{m_l} \epsilon_{lh}{}^{y_{glh}}. \tag{19.16}$$

The summation runs over all possible response vectors $\boldsymbol{y}_g$ consistent with raw score $r = \sum_{l=1}^{u} \sum_{h=1}^{m_l} h y_{glh}$, with y_{glh} defined analogous to the x_{vlh} for $S_v \in G_g$; the index g is necessary because $\sum_{h=0}^{m_l} x_{vlh} = \sum_{h=0}^{m_l} y_{glh} = 1$ for all I_l^* presented to persons $S_v \in G_g$, but $x_{vlh} = y_{glh} = 0$ for $h = 0, \dots, m_l$ if item I_l^* was not presented to $S_v \in G_g$. This ensures that no superfluous terms are generated in (19.16) and (19.15): if I_l^* was not presented to $S_v \in G_g$, $\epsilon_{lh}{}^{y_{glh}} = \epsilon_{lh}^0 = 1$ for $h = 0, \dots, m_l$ in (19.16), and $\epsilon_{lh}{}^{s_{glh}} = \epsilon_{lh}^0 = 1$ in (19.15), so that V-item I_l^* is correctly ignored.

Taking logarithms of (19.15),

$$\ln L = \sum_g \Big(\sum_{l=1}^{u} \sum_{h=1}^{m_l} s_{glh} \beta^*_{lh} - \sum_r n_{gr} \ln \gamma_{gr} \Big). \tag{19.17}$$

It is well-known that L in (19.17) is the log-likelihood function of an exponential family (cf. Chapter 15). This implies that, after proper normalization of the parameters and given sufficiently large samples, $\ln L$ will have a unique maximum.

Differentiating $\ln L$ with respect to the basic parameters α_j in (19.12) yields the CML estimation equations,

$$\frac{\partial \ln L}{\partial \alpha_j} = \sum_g \sum_{l=1}^{u} \sum_{h=1}^{m_l} w_{lhj} \Big(s_{glh} - \sum_r \frac{n_{gr}}{\gamma_{gr}} \frac{\partial \gamma_{gr}}{\partial \epsilon_{lh}} \epsilon_{lh} \Big) = 0, \tag{19.18}$$

for $j = 1, \dots, p$. To solve (19.18) numerically, suitable recursive procedures for the γ_{gr} and their derivatives are needed. Two properties of such procedures seem indispensable: (a) numerical accuracy, and (b) efficiency in terms of computing expenses. Both goals can be reached by mixing the following two recursive algorithms for the combinatorial functions γ.

---

[3]It would be consistent to add asterisks to ϵ_{lh} and γ_{gr}, since they refer to V-items and their parameters, but we drop these asterisks for the sake of simplicity of the formulae.

19.3.1 The Summation Algorithm

Let the subset of V-items presented to group G_g be denoted $J_1, \dots, J_a$ for simplicity. (The index a should be a_g, of course, but with this in mind we write 'a' for convenience. Moreover, we simply write $\gamma_{gr}(J_1, \dots, J_s)$, for $s = 1, \dots, a$, instead of $\gamma_{gr}(\epsilon_{10}, \dots, \epsilon_{1m_1}; \dots; \epsilon_{s0}, \dots, \epsilon_{sm_s})$.)

Then,

$$\gamma_{gr}(J_1, \dots, J_s) = \sum_{h=0}^{\min(r,m_s)} \gamma_{g,r-h}(J_1, \dots, J_{s-1})\epsilon_{sh} \tag{19.19}$$

is a recurrence relation for the combinatorial functions γ_{gr}. It is numerically stable because the sum in (19.19) contains only positive terms. Starting values for the recursion are $\gamma_{gh}(J_1) = \epsilon_{1h}$, for $h = 0, \dots, m_1$. In each iteration, all values of the raw score r obtainable on items $J_1, \dots, J_s$ are considered. This method was first suggested by Andersen (1972) and has since then been used successfully in several computer programs; it is denoted the 'summation algorithm'.

The partial derivative of a γ_{gr} with respect to ϵ_{tk} is

$$\frac{\partial \gamma_{gr}}{\partial \epsilon_{tk}} = \gamma_{g,r-k}(J_1, \dots, J_{t-1}, J_{t+1}, \dots, J_s) =: \gamma_{g,r-k}^{(t)}. \tag{19.20}$$

If $J_l \notin \{J_1, \dots, J_s\}$, $\gamma_{g,r-k}^{(l)} = 0$. A recursion for $\partial\gamma_{gr}/\partial\epsilon_{tk}$ is obtained by differentiating the right-hand side of (19.19) with respect to ϵ_{tk}, which yields

$$\frac{\partial \gamma_{gr}(J_1, \dots, J_s)}{\partial \epsilon_{tk}} = \sum_{h=0}^{\min(r,m_s)} \frac{\partial \gamma_{g,r-h}(J_1, \dots, J_{s-1})}{\partial \epsilon_{tk}} \epsilon_{sh} + \gamma_{g,r-h}(J_1, \dots, J_{s-1})\delta_{st}\delta_{hk}, \tag{19.21}$$

where $\delta_{st} = 1$ for $s = t$, and $\delta_{st} = 0$ for $s \neq t$ (the Kronecker symbol). Again, this recursion involves only summations of positive terms and is hence numerically stable.

19.3.2 The Difference Algorithm

The fastest algorithm known for computing the γ_r and their first derivatives in the dichotomous RM is the 'difference algorithm' used by Fischer and Allerup (1968; for details see Fischer, 1974, pp. 241–244). Since, however, it was found to be prone to numerical errors when tests are longer than, e.g., $k = 40$, it was later given up (for a discussion, see Gustafsson, 1980a,b). With a slight modification and suitable generalization to polytomous items, however, it affords a very expedient and yet sufficiently accurate technique for computing the first-order derivatives of the γ_{gr}. To obtain the highest possible accuracy of the results, we suggest mixing the summation and the difference algorithms as follows (see Fischer & Ponocny, 1994; Liou, 1994):

Let the γ_{gr} be computed by means of the summation method as above, using (19.19) for all a V-items given to group G_g. Regarding $\gamma^{(t)}_{g,r-k}$, observe that

$$\gamma_{g1} = \gamma^{(t)}_{g0}\epsilon_{t1} + \gamma^{(t)}_{g1},$$
$$\gamma_{g2} = \gamma^{(t)}_{g0}\epsilon_{t2} + \gamma^{(t)}_{g1}\epsilon_{t1} + \gamma^{(t)}_{g2},$$
$$\vdots \quad \vdots$$

so that

$$\begin{aligned} \gamma^{(t)}_{g1} &= \gamma_{g1} - \gamma^{(t)}_{g0}\epsilon_{t1}, \\ \gamma^{(t)}_{g2} &= \gamma_{g2} - \gamma^{(t)}_{g0}\epsilon_{t2} - \gamma^{(t)}_{g1}\epsilon_{t1}, \\ \vdots \quad & \quad \vdots \\ \gamma^{(t)}_{gr} &= \gamma_{gr} - \gamma^{(t)}_{g,r-1}\epsilon_{t1} - \ldots - \gamma^{(t)}_{g,r-q}\epsilon_{tq}, \end{aligned} \tag{19.22}$$

where $q = \min(r, m_t)$. Starting with $r = 1$, (19.22) affords a complete 'forward' recursion for $r = 1, \ldots, \tilde{r}$ with $\tilde{r} := r_{\max}$, where $r_{\max}$ denotes the largest possible raw score.

Clearly, since (19.22) involves subtractions, this method *may* be affected by numerical error; but if the recursion is applied separately to each item I_t, any numerical error occurring for one index value t cannot spread to other values of t. In order to safeguard against numerical error, a check of the accuracy is necessary, which can be carried out separately for each t. To that end, basically the same recursion can be run backwards also, starting with $r = \tilde{r}$ and reducing r by 1 in each step: from

$$\gamma_{g,\tilde{r}-l} = \gamma^{(t)}_{g,\tilde{r}-l-m_t}\epsilon_{tm_t} + \gamma^{(t)}_{g,\tilde{r}-l-m_t+1}\epsilon_{t,m_t-1} + \ldots + \gamma^{(t)}_{g,\tilde{r}-l}$$

one obtains the recurrence relation

$$\gamma^{(t)}_{g,\tilde{r}-l-m_t} = \epsilon^{-1}_{tm_t}\left(\gamma_{g,\tilde{r}-l} - \gamma^{(t)}_{g,\tilde{r}-m_t-(l-1)}\epsilon_{t,m_t-1} - \ldots - \gamma^{(t)}_{g,\tilde{r}-l}\right). \tag{19.23}$$

Therefore, a backward recursion consists in the application of the recurrence relation (19.23) for $l = 0, \ldots, \tilde{r} - m_t$, setting $\gamma^{(t)}_{gr} = 0$ for $r > \tilde{r} - m_t$:

$$\begin{aligned} \gamma^{(t)}_{g,\tilde{r}-m_t} &= \epsilon^{-1}_{tm_t}\gamma_{g,\tilde{r}}, \\ \gamma^{(t)}_{g,\tilde{r}-1-m_t} &= \epsilon^{-1}_{tm_t}(\gamma_{g,\tilde{r}-1} - \gamma^{(t)}_{g,\tilde{r}-m_t}\epsilon_{t,m_t-1}), \\ \gamma^{(t)}_{g,\tilde{r}-2-m_t} &= \epsilon^{-1}_{tm_t}(\gamma_{g,\tilde{r}-2} - \gamma^{(t)}_{g,\tilde{r}-m_t}\epsilon_{t,m_t-2} - \gamma^{(t)}_{g,\tilde{r}-m_t-1}\epsilon_{t,m_t-1}), \end{aligned}$$

and so forth.

Thus, in principle one might compute two values for each $\gamma^{(t)}_{gr}$ and then compare the elements of each pair. In practice, however, it suffices to apply the forward recursion, beginning with $r = 1$, until in (19.22) the differences become 'small', that is, until

$$\gamma^{(t)}_{gr} < \gamma^{(t)}_{g,r-q}\epsilon_{tq} \tag{19.24}$$

(at $r = \tilde{k}$, say), and then to apply the backward recursion (19.23) from $r = \tilde{r}$ to $r = \tilde{k}$; finally, to compare the two values obtained for $\gamma_{g\tilde{k}}^{(t)}$. If they are close enough, accept all $\gamma_{gr}^{(t)}$ for this value of t; if not, reject all $\gamma_{gr}^{(t)}$ for this particular t and recompute them by the more stable summation algorithm based on (19.21). (Our condition for accepting the trial values is that the relative error of $\gamma_{g\tilde{k}}^{(t)}$ be less than 0.00001.)

Simulation studies showed that this version of the difference algorithm indeed accelerates the estimation procedure, without any perceptible loss of accuracy. A typical example with 50 items and 4 categories per item was computed on a PC AT 486 within 4 mins using only the summation algorithm, whereas for the difference algorithm as described above, 3 mins were required; that is, the difference algorithm was about 25% faster, in spite of about 50% rejections and recalculations of trial values. All absolute differences between estimates of $\gamma_{gr}^{(t)}$ under the two algorithms were less than 0.0001. A reduction of 25% computing time may not seem convincing; yet it should be taken into account that this example required a large number of corrections entailing a second independent computation of $\gamma_{gr}^{(t)}$ by means of (19.21). In another simulation example with 35 items and 4 categories, an acceleration from 39 secs to 22 secs was achieved, that is 44%; in this example, only 14% of the $\gamma_{gr}^{(t)}$ had to be recomputed. It is remarkable that the advantage of the difference algorithm is even greater for the dichotomous RM: when no rejections occurred, the ratio of the run times for both algorithms was > 5 (namely, 5 mins 45 secs for the summation vs. 1 min 4 secs for the difference algorithm), again with an absolute difference in the results of less than 0.0001.

(Remark: The old algorithm for the dichotomous RM used by Fischer and Allerup (1968; described in detail in Fischer, 1974) employed both the forward and the backward recursions, but made only one overall check of the accuracy instead of one separate check for each t, and did not employ the summation method (19.19) to correct insufficiently accurate values. This explains why the old algorithm was not so successful in long tests.)

19.3.3 Iterative Solution of the CML Equations

The iterative procedure typically employed for solving the nonlinear CML equations is the well-known Newton-Raphson method. This, however, also requires the computation of the second-order partial derivatives of $\ln L$, for which recursions are obtained by differentiating (19.21) once more (cf. Fischer & Parzer, 1991, who applied this method to the LRSM). Since, however, the number of parameters in the LPCM is generally much larger than in the LRSM, this approach is not recommended in view of the relatively heavy computations required for the Hesse matrix and its inversion. The latter is entirely avoided by using a 'Quasi-Newton' method, namely, the Broyden-Fletcher-Goldfarb-Shanno

(BFGS) approximation (see, e.g., Churchhouse, 1981, pp. 493–496, or Kosmol, 1989, pp. 179–191).

The BFGS method of descent (for minimizing $-\ln L$) produces a sequence of matrices $\boldsymbol{B}_k$, $k = 1, 2, \ldots$, converging to $\boldsymbol{I}^{-1}$, the inverse of the information matrix evaluated at $\hat{\boldsymbol{\alpha}}$. In combination with a suitable step-length rule (e.g., the so-called 'Armijo rule', see below), the BFGS method guarantees that all $\boldsymbol{B}_k$ are positive definite and that the sequence $\boldsymbol{\alpha}_k$ converges to a minimum solution $\hat{\boldsymbol{\alpha}}$ of $-\ln L$ at least in Q-superlinear form; this means that there exists a sequence of non-negative integers c_k converging to zero such that, for all k,

$$||\boldsymbol{\alpha}_{k+1} - \hat{\boldsymbol{\alpha}}|| \leq c_k ||\boldsymbol{\alpha}_k - \hat{\boldsymbol{\alpha}}||. \tag{19.25}$$

To determine the step length, the 'Armijo rule with extended step length' (Armijo, 1966; see also Kosmol, 1989, p. 86) can be recommended. The principle of this rule is that the steepness of the secant between $\boldsymbol{\alpha}_k$ and $\boldsymbol{\alpha}_{k+1}$ should not become much smaller than the steepness of the tangent at $\boldsymbol{\alpha}_k$. The step-length rule is defined as follows: let two constants $d \in (0, 1)$ and $c \in (0, \frac{1}{2})$ be given (which may be used for 'tuning' the procedure), and let $-\ln L(\boldsymbol{\alpha}_k)$ be denoted by $f(\boldsymbol{\alpha}_k)$; choose the step length $\lambda_k = d^{n_k}$, where n_k is the smallest (possibly negative) integer satisfying

$$f(\boldsymbol{\alpha}_k) - f(\boldsymbol{\alpha}_k - d^{n_k} \boldsymbol{B}_k \bigtriangledown f_k) \geq c d^{n_k} \langle \bigtriangledown f_k, \boldsymbol{B}_k \bigtriangledown f_k \rangle, \tag{19.26}$$

but such that (19.26) is not satisfied by $n_k - 1$. (The symbol $\langle \cdot, \cdot \rangle$ denotes the inner vector product, and $\bigtriangledown f_k$ the gradient of f at $\boldsymbol{\alpha}_k$.)

This rule ensures that the descent (with respect to $f = -\ln L$) cannot become too small at the beginning of the iteration or in the neighborhood of the solution; the step length is variable and is automatically controlled by the procedure itself.[4]

Although the BFGS method requires a larger number of iterations than the Newton-Raphson method, each iteration is much faster, so that the total time consumed is less, particularly so in applications where the number of parameters is large. The BFGS algorithm has been seen to be very efficient for Rasch-type models, because these are exponential families, that is, the function $\ln L$ has a unique maximum whenever the model is in minimal form (has been normalized appropriately), and if the sample size is sufficiently large (cf. the uniqueness conditions in the dichotomous RM and LLTM discussed in Chapters 3 and 8). The BFGS method even works if the model is *not* in minimal form (e.g., when the weight matrix $\boldsymbol{W}^+$ of the LLTM fails to satisfy the critical rank condition), but then the sequence of matrices $\boldsymbol{B}_k$ does *not* converge to the asymptotic information matrix. In such cases, the parameter estimates still are correct (except

[4]The program 'LpcM' using the algorithms described in this section can be obtained from the authors. Please e-mail to GH.FISCHER@UNIVIE.AC.AT.

for proper normalization), but the estimates of their confidence intervals may be completely misleading. Hence, the user has to check whether his/her model is parameterized and normalized appropriately.

19.4 Some Typical Designs and Hypothesis Testing

Since the LPCM has been defined in a sufficiently general form in (19.12), it can now serve as a basis for formulating and testing a host of composite hypotheses about the V-item parameters β^*_{lh} and the basic parameters α_j (e.g., treatment effects η_j and a trend τ). Let H_0 be a restriction of a hypothesis H_1 where the latter is believed to be true, and assume that the LPCM is identifiable under both hypotheses (the parameters are properly normalized, i.e., the model is in minimal form, and there are sufficient data); if L_0 and L_1 are the maxima of the respective conditional likelihood functions, then

$$-2\ln\lambda = -2(\ln L_0 - \ln L_1) \overset{as.}{\sim} \chi^2_{(df)},$$

where the number of degrees of freedom, df, equals the difference of the numbers of parameters under H_1 and H_0, respectively (cf. Chapter 18). Special cases of such hypothesis tests arise if the sample of persons is split into subgroups as in Andersen's (1973b) goodness of fit test for the RM. Moreover, the asymptotic standard error of the parameter estimates $\hat{\alpha}_j$ can be computed from the inverse of the Hesse matrix, which results as a by-product of the BFGS method.

In what follows, we briefly discuss a few typical longitudinal designs and related hypotheses.

19.4.1 One unidimensional latent trait, two time points

Suppose that a unidimensional test of k items, each with $m+1$ response categories, is administered to persons at two time points, T_1 and T_2. Let the persons receive – possibly different – treatment combinations, including the 'null-combination' of no treatment (Control Group). All persons receiving the same treatment combination will be called a Treatment Group, G_g. In formulating a model for the measurement of treatment effects, we sometimes replace the abstract index 'l' in β^*_{lh} for convenience by an index combination g, i, and t, where $g = 1, \dots, d$ represents the Treatment Groups, $i = 1, \dots, k$ the items, and $t = 1, 2$ the time points. The data are of the form x_{viht}, for $v = 1, \dots, n$; $i = 1, \dots, k$; $h = 0, \dots, m$; $t = 1, 2$, as illustrated in Table 19.1 (for brevity here restricted to two groups, G_1 and G_2). Clearly, in terms of V-items, the design is structurally incomplete.

Let one treatment effect η_j be assigned to each treatment B_j, and let a trend parameter τ be the representative of change occurring independently of the treatments (e.g., change in the response behavior due to development or to repeated presentation of the same test). This design generates $(d+1)k$ V-items,

TABLE 19.1. Repeated measurement data in a design with k unidimensional items given to persons of $d = 2$ groups, G_1 and G_2, at two time points, T_1 and T_2. $I_1, \ldots, I_k$ are the k items of the test given at T_1; I'_i and I''_i, for $i = 1, \ldots, k$, are the same items given at T_2, i.e., after the treatments. These $3k$ V-items are denoted I^*_l, $l = 1, \ldots, 3k$. Persons S_v belong to group G_1, S_w to group G_2. The symbol (x_{viht}) denotes the response vector of elements x_{viht} with fixed indices v, i, t, for $h = 0, \ldots, m$.

		Time Point T_1	Time Point T_2	
	Items:	$I_1 \ \ldots \ I_k$	$I'_1 \ \ldots \ I'_k$	$I''_1 \ \ldots \ I''_k$
	V-Items:	$I^*_1 \ \ldots \ I^*_k$	$I^*_{k+1} \ \ldots \ I^*_{2k}$	$I^*_{2k+1} \ \ldots \ I^*_{3k}$
G_1	1 $\vdots$ n_1	Item Response Vectors (x_{vih1})	Item Response Vectors (x_{vih2})	No Observations
G_2	$n_1 + 1$ $\vdots$ $n_1 + n_2$	Item Response Vectors (x_{wih1})	No Observations	Item Response Vectors (x_{wih2})

e.g., in the case of $d = 2$ groups as shown in Table 19.1, k V-items I^*_i, $i = 1, \ldots, k$, which are identical to real items $I_1, \ldots, I_k$ given at T_1, plus $2k$ V-items representing the k items given after the treatments.

Now the effects of the treatments can be modeled, e.g., as follows:

$$\begin{aligned} \beta^*_{gih1} &= \beta_{ih} + hc && \text{for the } k \text{ V-items } I^*_i \ (k \text{ real items } I_i) \text{ presented at } T_1, \\ \beta^*_{gih2} &= \beta_{ih} + h \sum_j q_{gj} \eta_{jh} + h\tau + hc, && \text{for the } 2k \text{ V-items } I^*_{k+i} \text{ and } I^*_{2k+i} \text{ presented at } T_2, \end{aligned} \tag{19.27}$$

where

β_{ih} are the item parameters of the (real) items I_i;

β^*_{giht}, for all g and $t = 1, 2$, are the V-item parameters; notice that $\beta^*_{1ih1} = \beta^*_{2ih1}$ for all i, h;

q_{gj} is the dosage of treatment B_j, $j = 1, \ldots, b$, given to treatment group G_g between T_1 and T_2;

η_{jh} is the effect of treatment B_j on the preference for response category C_h, with $\eta_{j0} = 0$ for all B_j, so that normalization (19.7) is preserved;

τ is a trend effect (independent of the treatments);

c is a normalization constant needed to assure that the β^*_{giht} satisfy a normalization condition analogous to (19.8).

TABLE 19.2. Matrix of weights w_{lhj} in (19.12) specified for Model 1, however, given here only for one treatment group G_g. All unspecified matrix elements are zero. Note that the β^*_{giht} are subject to normalization conditions equivalent to those in (19.7)–(19.8); this is the reason why the table need not show the $\beta_{gi0t} = 0$, and why the first column has to be deleted to make $\beta_{11} = \beta^*_{g111} = 0$, see (19.8).

	β_{11}	$\beta_{12} \dots \beta_{1m}$	$\dots$	$\beta_{k1}\ \beta_{k2} \dots \beta_{km}$	$\eta_{11}\ \eta_{12} \cdots \eta_{1m}$	$\dots$	$\eta_{b1}\ \eta_{b2} \cdots \eta_{bm}$	τ
β^*_{g111}	1							
β^*_{g121}		1						
$\vdots$		$\ddots$						
β^*_{g1m1}		1						
$\vdots$			$\ddots$					
β^*_{gk11}				1				
β^*_{gk21}				1				
$\vdots$				$\ddots$				
β^*_{gkm1}				1				
β^*_{g112}	1				q_{g1}	$\dots$	q_{gb}	1
β^*_{g122}		1			$2q_{g1}$	$\dots$	$2q_{gb}$	2
$\vdots$		$\ddots$			$\ddots$	$\vdots$	$\ddots$	$\vdots$
β^*_{g1m2}		1			mq_{g1}	$\dots$	mq_{gb}	m
$\vdots$			$\ddots$			$\ddots$		$\vdots$
β^*_{gk12}				1	q_{g1}	$\dots$	q_{gb}	1
β^*_{gk22}				1	$2q_{g1}$	$\dots$	$2q_{gb}$	2
$\vdots$				$\ddots$	$\ddots$	$\vdots$	$\ddots$	$\vdots$
β^*_{gkm2}				1	mq_{g1}	$\dots$	mq_{gb}	m

The model defined in (19.27) will be denoted Model 1. The weights w_{lhj} in (19.12) specified for Model 1 are shown as elements of a matrix in Table 19.2.

A null-hypothesis saying that the treatments influence only the latent trait parameters θ_v, but do not interact with the response categories, can now be formulated as a special case of Model 1: for time point T_1, the parameters are as in (19.27), and for T_2 we hypothesize

$$\beta^*_{gih2} = \beta_{ih} + h\sum_j q_{gj}\eta_j + h\tau + hc, \quad \text{for the } k \text{ V-items given to persons in } G_g \text{ at } T_2. \tag{19.28}$$

This is equivalent to setting $\eta_{jh} = \eta_j$, for $h = 1, \dots, m$, in (19.27), reducing the bm parameters η_{jh} to b parameters η_j, so that the respective CLR test has $df = b(m-1)$. The corresponding weight matrix is obtained from that in Table

TABLE 19.3. Matrix of weights w_{lhj} in (19.12) specified for Model 2, i.e., assuming in Model 1 that $\eta_{jh} = \eta_j$, for $h = 1, \ldots, m$. All unspecified matrix elements are zeros. Again, the first column has to be deleted for normalization, making $\beta_{11} = 0$.

	β_{11}	β_{12}	$\ldots$	β_{1m}	$\ldots$	β_{k1}	β_{k2}	$\ldots$	β_{km}	η_1	$\ldots$	η_b	τ
β^*_{g111}	1												
β^*_{g121}		1											
$\vdots$			$\ddots$										
β^*_{g1m1}				1									
$\vdots$					$\ddots$								
β^*_{gk11}						1							
β^*_{gk21}							1						
$\vdots$								$\ddots$					
β^*_{gkm1}									1				
β^*_{g112}	1									q_{g1}	$\ldots$	q_{gb}	1
β^*_{g122}		1								$2q_{g1}$	$\ldots$	$2q_{gb}$	2
$\vdots$			$\ddots$							$\vdots$	$\vdots$	$\vdots$	$\vdots$
β^*_{g1m2}				1						mq_{g1}	$\ldots$	mq_{gb}	m
$\vdots$					$\ddots$					$\vdots$	$\ddots$	$\vdots$	$\vdots$
β^*_{gk12}						1				q_{g1}	$\ldots$	q_{gb}	1
β^*_{gk22}							1			$2q_{g1}$	$\ldots$	$2q_{gb}$	2
$\vdots$								$\ddots$		$\vdots$	$\vdots$	$\vdots$	$\vdots$
β^*_{gkm2}									1	mq_{g1}	$\ldots$	mq_{gb}	m

19.2 by collapsing the columns corresponding to parameters $\eta_{j1}, \ldots, \eta_{jm}$ into one column for η_j, see Table 19.3. This model is denoted Model 2.

Extensions of this design to any number of treatment groups and time points are obvious and need not be discussed in detail.

19.4.2 Multidimensional Items, Two Time Points

An attractive property of the linear logistic model with relaxed assumptions (LLRA) as compared to the unidimensional linear logistic test model (LLTM) or to the RM is the multidimensionality of the underlying latent traits. As was shown in Chapter 9, the LLRA is just a reinterpretation of a special LLTM with test length $k_v = 2$ and structurally incomplete data. Similarly, the LPCM can be reinterpreted in such a way that change in multidimensional item sets becomes describable. Let θ_{vi} be the latent trait parameter of S_v corresponding to item I_i. In order to condition the θ_{vi} of person S_v, for all $i = 1, \ldots, k$, out of the likelihood function, we maximize the latter under k marginal conditions

TABLE 19.4. Data for Model 3 arranged as responses of V-persons of two treatment groups, G_1 and G_2. For brevity, the table gives only the data of those k V-persons S_{vi}^* who represent one typical (real) person $S_v \in G_1$, and the k V-persons S_{wi}^* who represent one typical (real) person $S_w \in G_2$. The symbol (x_{viht}) denotes a response vector of elements x_{viht} with fixed indices v, i, t, for $h = 0, \ldots, m$. Real items $I_1, \ldots, I_k$ are identical to V-items $I_1^*, \ldots, I_k^*$ (presented at T_1).

Virtual Persons	Time Point T_1 V-Items			Time Point T_2 V-Items					
	I_1^*	$\cdots$	I_k^*	I_{k+1}^*	$\cdots$	I_{2k}^*	I_{2k+1}^*	$\cdots$	I_{3k}^*
$\vdots$	$\vdots$		$\vdots$	$\vdots$		$\vdots$			
S_{v1}^*	(x_{v1h1})	–	–	(x_{v1h2})	–	–	No		
$\vdots$	–	$\ddots$	–	–	$\ddots$	–	Obser-		
S_{vk}^*	–	–	(x_{vkh1})	–	–	(x_{vkh2})	vations		
$\vdots$	$\vdots$		$\vdots$	$\vdots$		$\vdots$			
$\vdots$	$\vdots$		$\vdots$				$\vdots$		$\vdots$
S_{w1}^*	(x_{w1h1})	–	–	No			(x_{w1h2})	–	–
$\vdots$	–	$\ddots$	–	Obser-			–	$\ddots$	–
S_{wk}^*	–	–	(x_{wkh1})	vations			–	–	(x_{wkh2})
$\vdots$	$\vdots$	$\vdots$	$\vdots$				$\vdots$	$\vdots$	$\vdots$

$$\sum_t \sum_h h x_{viht} = r_{vi} = \text{fixed}, \quad \text{for } i = 1, \ldots, k. \tag{19.29}$$

That is, the model treats the responses of S_v to any two different items, e.g., I_i and I_j, as if they stemmed from two different persons, S_v and S_v', respectively, with person parameters θ_{vi} and θ_{vj}. If the test consists of k (real) items, multidimensionality implies that each person S_v is technically represented by k 'virtual persons' (V-persons) S_{vi}^*, $i = 1, \ldots, k$. Rearranging the data of Table 19.1 accordingly leads to the data structure given in Table 19.4: each V-person responds to only two V-items, i.e., the test length is $k_v = 2$ for all S_v (in the case of two time points), and the data are highly incomplete. The model will be denoted Model 3.

Within the CML framework that is central to this discourse, the main difference between Model 3 and Model 1 lies in the marginal conditions (19.29), in contrast to fixing the raw scores r_v in (19.9). The weight matrix for Model 3 is obtained from that of Model 1, given in Table 19.2, by deleting the columns denoted by $\beta_{11}, \beta_{21}, \beta_{31}, \ldots, \beta_{k1}$, thus normalizing $\beta_{i1} = 0$, for $i = 1, \ldots, k$, and so anchoring all k latent scales. The H_0 stating that there are no treatment $\times$

category interactions is again formalized by collapsing the columns corresponding to $\eta_{j1}, \ldots, \eta_{jm}$ into one column for η_j, $j = 1, \ldots, b$, as was done for Model 2 in Table 19.3.[5]

Many obvious extensions and modifications suggest themselves, such as, e.g., replacing τ by τ_h, $h = 1, \ldots, m$, etc.

19.4.3 Modeling Response Style

Response style is a ubiquitous nuisance in applications of rating scales. Fischer and Ponocny (1994) found indications of response style in the self-ratings of patients' symptoms, in the sense that patients with a high overall score tended to overemphasize their complaints by choosing category C_m over-proportionally, whereas patients with few symptoms tended to underscore their feeling of ailment by noticeably preferring category C_0. To model this response bias, the authors replaced β_{im} by $\beta_{im} + \lambda$ for the high-scoring group and, similarly, β_{i0} by $\beta_{i0} + \lambda$ for the low-scoring group; regarding the latter, this reparameterization is equivalent, under normalization (19.7), to substituting β_{ih} by $\beta_{ih} - \lambda$, for $h = 1, \ldots, m$. Thus, λ measures the tendency of each group to choose one of the extreme response categories. – It is easy to think of other similar formalizations of response style effects. A more general approach to heterogeneity of respondents will be discussed in Chapter 20, where persons are allowed to belong to latent classes with respect to their response behavior.

19.4.4 Further Hypotheses on Treatment Effects

In principle, all the hypotheses listed in Chapter 9 in connection with the LLRA can be formalized and tested also within the LPCM. Particularly useful for assessing the fit of the LPCM are the hypotheses of the generalizability of effects over (a) subgroups of items and (b) of persons. Since all these hypothesis tests are completely analogous to those discussed in Chapter 9, we leave their eventual realization to the reader.

---

[5]The program 'LPCM', mentioned in Section 19.3, is capable of estimating all these models.

20

Polytomous Mixed Rasch Models

Matthias von Davier and Jürgen Rost[1]

ABSTRACT In this chapter, the discrete mixture distribution framework is developed for Rasch models for items with a polytomous ordered response format. Four of these models will be considered, based on different restrictions regarding the so-called threshold distances (Section 20.2). Since the concept of thresholds is basic for understanding this family of Rasch models, Section 20.1 elaborates the threshold concept in some detail. Section 20.3 focusses on parameter estimation, particularly describing some results on the computation of the γ-functions which correspond to the symmetric functions of the dichotomous Rasch model. Section 20.4 gives some references to applications of these models.

20.1 The Threshold Reparameterization

It has already been mentioned in Chapter 15 that the category parameters ω_h of the Rating Scale Model (RSM) in (15.2) and the item-category parameters of the unrestricted model (15.3) can be transformed into algebraically identical versions in such a way that the cumulative character of these parameters becomes obvious:

$$\omega_h = -\sum_{s=1}^{h} \tau_s \quad \text{and} \quad \beta_{ih} = -\sum_{s=1}^{h} \tau_{is}$$

for all i and $h \in \{1, \dots, m\}$, and $\omega_0 = \beta_{i0} = 0$ for all i, see (15.5) and (15.8). These parameters τ_h and τ_{ih} have a more staightforward interpretation than the cumulative parameters. As Andrich (1978a,b) has shown, they define those points on the latent continuum where the choice probabilities for adjacent categories are equally high. Expressed graphically, they define where the category characteristic curves of adjacent categories intersect (see Figure 20.1).

In order to derive this result, we introduce the concept of a threshold. Let P_{vih} denote the response probability of individual S_v for category C_h on item I_i; then, the probability of passing the h-th threshold, i.e., the threshold 'between' response categories C_{h-1} and C_h, is

$$q_{vih} := \frac{P_{vih}}{P_{vi,h-1} + P_{vih}} = P(X_{vi} = h | (X_{vi} = h-1) \vee (X_{vi} = h)),$$

for $h \in \{1, \dots, m\}$. So q_{vih} is the conditional probability of responding in the upper category, given that only two successive categories are considered. In the

[1]Institut für die Pädagogik der Naturwissenschaften an der Universität Kiel, Olshausenstraße 62, D-24118 Kiel; e-mail: VDAVIER@IPN.UNI-KIEL.DE or ROST@IPN.UNI-KIEL.DE

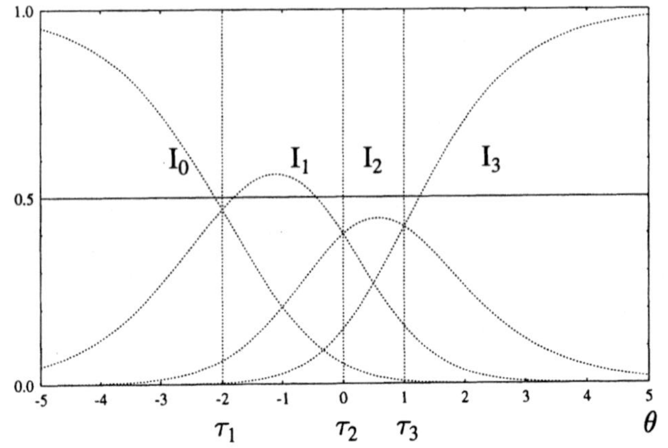

FIGURE 20.1. Category characteristic curves and the intersects

dichotomous case, this is the only threshold, namely, the threshold between categories C_0 and C_1, and its probability is defined by the logistic function of the Rasch model (RM). As a straightforward generalization to the polytomous case, Andrich (1978a) assumed the same function for all threshold probabilities,

$$q_{vih} = \frac{\exp(\theta_v - \tau_{ih})}{1 + \exp(\theta_v - \tau_{ih})}, \tag{20.1}$$

for $h \in \{1, \ldots, m\}$, with one item parameter τ_{ih} for each threshold h. For a discrete random variable X with $m + 1$ categories, there are m independent probabilities $P(X = h)$, so in the case of ordered categories, m independent threshold probabilities may be defined. From assumption (20.1) it can be derived that the response probabilities have the following form (Andrich, 1978a; Masters, 1982):

$$P(X_{vi} = h) = \frac{\exp(h\theta_v - \sum_{s=1}^{h} \tau_{is})}{\sum_{l=0}^{m} \exp(l\theta_v - \sum_{t=1}^{l} \tau_{it})} = \frac{\exp(h\theta_v + \beta_{ih})}{\sum_{l=0}^{m} \exp(l\theta_v + \beta_{il})}, \tag{20.2}$$

which is model (15.3) with cumulative parameters $\beta_{ih} = -\sum_{s=1}^{h} \tau_{is}$. In formula (20.2) the convenient definition $\sum_{i=1}^{0} x_i = 0$ is used to simplify notation. The integer counts h, i.e., the number of thresholds passed, are used here instead of the category coefficients w_h from Chapter 15. This simply follows from the requirement that these coefficients have to be equidistant in order to make conditional maximum likelihood estimation procedures feasible (Andersen, 1977). As shown above, these 'scoring coefficients' follow directly from the assumption of simple

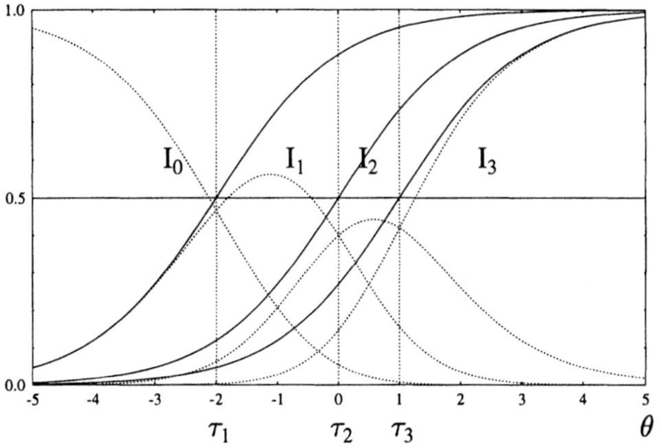

FIGURE 20.2. Category characteristic curves and threshold probabilities

logistic threshold probabilities for all thresholds. Figure 2 shows the threshold probabilities (q_{vih}, solid lines) as well as the response probabilities (P_{vih}, dotted lines) as a function of the person parameter θ.

It can be seen that the points of inflexion of the threshold probabilities and the intersection points of the corresponding category probabilities have the same location on the continuum. Moreover, these points define a partition of the latent continuum into m+1 intervals, namely,

$$I_{i0} =]-\infty, \tau_{i1}], \ \ I_{i1} =]\tau_{i1}, \tau_{i2}], \ldots, I_{im} =]\tau_{im}, +\infty[.$$

Within each of these intervals, one of the response categories has the highest choice probability; in some sense the response scale is mapped onto the latent scale; however, this is only the case if all thresholds are ordered, i.e., if their parameters increase with the category numbers. Andrich (1982b) has pointed out that ordered thresholds are a necessary condition for measuring a trait using items with an ordinal response format.

It is this interpretation of the model parameters that forms the basis for defining a family of restricted polytomous RMs with ordered response categories, of which the rating scale model is but one special case. These restricted models follow from different assumptions about the variation of the threshold distances across items and response categories and, regarding their mixture generalizations, across latent populations. We shall call polytomous RMs with ordered response categories 'ordinal RMs' for short.

20.2 The Mixed Ordinal RM and Three Restricted Models

The ordinal RM described by (20.2) can easily be generalized to a mixture distribution model (cf. Rost, 1991, and Chapter 14). Let $m+1$ be the number of categories, $x_i \in \{0, \ldots, m\}$ a person's response to item I_i, and $\boldsymbol{x} = (x_1, \ldots, x_k)$ the response vector. The general structure of the mixed ordinal RM (MORM) then is

$$P(\boldsymbol{X}_v = \boldsymbol{x}) = \sum_{c=1}^{C} \pi_c \prod_{i=1}^{k} \frac{\exp\left(x_i \theta_{vc} - \sum_{l=1}^{x_i} \tau_{ilc}\right)}{\sum_{h=0}^{m} \exp\left(h\theta_{vc} - \sum_{l=1}^{h} \tau_{ilc}\right)}, \tag{20.3}$$

with $\sum_{l=1}^{0} \tau_{ilc} = 0$. The item parameters τ_{ihc} are referred to as threshold parameters and can be cumulated to yield the β as shown in Section 20.1 and Chapter 15. However, all parameters, the trait parameters θ_{vc} and the threshold parameters τ_{ihc}, are class-specific and are estimated for each latent population. Like the item parameters in the ordinary polytomous RM, the threshold parameters τ_{ihc} are subject to a normalization condition, i.e., the sum of all threshold parameters within a latent class is set to zero,

$$\sum_{i=1}^{k} \sum_{h=1}^{m} \tau_{ihc} = 0 \quad \text{for all } c.$$

The π_c again parameterize the class sizes and hence add up to one,

$$\sum_{c} \pi_c = 1.$$

To estimate the item parameters in the latent classes, the trait parameters θ_{vc} can be conditioned out by introducing conditional score probabilities $\pi_{r|c}$ (cf. Chapter 14),

$$P(\boldsymbol{X} = \boldsymbol{x}) = \sum_{c=1}^{C} \pi_c \pi_{r|c} \frac{\exp\left(-\sum_{i=1}^{k} \sum_{l=1}^{x_i} \tau_{ilc}\right)}{\gamma_r(\boldsymbol{\alpha}_c)}. \tag{20.4}$$

The combinatorial $\gamma_r(\boldsymbol{\alpha}_c)$-functions of parameters $\alpha_{ixc} := \exp(-\sum_{l=1}^{x} \tau_{ilc})$ in class c are analogous to the elementary symmetric functions occurring in the dichotomous case. They are class-specific versions of the γ-functions already introduced in (15.20) in Chapter 15.

The conditional score probabilities $\pi_{r|c}$ add up to 1,

$$\sum_{r} \pi_{r|c} = 1 \quad \text{for all } c,$$

and can be made subject to further restrictions, e.g., the two-parameter logistic restriction introduced in Chapter 14.

Model (20.3) or (20.4), respectively, is the mixture generalization of the partial credit model (PCM; Masters, 1982) and represents the most general MORM. All other ordinal RMs, obtained by restricting the threshold distances in specific ways, are submodels of this general ordinal RM, which is also true for their mixture generalizations. In what follows, three submodels resulting from certain restrictions of the τ_{ihc} will be described. All of them can be estimated by the WINMIRA-program (von Davier, 1994).

- The restriction

$$\tau_{ihc} = \sigma_{ic} + \tau_{hc}, \tag{20.5}$$

  with $\sum_h \tau_{hc} = 0$ for all c, yields the rating scale model (RSM; Andrich, 1978a) or its mixture generalization, respectively; here, σ_{ic} denotes the item parameter of item I_i in class c. In this model, the same threshold distances between response categories are assumed for all items, defined by the τ parameters. In their mixture generalization, however, these threshold parameters may vary from class to class, as do the item parameters σ_{ic}.

- Similarly,

$$\tau_{ihc} = \sigma_{ic} + [(m+1) - 2h]\delta_{ic} \tag{20.6}$$

  defines the mixed dispersion model (Andrich, 1982a) which assumes equidistant thresholds for each item. The δ_{ic} parameterize half the distance between any two adjacent thresholds in class c, i.e., these distance parameters are also class-specific. The scoring coefficient $(m+1) - 2h$ defines the number of half-distances a category C_h departs from the midpoint of the rating scale $(m+1)/2$. For example, the first threshold is 1.5 distance units away from the midpoint of a 4-point rating scale, so that the scoring coefficient is $(m+1) - 2h = 3$.

  As a consequence of this definition, an item parameter σ_{ic} parameterizes the mean value of all thresholds τ_{ihc}, as they do in the mixed RSM (20.6).

- Finally, a combination of these two models, the so-called successive intervals model (Rost, 1988), can be defined. It has the decomposition

$$\tau_{ihc} = \sigma_{ic} + \tau_{hc} + [(m+1) - 2h]\delta_{ic}, \tag{20.7}$$

  with $\sum_h \tau_{hc} = 0$ and $\sum_i \delta_{ic} = 0$ for all c. This model can be considered as a RSM extended by an additional dispersion parameter δ_{ic} stretching or shrinking the threshold distances of the τ_{hc} parameters.

The restriction of the threshold parameters allows one to test specific hypotheses about the response process, e.g., about the 'sizes' of the response categories in terms of threshold distances, or about 'response sets' in terms of symmetric distortions of the distances towards the mean or towards one end of the rating scale. The mixture generalizations of these models allow one to test specific hypotheses about differences of threshold distances between (latent) groups of individuals. Some applications will be mentioned in Section 20.4.

20.3 CML Estimation for Restricted Models

Parameter estimation in these models can be done by means of the EM algorithm as described for the dichotomous MRM in Chapter 14, Section 14.3. In this procedure, conditional maximum likelihood (CML) estimates of all class-specific item-threshold parameters are required in each M-step. The conditional log-likelihood function for the unrestricted model is

$$\ln L_c = \sum_{\boldsymbol{x}} \hat{n}\left(\boldsymbol{x}|c\right) \left[\ln \pi_{r|c} + \sum_{i=1}^{k} \beta_{ih_i c} - \ln\left[\gamma_r(\boldsymbol{\alpha}_c)\right] \right], \tag{20.8}$$

see (12.14) in Chapter 12 for the dichotomous case.

The only problem in applying the CML estimation equations based on this likelihood is the calculation of the functions $\gamma_r(\boldsymbol{\alpha}_c)$ and their partial derivatives. Rost (1991) describes a recursive algorithm for the parameters of the unrestricted mixed ordinal RM; it is a generalization of the summation algorithm which has been elaborated by Andersen (1972), see also Chapters 15 and 19.

The recursion formulae are

$$\tilde{\gamma}_{r|1}(\boldsymbol{\alpha}_c) = \alpha_{1rc} \quad \text{for } r \in \{0, \ldots, m\}, \tag{20.9}$$

and

$$\tilde{\gamma}_{r|i+1}(\boldsymbol{\alpha}_c) = \sum_{s=0}^{m} \tilde{\gamma}_{(r-s)|i}(\boldsymbol{\alpha}_c)\alpha_{(i+1)sc}, \tag{20.10}$$

for $r \in \{0, \ldots, (i+1)m\}$, where $\tilde{\gamma}_{r|i}(\boldsymbol{\alpha}_c) := 0$ for all $r < 0$. The symbol $\tilde{\gamma}_{r|i}(\boldsymbol{\alpha}_c)$ refers to the i-th step of the recursive algorithm of the γ-function of order r in class c; at the same time, i is also the index of the item actually introduced in that step. For example, the γ-function for score r in class c in step two, $\tilde{\gamma}_{r|2}(\boldsymbol{\alpha}_c)$, is

$$\begin{aligned}\tilde{\gamma}_{r|2}(\boldsymbol{\alpha}_c) &= \sum_{s=0}^{m} \tilde{\gamma}_{(r-s)|1}(\boldsymbol{\alpha}_c)\alpha_{2sc} \\ &= \alpha_{1rc}\alpha_{20c} + \alpha_{1(r-1)c}\alpha_{21c} + \ldots + \alpha_{1(r-m)c}\alpha_{2mc} \,.\end{aligned}$$

Each additive term in step two consists of two (cumulative) item-threshold parameters, so that the sums of both category numbers, i.e., $r+0$, $(r-1)+1$,...,$(r-m)+m$, always equals r. In the case $r \geq 2m$, of course, the γ-function of order r is not defined in that step; however, that function is generated in one of the following steps, namely, as soon as $i > r/m$.

The algorithm has been shown to be fast and accurate even for moderate numbers of items, 30 items with 5 categories, say. It can also be used for computing the γ-functions of the restricted models (20.5) through (20.7), substituting the item parameters of the PCM by the linear decomposition of a particular submodel.

The computation of the first partial derivatives of the γ-functions as well as the entire parameter estimation procedure turns out to be somewhat different in these submodels, though. This is shown in the following for the successive intervals model, because this model contains all kinds of parameters and the estimation equations for the two more restrictive models (20.5) and (20.6) are obtained by simply skipping the steps for the τ- and δ-parameters.

To simplify the notation, the class index c will be dropped, since the estimation of parameters is class-specific. Let $\beta_s := \sum_{h=1}^{s} \tau_h$ denote the cumulated threshold parameters, and $g(x) := \sum_{s=1}^{x}(m+1-2s) = x(m-x)$ be the quadratic coefficient for the dispersion parameter δ; then,

$$\ln L_{\text{Cond}} = \sum_{v=1}^{N} \ln P(\boldsymbol{x}_v|r_v) = \sum_{v=1}^{N}\Big[-\sum_{i=1}^{k} x_{vi}\sigma_i + \beta_{x_{vi}} + g(x_{vi})\delta_i\Big] - \sum_{r=0}^{km} n(r)\ln\gamma_r. \tag{20.11}$$

The first partial derivative of this function with respect to any one of the parameters, $\vartheta \in \{\tau_i, \beta_s, \delta_i\}$, is

$$\frac{\partial \ln L_{\text{Cond}}}{\partial\vartheta} = -\sum_{v=1}^{N} f_\vartheta(\boldsymbol{x}_v, i, s) - \sum_{r=0}^{km} n(r)\frac{\partial\gamma_r}{\partial\vartheta}\gamma_r{}^{-1}, \tag{20.12}$$

where

$$\begin{aligned} f_{\tau_i}(\boldsymbol{x}, i, s) &:= x_i; \\ f_{\beta_s}(\boldsymbol{x}, i, s) &:= n_s(\boldsymbol{x}) := \textstyle\sum_{i=1}^{k} 1_{\{s\}}(x_i), \text{ the frequency of category } C_s \\ & \text{in pattern } \boldsymbol{x}, \text{ e.g., } n_0(0111) = 1,\ n_1(0111) = 3, \text{ etc.;} \\ f_{\delta_i}(\boldsymbol{x}, i, s) &:= g(x_i) = x_i(m - x_i). \end{aligned}$$

Therein, ϑ can be any type of parameter as defined in equations (20.5) to (20.7). The first partial derivatives of the γ-functions,

$$\gamma_r = \sum_{\boldsymbol{x}|r} \exp\Big[-\sum_{i=1}^{k} x_i\sigma_i + \beta_{x_i} + \delta_i g(x_i)\Big],$$

where the summation runs over all patterns $\boldsymbol{x}$ with score r, and

$$\beta_{ix} := -(x\sigma_i + \beta_x + g(x)\delta_i),$$

are

$$\frac{\partial\gamma_r}{\partial\sigma_j} = -\sum_{\boldsymbol{x}|r} x_j \exp\Big(\sum_{i=1}^{k}\beta_{ix_i}\Big), \tag{20.13}$$

$$\frac{\partial \gamma_r}{\partial \beta_s} = -\sum_{\boldsymbol{x}|r} n_s(\boldsymbol{x}) \exp\Big(\sum_{i=1}^{k} \beta_{ix_i}\Big), \tag{20.14}$$

$$\frac{\partial \gamma_r}{\partial \delta_j} = -\sum_{\boldsymbol{x}|r} g(x_j) \exp\Big(\sum_{i=1}^{k} \beta_{ix_i}\Big). \tag{20.15}$$

It turns out that computing the derivatives of the γ-functions results in multiplicative coefficients for each term of a γ-function, i.e., x_j for σ_j, $n_s(\boldsymbol{x})$ for β_s, and $g(x_j)$ for δ_j. This property can be used for devising an algorithm for the derivatives based on the summation algorithm described above.

For the first partial derivatives with respect to the σ- and δ-parameters, the summation algorithm defined by (20.9) and (20.10) can be carried out with a modified set of α-parameters: for the derivatives with respect to σ_i, define

$$\alpha^i_{js} := \begin{cases} \alpha_{js} & \text{for } j \neq i, \\ -s\alpha_{js} & \text{for } j = i, \end{cases} \tag{20.16}$$

for all $s \in \{0, \ldots, m\}$. By (20.13) the computation of γ with the modified parameters $\boldsymbol{\alpha}^i$ yields

$$\frac{\partial \gamma_r(\boldsymbol{\alpha})}{\partial \sigma_i} = \gamma_r(\boldsymbol{\alpha}^i). \tag{20.17}$$

The computation of the derivatives with respect to δ_i is carried out by defining

$$\text{for all } s \text{ let } \alpha^i_{js} := \begin{cases} \alpha_{js} & \text{for } j \neq i, \\ -g(s)\alpha_{js} & \text{for } j = i. \end{cases} \tag{20.18}$$

Applying (20.15) shows that the computation of the γ functions with $\boldsymbol{\alpha}^i$ is equivalent to the computation of the derivatives of the original $\gamma(\boldsymbol{\alpha})$ with respect to δ_i, i.e.,

$$\frac{\partial \gamma_r(\boldsymbol{\alpha})}{\partial \delta_i} = \gamma_r(\boldsymbol{\alpha}^i). \tag{20.19}$$

In the case of derivatives with respect to the threshold parameters β_s, see (20.14), all patterns $\boldsymbol{x}$ with the same number n of a particular category C_s, can be grouped. By defining

$$\tilde{\gamma}_{r,n,s} := \sum_{\{\boldsymbol{x}|r,n,s\}} \Big[\prod_{i=1}^{k} \exp\left[-\left(x_i\sigma_i + \beta_{x_i} + g(x_i)\delta_i\right)\right]\Big], \tag{20.20}$$

it follows that

$$\frac{\partial \gamma_r(\boldsymbol{\alpha})}{\partial \beta_s} = \sum_{n=0}^{k} n\tilde{\gamma}_{r,n,s} \,. \tag{20.21}$$

In the case of the first-order gamma function, any category C_s can occur only once, so the recursive computation starts with setting, for all $r \in \{0, \ldots, m\}$,

$$\tilde{\gamma}^{(1)}_{r,n,s} := \begin{cases} \alpha_{1s} & \text{for } r = s \text{ and } n = 1, \\ \alpha_{1r} & \text{for } r \neq s \text{ and } n = 0, \\ 0 & \text{otherwise.} \end{cases} \tag{20.22}$$

The general recursion then becomes

$$\tilde{\gamma}^{(i+1)}_{r,n,s} = \tilde{\gamma}^{(i)}_{r-s,n-1,s}\alpha_{(i+1)s} + \sum_{x \neq s} \tilde{\gamma}^{(i+1)}_{r-x,n,s}\alpha_{(i+1)x}, \tag{20.23}$$

for $n \in \{0, \ldots, i+1\}$.

Stated verbally, the γ-function for $i + 1$ items is obtained by summing over the γ-functions for i items, where the multiplication with the $(i + 1)$-th item parameter depends on the threshold number s. If $x = s$, the multiplication has to be done with a γ-function with a threshold count $n - 1$ different from that in the case $x \neq s$.

These recursions enable one to compute the γ-functions as well as their derivatives, and therefore also the derivatives of the conditional likelihood function. In contrast to the mixed PCM, where a simple iterative procedure for parameter estimation is feasible (cf. Rost, 1991), parameter estimation for the restricted models has to be done by a generalized gradient method or some other standard method for maximizing functions.

20.4 Applications

The MORMs have their main domain of applications in the assessment of personality traits and the measurement of attitudes (Rost, 1990b) or interests (Rost, 1991).

Rost and Georg (1991) analyzed an attitude scale designed to measure 'adolescent centrism' and identified two latent subpopulations. The sample was fitted best by a two-class solution with one 80% and one 20% class differing in their use of the response format. There is some evidence that the smaller class consists of persons with a somewhat strange response behavior, denoted the 'unscalables'. Measures of item-(mis)fit and the absence of ordered thresholds indicate that this class did not use the scale in a systematic way to express their attitude.

Rost (1995) discusses the identification of subpopulations with different response sets, e.g., the avoidance of some category. As he points out, a tendency not to choose the indifference category in the middle of the response format can be assumed whenever the adjacent thresholds are reversed in order within some subpopulation. Accordingly, the avoidance of extreme answers in some subpopulation is indicated by a very easy (i.e., large positive) first threshold and a very difficult (i.e., large negative) last threshold.

In Retrospect

21

What Georg Rasch Would Have Thought about this Book

Erling B. Andersen[1]

Generally, I believe that Georg Rasch would have been very pleased with the book – even proud to see the amount of scientific work on a high academic level which has been carried out concerning the Rasch model (RM) over the last 15 years. He would, at the same time, have been surprised by many developments and have looked in vain for a discussion of some issues which he would have thought to be natural and necessary to take up; he would finally be critical about many results and many approaches. All these opinions are for obvious reasons only qualified guesses on my part, and opinions that are entirely my own, i.e., I have not discussed them with other researchers or the editors of this book.

On the other hand, I worked with Georg Rasch for almost 20 years. Not equally intensely during all these years, but we did have serious and deep discussions of the model – and many other scientific issues – on many occasions. Georg Rasch was a very rich person as a human being and as a scientist. Any discussion with Georg Rasch was, therefore, an occasion to be cherished, and an occasion to enrich your knowledge.

Firstly and primarily, I think that Georg Rasch would have been pleased with the book because a group of scientists have undertaken the task of describing the *status quo* for the RM, and because one of his very first followers on the international scene, Gerhard Fischer, was the person who took the initiative. At the same time, I am convinced that his first surprise would be the list of authors, with a large majority of Dutch authors, no contributions from the United States, and only one author from Denmark. It was clear already when he was still alive that there was a strong interest in the Rasch model in the Netherlands and in Germany (in the latter case due to the fact that some of Gerhard Fischer's students attained university positions in Germany). But I do not think that Georg Rasch foresaw the enormous interest in the RM in the Netherlands, and the fact that so many of the new developments in the field are due to Dutch researchers. That so few contributions in recent years have come from Denmark is due to structural problems in our university system, which

[1]Institute of Statistics, University of Copenhagen, Studiestræde 6, DK-1455 Copenhagen K; e-mail: USIEBA@PC.IBT.DK

were in part already evident during the 10 years Georg Rasch was professor at the University of Copenhagen. The missing authors from the United States and Australia, of course, do not reflect any lack of important contributions to the theory of the RM from the US or Australia. This is clear from the list of references, which I am sure the editors would have pointed out to Georg Rasch.

In the following, I shall try to comment on some selected general issues. Space forces me to be selective. I hope that none of the authors whose chapters I do not mention, will feel offended by this. I shall, however, start by a few words on the very name: 'Rasch model'.

At some time during 1969 or 1970, I discussed the title of my dissertation with Georg Rasch. On this occasion, he was very eager not to call the model the 'RM'. Instead, he suggested the name 'models for measurement'. I felt at the time that it was an awkward name and sounded too much like 'measurement models', which was commonly used for quite different types of models. At first I thought it was a trait of modesty in Georg Rasch's personality. But, on second thought I did not find this trait of modesty consistent with my general impression of his personality. In subsequent discussions I became more and more convinced that it was not modesty, but that his suggested name had the clear purpose of stressing the most important property of the model: that it solved a basic measurement problem in the social sciences, or as it became clearer later, in Georg Rasch's opinion in all sciences. As a consequence of these discussions, my dissertation carried the title: Conditional Inference and Models for Measurement. Time has proved that Georg Rasch was not successful in this attempt to name the models 'models for measurement'. I think, by the way, that the plural form referred both to the extensions to the polytomous case and to the multiplicative Poisson model, which is as central in his 1960 book as the model we now know as the RM.

In general, I believe that Georg Rasch would have liked Chapter 1 very much. He would have been pleased to note that the model is based on the idea that the items can be regarded as 'instruments' for measurement, and that the statistical analysis aims at improving the quality of this instrument. Many of the basic ideas Georg Rasch developed to define the framework of his model were based on analogues to the way instruments are used in experimental natural sciences.

There is one issue which was never completely clarified before Georg Rasch died, namely, the extent to which he accepted the idea of a probability distribution of the latent trait over its range space. I still believe that he never accepted that a latent density be built into the model. My own first paper including a latent density in the model, was not published until 1977, a few years before he died. There are two possible explanations for this. The first one is strictly scientific. Georg Rasch was not occupied with describing populations. He regarded the analysis based on his model as a way to measure differences between individuals as defined in relation to the given items, and by a basic principle, to be independent of which items were used for the measurement. As regards the individuals, he was quite content that the score was a sufficient statistic for the latent parameter. The second explanation is a more complicated one, but in my opinion carries some weight. Georg Rasch had a very obvious animosity towards the normal distribution. At certain occasions, when we had all consumed a generous amount of alcohol, he would invite all persons present to a party on his front lawn to burn all books containing the word 'normal distribution'. This animosity came from two applications of the normal distribution, which Georg Rasch felt was completely unjustified. The first one was standardization of educational or psychological tests, in particular intelligence tests, where individuals were classified by their position in terms of percentiles of a normal distribution describing the variation of the test score over a 'standard population'. The second one was the use of the cumulative distribution function of the normal distribution as a model for the item characteristic curve instead of the inverse logistic curve underlying the RM. He was aware that it was important to be able to compare populations, or at least groups of individuals, for example, at different points in time. But it never took the form of a standardization. Instead, he spent much time on a theory for 'building bridges' between tests administered at different time points to different individuals.

In Chapter 2 the RM is derived from a continuous monotonely increasing ICC from 0 to 1, local independence, and sufficiency of the score as an estimate for the latent trait . It is clear from the 1960 book, and also from the way Georg Rasch introduced me to both the dichotomous and polytomous models in the spring of 1961, that this was the way he originally arrived at the model himself. (This introduction took place every Tuesday morning, when Georg Rasch spent two to three hours in his private home, with me as the only student present, going over the theory of the models.) During these morning sessions he also, however, presented his earliest attempts to derive the model from basic measurement requirements. Those were the same ideas he had presented at the Berkeley symposium in the summer of 1960, and which he, as a measurement principle, called specific objectivity. This is described in Section 2.3, which Georg Rasch would have read with great interest and enjoyed the clear and informative presentation.

There is a historical note, which is important to include. Georg Rasch was a very lazy journal and book reader, as can be seen, e.g., from the sparseness of references in his papers. He once admitted quite frankly to me that he never really read papers even in major statistical journals. This, unfortunately, meant that he was not familiar with the modern theory of functional equations or with group theory. As a consequence, he spent, as far as I could judge at the time, far too much time establishing results concerning the parametric structure of the model by old-fashioned methods, which could have been much more easily derived, and in much more general forms, by well-known results from the just-mentioned mathematical disciplines. At one point in time he got hold of a book on Lie groups, which he read with great interest. Some time later he informed me that it was a great relief to him that his own calculations were very similar to the ones used in Lie group theory. The fact that applications of Lie group theory would have made life easier somehow escaped him, as far as I could figure out. In fact, Georg Rasch was a very clever and capable mathematician. My father, who was also a mathematician, once told me that Georg Rasch was quite famous already as a graduate student for being able to solve complicated differential equations. One story, which my father claimed was true, went that one of the famous mathematics professors showed up one morning at Georg Rasch's dormitory, while he was shaving, to unravel a riddle connected with a complicated differential equation.

Estimation problems are treated in Chapter 3. It is quite common now to justify the CML method by inconsistency of the joint ML estimates for the item parameters. This was never a motivation for Georg Rasch when he suggested the CML method. For him, separability was the key issue. Thus the existence of a probability distribution, not dependent on the latent trait, was the main property. Strangely also, he never seemed to be aware of the importance of the conditionality principle in other statistical problems, e.g., in testing composite hypotheses. I shall return to his attitude towards statistical tests later. It is also important to observe, as a historical note, that he never accepted asymptotic results. He was convinced that all statistical arguments could be made precise and that we do not have to rely on asymptotics.

The method now known as MML estimation was never discussed by Rasch. I have noted earlier that he somehow felt uneasy about introducing a random variation of the latent trait over the population in the model. This, to a large extent, was due to the fact that he had discovered a 'population-free' method to estimate the item parameters and to check the model. As regards the latent trait, he seems to have felt content with having a sufficient statistic (the score) for this parameter. The only two places I recall having seen him mention a distribution of the latent trait, is on the second page of chapter X in his 1960 book, where he uses a distribution H to describe properties of the variation of the item totals, and in his derivation of a model later called the RC-association model by Leo Goodman. I briefly mention this derivation in Section 15.6.

I am convinced that Georg Rasch would have been very excited about the papers by Follman (1988), Lindsay, Clogg, and Grego (1991), De Leeuw and Verhelst (1986), and Holland (1981), all concerned with basic properties of the model, especially its identifiability. I am sure he would have seen one of the main results, namely, that the latent density is only identifiable at a certain number of θ-points (around the maximum score divided by 2), as justification for using the score as an estimate for θ and saying nothing about a latent density.

He had met Paul Lazarsfeld on several occasions and sometimes briefly referred to his work on the latent class model. But I do not remember him making any direct hints of connections to his own model. And he never actually referred to the Rasch models as a type of latent structure model. I am sure, however, that he would have enjoyed reading Chapter 13, which discusses the connection to

Lazarsfeld's work.

I would very much like to say something about the chapters covering tests for fit of the RM, in particular Chapters 5 and 18. It is on the other hand very difficult to predict exactly what would have been Georg Rasch's reaction to the methods discussed by Glas and Verhelst. Overall, Rasch was against all kinds of statistical tests. He called them 'destructive', as opposed to 'constructive' because they either accepted or rejected a hypothesis, or, in this case, a model, and did not provide the researcher with any form of information about the reason for a rejection, or what modifications one could imagine to improve the model. In addition, he was not very comfortable with any form of asymptotic results. If one checks the main published work by Georg Rasch, the only place where a limiting χ^2-distribution is mentioned is in the 1960 book, Section VIII.6, where the possibility of an approximate χ^2-test is mentioned very briefly "using ordinary approximations", as he writes. But this is in connection with the Poisson model, and he gives very little details. For the RM, Rasch claimed that model checks should, in principle, be based on the conditional distribution of the whole data set given both the item totals (which are sufficient for the item parameters) and the scores (which are sufficient for the latent parameters). This probability is very complicated algebraically, and Rasch never mastered any model check procedure based on it. With these remarks in mind, I do not believe Rasch would have been very excited about Chapters 5 and 18. I think he in particular would have been disappointed that nothing had happend over the years regarding the usefulness of his parameter free probability, just described, as a tool for checking the model. (There is some very recent work by Claus Holst in Copenhagen, by the way, on this issue). Actually, Georg Rasch put a lot of emphasis on checking the fit of the model, including whether to omit items or divide items in homogeneous subgroups. The procedures suggested by Rasch never essentially deviated from the graphical techniques suggested in the 1960 book, and he never made serious attempts to completely solve how large deviations from the identity line are permitted under the model. His attitude towards the test suggested in Andersen (1973b) (actually, at a conference he attended, several years earlier) was mild interest and certainly not any form of enthusiasm.

I may, however, be very wrong about how Georg Rasch would have felt about Chapters 5 and 18. It is, indeed, likely that his interest would have been awoken by the basic philosophy of these chapters, namely, to look for variations of the basic χ^2-test, where various forms of deviations from the model are emphasized. I think he would have liked the idea that a test is not just 'destructive', as he called it, but actually points towards reasons for the lack of fit and hence towards possible model reformulations.

A general comment, which applies to Chapters 5 and 18, and indeed to the book as a whole, is concerned with the use of graphs, in particular when testing goodness of fit. Georg Rasch liked graphs, and he liked to plot as many of the original data points as possible, and in a way which revealed as much information as possible about the way these data points deviated from what should be expected under the model. He was an expert in this area. From my earliest work as a student in the Military Psychology Group, where I spent days producing page after page of logistically transformed numbers plotted against square roots of marginals, or whatever new idea Rasch had arrived at, I have always been impressed by how clever he was at selecting a subdata set, performing a data splitting, or making a transformation that made the model fit the data. Many of these ideas are clear analogues to what today generally goes under the name 'diagnostics'. The present book contains very few graphs indeed: according to a quick count based on my copy of the manuscript, 8 altogether. It is true that further developments of the way Georg Rasch used graphs have been rare events. And the book, of course, reflects this fact. But Rasch would have wondered about what happened to the use of graphs. And I think he would have been quite justified in this. Could it be that we have used computers in a wrong way? Since Rasch retired from active duty, have we emphasized the power of computers to do complicated calculations and solving complicated equations over the power of the computers to make nice and illustrative graphs?

The next comments are related to several chapters, namely 8, 9, 10, 14, 15, 19, and 20, all of which deal with extensions of the Rasch model. Rasch would have read all these chapters with great interest. They would have shown him what a rich class of models he arrived at, when he, in the now famous publications of 1960 and 1961, presented the dichotomous and the polytomous RMs. I think he would also have been very satisfied that many (if not all) of these models were suggested by his follower Ben Wright in Chicago, and his students, primarily Geoffrey Masters and David Andrich, and by Gerhard Fischer in Vienna and his students. He would also have been very pleased to see the range of applications published by these two circles of authors.

It is tempting to make more qualified guesses on Georg Rasch's reaction to the rating scale model, like whether equidistant scoring of categories is justified, and how critical lack of equidistance is, or his response to the partial credit model. But I shall refrain from doing so.

Recently, I read the Berkeley symposium paper (Rasch, 1961) again. One of the things that struck me was the generality of the polytomous model as presented in formula (4.2). I am sure that Georg Rasch himself would have claimed for many of the polytomous models presented and discussed in this book that they are actually just special cases of his basic model of 1961. In many cases he would be right, but, of course, wrong in others.

One of the features of the CML estimation procedure, which is basic to the RM, is the need for calculations of the so-called γ-function ratios. A substantial part of the number of hours – and the amount of sweat – spent working with the RM, has been in connection with derivation and computation of γ-ratios. The problems with computation of γ-ratios are discussed, or at least mentioned, in Sections 3.3, 8.2, 15.3, and 19.3. Rasch was very much aware of this problem. He never mastered the use of a computer himself, and so relied on assistants to do the work. Hence, the first computer algorithms were developed in his department at the University of Copenhagen, and at the Danish Institute for Educational Research. When I worked with him from 1961–63 as a student and during my military service in 1963–64, there were only few computers in Denmark, and as late as in my Ph.D. dissertation of 1965, published in 1966, I wrote "the γ-functions depend on the items parameters in so complicated a fashion that an estimation procedure cannot be carried out in practice". But things happened very fast. When I visited Ben Wright in Chicago in the spring of 1967, he and his graduate students were working hard on a Newton-Raphson procedure involving a recursive calculation of the γ-ratios. At that time it was the method described on p. 134 as the 'difference method'. Also during the spring of 1967 Gerhard Fischer was visiting Copenhagen, and worked on the problem together with Peter Allerup. In subsequent years, the 'rounding error problem' inherent in the difference methods was successfully tackled. And since the early 70s we have had fast and efficient procedures for the calculation of the γ-functions.

Whether Rasch would have read the sections on numerical problems, I do not know. It is a fact, however, that he one day in 1964, when I still believed that γ-ratios could not be computed, outlined the difference method to me on the reverse side of a model fit plot. So, maybe Georg Rasch would have read all the more technical matters and descriptions of computer algorithms with great interest. He, in fact, foresaw the potential of computers very early, and actually made arrangements in 1962, when there was only one computer (still based on radio tubes) in Copenhagen, to have some of the data from a military test analysed on the computer.

When I was preparing this chapter, I re-read most of Georg Rasch's writings from the 1960 book up through the 60s and 70s. During this reading, it occurred to me that the over-all answer to the question posed in the heading of this chapter would have been quite different in 1961, around 1968, and around 1977. In 1961, when Georg Rasch still did not hold a permanent position at the University of Copenhagen, his overall reaction would have been amazement at the amount of research based on his models, the range of their applications, and the many developments of statistical tools the models had fostered. In 1968, when he was an established professor at the university, his model widely recognized, and he himself acknowledged as a leading and innovative statistician, he would have been pleased with many chapters, especially those concerned with model extensions, but he would at the same time have been critical of many other chapters, especially those concerned with statistical tests and models for a population variation of the latent parameter. In 1977, when he was primarily occupied with basic issues of philosophy of science, he would, I think, have been very disappointed that the developments up through the 80s and early 90s have been so much concerned with statistical techniques, while so few scientists have worked on basic philosophical issues. But let me emphasize once more that all I have written in this chapter is not – and cannot be – more than qualified guesses.

References

Ackerman, T. A. (1989). *An alternative methodology for creating parallel test forms using the IRT information function.* Paper presented at the NCME Annual Meeting, March 1989, San Francisco, CA.

Aczél, J. (1966). *Lectures on functional equations and their applications.* New York: Academic Press.

Adema, J. J. (1988). *A note on solving large-scale zero-one programming problems.* (Research Report 88-4.) Enschede: University of Twente.

Adema, J. J. (1990a). The construction of customized two-stage tests. *Journal of Educational Measurement, 27,* 241–253.

Adema, J. J. (1990b). *Models and algorithms for the construction of achievement tests.* (Doctoral Thesis.) Enschede: University of Twente.

Adema, J. J. (1992a). Methods and models for the construction of weakly parallel tests. *Applied Psychological Measurement, 16,* 53–63.

Adema, J. J. (1992b). Implementation of the branch-and-bound method for test construction problems. *Methodika, 6,* 99–117.

Adema, J. J., Boekkooi-Timminga, E., & Gademann, A. J. R. M. (1992). Computerized test construction. In M. Wilson (Ed.), *Objective measurement: Theory into practice, Vol. I* (pp. 261–273). Norwood, NJ: Ablex.

Adema, J. J., Boekkooi-Timminga, E., & Van der Linden, W. J. (1991). Achievement test construction using 0-1 linear programming. *European Journal of Operational Research, 55,* 103–111.

Adema, J. J., & Van der Linden, W. J. (1989). Algorithms for computerized test construction using classical item parameters. *Journal of Educational Statistics, 14,* 279–290.

Agresti, A. (1993). Computing conditional maximum likelihood estimates for generalized Rasch models using simple loglinear models with diagonals parameters. *Scandinavian Journal of Statistics, 20,* 63–71.

Aitchison, J., & Silvey, S. D. (1958). Maximum likelihood estimation of parameters subject to restraints. *Annals of Mathematical Statistics, 29,* 813–828.

Allen, M. J., & Yen W. M. (1979). *Introduction to measurement theory.* Monterey: Brooks-Cole.

Amthauer, R. (1970). Intelligenzstrukturtest IST-70. Göttingen: Hogrefe.

Andersen, E. B. (1966). *Den diskrete målingsmodel af endelig orden med anvendelse på et socialpsykologisk materiale.* [The discrete measurement model of finite order with an application to data from social psychology.] Copenhagen: Statens Trykningskontor.

Andersen, E. B. (1970). Asymptotic properties of conditional maximum likelihood estimators. *Journal of the Royal Statistical Society, Series B, 32,* 283–301.

Andersen, E. B. (1971a). *Conditional inference for multiple choice questionnaires.* (Report No. 8.) Copenhagen: Copenhagen School for Economics and Business Administration.

Andersen, E. B. (1971b). Asymptotic properties of conditional likelihood ratio tests. *Journal of the American Statistical Association, 66,* 630–633.

Andersen, E. B. (1972). The numerical solution of a set of conditional estimation equations. *Journal of the Royal Statistical Society, Series B, 34,* 42–54.

Andersen, E. B. (1973a). Conditional inference for multiple-choice questionnaires. *British Journal of Mathematical and Statistical Psychology, 26,* 31–44.

Andersen, E. B. (1973b). A goodness of fit test for the Rasch model. *Psychometrika, 38,* 123–140.

Andersen, E. B. (1973c). *Conditional inference and models for measuring.* Copenhagen: Mentalhygiejnisk Forlag.

Andersen, E. B. (1977). Sufficient statistics and latent trait models. *Psychometrika, 42,* 69–81.

Andersen, E. B. (1980a). *Discrete statistical models with social science applications.* Amsterdam: North-Holland.

Andersen, E. B. (1980b). Comparing latent distributions. *Psychometrika, 45,* 121–134.

Andersen, E. B. (1983). A general latent structure model for contingency table data. In H. Wainer & S. Messik (Eds.), *Principals of modern psychological measurement* (pp. 117–139). Hillsdale, NJ: Erlbaum.

Andersen, E. B. (1985). Estimating latent correlations between repeated testings. *Psychometrika, 50,* 3–16.

Andersen, E. B. (1991). *The statistical analysis of categorical data* (2nd ed.). Heidelberg: Springer-Verlag.

Andersen, E. B., & Madsen, M. (1975). *A computer program for parameter estimation and model check for a Rasch-model* (Research Report No. 6). Copenhagen: Universitetets Statistiske Institut.

Andersen, E. B., & Madsen, M. (1977). Estimating the parameters of the latent population distribution. *Psychometrika, 42,* 357–374.

Andrich, D. (1978a). A rating formulation for ordered response categories. *Psychometrika, 43,* 561–573.

Andrich, D. (1978b). Application of a psychometric rating model to ordered categories which are scored with successive integers. *Applied Psychological Measurement, 2,* 581–594.

Andrich, D. (1982a). An extension of the Rasch model for ratings providing both location and dispersion parameters. *Psychometrika, 47,* 105–113.

Andrich, D. (1982b). Using latent trait measurement models to analyse attitudinal data: A synthesis of viewpoints. In Spearritt, D. (Ed.), *The improvement of measurement in education and psychology* (pp. 89–128). Hawthorn, Victoria: Australian Council for Educational Research (ACER).

Andrich, D. (1989a). Distinctions between assumptions and requirements in measurement in the social sciences. In J. A. Keats, R. Taft, R. A. Heath, & S. H. Lovibond (Eds.), *Mathematical and Theoretical systems.* North Holland: Elsevier Science Publishers.

Andrich, D. (1989b). Distinctions between assumptions and requirements in measurement in the social sciences. *Bulletin of the International Test Commission, 28/29,* 46–66.

Armijo, L. (1966). Minimization of functions having Lipschitz-continuous first partial derivatives. *Pacific Journal of Mathematics, 16,* 1–3.

Armstrong, R. D., & Jones, D. H. (1992). Polynomial algorithms for item matching. *Applied Psychological Measurement, 16,* 365–371.

Armstrong, R. D., Jones, D. H., & Wu, I. L. (1992). An automated test development of parallel tests from a seed test. *Psychometrika, 57,* 271–288.

Bahadur, R. R. (1954). Sufficiency and statistical decision functions. *Annals of Mathematical Statistics, 25,* 423–462.

Baker, F. B. (1992). *Item response theory. Parameter estimation techniques.* New York: Marcel Dekker.

Baker, F. B. (1993). Sensitivity of the linear logistic test model to misspecification of the weight matrix. *Applied Psychological Measurement, 17,* 201–210.

Baker, F. B., Cohen, A. S., & Barmish, B. R. (1988). Item characteristics of tests constructed by linear programming. *Applied Psychological Measurement, 12,* 189–199.

Barisch, S. (1989). *Einstellung zur Epilepsie und Einstellungsänderung durch Information.* [Attitudes towards epilepsy and attitude change caused by information.] (Master's Thesis.) Vienna: University of Vienna.

Barndorff-Nielsen, O. (1978). *Information and exponential families in statistical theory.* New York: J. Wiley.

Bartholomew, D. J. (1980). Factor analysis for categorical data. *Journal of the Royal Statistical Society, Series B, 42,* 293–321.

Bickel, P. J., & Freedman, D. A. (1981). Some asymptotic theory for the bootstrap. *The Annals of Statistics, 9,* 1196–1217.

Birch, M. W. (1963). Maximum likelihood in three-way contingency tables. *Journal of the Royal Statistical Society, Series B, 25,* 220–233.

Birch, M. W. (1964). A new proof of the Pearson-Fisher theorem. *Annals of Mathematical Statistics, 35,* 718–824.

Birnbaum, A. (1968). Some latent trait models and their use in inferring an examinee's ability. In F. M. Lord & M. R. Novick (Eds.), *Statistical theories of mental test scores* (pp. 395–479). Reading, MA: Addison-Wesley.

Bishop, Y. M. M., Fienberg, S. E., & Holland, P. W. (1975). *Discrete Multivariate Analysis.* Cambridge, MA: MIT Press.

Bock, R. D., & Aitkin, M. (1981). Marginal maximum likelihood estimation of item parameters: An application of an EM algorithm. *Psychometrika, 46,* 443–459.

Bock, R. D., & Lieberman, M. (1970). Fitting a response model for n dichotomously scored items. *Psychometrika, 35,* 179–197.

Boekkooi-Timminga, E. (1987). Simultaneous test construction by zero-one programming. *Methodika, 1,* 102–112.

Boekkooi-Timminga, E. (1989). *Models for computerized test construction.* (Doctoral Thesis, University of Twente.) De Lier, NL: Academisch Boeken Centrum.

Boekkooi-Timminga, E. (1990a). The construction of parallel tests from IRT-based item banks. *Journal of Educational Statistics, 15,* 129–145.

Boekkooi-Timminga, E. (1990b). A cluster-based method for test construction. *Applied Psychological Measurement, 14,* 341–354.

Boekkooi-Timminga, E., & Adema, J. J. (1992). An interactive approach to modifying infeasible 0-1 linear programming models for test construction. (Submitted for publication.)

Boekkooi-Timminga, E., & Sun, L. (1992). Contest: A computerized test construction system. In J. Hoogstraten & W. J. Van der Linden (Eds.), *Methodologie: Onderwijsresearchdagen '91* (pp. 69–76). Amsterdam: SCO.

Boomsma, A. (1991). *BOJA, a program for bootstrap and jackknife analysis.* Groningen: iec ProGAMMA.

Bradley, R. A. (1984). Paired comparisons. Some basic procedures and examples. In P. R. Krishnaiah & P. K. Sen (Eds.), *Handbook of statistics, Vol. IV* (pp. 299–326). New York: Elsevier Science Publishers.

Bradley, R. A., & El-Helbawy, A. T. (1976). Treatment contrasts in paired comparisons: Basic procedures with application to factorials. *Biometrika, 63,* 255–262.

Bradley, R. A., & Terry, M. E. (1952). Rank analysis of incomplete block designs, I. The method of pair comparisons. *Biometrika, 39,* 324–345.

Bügel, K., & Glas, C. A. W. (1991). Item specifieke verschillen in prestaties tussen jongens en meisjes bij tekstbegrip examens moderne vreemde talen. *Tijdschrift voor Onderwijsresearch, 16,* 337–351.

Buse, A. (1982). The likelihood ratio, Wald, and Lagrange multiplier tests: An expository note. *The American Statistician, 36,* 153–157.

Bush, R. R., & Mosteller, F. (1951). A mathematical model for simple learning. *Psychological Review, 58,* 313–323.

Campbell, N. R. (1928). *An account of the principles of measurement and calculation.* London: Longmans, Green, & Co.

Chang, H.-H., & Stout, W. (1993). The asymptotic posterior normality of the latent trait in an IRT model. *Psychometrika, 58,* 37–52.

Christofides, N. (1975). *Graph theory. An algorithmic approach.* New York: Academic Press.

Churchhouse, R. F. (1981). *Handbook of applicable mathematics, Vol. III.* Chichester-New York: J. Wiley.

Clogg, C. C. (1988). Latent class models for measuring. In R. Langeheine & J. Rost (Eds.), *Latent Trait and Latent Class Models* (pp. 173–206). New York: Plenum.

Colonius, H. (1979). Zur Eindeutigkeit der Parameter im Rasch-Modell. [On the uniqueness of parameters in the Rasch model.] *Psychologische Beiträge, 21,* 414–416.

Cox, D. R. (1970). *The analysis of binary data.* London: Methuen.

Cressie, N., & Holland, P. W. (1983). Characterizing the manifest probabilities of latent trait models. *Psychometrika, 48,* 129–141.

David, H. A. (1963). *The method of paired comparisons.* London: Griffin.

Davier v., M. (1994). WINMIRA: A Windows 3.x Program for Analyses with the Rasch Model, with the Latent Class Analysis, and with the Mixed Rasch Model. Kiel: Institute for Science Education (IPN).

Daellenbach, H. G., George, J. A., & McNickle, D. C. (1978). *Introduction to operations research techniques* (2nd ed.). Boston: Allyn & Bacon.

De Gruijter, D. N. M. (1990). Test constructions by means of linear programming. *Applied Psychological Measurement, 14,* 175–181.

De Leeuw, J., & Verhelst, N. D. (1986). Maximum likelihood estimation in generalized Rasch models. *Journal of Educational Statistics, 11,* 183–196.

Dempster, A. P., Laird, N. M., & Rubin, D. B. (1977). Maximum likelihood from incomplete data via the EM algorithm (with discussion). *Journal of the Royal Statistical Society, Series B, 39,* 1–38.

Doff, B. (1992). *Laufen und psychische Gesundheit.* [Jogging and psychic health.] (Master's Thesis.) Vienna: University of Vienna.

Ducamp, A., & Falmagne, J.-C. (1969). Composite measurement. *Journal of Mathematical Psychology, 6,* 359–390.

Duncan, O. D. (1984). Rasch measurement: Further examples and discussion. In C. F. Turner & E. Martin (Eds.), *Surveying subjective phenomena, Vol. II* (pp. 267–403). New York: Russell Sage Foundation.

Duncan, O. D., & Stenbeck, M. (1987). Are Likert scales unidimensional? *Social Science Research, 16,* 245–259.

Efron, B. (1982). *The jackknife, the bootstrap, and other resampling plans.* Philadelphia: Society for Industrial and Applied Mathematics.

Eggen, T. J. H. M., & Van der Linden, W. J. (1986). *The use of models for paired comparisons with ties.* (Research Report 86-8, Department of Education). Enschede: University of Twente.

El-Helbawy, A. T., & Bradley, R. A. (1977). Treatment contrasts in paired comparisons: Convergence of a basic iterative scheme for estimation. *Communications in Statistics – Theory and Methods, 6,* 197–207.

Elliott, C. D., Murray, D. J., & Saunders, R. (1971). *Goodness of fit to the Rasch model as a criterion of test unidimensionality.* Manchester: University of Manchester.

Ellis, J. L., & Van den Wollenberg, A. L. (1993). Local homogeneity in latent trait models. A characterization of the homogeneous monotone IRT model. *Psychometrika, 58,* 417–429.

Embretson, S. E. (1985). *Test design: Developments in psychology and psychometrics.* New York: Academic Press.

Embretson, S. E. (1991). A multidimensional latent trait model for measuring learning and change. *Psychometrika, 56,* 495–515.

Engelen, R. J. H. (1989). *Parameter estimation in the logistic item response model.* (Doctoral Thesis.) Enschede: University of Twente.

Estes, W. K. (1972). Research and theory on the learning of probabilities. *Journal of the American Statistical Association, 67,* 81–102.

Everitt, B. S., & Hand, D. J. (1981). *Finite Mixture Distributions.* London: Chapman & Hall.

Falmagne, J.-C. (1970). Unpublished manuscript cited in H.-Ch. Micko (1970).

Feldman, M. W., & Lewontin, R. C. (1975). The heritability hang-up. *Science, 190,* 1163–1168.

Feuerman, M., & Weiss, H. (1973). A mathematical programming model for test construction and scoring. *Management Science, 19,* 961–966.

Fischer, G. H. (1968). Neue Entwicklungen in der psychologischen Testtheorie. [New developments in mental test theory.] In G. H. Fischer (Ed.), *Psychologische Testtheorie* (pp. 15–158). Berne: Huber.

Fischer, G. H. (1972a). A measurement model for the effect of mass-media. *Acta Psychologica, 36,* 207–220.

Fischer, G. H. (1972b). *A step towards dynamic test theory.* (Research Bulletin No. 10.) Vienna: Department of Psychology, University of Vienna.

Fischer, G. H. (1973). The linear logistic test model as an instrument in educational research. *Acta Psychologica, 37,* 359–374.

Fischer, G. H. (1974). *Einführung in die Theorie psychologischer Tests.* [Introduction to mental test theory.] Berne: Huber.

Fischer, G. H. (1976). Some probabilistic models for measuring change. In: D. N. M. De Gruijter & L. J. Th. Van der Kamp (Eds.), *Advances in psychological and educational measurement* (pp. 97–110). New York: J. Wiley.

Fischer, G. H. (1977a). Linear logistic trait models: Theory and application. In H. Spada & W. F. Kempf (Eds.), *Structural models of thinking and learning* (pp. 203–225). Berne: Huber.

Fischer, G. H. (1977b). Some probabilistic models for the description of attitudinal and behavioral changes under the influence of mass communication. In W. F. Kempf & B. H. Repp (Eds.), *Mathematical models for social psychology* (pp. 102–151). Berne: Huber.

Fischer, G. H. (1981). On the existence and uniqueness of maximum-likelihood estimates in the Rasch model. *Psychometrika, 46*, 59–77.

Fischer, G. H. (1982). *Logistic latent trait models with linear constraints: Formal results and typical applications.* (Research Bulletin No. 24.) Vienna: Department of Psychology, University of Vienna.

Fischer, G. H. (1983a). Logistic latent trait models with linear constraints. *Psychometrika, 48*, 3–26.

Fischer, G. H. (1983b). Some latent trait models for measuring change in qualitative observations. In D. J. Weiss (Ed.), *New horizons in testing* (pp. 309–329). New York: Academic Press.

Fischer, G. H. (1983c). *Zum Problem der Validierung diagnostischer Entscheidungen in der Verkehrspsychologie.* [On the problem of validating diagnostic decisions in traffic psychology.] Unpublished manuscript. Vienna: Department of Psychology, University of Vienna.

Fischer, G. H. (1987a). Applying the principles of specific objectivity and generalizability to the measurement of change. *Psychometrika, 52*, 565–587.

Fischer, G. H. (1987b). Heritabilität oder Umwelteffekte? Zwei verschiedene Ansätze zur Analyse von Daten getrennt aufgewachsener eineiiger Zwillinge. [Heritability or environmental effects? Two approaches to analyzing data of monozygotic twins reared apart.] In E. Raab & G. Schulter (Eds.), *Perspektiven Psychologischer Forschung* (pp. 37–55). Vienna: Deuticke.

Fischer, G. H. (1988). Spezifische Objektivität: Eine wissenschaftstheoretische Grundlage des Rasch-Modells. [Specific Objectivity: An epistimological foundation of the Rasch model.] In K. D. Kubinger (Ed.), *Moderne Testtheorie* (pp. 87–111). Weinheim: Beltz.

Fischer, G. H. (1989). An IRT-based model for dichotomous longitudinal data. *Psychometrika, 54*, 599–624.

Fischer, G. H. (1991a). A new methodology for the assessment of treatment effects. *Evaluación Psicológica - Psychological Assessment, 7*, 117–147.

Fischer, G. H. (1991b). On power series models and the specifically objective assessment of change in event frequencies. In J.-C. Falmagne & J.-P. Doignon (Eds.), *Mathematical Psychology: Current developments* (pp. 293–310). New York: Springer-Verlag.

Fischer, G. H. (1992). The 'Saltus model' revisited. *Methodika, 6*, 87–98.

Fischer, G. H. (1993a). The Measurement of environmental effects: An alternative to the estimation of heritability in twin data. *Methodika, 7*, 20–43.

Fischer, G. H. (1993b). Notes on the Mantel-Haenszel procedure and another chi-squared test for the assessment of DIF. *Methodika, 7*, 88-100.

Fischer, G. H., & Allerup, P. (1968). Rechentechnische Fragen zu Raschs eindimensionalem Modell. In G. H. Fischer (Ed.), *Psychologische Testtheorie* (pp. 269–280). Berne: Huber.

Fischer, G. H., & Formann, A. K. (1972). *An algorithm and a FORTRAN program for estimating the item parameters of the linear logistic test model.* (Research Bulletin No. 11.) Vienna: Department of Psychology, University of Vienna.

Fischer, G. H., & Formann, A. K. (1981). Zur Schätzung der Erblichkeit quantitativer Merkmale. [Estimation of the heritability of quantitative variables.] *Zeitschrift für Differentielle und Diagnostische Psychologie, 2,* 189–197.

Fischer, G. H., & Formann, A. K. (1982a). Some applications of logistic latent trait models with linear constraints on the parameters. *Applied Psychological Measurement, 4,* 397–416.

Fischer, G. H., & Formann, A. K. (1982b). Veränderungsmessung mittels linear-logistischer Modelle. [Measuring change by means of linear logistic models.] *Zeitschrift für Differentielle und Diagnostische Psychologie, 3,* 75–99.

Fischer, G. H., & Parzer, P. (1991a). An extension of the rating scale model with an application to the measurement of treatment effects. *Psychometrika, 56,* 637–651.

Fischer, G. H., & Parzer, P. (1991b). LRSM: Parameter estimation for the linear rating scale model. *Applied Psychological Measurement, 15,* 138.

Fischer, G. H., & Pendl, P. (1980). Individualized testing on the basis of the dichotomous Rasch model. In L. J. Th. Van der Kamp, W. F. Langerak, & D. N. M. De Gruijter (Eds.), *Psychometrics for educational debates* (pp. 171–188). New York: J. Wiley.

Fischer, G. H., & Ponocny, I. (1994). An extension of the partial credit model with an application to the measurement of change. *Psychometrika, 59,* 177–192.

Fischer, G. H., & Scheiblechner, H. H. (1970). Algorithmen und Programme für das probabilistische Testmodell von Rasch. [Algorithms and programs for Rasch's probabilistic test model.] *Psychologische Beiträge, 12,* 23–51.

Fischer, G. H., & Spada, H. (1973). *Die psychometrischen Grundlagen des Rorschachtests und der Holtzman Inkblot Technique.* [The psychometric fundament of the Rorschach Test and the Holtzman Inkblot Technique.] Berne: Huber.

Fischer, G. H., & Tanzer, N. (1994). Some LLTM and LBTL relationships. In G. H. Fischer & D. Laming (Eds.), *Contributions to Mathematical Psychology, Psychometrics, and Methodology* (pp. 277–303). New York: Springer-Verlag.

Follmann, D. A. (1988). Consistent estimation in the Rasch model based on nonparametric margins. *Psychometrika, 53,* 553–562.

Ford, L. R. Jr. (1957). Solution of a ranking problem from binary comparisons. *American Mathematical Monthly, 64,* 28–33.

Formann, A. K. (1973). *Die Konstruktion eines neuen Matrizentests und die Untersuchung des Lösungsverhaltens mit Hilfe des linear logistischen Testmodells.* [The construction of a new matrices test and the investigation of the response behavior by means of the linear logistic test model.] (Doctoral Thesis.) Vienna: University of Vienna.

Formann, A. K. (1982). Linear logistic latent class analysis. *Biometrical Journal, 24,* 171–190.

Formann, A. K. (1984). *Die Latent-Class-Analyse.* Weinheim and Basel: Beltz Verlag.

Formann, A. K. (1985). Constrained latent class models: Theory and applications. *British Journal of Mathematical and Statistical Psychology, 38,* 87–111.

Formann, A. K. (1986). A note on the computation of the second order derivatives of the elementary symmetric functions in the Rasch model. *Psychometrika, 51,* 335–339.

Formann, A. K. (1989). Constrained latent class models: Some further applications. *British Journal of Mathematical and Statistical Psychology, 42,* 37–54.

Formann, A. K. (1992). Linear logistic latent class analysis for polytomous data. *Journal of the American Statistical Association, 87,* 476–486.

Formann, A. K., & Piswanger, K. (1979). *Wiener Matrizen-Test. Ein Rasch-skalierter sprachfreier Intelligenztest.* [The Viennese Matrices Test. A Rasch-scaled culture-fair intelligence test.] Weinheim: Beltz Test.

Gardner, H. (1983). *Frames of mind: The theory of multiple intelligences.* New York: Basic Books.

Gittler, G. (1991). *Dreidimensionaler Würfeltest (3DW). Ein Rasch-skalierter Test zur Messung des räumlichen Vorstellungsvermögens.* [The three-dimensional cubes-test (3 DW). A Rasch-scaled test for spatial ability.] Weinheim: Beltz Test.

Gittler, G. (1992). *Testpsychologische Aspekte der Raumvorstellungsforschung – Kritik, Lösungsansätze und empirische Befunde.* [Test-psychological aspects of research on spatial ability – critique, tentative solutions, and empirical findings.] (Habilitations-Thesis.) Vienna: University of Vienna.

Gittler, G. (1994). Intelligenzförderung durch Schulunterricht: Darstellende Geometrie und räumliches Vorstellungsvermögen. In G. Gittler, M. Jirasko, U. Kastner-Koller, C. Korunka, & A. Al-Roubaie (Eds.), *Die Seele ist ein weites Land.* Vienna: WUV-Universitätsverlag.

Gittler, G., & Wild, B. (1988). Der Einsatz des LLTM bei der Konstruktion eines Itempools für das adaptive Testen. [Usage of the LLTM for the construction of an item pool for adaptive testing.] In K. D. Kubinger (Ed.), *Moderne Testtheorie* (pp. 115–139). Weinheim: Beltz.

Glas, C. A. W. (1981). *Het Raschmodel bij data in een onvolledig design.* [The Rasch model for data in an incomplete design]. (PSM-Progress Reports 81-1). Utrecht: Vakgroep PSM van de subfaculteit Psychologie.

Glas, C. A. W. (1988a). The derivation of some tests for the Rasch model from the multinomial distribution. *Psychometrika, 53,* 525–546.

Glas, C. A. W. (1988b). The Rasch model and multi-stage testing. *Journal of Educational Statistics, 13,* 45–52.

Glas, C. A. W. (1989). *Contributions to estimating and testing Rasch models.* (Doctoral Thesis.) Enschede: University of Twente.

Glas, C. A. W. (1992). A Rasch model with a multivariate distribution of ability. In M. Wilson (Ed.), *Objective measurement: Theory into practice, Vol. I.* (pp. 236–258). New Jersey: Ablex Publishing Corporation.

Glas, C. A. W., & Ellis, J. L. (1993). *RSP, Rasch Scaling Program: Computer program and user's manual.* Groningen: ProGAMMA.

Glas, C. A. W., & Verhelst, N. D. (1989). Extensions of the partial credit model. *Psychometrika, 54,* 635–659.

Glatz, E.-M. (1977). *Die Wirksamkeit eines verhaltenstherapeutischen Eß-Trainings bei geistig retardierten Kindern.* [The efficacy of behavior modification of eating behavior in mentally retarded children.] (Doctoral Thesis.) Vienna: University of Vienna.

Goodman, L. A. (1974). Exploratory latent structure analysis using both identifiable and unidentifiable models. *Biometrika, 61,* 215–231.

Goodman, L. A. (1978). *Analyzing qualitative/categorical data: Loglinear models and latent structure analysis.* London: Addison Wesley.

Goodman, L. A. (1979). Simple methods for the analysis of association in cross-classifications having ordered categories. *Journal of the American Statistical Association, 74,* 537–552.

Goodman, L. A. (1981). Association models and canonical correlation in the analysis of cross-classifications having ordered categories. *Journal of the American Statistical Association, 76,* 320–334.

Guilford, J. P. (1967). *The nature of human intelligence.* New York: McGraw-Hill.

Gulliksen, H. (1950). *Theory of mental tests.* New York: J. Wiley.

Gustafsson, J. E. (1977). *The Rasch model for dichotomous items: Theory, applications and a computer program.* (Reports from the Institute of Education, No. 63). Göteborg: University of Göteborg.

Gustafsson, J. E. (1979). *PML: A computer program for conditional estimation and testing in the Rasch model for dichotomous items.* (Reports from the Institute of Education, No. 85). Göteborg: University of Göteborg.

Gustafsson, J. E. (1980a). A solution of the conditional estimation problem for long tests in the Rasch model for dichotomous items. *Educational and Psychological Measurement, 40,* 337–385.

Gustafsson, J. E. (1980b). Testing and obtaining fit of data to the Rasch model. *British Journal of Mathematical and Statistical Psychology, 33,* 205–233.

Haberman, S. J. (1974). *The analysis of frequency data.* Chicago: University of Chicago Press.

Haberman, S. J. (1977). Maximum likelihood estimates in exponential response models. *The Annals of Statistics, 5,* 815–841.

Haberman, S. J. (1978). *Analysis of qualitative data. Introductory topics, Vol. I.* New York: Academic Press.

Haberman, S. J. (1979). *Analysis of qualitative data. New developments, Vol. II.* New York: Academic Press.

Hambleton, R. K. (1992). Measurement advances to address educational policy questions. In Plomp, T., Pieters, J., & Feteris, A. (Eds.), *European Conference on Educational Research* (pp. 681–684). Enschede: Department of Education, University of Twente.

Hambleton, R. K., and Rogers, H. J. (1989). Detecting potentially biased test items: Comparison of IRT area and Mantel-Haenszel methods. *Applied Measurement in Education, 2,* 313–334.

Hambleton, R. K., & Swaminathan, H. (1985). *Item response theory: Principles and applications.* Boston: Kluwer-Nijhoff.

Hamburger, H. (1920). Über eine Erweiterung des Stieltjesschen Momentproblems. [On an extension of the Stieltjes moment problem.] *Mathematische Annalen, 81,* 235–319.

Hamerle, A. (1979). Über die meßtheoretischen Grundlagen von Latent-Trait-Modellen. [On the measurement-theoretic foundation of latent trait models.] *Archiv für Psychologie, 132,* 19–39.

Hamerle, A. (1982). *Latent-Trait-Modelle.* [Latent trait models.] Weinheim: Beltz.

Hammer, H. (1978). *Informationsgewinn und Motivationseffekt einer Tonbildschau und eines verbalen Lehrvortrages.* [Information gain and motivational effects via a slide-and-sound show and via a teacher's presentation.] (Doctoral Thesis.) Vienna: University of Vienna.

Harary, F., Norman, R. Z., & Cartwright, D. (1965). *Structural models: An introduction to the theory of directed graphs.* New York: J. Wiley.

Heckl, U. (1976). *Therapieerfolge bei der Behandlung sprachgestörter Kinder.* [Effects of therapy in speech-handicapped children.] (Doctoral Thesis.) Vienna: University of Vienna.

Hoijtink, H., & Boomsma, A. (1991). *Statistical inference with latent ability estimates.* (Heymans Bulletins Psychologische Instituten R. U. Groningen, HB-91-1045-EX.) Groningen: Vakgroep S&M FPPSW, University of Groningen.

Holland, P. W. (1981). When are item response theory models consistent with observed data? *Psychometrika, 46,* 79–92.

Holland, P. W. (1990). On the sampling theory foundations of item response theory models. *Psychometrika, 55,* 577–601.

Holland, P. W., and Thayer, D. T. (1988). Differential item functioning and the Mantel-Haenszel procedure. In H. Wainer & H. I. Braun (Eds.), *Test validity.* Hillsdale, NJ: Lawrence Erlbaum.

Holtzman, W. H., Thorpe, J. S., Swartz, J. D., & Heron, E. W. (1961). *Inkblot perception and personality.* Austin: University of Texas Press.

Hommel, G. (1983). Tests of the overall hypothesis for arbitrary dependence structures. *Biometrical Journal, 25,* 423–430.

Hornke, L. F., & Habon, M. W. (1986). Rule-based item bank construction and evaluation within the linear logistic framework. *Applied Psychological Measurement, 10,* 369–380.

Hornke, L. F., & Rettig, K. (1988). Regelgeleitete Itemkonstruktion unter Zuhilfenahme kognitionspsychologischer Überlegungen. [Rule-based item construction using concepts of cognitive psychology.] In K. D. Kubinger (Ed.), *Moderne Testtheorie* (pp. 140–162). Weinheim: Beltz.

Hout, M., Duncan, O. D., & Sobel, M. E. (1987). Association and heterogeneity: Structural models of similarities and differences. *Sociological Methodology, 17,* 145–184.

Huber, P. J. (1981). *Robust statistics.* New York: J. Wiley.

Hubert, L., Golledge, R. G., Costanzo, C. M., Gale, N., & Halperin, W. C. (1984). Nonparametric tests for directional data. In G. Bahrenberg, M. M. Fischer, & P. Nijkamp (Eds.), *Recent developments in spatial data analysis* (pp. 171–189). Brookfield VT: Gower.

Hubert, L., & Schultz, J. (1976). Quadratic assignment as a general data analysis strategy. *British Journal of Mathematical and Statistical Psychology, 29,* 190–241.

Iby, M. (1987). *Die Effektivität von Kommunikationsseminaren in der innerbetrieblichen Ausbildung der Zentralsparkasse.* [The efficacy of communication seminars for employees of the 'Zentralsparkasse'.] (Doctoral Thesis.) Vienna: University of Vienna.

Irtel, H. (1987). On specific objectivity as a concept in measurement. In E. E. Roskam & R. Suck (Eds.), *Progress in mathematical psychology* (pp. 35–45). Amsterdam: North-Holland.

Irtel, H. (1994). The uniqueness structure of simple latent trait models. In G. H. Fischer & D. Laming (Eds.), *Contributions to mathematical psychology, psychometrics, and methodology* (pp. 265–275). New York: Springer-Verlag.

Jacobsen, M. (1989). Existence and unicity of MLEs in discrete exponential family distributions. *Scandinavian Journal of Statistics, 16,* 335–349.

Jannarone, R. J. (1986). Conjunctive item response theory kernels. *Psychometrika, 51,* 357–373.

Jansen, P. G. W. (1984). Computing the second-order derivatives of the symmetric functions in the Rasch model. *Kwantitatieve Methoden, 13,* 131–147.

Johnson, N. L., & Kotz, S. (1969). *Distributions in statistics: Discrete distributions, Vol. I.* Boston: Houghton Mifflin.

Junker, B. W. (1991). Essential independence and likelihood-based ability estimation for polytomous items. *Psychometrika, 56,* 255–278.

Karlin, S., & Studden, W. J. (1966). *Tchebycheff Systems: With applications to analysis and statistics.* New York: J. Wiley.

Kelderman, H. (1984). Loglinear Rasch model tests. *Psychometrika, 49,* 223–245.

Kelderman, H. (1988). Common item equating using the loglinear Rasch model. *Journal of Educational Statistics, 13,* 313–318.

Kelderman, H. (1989). *Loglinear multidimensional IRT models for polytomously scored items.* Paper presented at the Fifth International Objective Measurement Workshop, Berkeley. (ERIC document Reproduction Service No. ED 308 238.)

Kelderman, H. (1992). Computing maximum likelihood estimates of loglinear models from marginal sums with special attention to loglinear item response theory. *Psychometrika, 57,* 437–450.

Kelderman, H. (in press). Loglinear multidimensional latent trait models for polytomously scored items. In R. Hambleton & W. J. Van der Linden (Eds.), *Handbook of modern item response theory.* New York: Springer-Verlag.

Kelderman, H., Mellenbergh, G. J., & Elshout, J. J. (1981). Guilford's facet theory of intelligence: An empirical comparison of models. *Multivariate Behavioral Research, 16,* 37–62.

Kelderman, H., & Rijkes, C. P. M. (1994). Loglinear multidimensional IRT models for polytomously scored items. *Psychometrika, 59,* 149–176.

Kelderman, H., & Steen, R. (1988). *LOGIMO I: Loglinear item response theory modeling.* Program manual. Groningen: ProGAMMA.

Kempf, W. (1972). Probabilistische Modelle experimentalpsychologischer Versuchssituationen. *Psychologische Beiträge, 14,* 16–37.

Kempf, W. (1974) Dynamische Modelle zur Messung sozialer Verhaltenspositionen. [Dynamic models for measuring social relationships.] In W. Kempf (Ed.), *Probabilistische Modelle in der Sozialpsychologie* (pp. 13–55). Berne: Huber.

Kempf, W. (1977a). A dynamic test model and its use in the micro-evaluation of instrumental material. In H. Spada & W. Kempf (Eds.), *Structural models for thinking and learning* (pp. 295–318). Berne: Huber.

Kempf, W. (1977b). Dynamic models for the measurement of 'traits' in social behavior. In W. Kempf & B. H. Repp (Eds.), *Mathematical models for social psychology* (pp. 14–58). Berne: Huber.

Kiefer, J., & Wolfowitz, J. (1956). Consistency of the maximum likelihood estimator in the presence of infinitely many nuisance parameters. *Annals of Mathematical Statistics, 27,* 887–906.

Klauer, K. C. (1988). Tests von Latent-trait-Annahmen: Ein Schluß von der Population auf die Person. [Tests of latent trait assumptions: Inferences about individuals on the basis of population data.] *Zeitschrift für Differentielle und Diagnostische Psychologie, 9,* 97–104.

Klauer, K. C. (1990). Asymptotic properties of the ML estimator of the ability parameter when item parameters are known. *Methodika, 4,* 23–26.

Klauer, K. C. (1991a). Exact and best confidence intervals for the ability parameter of the Rasch model. *Psychometrika, 56,* 535–547.

Klauer, K. C. (1991b). An exact and optimal standardized person test for assessing consistency with the Rasch model. *Psychometrika, 56,* 213–228.

Klauer, K. C., & Rettig, K. (1990). An approximately standardized person test for assessing consistency with a latent trait model. *British Journal of Mathematical and Statistical Psychology, 43,* 193–206.

Klinkenberg, E. (1992). *Estimating latent ability: A comparison of the WLE and EAP estimator.* (Heymans Bulletins Psychologische Instituten R. U. Groningen, HB-92-1079-SW.) Groningen: Vakgroep S & M FPPSW, University of Groningen.

Köller, O., Rost, J., & Köller, M. (1994). Individuelle Unterschiede beim Lösen von Raumvorstellungsaufgaben aus dem IST-70 bzw. IST-70 Untertest "Würfelaufgaben". *Zeitschrift für Psychologie, 202,* 65–85.

Kok, F. G., Mellenbergh, G. J., & Van der Flier, H. (1985). Detecting experimentally induced item bias using the iterative logit method. *Journal of Educational Measurement, 22,* 295–303.

Koschier, A. (1993). Wirksamkeit von Kommunikationstrainings. [The efficacy of communication training.] (Master's Thesis.) Vienna: University of Vienna.

Kosmol, P. (1989). *Methoden zur numerischen Behandlung nichtlinearer Gleichungen und Optimierungsaufgaben.* Stuttgart: Teubner.

Krantz, D. H., Luce, R. D., Suppes, P., & Tversky, A. (1971). *Foundations of measurement, Vol. I.* New York: Academic Press.

Krein, M. G., & Nudel'man, A. A. (1977). *The markov moment problem and extremal problems.* Providence, RI: American Mathematical Society.

Kropiunigg, U. (1979a). *Wirkungen einer sozialpolitschen Medienkampagne.* [Attitudinal effects of a mass media campaign on social policies.] (Doctoral Thesis.) Vienna: University of Vienna.

Kropiunigg, U. (1979b). Einstellungswandel durch Massenkommunikation. [Attitude change via mass communication.] *Österreichische Zeitschrift für Soziologie, 4,* 67–71.

Kubinger, K. D. (1979). Das Problemlöseverhalten bei der statistischen Auswertung psychologischer Experimente. Ein Beitrag hochschuldidaktischer Forschung. [The task-solving behavior in the statistical analysis of psychological experiments. An example of research in didactics.] *Zeitschrift für Experimentelle und Angewandte Psychologie, 26,* 467–495.

Kuhn, T. S. (1961/1977). The function of measurement in modern physical science. *ISIS, 52,* 161–190. Reproduced in T. S. Kuhn, (1977), *The essential tension.* Chicago: The University of Chicago Press.

Layzer, D. (1974). Heritability analyses of IQ scores: Science or numerology? *Science, 183,* 1259–1266.

Lazarsfeld, P. F. (1950). The logical and mathematical foundation of latent structure analysis. In S. A. Stouffer, L. Guttman, E. A. Suchman, P. F. Lazarsfeld, S. A. Star, & J. A. Clausen (Eds.), *Studies in Social Psychology in World War II, Vol. IV: Measurement and Prediction* (pp. 362–412). Princeton, NJ: Princeton University Press.

Lazarsfeld, P. F. (1959). Latent structure analysis. In S. Koch (Ed.), *Psychology: A study of a science, Vol. III.* New York: McGraw-Hill.

Lazarsfeld, P. F., & Henry, N. W. (1968). *Latent Structure Analysis.* Boston: Houghton Mifflin.

Lehmann, E. L. (1983). *Theory of point estimation.* New York: J. Wiley.

Lehmann, E. L. (1986). *Testing statistical hypotheses* (2nd ed.). New York: J. Wiley.

Lenstra, J. K., & Rinnooy Kan, A. H. G. (1979). Computational complexity of discrete optimization problems. In P. L. Hammer, E. L. Johnson, & B. H. Korte (Eds.), *Discrete optimization I.* New York: North-Holland.

Levine, M. V., & Drasgow, F. (1983). Appropriateness measurement: Validity studies and variable ability models. In D. J. Weiss (Ed.), *New horizons in testing* (pp. 109–131). New York: Academic Press.

Levine, M. V., & Drasgow, F. (1988). Optimal appropriateness measurement. *Psychometrika, 53,* 161–176.

Likert, R. (1932). A technique for the measurement of attitude. *Archives of Psychology, No. 140,* 1–55.

Lindsay, B. G. (1983). The geometry of mixture likelihoods: A general theory. *The Annals of Statistics, 11,* 86–94.

Lindsay, B. G., Clogg, C. C., & Grego, J. (1991). Semiparametric estimation in the Rasch model and related exponential response models, including a simple latent class model for item analysis. *Journal of the American Statistical Association, 86,* 96–107.

Liou, M. (1994). More on the computation of higher-order derivatives of the elementary symmetric functions in the Rasch model. *Applied Psychological Measurement, 18,* 53–62.

Liou, M., & Yu, L. (1991). Assessing statistical accuracy in ability estimation: A bootstrap approach. *Psychometrika, 56,* 55–67.

Liou, M., & Chang, C.-H. (1992). Constructing the exact significance level for a person fit statistic. *Psychometrika, 57,* 169–181.

Little, R. J. A., & Rubin, D. B. (1987). *Statistical analysis with missing data.* New York: J. Wiley.

Lord, F. M. (1952). A theory of test scores. *Psychometric Monograph, No. 7.* The Psychometric Society.

Lord, F. M. (1974). Estimation of latent ability and item parameters when there are omitted responses. *Psychometrika, 39,* 247–264.

Lord, F. M. (1980). *Applications of item response theory to practical testing problems.* Hillsdale, NJ: Lawrence Erlbaum.

Lord, F. M. (1983a). Unbiased estimators of ability parameters, of their variance, and of their parallel-forms reliability. *Psychometrika, 48,* 233–245.

Lord, F. M. (1983b). Maximum likelihood estimation of item response parameters when some responses are omitted. *Psychometrika, 48,* 477–482.

Lord, F. M., & Novick, M. R. (1968). *Statistical theories of mental test scores.* Reading, MA: Addison-Wesley.

Luce, R. D. (1959). *Individual choice behavior: A theoretical analysis.* New York: J. Wiley.

Luce, R. D., Bush, R. R., & Galanter E. (Eds.) (1963). *Handbook of mathematical psychology, Vol. III.* New York: J. Wiley.

Luecht, R. M., & Hirsch, T. M. (1992). Item selection using an average growth approximation of target information functions. *Applied Psychological Measurement, 16,* 41–51.

Mandanski, A. (1963). Tests of homogeneity for correlated samples. *Journal of the American Statistical Association, 58,* 97–119.

Martin-Löf, P. (1973). *Statistiska modeller.* [Statistical models.] Anteckningar från seminarier lasåret 1969-1970, utarbetade av Rolf Sundberg. Obetydligt ändrat nytryck, Oktober 1973. Stockholm: Institutet för Försäkringsmatematik och Matematisk Statistisk vid Stockholms Universitet.

Martin-Löf, P. (1974). The notion of redundancy and its use as a quantitative measure of the discrepancy between a statistical hypothesis and a set of observational data. *Scandinavian Journal of Statistics, 1,* 3–18.

Masters, G. N. (1982). A Rasch model for partial credit scoring. *Psychometrika, 47,* 149–174.

McCullagh, P., & Nelder, J. A. (1983). *Generalized linear models.* London: Chapman and Hall.

McHugh, R. B. (1956). Efficient estimation and local identification in latent class analysis. *Psychometrika, 21,* 331–347.

Medina-Díaz, M. (1993). Analysis of cognitive structure. *Applied Psychological Measurement, 17,* 117–130.

Mellenbergh, G. J. (1982). Contingency table models for assessing item bias. *Journal of Educational Statistics, 7,* 105–118.

Mellenbergh, G. J. (1983). Conditional item bias methods. In S. H. Irvine & W. J. Berry (Eds.), *Human assessment and cultural factors.* New York: Plenum Press.

Meredith, W., & Kearns, J. (1973). Empirical Bayes point estimates of latent trait scores without knowledge of the trait distribution. *Psychometrika, 38,* 533–554.

Micko, H.-C. (1969). A psychological scale for reaction time measurement. *Acta Psychologica, 30,* 324.

Micko, H.-C. (1970). Eine Verallgemeinerung des Messmodells von Rasch mit einer Anwendung auf die Psychophysik der Reaktionen. [A generalization of Rasch's measurement model with an application to the psychophysics of reactions.] *Psychologische Beiträge, 12,* 4–22.

Miller, R. G. (1974). A thrustworthy jackknife. *The Annals of Mathematical Statistics, 35,* 1594–1605.

Mislevy, R. J. (1981). *A general linear model for the analysis of Rasch item threshold estimates.* (Doctoral Thesis.) Chicago: University of Chicago.

Mislevy, R. J. (1984). Estimating latent distributions. *Psychometrika, 49,* 359–381.

Mislevy, R. J. (1985). Estimation of latent group effects. *Journal of the American Statistical Association, 80,* 993–997.

Mislevy, R. J. (1986). Bayes modal estimation in item response models. *Psychometrika, 51,* 177–195.

Mislevy, R. J. (1988). Exploiting auxiliary information about items in the estimation of Rasch item difficulty parameters. *Applied Psychological Measurement, 12,* 281–296.

Mislevy, R. J., & Bock, R. D. (1982). Biweight estimates of latent ability. *Educational and Psychological Measurement, 42,* 725–737.

Mislevy, R. J., & Bock, R. D. (1986). *PC-Bilog: Item analysis and test scoring with binary logistic models.* Mooresville: Scientific Software.

Mislevy, R. J., & Verhelst, N. D. (1990). Modeling item responses when different subjects employ different solution strategies. *Psychometrika, 55,* 195–215.

Mislevy, R. J., Wingersky, M. S., Irvine, S. H., & Dann, P. L. (1991). Resolving mixtures of strategies in spatial visualization tasks. *British Journal of Mathematical and Statistical Psychology, 44,* 265–288.

Mislevy, R. J., & Wu, P.-K. (1988). *Inferring examinee ability when some item responses are missing.* (Research Report RR-88-48-ONR.) Princeton, NJ: Educational Testing Service.

Mohammadzadeh-Koucheri, F. (1993). *Interkultureller Vergleich mit einer variierten Form des Matrizentests von Formann.* [Cross-cultural comparisons using a variation of the matrices test of Formann.] (Master's Thesis.) Vienna: University of Vienna.

Molenaar, I. W. (1983). Some improved diagnostics for failure in the Rasch model. *Psychometrika, 48,* 49–72.

Molenaar, I. W., & Hoijtink, H. (1990). The many null distributions of person fit indices. *Psychometrika, 55,* 75–106.

Mood, A. M., Graybill, F. A., & Boes, D. C. (1974). *Introduction to the theory of statistics.* London: McGraw-Hill.

Mutschlechner, R. (1987). *Der Patient im Krankenhaus. – Ein Versuch, die Wirksamkeit einer psychologischen Betreuung nachzuweisen.* [The patient in hospital. – An attempt at assessing the effects of psychological treatments.] (Doctoral Thesis.) Vienna: University of Vienna.

Naas, J., & Schmid, H. L. (1961). *Mathematisches Wörterbuch, Vol. II.* [Mathematical Dictionary.] Berlin: Akademie-Verlag.

Nährer, W. (1977). *Modellkontrollen bei der Anwendung des linearen logistischen Modells in der Psychologie.* [Tests of fit in the application of linear logistic models in psychology.] (Doctoral Thesis.) Vienna: University of Vienna.

Nährer, W. (1980). Zur Analyse von Matrizenaufgaben mit dem linearen logistischen Testmodell. [On the analysis of matrices items by means of the linear logistic test model.] *Zeitschrift für Experimentelle und Angewandte Psychologie, 27,* 553–564.

Nap, R. E. (1994). *Variants for item parameter estimation in the Rasch model.* (Heymans Bulletin HB-94-1153-SW.) Groningen: Vakgroep S&M FPPSW, University of Groningen.

Nelder, J. A., & Wedderburn, R. W. M. (1972). Generalized linear models. *Journal of the Royal Statistical Society, Series A, 135,* 370–384.

Nemhauser, G. L., & Wolsey, L. A. (1988). *Integer and combinatorial optimization.* New York: J. Wiley.

Neuwirth, E. (1988). *Some results on latent additive functions for statistical models.* (Unpublished paper.) Vienna: Department of Statistics, University of Vienna.

Neyman, J., & Scott, E. L. (1948). Consistent estimates based on partially consistent observations. *Econometrica, 16,* 1–32.

Noack, A. (1950). A class of random variables with discrete distributions. *Annals of Mathematical Statistics, 21,* 127–132.

Ogasawara, T., & Takahashi, M. (1951). Independence of quadratic forms in normal system. *J. Sci. Hiroshima University, 15,* 1–9. Cited in Rao, C. R. (1973), *Linear Statistical Inference and its Applications.* New York: J. Wiley.

Papadimitriou, C. H., & Steiglitz, K. (1982). *Combinatorial optimization: Algorithms and complexity.* Englewood Cliffs, NJ: Prentice Hall.

Patil, G. P. (1965). On the multivariate generalized power series distribution and its application to the multinomial and negative multinomial. In G. P. Patil (Ed.), *Classical and contagious discrete distributions* (pp. 183–194). London: Pergamon Press.

Pfanzagl, J. (1971). *Theory of measurement.* New York: J. Wiley.

Pfanzagl, J. (1993). A case of asymptotic equivalence between conditional and marginal maximum likelihood estimators. *Journal of Statistical Planning and Inference, 35,* 301–307.

Pfanzagl, J. (1994). On item parameter estimation in certain latent trait models. In G. H. Fischer & D. Laming, (Eds.), *Contributions to Mathematical Psychology, Psychometrics, and Methodology* (pp. 249–263). New York: Springer-Verlag.

Piswanger, K. (1975). *Interkulturelle Vergleiche mit dem Matrizentest von Formann.* [Cross-cultural comparisons using the matrices test of Formann.] (Doctoral Thesis.) Vienna: University of Vienna.

Rao, C. R. (1948). Large sample tests of statistical hypotheses concerning several parameters with applications to problems of estimation. *Proceedings of the Cambridge Philosophical Society, 44,* 50–57.

Rao, C. R. (1965/1973). *Linear statistical inference and its applications.* New York: J. Wiley.

Rao, S. S. (1985). *Optimization theory and applications* (2nd ed.). New Delhi: Wiley Eastern Limited.

Rasch, G. (1960). *Probabilistic models for some intelligence and attainment tests.* Copenhagen: The Danish Institute of Educational Research. (Expanded edition, 1980. Chicago: The University of Chicago Press.)

Rasch, G. (1961). On general laws and the meaning of measurement in psychology. *Proceedings of the IV. Berkeley Symposium on mathematical statistics and probability, Vol. IV* (pp. 321–333). Berkeley: University of California Press.

Rasch, G. (1965). *Statistisk seminar.* [Statistical seminar]. (Notes taken by J. Stene.) Copenhagen: Department of Statistics, University of Copenhagen.

Rasch, G. (1966). *The theory of statistics.* (Lecture notes taken by Ulf Christiansen.) Copenhagen: Department of Statistics, University of Copenhagen.

Rasch, G. (1967). An informal report an a theory of objectivity in comparisons. In L. J. Th. Van der Kamp & C. A. J. Vlek (Eds.), *Psychological Measurement theory* (pp. 1–19). Proceedings of the NUFFIC international summer session in science at "Het Oude Hof", The Hague, July 14-28, 1966. Leyden: University of Leyden.

Rasch, G. (1968). *A mathematical theory of objectivity and its consequences for model construction.* Paper presented at the European Meeting on Statistics, Econometrics, and Management Science, Amsterdam, September 2-7, 1968.

Rasch, G. (1972). Objectivitet i samfundsvidenskaberne et metodeproblem. [Objectivity in the social sciences as a methodological problem.] *Nationaløkonomisk Tidsskrift, 110,* 161–196.

Rasch, G. (1973). *Two applications of the multiplicative Poisson models in road accidents statistics.* Invited paper presented at the 1973 Meeting of the International Statistical Institute in Vienna.

Rasch, G. (1977). On specific objectivity. An attempt at formalizing the request for generality and validity of scientific statements. In M. Blegvad (Ed.), *The Danish Yearbook of Philosophy* (pp. 58–94). Copenhagen: Munksgaard.

Rella, E. (1976). *Trainierbarkeit des Antizipierens von Gefahrensituationen im Straßenverkehr.* [The trainability of the anticipation of dangers in traffic.] (Doctoral Thesis.) Vienna: University of Vienna.

Rop, I. (1977). The application of a linear logistic model describing the effects of pre-school curricula on cognitive growth. In H. Spada & W. F. Kempf (Eds.), *Structural models of thinking and learning* (pp. 281–293). Berne: Huber.

Rorschach, H. (1921). *Psychodiagnostik.* Berne: Huber.

Roskam, E. E. (1983). Allgemeine Datentheorie. [General data theory.] In H. Feger & J. Bredenkamp (Eds.), *Messen und Testen. Enzyklopädie der Psychologie, Forschungsmethoden, Vol. 3.* (pp. 1–135). Göttingen: Verlag für Psychologie Hogrefe.

Roskam, E. E., & Jansen, P. G. W. (1984). A new derivation of the Rasch model. In E. Degreef & J. Van Buggenhaut (Eds.), *Trends in Mathematical Psychology* (pp. 293–307). Amsterdam: Elsevier Science Publishers.

Rost, J. (1988). Measuring attitudes with a threshold model drawing on a traditional scaling concept. *Applied Psychological Measurement, 12,* 397–409.

Rost, J. (1990a). Rasch models in latent classes: An integration of two approaches to item analysis. *Applied Psychological Measurement, 3,* 271–282.

Rost, J. (1990b). Einstellungsmessung in der Tradition von Thurstone's Skalierungsverfahren. [Attitude measurement in the tradition of Thurstonian scaling.] *Empirische Pädagogik, 4,* 83–92.

Rost, J. (1991). A logistic mixture distribution model for polychotomous item responses. *The British Journal of Mathematical and Statistical Psychology, 44,* 75–92.

Rost, J. (1995). Applications of the mixed Rasch model to personality questionnaires. In J. Rost & R. Langeheine (Eds.), *Applications of latent trait and latent class models in the social sciences* (in preparation).

Rost, J., & Davier v., M. (1992). MIRA. *A PC-program for the Mixed Rasch model – User Manual.* Kiel: Institute for Science Education (IPN).

Rost, J., & Davier v., M. (1993). Measuring different traits in different populations with the same items. In R. Steyer, K. F. Wender, & K. F. Widaman (Eds.), *Psychometric Methodology. Proceedings of the 7th European Meeting of the Psychometric Society in Trier.* Stuttgart: Gustav Fischer Verlag.

Rost, J., & Georg, W. (1991). Alternative Skalierungsmöglichkeiten zur klassischen Testtheorie am Beispiel der Skala 'Jugendzentrismus'. [Alternatives to classical test theory illustrated by a scale for 'Jugendzentrismus' as an example.] *ZA-Information, 28,* 52–75.

Rubin, D. B. (1987). *Multiple imputation for nonresponse in surveys.* New York: J. Wiley.

Samejima, F. (1977). Weakly parallel tests in latent trait theory with some criticism of classical test theory. *Psychometrika, 42,* 193–198.

Samejima, F. (1993). An approximation for the bias function of the maximum likelihood estimate of a latent variable for the general case where the item responses are discrete. *Psychometrika, 58,* 119–138.

Sanathanan, L., & Blumenthal, S. (1978). The logistic model and estimation of latent structure. *Journal of the American Statistical Association, 73,* 794–799.

Scheiblechner, H. (1971a). *CML-parameter-estimation in a generalized multifactorial version of Rasch's probabilistic measurement model with two categories of answers.* (Research Bulletin No. 4.) Vienna: Department of Psychology, University of Vienna.

Scheiblechner, H. (1971b). *A simple algorithm for CML-parameter-estimation in Rasch's probabilistic measurement model with two or more categories of answers.* (Research Bulletin No. 5.) Vienna: Department of Psychology, University of Vienna.

Scheiblechner, H. (1972). Das Lernen und Lösen komplexer Denkaufgaben. [The learning and solving of complex reasoning items.] *Zeitschrift für Experimentelle und Angewandte Psychologie, 3,* 456–506.

Schmied, C. (1987). *Die Effektivität von Managementseminaren – durchgeführt bei den Österreichischen Bundesbahnen.* [The efficacy of management seminars for employees of the Austrian Federal Railways]. (Doctoral Thesis.) Vienna: University of Vienna.

Scholz, A., & Schoeneberg, B. (1955). *Einführung in die Zahlentheorie.* [Introduction to number theory.] Sammlung Göschen, Band 1131. Berlin: Walter de Gruyter.

Schumacher, M. (1980). Point estimation in quantal response models. *Biometrical Journal, 22,* 315–334.

Serfling, R. J. (1980). *Approximation theorems of mathematical statistics.* New York: J. Wiley.

Sharda, R. (1992). *Linear programming software for personal computers: 1992 survey.* OR/MS Today, June 1992, 44–60.

Shealy, R. S., & Stout, W. (1993). A model-based standardization approach that separates true bias/DIF from group ability differences and detects test bias/DTF as well as item bias/DIF. *Psychometrika, 58,* 159–194.

Shohat, J. A., & Tamarkin, J. D. (1943). *The Problem of Moments.* New York: American Mathematical Society.

Smith, R. M., Kramer, G. A., & Kubiak, A. T. (1992). Components of difficulty in spatial ability test items. In M. Wilson (Ed.), *Objective measurement: Theory into practice, Vol. I* (pp. 157–174). Norwood, NJ: Ablex Publishing Corporation.

Sommer, K. (1979). *Der Effekt kindzentrierter Gruppengespräche auf das Verhalten von Kindern im Vorschulalter.* [The effects of a child-centered group therapy on the behavior of preschool-children.] (Doctoral Thesis.) Vienna: University of Vienna.

Spada, H. (1976). *Modelle des Denkens und Lernens.* [Models of thinking and learning.] Berne: Huber.

Spada, H., & Kluwe, R. (1980). Two models of intellectual development and their reference to the theory of Piaget. In R. Kluwe & H. Spada (Eds.), *Developmental models of thinking* (pp. 1–30). New York: Academic Press.

Spada, H., & May, R. (1982). The linear logistic test model and its application in educational research. In D. Spearritt (Ed.), *The improvement of measurement in education and psychology* (pp. 67–84). Hawthorn, Victoria: The Australian Council for Educational Research.

Staphorsius, G. (1992). *CLIB-toetsen.* [CLIB-tests]. Arnhem: CITO.

Stelzl, I. (1979). Ist der Modelltest des Rasch-Modells geeignet, Homogenitätshypothesen zu prüfen? Ein Bericht über Simulationsstudien mit inhomogenen Daten. [Is the test of fit of the Rasch model effective in testing homogeneity hypotheses? A report on simulation studies based on inhomogeneous data.] *Zeitschrift für Experimentelle und Angewandte Psychologie, 26,* 652–672.

Stene, J. (1968). Einführung in Raschs Theorie psychologischer Messung. [Introduction to Rasch's theory of measurement in psychology.] In G. H. Fischer (Ed.), *Psychologische Testtheorie* (pp. 229–268). Berne: Huber.

Sternberg, S. H. (1959). A path dependent linear model. In R. R. Bush & W. K. Estes (Eds.), *Studies in mathematical learning theory* (pp. 308–339). Stanford: Stanford University Press.

Sternberg, S. H. (1963). Stochastic learning theory. In R. D. Luce, R. R. Bush, & E. Galanter (Eds.), *Handbook of mathematical psychology, Vol. II* (pp. 1–120). New York: J. Wiley.

Stieltjes, T. J. (1894). Recherches sur les fractions continues. *Annales de la Faculté des Sciences, Toulouse, 8,* 1–122.

Stocking, M. L., & Swanson, L. (1993). A method for severly constrained item selection in adaptive testing. *Applied Psychological Measurement, 17,* 277–292.

Stokman, F. N. (1977). *Roll calls and sponsorship.* A methodological analysis of third world group formation in the United Nations. Leiden: Sijthoff.

Stouffer, S. A., & Toby, J. (1951). Role conflict and personality. *American Journal of Sociology, 56,* 395–406.

Stout, W. F. (1990). A new item response theory modeling approach with applications to unidimensionality assessment and ability estimation. *Psychometrika, 55,* 293–326.

Swaminathan, H., & Gifford, J. A. (1982). Bayesian estimation in the Rasch model. *Journal of Educational Statistics, 7,* 175–192.

Swanson, L., & Stocking, M. L. (1993). A model and heuristic for solving very large item selection problems. *Applied Psychological Measurement, 17,* 151–166.

Tanzer, N. (1984). *On the existence of unique joint maximum-likelihood estimates in linear logistic latent trait models for incomplete dichotomous data.* (Research Bulletin No. 25.) Vienna: Department of Psychology, University of Vienna.

Tatsuoka, K. K. (1984). Caution indices based on item response theory. *Psychometrika, 49,* 95–110.

Teicher, H. (1963). Identifiability of finite mixtures. *Annals of Mathematical Statistics, 34,* 1265–1269.

Ten Have, T. R. (1991). *Parameter symmetric and exchangeable log-linear models.* (Doctoral Thesis.) Michigan: Department of Biostatistics, University of Michigan.

Ten Have, T. R., & Becker, M. P. (1991). A hierarchy of first order parameter symmetric and exchangeable log-linear models. In *Proceedings of the Sixth International Workshop on Statistical Modelling.* Utrecht, The Netherlands: University of Utrecht.

Theunissen, T. J. J. M. (1985). Binary programming and test design. *Psychometrika, 50,* 411–420.

Theunissen, T. J. J. M. (1986). Some applications of optimization algorithms in test design and adaptive testing. *Applied Psychological Measurement, 10,* 381–389.

Thissen, D. (1982). Marginal maximum likelihood estimation for the one-parameter logistic model. *Psychometrika, 47,* 175–186.

Thurstone, L. L. (1925). A method of scaling psychological and educational tests. *Journal of Educational Psychology, 16,* 433–451.

Timminga, E., & Adema, J. J. (in press). An interactive approach to modifying infeasible 0-1 linear programming models for test construction. In G. Engelhard & M. Wilson (Eds.), *Objective measurement: Theory into practice, Vol. III.* Norwood, N. J.: Ablex Publishing Corporation.

Titterington, D. M., Smith, A. F. M., & Makov, U. E (1985). *Statistical analysis of finite mixture distributions.* Chichester: Wiley.

Tjur, T. (1982). A connection between Rasch's item analysis model and a multiplicative Poisson model. *Scandinavian Journal of Statistics, 9,* 23–30.

Trabin, T. E., & Weiss, D. J. (1983). The person response curve: Fit of individuals to item response theory models. In D. J. Weiss (Ed.), *New horizons in testing* (pp. 83–108). New York: Academic Press.

Tsutakawa, R. K., & Soltys, M. J. (1988). Approximation for Bayesian ability estimation. *Journal of Educational Statistics, 13,* 117–130.

Tsutakawa, R. K., & Johnson, J. C. (1990). The effect of uncertainty of item parameter estimation on ability estimates. *Psychometrika, 55,* 371–390.

Tutz, G. (1989). *Latent Trait-Modelle für ordinale Beobachtungen.* [Latent trait models for ordinal data.] Berlin: Springer-Verlag.

Van der Linden, W. J., & Boekkooi-Timminga, E. (1988). A zero-one programming approach to Gulliksen's matched random subsets method. *Applied Psychological Measurement, 12,* 201–209.

Van der Linden, W. J., & Boekkooi-Timminga, E. (1989). A maximin model for test design with practical constraints. *Psychometrika, 54,* 237–247.

Van Maanen, L., Been, P., & Sijtsma, K. (1989). The linear logistic test model and heterogeneity of cognitive strategies. In Roskam, E. E. (Ed.), *Mathematical psychology in progress* (pp. 267–287). New York: Springer-Verlag.

Van de Vijver, F. J. R. (1988). Systematizing item content in test design. In R. Langeheine & J. Rost (Eds.), *Latent trait and latent class models* (pp. 291–307). New York: Plenum.

Van den Wollenberg, A. L. (1982). Two new test statistics for the Rasch model. *Psychometrika, 47,* 123–139.

Verhelst, N. D. (1993). *On the standard errors of parameter estimators in the Rasch model.* (Measurement and Research Department Reports, 93-1.) Arnhem: CITO.

Verhelst, N. D., & Eggen, T. J. H. M. (1989). *Psychometrische en statistische aspecten van peilingsonderzoek* [Psychometric and statistical aspects of assessment research.] (PPON-rapport, 4) Arnhem: CITO.

Verhelst, N. D., & Glas C. A. W. (1993). A dynamic generalization of the Rasch model. *Psychometrika, 58,* 395–415.

Verhelst, N. D., Glas, C. A. W., & Van der Sluis, A. (1984). Estimation problems in the Rasch model: The basic symmetric functions. *Computational Statistics Quarterly, 1,* 245–262.

Verhelst, N. D., Glas, C. A. W., & Verstralen, H. H. F. M. (1994). *OPLM: Computer program and manual.* Arnhem: CITO.

Verhelst, N. D., & Molenaar, I. W. (1988). Logit based parameter estimation in the Rasch model. *Statistica Neerlandica, 42,* 273–295.

Verhelst, N. D., & Veldhuijzen, N. H. (1991). *A new algorithm for computing elementary symmetric functions and their first and second derivatives.* (Measurement and Research Department Reports, 91-1.) Arnhem: CITO.

Verhelst, N. D., Verstralen, H. H. F. M., & Eggen, T. J. H. M. (1991). *Finding starting values for the item parameters and suitable discrimination indices in the one parameter logistic model.* (Measurement and Research Department reports, 91-10.) Arnhem: CITO.

Verschoor, A. (1991). *OTD (optimal test design).* (Computer program.) Arnhem: CITO.

Wainer, H., & Wright, B. D. (1980). Robust estimation of ability in the Rasch model. *Psychometrika, 45,* 373–391.

Wald, A. (1943). Tests of statistical hypotheses concerning several parameters when the number of observations is large. *Transactions of the American Mathematical Society, 54,* 426–482.

Warm, T. A. (1989). Weighted likelihood estimation of ability in item response models. *Psychometrika, 54,* 427–450.

Weiss, D. J. (Ed.) (1982). *New horizons in testing: Latent trait theory and computerized adaptive testing.* New York: Academic Press.

Whitely, S. E., & Schneider, L. M. (1981). Information structure for geometric analogies: A test theory approach. *Applied Psychological Measurement, 5,* 383–397.

Widowitz, E. (1987). *Der Effekt autogenen Trainings bei funktionellen Erkrankungen.* [The effect of a relaxation training on the functional syndrome]. (Doctoral Thesis.) Vienna: University of Vienna.

Williams, H. P. (1985). *Model building in mathematical programming.* New York: J. Wiley.

Wilson, M. (1989). Saltus: A psychometric model of discontinuity in cognitive development. *Psychological Bulletin, 105,* 276–289.

Wilson, M., & Adams, R. A. (1993). Marginal maximum likelihood estimation for the ordered partition model. *Journal of Educational Statistics, 18,* 69–90.

Wingersky, M. S., Barton, M. A., & Lord, F. M. (1982). *LOGIST user's guide.* Princeton, NJ: Educational Testing Service.

Wilson, M., & Masters, G. N. (1993). The partial credit model and null categories. *Psychometrika, 58,* 87–99.

Witek, J. (1979). *Die Effektivität des gruppendynamischen Sensitivity Trainings.* [The efficacy of a group-dynamic sensitivity training.] (Doctoral Thesis.) Vienna: University of Vienna.

Witting, H. (1978). *Mathematische Statistik* (3rd ed.). [Mathematical Statistics.] Stuttgart: Teubner.

Wottawa, H. (1980). *Grundriß der Testtheorie.* [Compendium of test theory.] München: Juventa.

Wright, B. D. (1977). Solving measurement problems with the Rasch model. *Journal of Educational Measurement, 14,* 97–115.

Wright, B. D., & Masters, G. N. (1982). Rating scale analysis: Rasch measurement. Chicago: Mesa Press.

Wu, C. F. J. (1983). On the convergence properties of the EM algorithm. *The Annals of Statistics, 11,* 95–103.

Yen, W. M. (1981). Using simulation results to choose a latent trait model. *Applied Psychological Measurement, 5,* 245–262.

Yen, W. M. (1983). *Use of the three-parameter model in the development of a standardized achievement test.* In R. K. Hambleton (Ed.), Applications of item response theory (pp. 123–141). Vancouver: Educational Institute of British Columbia.

Zeman, M. (1976). *Die Wirksamkeit der mathematischen Früherziehung.* [The efficacy of early mathematics training.] (Doctoral Thesis.) Vienna: University of Vienna.

Zermelo, E. (1929). Die Berechnung der Turnierergebnisse als ein Maximumproblem der Wahrscheinlichkeitsrechnung. [The computation of tournament results as a maximum-probability problem.] *Mathematische Zeitschrift, 29,* 436–460.

Zimprich, H. (1980). Behandlungskonzepte und -resultate bei psychosomatischen Erkrankungen im Kindesalter. [Treatment designs and results in psychosomatic diseases in children.] *Pädiatrie und Pädologie, Supplementum 6,* 131–198.

Zwinderman, A. H. (1991a). *Studies of estimating and testing Rasch models.* (NICI Technical Report 91-02) (Doctoral Thesis.) Nijmegen: University of Nijmegen.

Zwinderman, A. H. (1991b). A generalized Rasch model for manifest predictors. *Psychometrika, 56,* 589–600.

Zwinderman, A. H. (1991c). A two stage Rasch model approach to dependent item responses: An application of constrained latent trait models. *Methodika, 5,* 33–46.

Author Index

Subject Index

Abbreviations

ANOVA	Analysis of Variance
AS	Alternative System
BFGS	Broyden-Fletcher-Goldfarb-Shanno Algorithm
BAN	Best Asymptotic Normal
BME	Bayes Modal Estimator
BTL	Bradley-Terry-Luce Model
CLR	Conditional Likelihood Ratio
CMI	Complete Measurement Interchangeability
CML	Conditional Maximum Likelihood
CONTEST	program name, see Boekkooi-Timminga & Sun, 1992
DIF	Differential Item Functioning
DTF	Differential Test Functioning
EAP	Bayes Expected A Posteriori
EM	Expected Marginals
GLM	Generalized Linear Model
H_0	Null Hypothesis
H_1	Alternative Hypothesis
ICC	Item Characteristic Curve
IPF	Iterated Proportional Fitting
IRF	Item Response Function
IRT	Item Response Theory
IST	Intelligenz-Struktur-Test
JML	Joint Maximum Likelihood
LBTL	Linear Bradley-Terry-Luce Model
LCA	Latent Class Analysis
LCALIN	Linear Logistic Latent Class Analysis
LC/RM	Latent Class Rasch Model
LLM	Loglinear Model
LLRA	Linear Logistic Model with Relaxed Assumptions
LLTM	Linear Logistic Test Model

LM	Langrange Multiplier
LOGIMO	Loglinear IRT Modeling; program name, see Kelderman and Steen, 1988
LOGIST	program name, see Wingersky, Barton, and Lord (1982)
LP	Linear Program
LPCM	Linear Partial Credit Model
LPCM	Linear Partial Credit Model; program name, see Fischer & Ponocny (1994)
LR	Likelihood Ratio
LRSM	Linear Rating Scale Model
MDM	Mixed Distribution Model
MINCHI	Minimum Chi-Square
MIRA	Mixed Rasch Model
ML	Maximum Likelihood
MLE	Maximum Likelihood Estimator
MLR	Marginal Likelihood Ratio
MML	Marginal Maximum Likelihood
MMScore	Marginal Mixture of the score groups
MORM	Mixed Ordinal Rasch Model
MRM	Mixed Rasch Model
MZTA	Monozygotic Twins Reared Apart
OPLM	One Parameter Logistic Model
OTD	Optimal Test Design; program name, see Verschoor (1991)
PCM	Partial Credit Model
QMI	Quasi Measurement Interchangeability
RM	Rasch Model
RSM	Rating Scale Model
RSP	Rasch Scaling Program
SO	Specific Objectivity
2PLM	Two Parameter Logistic Model
UML	Unconditional Maximum Likelihood
UMP	Uniformly Most Powerful
WINMIRA	program name, see von Davier (1994)
WLE	Warm Weighted Likelihood Estimator
WMT	Wiener Matrizentest

Printed in the United States
69229LVS00001BB/14

9 780387 944999